Advances in Intelligent Systems and Computing

Volume 408

Series editor

Janusz Kacprzyk, Polish Academy of Sciences, Warsaw, Poland
e-mail: kacprzyk@ibspan.waw.pl

About this Series

The series "Advances in Intelligent Systems and Computing" contains publications on theory, applications, and design methods of Intelligent Systems and Intelligent Computing. Virtually all disciplines such as engineering, natural sciences, computer and information science, ICT, economics, business, e-commerce, environment, healthcare, life science are covered. The list of topics spans all the areas of modern intelligent systems and computing.

The publications within "Advances in Intelligent Systems and Computing" are primarily textbooks and proceedings of important conferences, symposia and congresses. They cover significant recent developments in the field, both of a foundational and applicable character. An important characteristic feature of the series is the short publication time and world-wide distribution. This permits a rapid and broad dissemination of research results.

Advisory Board

Chairman

Nikhil R. Pal, Indian Statistical Institute, Kolkata, India
e-mail: nikhil@isical.ac.in

Members

Rafael Bello, Universidad Central "Marta Abreu" de Las Villas, Santa Clara, Cuba
e-mail: rbellop@uclv.edu.cu

Emilio S. Corchado, University of Salamanca, Salamanca, Spain
e-mail: escorchado@usal.es

Hani Hagras, University of Essex, Colchester, UK
e-mail: hani@essex.ac.uk

László T. Kóczy, Széchenyi István University, Győr, Hungary
e-mail: koczy@sze.hu

Vladik Kreinovich, University of Texas at El Paso, El Paso, USA
e-mail: vladik@utep.edu

Chin-Teng Lin, National Chiao Tung University, Hsinchu, Taiwan
e-mail: ctlin@mail.nctu.edu.tw

Jie Lu, University of Technology, Sydney, Australia
e-mail: Jie.Lu@uts.edu.au

Patricia Melin, Tijuana Institute of Technology, Tijuana, Mexico
e-mail: epmelin@hafsamx.org

Nadia Nedjah, State University of Rio de Janeiro, Rio de Janeiro, Brazil
e-mail: nadia@eng.uerj.br

Ngoc Thanh Nguyen, Wroclaw University of Technology, Wroclaw, Poland
e-mail: Ngoc-Thanh.Nguyen@pwr.edu.pl

Jun Wang, The Chinese University of Hong Kong, Shatin, Hong Kong
e-mail: jwang@mae.cuhk.edu.hk

More information about this series at http://www.springer.com/series/11156

Suresh Chandra Satapathy · Amit Joshi
Nilesh Modi · Nisarg Pathak
Editors

Proceedings of International Conference on ICT for Sustainable Development

ICT4SD 2015 Volume 1

 Springer

Editors
Suresh Chandra Satapathy
Department of Computer Science and
  Engineering
Anil Neerukonda Institute of Technology
  and Sciences
Visakhapatnam, Andhra Pradesh
India

Amit Joshi
Sabar Institute of Technology
Sabarkantha, Gujarat
India

Nilesh Modi
Narsinhbhai Institute of Computer Studies
  and Management
Kadi, Gujarat
India

Nisarg Pathak
Narsinhbhai Institute of Computer Studies
  and Management
Kadi, Gujarat
India

ISSN 2194-5357   ISSN 2194-5365   (electronic)
Advances in Intelligent Systems and Computing
ISBN 978-981-10-0127-7   ISBN 978-981-10-0129-1   (eBook)
DOI 10.1007/978-981-10-0129-1

Library of Congress Control Number: 2015955863

Printed on acid-free paper

This Springer imprint is published by SpringerNature
The registered company is Springer Science+Business Media Singapore Pte Ltd.

Preface

These AISC volumes contain the papers presented at the ICT4SD 2015: International Conference on Information and Communication Technology for Sustainable Development. The conference was held during July 3–4, 2015 at Hotel Pride Ahmedabad, India, and communally organized by ASSOCHAM Gujarat Chapter, ACM Professional Chapter, GESIA and Sabar Institute of Technology, Gujarat and Computer Society of India, as Knowledge Partner. The objective of this international conference was to provide an opportunity for researchers, academicians, industry persons, and students to interact and exchange ideas, experience, and expertise in the current trend and strategies for Information and Communication Technologies. Besides this, participants were also enlightened about the vast avenues, current and emerging technological developments in the field of ICT in this era, and its applications. The conference attracted a large number of high-quality submissions and stimulated cutting-edge research discussions among many academic pioneering researchers, scientists, industrial engineers, and students from all over the world and provided a forum for researchers. Research submissions in various advanced technology areas were received and after a rigorous peer-review process with the help of program committee members and external reviewers, 154 (Vol-I: 77, Vol-II: 77) papers were accepted with an acceptance ratio of 0.43. The conference featured many distinguished personalities such as Mr. Job Glas, Head of Mission, NBSO, Netherlands, Mr. Volkmar Blech, Zera Gmbh, Germany, Dr. Mukesh Kumar, TITS, Bhiwani, Dr. Vipin Tyagi, Jaypee University, Guna, Prof. Pravesh Bhadviya, Director, Sabar Education, Mr. Bipin V. Mehta, President CSI, Mr. Hemal Patel, MD, Cyberoam, Ms. Bhagyesh Soneji, Chairperson, ASSOCHAM, and Mr. Jay Ruparel, President, GESIA. Separate invited talks were organized on industrial and academia tracks on both days. The conference also hosted a few tutorials and workshops for the benefit of participants. We are indebted to ASSOCHAM Gujarat Chapter, Sabar Institute of Technology and Computer Society of India, ACM Professional Chapter for their immense support to make this conference possible on such a grand scale. A total of 18 sessions were organized as a part of ICT4SD including 15 technical, 2 plenary, and 1 inaugural session. The Session Chairs for the technical sessions included Dr. Chirag Thaker, GEC, Bhavnagar, India, Dr. Vipin Tyagi, Jaypee University, MP, India, Dr. Munesh Trivedi, ABES Engineering College, Ghaziabad, India, Dr. Ramesh Thakur, DAVV, Indore, India, Dr. Dilip Kumar Sharma, GLA University, Mathura, India, Dr. Bhushan Trivedi, GLS University, Ahmedabad, India, Dr. S.M. Shah, KSV University, India, Dr. Nikita Vats Doohan, Indore, MP, India, Dr. Harshal Arolkar, GlS University, Ahmedabad, India, Dr. Priyanka Sharma, Raksha Shakti University, Ahmedabad, India, Dr. Nilesh Modi, KSV, Ahmedabad, India, Dr. Satyen Parikh, Ganpat University, India, Dr. Sakshi Kaushal, UIET, Punjab University, India, Dr. S.C. Satapathy, Visakhapatnam, India, and Dr. Nisarg Pathak, KSV, Ahmedabad, India.

We express our sincere thanks to the members of the technical review committee for their valuable support in doing critical reviews to enhance the quality of all accepted papers. Our heartfelt thanks are due to the National and International Advisory Committee and CSI Execomm Members for their support in making this a grand success. Our authors deserve big thanks since it is due to them that the conference was such a huge success.

Our sincere thanks to all the sponsors, press, print, and electronic media for their excellent coverage of this convention.

July 2015

Suresh Chandra Satapathy\
Amit Joshi\
Nilesh Modi\
Nisarg Pathak

Contents

Committee

Advisory Committee

Mr. H.R. Mohan, Past President, CSI
Prof. Bipin Mehta, President, CSI
Mr. P.N. Jain, Add. Sec., R&D, Government of Gujarat, India
Dr. Srinivas Padmanabhuni, President ACM India
Dr. Anirban Basu, Vice President, CSI
Prof. R.P. Soni, RVP, Region III, CSI
Dr. Malay Nayak, Director-IT, London
Mr. Chandrashekhar Sahasrabudhe, ACM India
Dr. Pawan Lingras, Saint Mary's University, Canada
Prof. (Dr.) P. Thrimurthy, Past President, CSI
Dr. Shayam Akashe, ITM, Gwalior, MP, India
Dr. S.C. Sathapathy, Visakhapatnam, India
Dr. Dharm Singh, Windhoek, Namibia
Prof. S.K. Sharma, Pacific University, Udaipur, India
Prof. H.R. Vishwakarma, VIT, Vellore, India
Prof. Pravesh Bhadviya, Director, Sabar Education, India
Mr. Mignesh Parekh, Ahmedabad, India
Dr. Muneesh Trivedi, ABES, Gaziabad, India
Dr. Chandana Unnithan, Victoria University, Australia
Prof. Deva Ram Godara, Bikaner, India
Dr. Y.C. Bhatt, Chairman, CSI Udaipur Chapter
Dr. B.R. Ranwah, Past Chairman, CSI Udaipur Chapter
Dr. Arpan Kumar Kar, IIT Delhi, India

Organizing Committee

General Chairs
Ms. Bhagyesh Soneji, Chairperson, ASSOCHAM Gujarat
Mr. Jay Ruparel, President, GESIA
Mr. Bharat Patel, COO, Yudiz Solutions

Organizing Chairs
Dr. Durgesh Kumar Mishra, Chairman, Division IV, CSI
Dr. Rajveer Shekhawat, Chairman, ACM Udaipur Chapter

Organizing Co-chair
Dr. Harshal Arolkar, Associate Professor, GLS Ahmedabad

Members
Dr. Vimal Pandya, Ahmedabad, India
Dr. G.N. Jani, Ahmedabad, India
Mr. Nilesh Vaghela, Electromech, Ahmedabad, India
Mr. Vinod Thummar, SITG, Gujarat, India
Dr. Chirag Thaker, GEC, Bhavnagar, Gujarat, India
Mr. Maulik Patel, SITG, Gujarat, India
Mr. Nilesh Vaghela, Electromech Corp., Ahmedabad, India
Dr. Savita Gandhi, GU, Ahmedabad, India
Mr. Nayan Patel, SITG, Gujarat, India
Dr. Jyoti Parikh, Associate Professor, CE, GU, Ahmedabad, India
Dr. Vipin Tyagi, Jaypee University, Guna, India
Prof. Sanjay Shah, GEC, Gandhinagar, India
Dr. Chirag Thaker, GEC, Bhavnagar, Gujarat, India
Mr. Mihir Chauhan, VICT, Gujarat, India
Mr. Chetan Patel, Gandhinagar, India

Program Committee

Program Chair
Dr. Nilesh Modi, Professor and Head, NICSM, Kadi

Program Co-chair
Dr. Nisarg Pathak, SSC, CSI, Gujarat

Members
Dr. Mukesh Sharma, SFSU, Jaipur
Dr. Manuj Joshi, SGI, Udaipur, India
Dr. Bharat Singh Deora, JRNRV University, Udaipur
Prof. D.A. Parikh, Head, CE, LDCE, Ahmedabad, India
Prof. L.C. Bishnoi, GPC, Kota, India

Mr. Alpesh Patel, SITG, Gujarat
Dr. Nisheeth Joshi, Banasthali University, Rajasthan, India
Dr. Vishal Gaur, Bikaner, India
Dr. Aditya patel, Ahmedabad University, Gujarat, India
Mr. Ajay Choudhary, IIT Roorkee, India
Dr. Dinesh Goyal, Gyan Vihar, Jaipur, India
Mr. Nirav Patel, SITG, Gujarat
Dr. Muneesh Trivedi, ABES, Gaziabad, India
Mr. Ajit Pujara, SITG, Gujarat, India
Dr. Dilip Kumar Sharma, Mathura, India
Prof. R.K. Banyal, RTU, Kota, India
Mr. Jeril Kuriakose, Manipal University, Jaipur, India
Dr. M. Sundaresan, Chairman, CSI Coimbatore Chapter
Prof. Jayshree Upadhyay, HOD-CE, VCIT, Gujarat
Dr. Sandeep Vasant, Ahmedabad University, Gujarat, India

About the Editors

Dr. Suresh Chandra Satapathy is currently working as Professor and Head, at the Department of CSE at Anil Neerukonda Institute of Technology and Sciences (ANITS), Andhra Pradesh, India. He obtained his Ph.D. in Computer Science and Engineering from JNTU Hyderabad and his M.Tech. in CSE from NIT, Rourkela, Odisha, India. He has 26 years of teaching experience. His research interests include data mining, machine intelligence, and swarm intelligence. He has acted as program chair of many international conferences and edited six volumes of proceedings from Springer LNCS and AISC series. He is currently guiding eight scholars for Ph.Ds. Dr. Satapathy is also a senior member of IEEE.

Er. Amit Joshi has experience of around 6 years in academic and industry in prestigious organizations in Rajasthan and Gujarat. Currently, he is working as Assistant Professor in the Department of Information Technology at Sabar Institute in Gujarat. He is an active member of ACM, CSI, AMIE, IEEE, IACSIT-Singapore, IDES, ACEEE, NPA, and many other professional societies. Currently, he is Honorary Secretary of CSI Udaipur Chapter and Honorary Secretary for ACM Udaipur Chapter. He has presented and published more than 40 papers in national and international journals/conferences of IEEE, Springer and ACM. He has also edited three books on diversified subjects including Advances in Open Source Mobile Technologies, ICT for Integrated Rural Development, and ICT for Competitive Strategies. He has also organized more than 25 national and international conferences and workshops including International Conference ETNCC 2011 at Udaipur through IEEE, international conference ICTCS—2014 at Udaipur through ACM, international conference ICT4SD 2015—by Springer recently. He has also served on Organizing and Program Committees of more than 50 conferences/seminars/workshops throughout the world and presented six invited talks at various conferences. For his contribution towards society he has been awarded by The Institution of Engineers (India), ULC, the Appreciation Award on the celebration of Engineers, 2014 and by SIG-WNs Computer Society of India on ACCE, 2012.

Dr. Nilesh Modi has rich experience of around 13 years in academics and in the IT industry. He holds a doctorate in e-Security (Computer Science and Application). Continuing his research on cybersecurity, presently he is pursuing postdoctoral research on Wireless Communication and Security and certification as Ethical Hacking. He has a good number of research papers in his name and has presented more than 75 research papers in international and national journals and conferences. He has delivered a number of expert talks on e-Security and hacking in national and international conferences. Dr. Modi, a person with vibrancy, is an active life member of CSI, ACM IEEE, IACSIT, IACSI, and IAEng apart from his academic and industrial careers. As a consultant, he contributes to different system development projects with the IT industry and has carried out different government projects.

Dr. Nisarg Pathak is an astute and result-oriented professional with 10 years of experience in teaching and carving state, national, and international events like workshops, seminars, and conferences. Being a mathematics scholar and a computer science professional, he is actively involved in the research of data mining and big data analytics. He has a strong list of national and international publications to his name. Dr. Pathak is currently Associate Professor of Computer Science and Application at Narsinhbhai Institute of Computer Studies and Management affiliated to Kadi Sarva Vishvavidyalaya. He received his Ph.D. from Hemchandracharya North Gujarat University in Computer Science and his Master's in Mathematics from the same university. His other tenures include research fellow at Indian Statistical Institute, Kolkata, and Indian Institute of Science, Bengaluru.

A Compact Triple and Quadruple Band Antenna for Bluetooth/ WiMAX/WLAN/X-Band Applications

Nirma Kumawat, Sanjay Gurjar and Sunita

Abstract In this paper, a compact antenna is presented with triple and quadruple bands. The antenna is designed on FR4 substrate using dielectric constant 4.4 and loss tangent is 0.02 with thickness 1.6 mm. The proposed antenna consists of a split ring inside a square patch with two parasitic element and a semicircle slot in the ground plane to create three bands and two more vertical slots inserted into the partial ground plane to introduce one more band. The proposed antenna covers Bluetooth, WiMAX, WLAN, and X-band applications with resonating frequencies at 2.47, 3.2, 5.39, 8.78 GHz, respectively. The overall dimension of the antenna is $25 \times 38 \times 1.6$ mm^3. Microstrip feed line is used to achieve 50 Ω impedance matching. All results of the proposed antenna are simulated using computer simulation technology software and show good acceptance with the theoretical concepts.

Keywords Parasitic element · Triple and quadruple band · Split ring

1 Introduction

Recently, many research papers have been reviewed on multiband antennas. The demand for multi-frequency application on a single portable device is increasing rapidly. A wideband UWB antenna has been developed which covers several wireless applications and FCC defined the 3.1–10.6 GHz frequency range for UWB antennas [1]. However, they have problems such as signal interference and constant gain. To resolve these problems, multiband antenna is designed having least

N. Kumawat (✉) · S. Gurjar
Ajmer Institute of Technology, Ajmer, India
e-mail: kumawatnirma@gmail.com

S. Gurjar
e-mail: ersanjay86@yahoo.in

Sunita
Government Women Engineering College, Ajmer, India
e-mail: olsunita30@gmail.com

© Springer Science+Business Media Singapore 2016
S.C. Satapathy et al. (eds.), *Proceedings of International Conference on ICT for Sustainable Development*, Advances in Intelligent Systems and Computing 408, DOI 10.1007/978-981-10-0129-1_1

interference with better gain. Over the past few years different and new mechanisms are being developed in multiband antenna design. In [2], fractal antenna is proposed for multiband applications and their design based on iteration method; after that a frequency reconfigurable multiband antenna was in trend but the structures of the reconfigurable antennas are complex and they have switching elements and biasing mechanism [3]. Multiband antennas are the alternate solution for complexity and interference.

Several multiband antennas have been reported in the literature [4–10]. In [4], a triple band antenna is presented using a split ring enclosed inside a rectangular patch and a U slot is cut from the partial ground plane. It covers WiMAX and WLAN applications. In [5], a compact antenna design is proposed by etching the open ended crossed double T-shaped slot in the ground plane and L-shaped line feed used to introduce WiMAX and WLAN applications. In [6, 7], a triple band antenna is designed using the defected ground structures for Bluetooth, WiMAX, and WLAN applications. In [8], cactus-shaped printed monopole antenna covers the triple bands. In [9] a multiband antenna is present in which right angled slots are used: two for dual band, three for triple band, and four for quad band. In [10], CPW-fed printed monopole antenna is used.

In this paper, a triple and quadruple band antenna is proposed. It has a simple and compact structure, which covers Bluetooth, WiMAX, WLAN, and X-band applications. Antenna consists of a square patch of side 18 mm. A split ring with two parasitic elements is inserted into the square patch to create the triple band. After that, two slots of dimension 1×4 mm^2 are cut from the partial ground plane and a fourth band is obtained. The antennas have stable radiation performance and VSWR < 2 in the desirable bands. Antenna design is described in Sect. 2. Section 3 gives the simulation results of the proposed antenna. Computer simulation technology (CST) is used as an electromagnetic solver to compute the results. Finally the paper is concluded in Sect. 4.

2 Antenna Design

The proposed antenna is fabricated on FR4 substrate having dielectric constant $\mathcal{E}r = 4.4$, with loss tangent $\delta = 0.02$ and thickness 1.6 mm. The optimized antenna structure is designed in two stages. In the first stage, triple band antenna is proposed while in the second stage optimized quadruple band antenna design is described.

Stage I: Triple Band Antenna

The proposed antenna consists of a square patch of side 18 mm. A split ring of 0.5 mm width is inserted inside the square patch and two parasitic elements are cut from inside the split ring to create the band for Bluetooth application. The band covers a frequency range from 2.44 to 2.55 GHz and resonance at 2.47 GHz. The microstrip feed line is used for perfect impedance matching. VSWR is less than 2 over the entire desirable band. Impedance bandwidths of the operating bands are

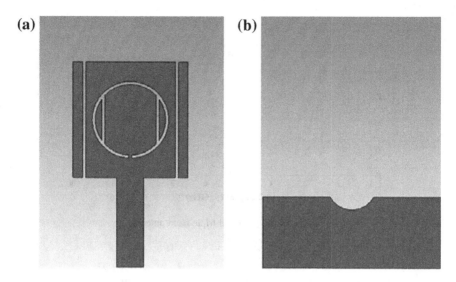

Fig. 1 Proposed antenna design for triple band. **a** *Top view*, **b** *bottom view*

Fig. 2 Structure of the
optimized antenna

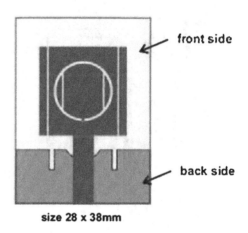

4.45 % for Bluetooth, 31.13 % for WiMAX, and 12.67 % for WLAN operating
band (Figs. 1 and 2).

Figure 3 shows the return loss curve for the proposed antenna. It is seen that the
value of return loss is below −10 dB for all the three operating bands corresponding
to Bluetooth, WiMax, and WLAN. Better impedance bandwidth is observed in the
application bands. Bandwidth for Bluetooth is 110 MHz, for WiMAX 1100 MHz,
and for WLAN 710 MHz.

Stage II: Optimized Quadruple Band Antenna

The optimized antenna presents the four bands by small modification in the triple
band antenna. Geometry of the optimized antenna is shown in Fig. 4. The two

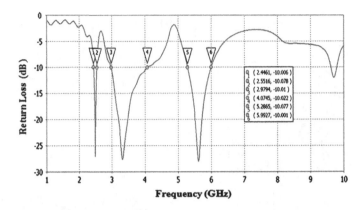

Fig. 3 Simulated return loss curve of the proposed triple band antenna

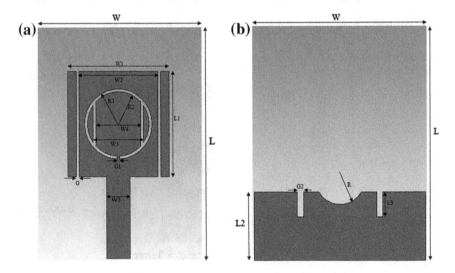

Fig. 4 Geometry of the optimized antenna. **a** *Top view*, **b** *bottom view*

vertical slots are etched from the partial ground plane to create the quad band. All the remaining dimensions of the antenna and substrate material is unchanged. The dimensions of the optimized antenna are as $W = 28$ mm, $L = 38$ mm, $W1 = 18$ mm, $L1 = 18$ mm, $W2 = 14$ mm, $W3 = 4.22$ mm, $W4 = 8$ mm, $W5 = 9$ mm, $R1 = 6$ mm, $R2 = 5.5$ mm, $G = 0.5$ mm, $G1 = 0,5$ mm, $L2 = 11$ mm, $R = 4$ mm, $G2 = 1$ mm, $L3 = 4$ mm. The fabricated antenna with dimension of 28×38 mm^2 is shown in Fig. 5.

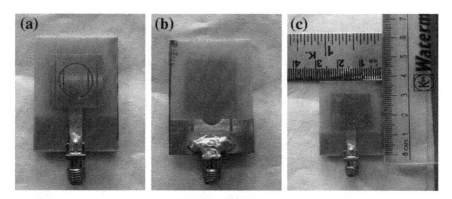

Fig. 5 Fabricated antenna. **a** *Top view*, **b** *bottom view*, **c** antenna with measurement with respect to cm scale

3 Results and Discussion

The CST electromagnetic solver is used to analyse the results and optimize antenna design. Simulated return loss curve of optimized antenna presents the multibands as shown in Fig. 6. Figure 6 shows four resonant bands at frequencies of 2.47 GHz, 3.2 GHz, 5.39 GHz, and 8.87 GHz. These bands are defined with bandwidth for −10 dB return loss, of about 110 MHz (2.44–2.55 GHz), 910 MHz (2.89–3.80), 480 MHz (5.18–5.66), and 1620 MHz (8.28–9.90 GHz) corresponding to Bluetooth, WiMAX, WLAN, and X-band applications, respectively. The impedance bandwidths of these four bands are 4.45, 28.4, 8.95, and 18.31 % with respect to the resonating frequencies of their operating bands.

In Fig. 7 measured return loss curve of the fabricated antenna is shown. The comparison between simulated and measured result is discussed in Table 1.

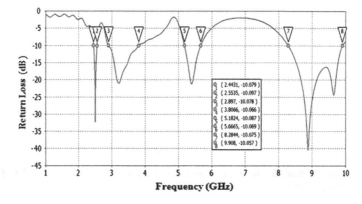

Fig. 6 Simulated return loss of the optimized antenna

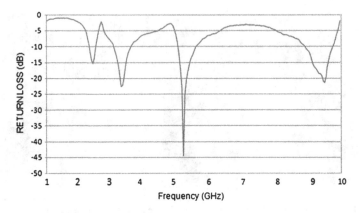

Fig. 7 Measured return loss of the fabricated antenna

Table 1 Comparison between simulated and measured results of the optimized antenna

Band	Simulated results		Measured results	
	Resonant frequency (GHz)	Bandwidth (MHz)	Resonant frequency (GHz)	Bandwidth (MHz)
First	2.47	110	2.48	190
Second	3.2	910	3.39	580
Third	5.39	480	5.37	570
Fourth	8.87	1620	9.87	1010

Radiation pattern of the optimized antenna is illustrated in Fig. 8. Radiation characteristics of an antenna measured radiation pattern, gain, and directivity in a far-field region. The simulation Fig. 8 of optimized antenna shows bidirectional radiation pattern and good directive gain at resonating frequencies over the operating bands (Table 2).

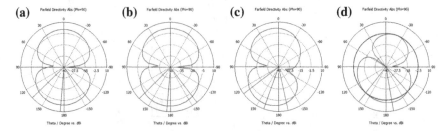

Fig. 8 Simulated far field directivity of optimized antenna. **a** At 2.47 GHz, **b** at 3.2 GHz, **c** at 5.39 GHz, **d** at 8.87 GHz

Table 2 Simulated directive gain of antenna at the operating frequency

Frequency (GHz)	Directive gain (dBi)
2.47	2.6
3.2	2.4
5.39	4.1
8.87	3

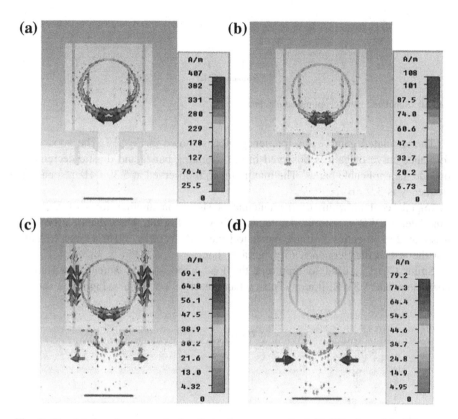

Fig. 9 Simulated surface current of optimized antenna. **a** At 2.47 GHz, **b** at 3.2 GHz, **c** at 5.39 GHz, **d** at 8.87 GHz

Surface current is another characteristic to understand the excitation behavior of the antenna. The surface current distribution for resonating frequencies at 2.47, 3.2, 5.39, and 8.87 GHz is shown in Fig. 9a–d.

In Fig. 9a the maximum current distribution is seen around the split ring at lower resonating frequency. In Fig. 9b, maximum splitting point of the ring is for 3.2 GHz. The next stronger current density is observed on the outside parasitic element of the patch at 5.39 GHz in Fig. 9c. In Fig. 9d around the partial ground plane slot, strong current distribution for upper resonating frequency is observed.

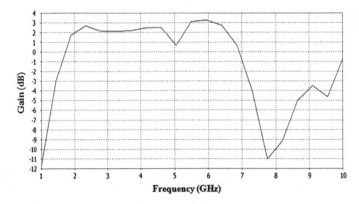

Fig. 10 Simulated gain of optimized antenna

The simulated gain versus frequency plot is shown in Fig. 10. The good performance curve of gain is observed in the operating bands and drastic decrement
outside the applicable bands. The maximum gain observed at 5.39 GHz resonating
frequency is 3.9 dB.

Impedance bandwidth of the antenna is shown in another form of VSWR.
Simulated VSWR of optimized antenna is shown in Fig. 11. The value of VSWR is
less than 2 in all entire desirable bands to Bluetooth, WiMax, WLAN, and X-band.
For the rest of the frequencies it is greater than 2.

Smith Chart is used to represent the load impedance of antenna, reflection
coefficient, and VSWR. It represents real and reactive load values. Figure 12 shows

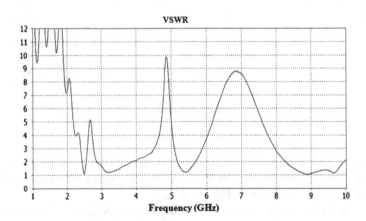

Fig. 11 Simulated VSWR of optimized antenna

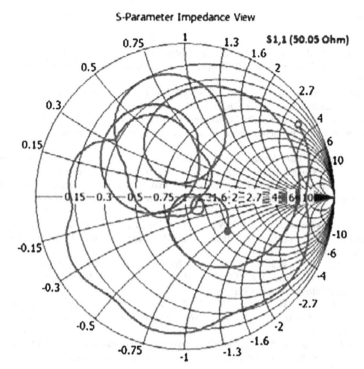

Fig. 12 Smith chart of optimized antenna

the smith chart for the optimized antenna. It is seen that the impedance of the antenna is good for perfect matching. The 50 Ω impedance is achieved by varying the width W3 of the microstrip line.

4 Conclusion

A compact antenna with triple and quadruple band characteristics has been presented in this paper. We illustrate it by inserting a split ring with two parasitic elements and two slots in ground plane, by which a multiband antenna could be designed. The radiation pattern shows good bidirectional behavior over the entire operating bands and the gain curve shows a drastic decrement outside the operating bands. The microstrip feed line impedance is well matched to 50 Ω. It is a simple, compact multiband band antenna for Bluetooth/WiMAX/WLAN/X-band wireless standards with VSWR < 2.

Acknowledgments I would like to thank the principal and H. O. D of E. C. E. department of Govt. Engineering College, Ajmer for helping to take the result of the fabricated antenna on PNA Vector Network Analyzer.

References

1. Abdollahvand, M., Dadashzadeh, G., & Mostafa, D. (2010). Compact dual band-notched printed monopole antenna for UWB application. *IEEE Antenna and Wireless Propagation Letters. 9.*
2. Choukiker, Y. K., & Behera, S. K. (2010). CPW-fed compact multiband sierpinski triangle antenna. In *Annual IEEE India Conference (INDICON).*
3. Pazin, L. & Leviatan, Y. (2013). Reconfigurable slot antenna for switchable multiband operation in a wide frequency range. *IEEE Antennas Wireless Propagation Letters, 12,* 329–332.
4. Zhai, H., Ma, Z., Han, Y. & Liang, C. (2013). A compact printed antenna for triple band WLAN/WiMAX applications. *IEEE Antenna and Wireless Propagation Letters 12,* 65, 68.
5. Ren, F. C., Zhang, F. S., Bao, J. H., Chen, B., & Jiao, Y. C. (2011). Compact triple frequency slot antenna for wlan/wimax operation. *Progress In Electromagnetic Research Letters, 26,* 21–30.
6. Liu, W.-C., *Senior Member, IEEE*, Wu, C.-M., & Dai, Y., (2011). Design of triple-frequency microstrip-fed monopole antenna using defected ground structure. *IEEE Transactions on Antennas and Propagation 59*(7).
7. Pei, J., Wang, A.-G., Gao, S., & Leng, W. (2011). Miniaturized triple-band antenna with a defected ground plane for WLAN/WiMAX applications. *IEEE Antennas Wireless Propagation Letters, 10,* 298–301.
8. Wong, S. H., Mok, W. C., Luk, K. M., & Lee, K. F. (2013). Single-layer single-patch dual-band and triple-band patch antenna. *IEEE Transactions on Antennas and Propagation, 61*(8), 4341–4344.
9. Rakluea, P., Anantrasirichai, N., Janchitrapongvej, K., & Wakabayashi, T. (2009). Multiband microstrip-fed right angle slot antenna design for wireless communication systems. *ETRI Journal 31*(3).
10. Liu, W.-C., Wu, C.-M., & Chu, N.-C. (2010). A compact CPW-fed slotted patch antenna for dual-band operation. *IEEE Antennas Wireless Propagation Letters, 9,* 110–113.

DDA: An Approach to Handle DDoS (Ping Flood) Attack

Virendra Kumar Yadav, Munesh Chandra Trivedi and B.M. Mehtre

Abstract Distributed denial of service attack (DDoS) is an attempt by malicious hosts to overload website, network, e-mail servers, applications, network resources, bandwidth, etc. Globally DDoS attacks affected four out of ten organizations (around 41 %) over the past few years. Challenges involved in taking counter measures against DDoS attacks are network infrastructure, identifying legitimate traffic from polluted traffic, attacker anonymity, large problem space, nature of attacks, etc. Several approaches proposed in the past few years to combat the problem of DDoS attacks. These approaches suffer for many limitations. Some of the limitations include: implementing filtering at router (firewall enabled) will create bottleneck, additional traffic, no means of sending alert to an innocent host acting as a bot, etc. Ping flood attack is one kind of DDoS attack. In this paper, ping flood attack is analyzed and a new approach, distributed defence approach (DDA) is proposed to mitigate ping flood attack. Distributed defence is applied with the help of routers connected to network when count of PING request crosses a threshold limit or packet size is greater than normal ping packet size. Concept of the proposed approach is to help the end router by putting less load during filtering attack packets, enhancing the speed of processing and informing the innocent host acting as bot simultaneously making the DDoS attack ineffective.

Keywords Distributive defence approach (DDA) · PING · Intrusion prevention system (IPS) · Message to source address (MtSA) · Next reset time (NRT)

V.K. Yadav (✉) · M.C. Trivedi
CSE Department, ABES Engineering College, Ghaziabad, Uttar Pradesh, India
e-mail: virendrashines@gmail.com

M.C. Trivedi
e-mail: Munesh.trivedi@gmail.com

B.M. Mehtre
IDRBT, Hyderabad, Andhra Pradesh, India
e-mail: bmmehtre@idrbt.ac.in

© Springer Science+Business Media Singapore 2016
S.C. Satapathy et al. (eds.), *Proceedings of International Conference on ICT for Sustainable Development*, Advances in Intelligent Systems and Computing 408, DOI 10.1007/978-981-10-0129-1_2

11

1 Introduction

DDoS attack is a kind of attack in which the attacker target the victim network resources such as bandwidth, memory, etc., so that victim may stop responding to legitimate users. The flood created by attacker forces victim to shut down for its legitimate user thus causing denial of service to its legitimate user. DDoS attack is also known as bandwidth attacks. DDoS attack can target many different network components such as firewalls, routers, ISPs, data centers, servers, appliances, etc. In DDoS attack, attacker creates the networks of bots also known as zombies by spreading malicious softwares. Sources of spreading malicious software could be emails, social media, Trojan viruses, malware, etc. Once infected, the machine will act as bot following attacker instructions remotely without their owner's knowledge. The collection of bots is commonly known as botnets. If number of bots involved during the DDoS attack is high, situation become more complex. Bots amplifies the power of attacker simultaneously making defence more complicated.

PING stands for Packet InterNet Groper. Mike Muuss has written the PING program to check the reachability of another host. PING uses two ICMP query messages: ICMP (ECHO request) and ICMP (ECHO reply). When a source make ICMP (ECHO (PING) request)) to another host, according to RFC 0792 guidelines, that host must reply with ICMP (ECHO (PING) reply)) after receiving the request from source. In Ping flood attack, attacker with the help of bots send several ICMP echo requests to victim without waiting for reply (Fig. 1). Now victim according to guidelines of RFC 0791 after receiving the ICMP echo request tries to reply with ICMP echo reply packets to source. Attacker sends request packets as fast as possible to consume bandwidth or network resources of victim, forcing victim to shut down or slowdown.

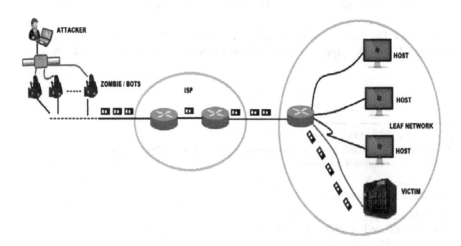

Fig. 1 DDoS attack overview

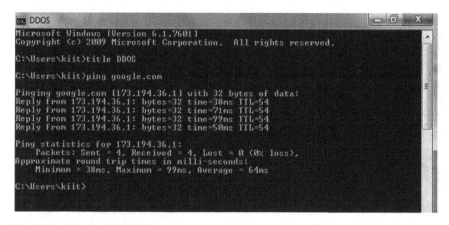

Fig. 2 Ping command overview

By default, the ICMP echo request packet contains data of 32 bytes under Windows (Fig. 2) and 56 bytes under Linux. But attacker can send data which can be greater than 32 bytes.

By default, the ICMP echo request packet contains data of 32 bytes under Windows (Fig. 2) and 56 bytes under Linux. But attacker can send data which can be greater than 32 bytes.

The length of an IP packet is 1514 bytes, maximum packet size supported by Ethernet. If attacker sends data in ICMP echo request which is more than 1500 bytes then sender or router will make fragments of this packet and set flag equal to one in flags field (Fig. 3). For example, suppose an attacker send 65,500 bytes data in each ICMP echo packet then it will get fragmented and receiver host (victim) should reassemble these IPv4 packets at their side. Sometimes when attacker sends

```
62 2.099681000 192.168.3.13 192.168.3.65 IPv4 1514 Fragmented IP protocol (proto=ICMP 1, off=0, ID=6b15) [Reassembled in #106]
⊞ Frame 62: 1514 bytes on wire (12112 bits), 1514 bytes captured (12112 bits) on interface 0
⊞ Ethernet II, Src: HonHaiPr_14:e2:1c (c0:18:85:14:e2:1c), Dst: IntelCor_8e:dd:a4 (a0:88:b4:8e:dd:a4)
⊟ Internet Protocol Version 4, Src: 192.168.3.13 (192.168.3.13), Dst: 192.168.3.65 (192.168.3.65)
    Version: 4
    Header length: 20 bytes
  ⊞ Differentiated Services Field: 0x00 (DSCP 0x00: Default; ECN: 0x00: Not-ECT (Not ECN-Capable Transport))
    Total Length: 1500
    Identification: 0x6b15 (27413)
  ⊟ Flags: 0x01 (More Fragments)
      0... .... = Reserved bit: Not set
      .0.. .... = Don't fragment: Not set
      ..1. .... = More fragments: Set
    Fragment offset: 0
    Time to live: 128
    Protocol: ICMP (1)
  ⊞ Header checksum: 0x226d [validation disabled]
    Source: 192.168.3.13 (192.168.3.13)
    Destination: 192.168.3.65 (192.168.3.65)
    [Source GeoIP: Unknown]
    [Destination GeoIP: Unknown]
    Reassembled IPv4 in frame: 106
⊟ Data (1480 bytes)
    Data: 080072c4000100b861626364656667686696a6b6c6d6e6f70...
    [Length: 1480]
```

Fig. 3 Captured packets (ICMP echo request)

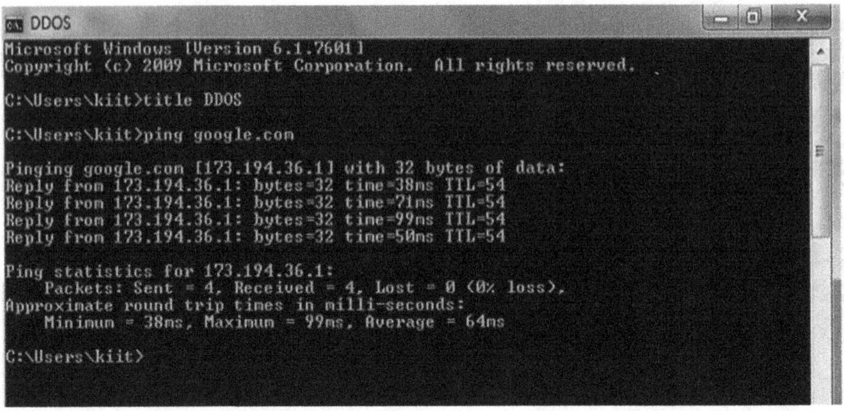

Fig. 4 Windows ping valid range (screen shots)

data which is more than victim machine can handle, then victim machine can crash (ping of death). For security reasons, Windows has fixed this size up to 65,500 bytes (Fig. 4).

1.1 Additional Information Required by the Volume Editor

If you have more than one surname, please make sure that the Volume Editor knows how you are to be listed in the author index.

1.2 Copyright Forms

The copyright form may be downloaded from the For Authors section of the LNCS Webpage: www.springer.com/lncs. Please send your signed copyright form to the Contact Volume Editor, either as a scanned pdf or by fax or by courier. One author

The important question is what are the consequences of DDoS attack? The consequences depend upon the intensions of the attacker and his success rate. According to Ponemon institute study, the average cost due to DDoS attack is $22,000 when downtime is equal to 1 min [1]. There are several variables to determine these costs; for example, volume of online business, brand value, competitors and business segment. Latest impact of DDoS attack is published in the South China morning post Hong Kong titled 'Cyberattack threatens to derail Hong Kong's unofficial vote on universal suffrage' (Fig. 5) [2].

The work in this paper presents distributed defence approach to prevent ping flood attack with the help of neighbouring routers simultaneously maintaining the efficiency of end router of victim network. The rest of the paper is organized as

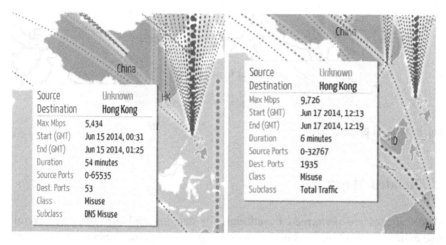

Fig. 5 Ping flood attack at Hong Kong [3]

follows: Sect. 2 contains the related works, Sect. 3 contains the proposed concept, Sect. 4 contains the proposed algorithm and Sect. 7 contains the conclusion of the proposed work (Figs. 6, 7 and 8).

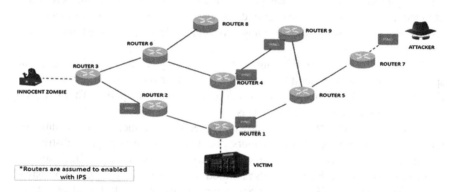

Fig. 6 Ping flood attack

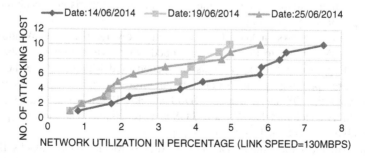

Fig. 7 Bandwidth consumption during PING flood attack

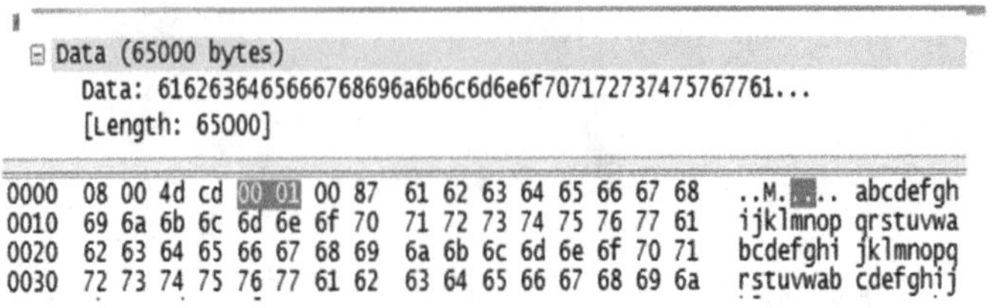

```
⊟ Data (65000 bytes)
    Data: 6162636465666768696a6b6c6d6e6f7071727374757677 61...
    [Length: 65000]

0000  08 00 4d cd 00 01 00 87  61 62 63 64 65 66 67 68   ..M.... abcdefgh
0010  69 6a 6b 6c 6d 6e 6f 70  71 72 73 74 75 76 77 61   ijklmnop qrstuvwa
0020  62 63 64 65 66 67 68 69  6a 6b 6c 6d 6e 6f 70 71   bcdefghi jklmnopq
0030  72 73 74 75 76 77 61 62  63 64 65 66 67 68 69 6a   rstuvwab cdefghij
```

Fig. 8 Wireshark screen shot-1

2 Related Work

DDoS attacks are performed by attacker for denying end-user service. Several researchers are working in this domain since past few years. Authors in the paper titled 'Towards User Centric Metrics for Denial of Service Measurement' introduced the concept of DoS impact metrics. With the help of these metrics, one can measure the QoS that is experienced by end user when it is receiving DDoS attack [1]. According to authors, measurement approaches that are followed during DDoS attacks are imprecise and incomplete. Percentage of failed transactions (PFT) is the main impact measure for each category of applications. PFT means the percentage of failed transactions that have occurred during the DDoS attack. In this paper, they also define the threshold model whose concept is developed, based on the past findings.

Concept of Hop-Count Filtering was proposed by authors to distinguish the spoofed IP packets from legitimate packets [2]. Server on the basis of time-to-live (TTL) value in the IP header and IP addresses, creates mapping. This mapping helps to identify the spoofed IP packets.

In the article, titled 'Survey of network-based defence mechanisms countering the DoS and DDoS problems', authors conducted the survey about distributed denial of service attack. They discussed the various kinds of DDoS attack such as protocol-based bandwidth attacks (SYN flood and ICMP flood), application-based bandwidth attack (HTTP flood and SIP flood), distributed reflector attacks, DNS amplification attacks and infrastructure attacks. The article also presents methods or strategies available to defend against the DDoS attack. Comparison of each method has also provided by authors [3].

Apple and Windows in late 2000s released Snow Leopard and Windows 7 respectively. According to them, the developed OS provides user a reliable and safer operating system. No experiments were conducted which evaluate the reliability of these operating system. So authors put efforts in conducting experiments, i.e. how both the OS faced against DDoS attacks. Based on the experimental results, authors concluded that Window 7 OS is more reliable than Snow Leopard in limiting adverse effects of DDoS attacks [4].

Authors in the paper, titled 'Defending Against Meek DDoS Attacks By IP Traceback-based Rate Limiting analyses a rate limit algorithm', i.e. maxmin-based rate limit algorithm. They try to put the algorithm fairness during the meek DDoS attack. Meek DDoS attack takes place when the bot or zombie behaves like a legitimate user. In that case, it is quite difficult to differentiate the legitimate traffic with the polluted traffic. Based on the analysis, authors proposed IP traceback-based rate limiting algorithm [5].

Subramani Rao and Sridhar Rao in their paper concluded with the help of experimental analysis about the role of network topology. According to them, topology of networks decides few important things such as: traffic amount passing through it, number of network elements, rate limiting, etc. [6]. Authors on the page number 39 mentioned one important line, i.e. in case of IPv6, DDoS attack would be stronger (88 %) in comparison with IPv4 [6].

Udaya Kiran Tupakula, Vijay Varadharajan in their paper analyzed the popular traceback technique. They also consider the real-time situation in which they raise certain issue such as how the attacker remains anonymous and remain untraced if any of these traceback techniques have been applied. Some of the IP traceback techniques considered in this paper are: single-packet IP traceback, IP packet marking technique and ICMP traceback technique, etc. [7].

Authors in this survey presented the study about the botnet. Nowadays, the 40 % of the host which are on the internet are infected and follow the instructions given by the attacker. Many of them are also unaware that they are acting as bot. In the study of botnet, focus has been put mainly in three areas: botnet understanding, tracking, detecting botnet and countering against botnet [8].

The authors in the paper proposed the system DoSTRACK. DoSTRACK system according to authors can handle the TCP SYN and reflection DDoS attacks [9].

Although there are several techniques available in the today's world, to mitigate with the DDoS attack or to prevent attack, still they have certain limitations.

- Implementing certain techniques will result in boosting DDoS attack traffic, for example, ICMP traceback.
- Implementing certain technique will cause the traffic when there is no DDoS attack, which is not encouraging, for example: single IP traceback.
- DoSTRACK approach seems to encouraging but it also has some limitations such as
- It works well for spoofed address but what will the case when attacker does not use the spoofed address. As nowadays, attacker needs not to spoof the source address.
- Also victim has to wait for certain threshold to initiate the attack prevention. Generally, ICMP (ECHO (PING) request) packet by default contains 32 bytes of data. If we receive the ping request packet which is carrying a payload for example say 1000 bytes, it cannot be considered as normal ping packet. So after receiving the ping packet of such payload, the immediate action is required without waiting for the certain threshold.
- How to send alert to a host who is unknowingly acting as bot?

The proposed technique will focused on ICMP ping flood attack. The technique tries to overcome the limitations of previous proposed concepts.

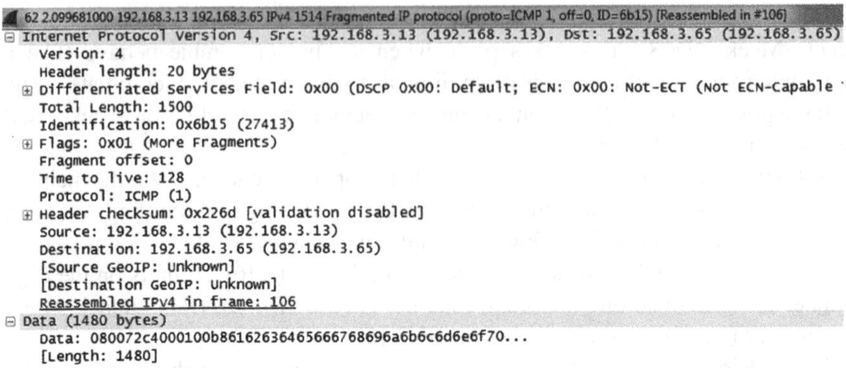

```
62 2.099681000 192.168.3.13 192.168.3.65 IPv4 1514 Fragmented IP protocol (proto=ICMP 1, off=0, ID=6b15) [Reassembled in #106]
☐ Internet Protocol Version 4, src: 192.168.3.13 (192.168.3.13), Dst: 192.168.3.65 (192.168.3.65)
    Version: 4
    Header length: 20 bytes
  ☐ Differentiated Services Field: 0x00 (DSCP 0x00: Default; ECN: 0x00: Not-ECT (Not ECN-Capable
    Total Length: 1500
    Identification: 0x6b15 (27413)
  ☐ Flags: 0x01 (More Fragments)
    Fragment offset: 0
    Time to live: 128
    Protocol: ICMP (1)
  ☐ Header checksum: 0x226d [validation disabled]
    Source: 192.168.3.13 (192.168.3.13)
    Destination: 192.168.3.65 (192.168.3.65)
    [Source GeoIP: Unknown]
    [Destination GeoIP: Unknown]
    Reassembled IPv4 in frame: 106
☐ Data (1480 bytes)
    Data: 080072c4000100b8616263646566676869a6b6c6d6e6f70...
    [Length: 1480]
```

Fig. 9 Wireshark screen shot-2

3 Proposed Work

This section contains the description of our proposed concept. Approach is to follow the distributive defence, when victim is receiving ping flood attack (DDoS attack). To understand the proposed work let us consider the Fig. 9 given below:

Before we start discussing the proposed model, consider the following assumptions:

- Routers are not compromised.
- Routers have capability to inspect the packet or routers are enabled with intrusion prevention system (IPS).

3.1 Case 1

When victim started receiving several ICMP echo request.

Subcase-1: when attacker does not spoofed the source address of botnets.

Suppose victim started receiving the ping flood attack. The end router-1 (R_1) which is connected with victim will generate an ICMP echo reply which will contain an alert message. R_1 will generate an alert message depending on certain conditions such as if it crosses a threshold of count, i.e. T_{count} or packet size is greater than normal ping packet size (by default windows ping with 32 bytes of data). R_1 will write three things in alert message, i.e. ACTION, message to source address (MtSA) and next reset time (NRT). R_1 will write these three things in

ICMP echo reply (data portion of the packet which contains default data. For example, suppose we are using windows OS, when we ping, the default data which a ping packet contains is abcd...w, repeatedly). In the data portion of ping packet, we can write these three things. R_1 will issue an alert message to all those router through which it is receiving the attack. Here in this case R_1 will issue an alert message to R_2, R_4 and R_5.

After receiving the alert message from R_1, R_2 will start dropping the ping packets whose destination address is victim IP address (VIP). R_2 will forward this packet to its upstream router, i.e. R_3. R_3 after receiving the ICMP echo reply packet (which contains the alert message) will start dropping the ping packets whose destination address is same as VIP. From the above example, it is clear that one or more innocent bots are connected to this router, i.e. R_3. Suppose IP address of bots (host) is not spoofed by the attacker. It means here we can assume that ICMP echo reply will reach to these bot or their end router. In this case, R_3. R_3 which is IPS enabled, will inspect the data field ICMP echo reply packet and read the ACTION, MtSA, NRT. R_3 will take necessary action that will ensure that in future or at least at that time the host will be prevented from acting as bot. The ICMP echo reply (alert message) will reach to its destination only, if message generating host, i.e. R_1 (in this case), how it chooses TTL field.

Subcase-2: when attacker spoofed the source address of botnets.

Here we have assumed that attacker does not spoofed IP addresses of its bots. Now consider the case, if addresses of bots are spoofed by the attacker; in this case, it is difficult to locate the bot and MtSA is of no use and ICMP echo reply will reach to spoofed address which is not participating in attack. Since ICMP echo reply is small packet, will not create trouble for the spoofed address who receive this. But it will alert the routers through which it pass to drop ping packets.

In similar fashion, all the routers will follow the same strategy during the ping flood attack.

3.2   Case 2

When victim started receiving several ICMP echo reply (reflector attacks).

This is a case in which it is receiving ICMP echo reply from several bots. In such cases, router R_1 will follow the same procedure as in case 1, but here it has to send an extra packet as alert message.

4 Proposed Algorithm

4.1 Algorithm-1 Victim Router

```
if ((length PING (ICMP(echo request)))≥ 1500 bytes OR
      COUNT = T_HIGH)
   {set (flag = ALERT)

   ICMP(echo reply(data))← write (ACTION, MtSA, NRT)
                  // this is one time reply send by route to alert
                                           neighbouring nodes
   DROP_ICMP echo request
   }
Else {echo reply}
```

4.2 Algorithm-2 Intermediate Router

```
if (ICMP echo reply (flag == ALERT))
   {  INSPECT ICMP echo reply (data)
            // check for spoofed source address (Egress Filtering)
If(INTERFACE_ROUTER (ICMP echo request=INTERFACE_ROUTER
   allowed))
   {FORWARD ICMP echo reply
    perform ACTION
          // perform action according to mentioned in the data field
                             of ICMP echo reply packet
      }
else { DROP
      // ICMP echo reply
      }}
else
     {  FORWARD ICMP echo reply  }
```

4.3 Algorithm-3 Attack Machine

```
if RECEIVED ICMP echo reply (flag == ALERT)
    {
        INSPECT ICMP echo reply (data)
        take ACTIONS
    }
else do nothing
```

5 Experimental Observations

To better understand the problem of ping flood attack, experiments have been conducted in MBS lab with the help of ten hosts in controlled environment.

The practice of ping flood is performed on three different days (Fig. 4). From the above graph, it is clear that on increasing the number of attacking hosts on y-axis, increase in network utilization is observed on x-axis. It means if the number of attacking host will increase, results in bandwidth consumption of victim host or server will increase.

We also recorded some more observation with the help of Wireshark. Some of the interesting observations are:

- The data which an ICMP echo request carries is default data which contains no meaningful information, for example: a, b, c, d…repeatedly.
- If attacker sends the packet which is of larger size say 65,500 bytes, then it gets fragmented and the ICMP request packet will contain the information about the frame number after which all packets will get reassemble. Consider the screen shots given below:

Conclusion of point number 2 is sending big volume of packet by the attacker to victim, to consume the processor time in rearrangements of packets.

Based on the observations which was obtained through experiments, we have developed some concepts, already been discussed in proposed work.

6 Efficiency (in Terms of Time, Bandwidth and Buffer Size)

Suppose number of packets reaching to end leaf router are $= n$. Let us assume that each packet take t time to drop. Total time consumed by these packets $= n \times t$. Consider two one step upstream router are started dropping. Suppose router one and two drop x and y packets, respectively, then number of packets reaching to end leaf router are $n - (x + y)$

Conclusion:

- $n > n - (x + y)$//number of packets now end router has to process is less.
- less packets $\rightarrow$ less processing time
- less packets $\rightarrow$ less buffer size needed to store, avoid legitimate packet to drop
- also alert bot to take security measures.
- No extra packet is needed

7 Conclusion

Conclusions are as follows:

- Approach is distributive, as we are not creating bottleneck at router R1 (considered example) during ping flood. As in some initial approaches, all the filtering is applied at end router.
- We are not creating any extra packet for generating the alert message, i.e. sending alert message in ICMP echo reply (data field).
- If address is not spoofed, then alert message will reach to bot who is unaware about its activity. The bot IPS or end router which is IPS enabled will take certain steps not to participate in attack at present, also in future.
- In case if address of bot is spoofed to fake source address, then it seems that alert message is of no use. But in that case it will also inform the router through which it passes, to drop the ping packet whose destination address is same as VIP. Thus helping in mitigating against the DDoS attack.

Acknowledgments I would like to thanks my guru Dr. B.M. Mehtre whose is also co-author of this paper, for his kind nature, faith and of course the guidance from time-to-time during the fellowship program. I would also like to thanks my parents, whose blessings are constant inspiration to me and also to Varsha Yadav for sharing his valuable time to share her knowledge.

References

1. Mirkovic, J., Hussain, A., Wilson, B., Fahmy, S., Reiher, P., Thomas, R., Yao, W., & Schwab, S. (2007). Towards user centric metrics for denial of service measurement. ExpCS'07, San Diego, CA. 13–14, June 2007. Copyright 2007 ACM.
2. Jin, C., Wang, H., & Shin, K.G. (2003). Hop-count filtering: an effective defense against spoofed DDos traffic. CCS'03, October 27–31, 2003, Washington, DC, USA. Copyright 2003 ACM 1-58113-738-9/03/0010.
3. Peng, T., Leckie, C., & Ramamohanarao, K. (2007). Survey of network-based defense mechanisms countering the DoS and DDos problems. *ACM Computing Surveys, 39*(1), Article 3 (April 2007), p. 42. doi:10.1145/1216370.1216373.

4. Kumar, S., & Surisetty, S. Microsoft versus apple: resilience against distributed denial-of-service attacks. *IEEE Security and Privacy, 10*(2).
5. Jing, Y., Wang, X., Xiao, X., & Zhang, G. (2006). Defending against meek DDos attacks by IP traceback-based rate limiting. *Global Telecommunications Conference, 2006. GLOBECOM '06*. IEEE, November 27 2006–December 1 2006.
6. Rao, S., & Rao, S. (2011). Denial of service attacks and mitigation techniques: Real time implementation with detailed analysis. This paper is from the SANS Institute Reading Room site©.
7. Kiran Tupakula, U., & Varadharajan, V. (2006). Analysis of traceback techniques. This paper appeared at the Fourth Australasian Information Security Workshop (AISW-NetSec 2006), Hobart, Australia. Conferences in Research and Practice in Information Technology (CRPIT), vol. 54.
8. Zhu, Z., Lu, G., Chen, Y., Judy Fu, Z., Roberts, P., & Han, K. (2008). Botnet research survey. *Annual IEEE International Computer Software and Applications Conference*, © 2008 IEEE.
9. Kiran Tupakula, U., Varadharajan, V., & Pandalaneni, S. R. (2009). DoSTRACK: a system for defending against DoS attacks. *SAC '09 Proceedings of the 2009 ACM Symposium on Applied Computing* (pp. 47–53). New York, NY, USA: ACM: ©2009.
10. Yuan, D., & Zhong, J. (2008). A lab implementation of SYN flood attack and defense. *SIGITE '08 Proceedings of the 9th ACM SIGITE Conference on Information Technology Education* (pp. 57–58). New York, NY, USA: ACM ©2008.
11. Bolz, C., Romney, G. W., & Rogers, B. L. (2004). Safely train security engineers regarding the dangers presented by denial of service attacks. *Proceeding CITC5 '04 Proceedings of the 5th Conference on Information Technology Education* (pp. 66–72) New York, NY, USA: ACM ©2004.
12. Batham, S., Yadav, V. K., & Kumar Mallik, V. K. (2014). ICSECV: An efficient approach of video encryption. In *Proceedings Contemporary Computing (IC3), 2014 Seventh International Conference*, 7–9 August 2014, pp. 425–430 (Available at IEEE Xplorer and DBLP, indexed by SCOPUS).
13. Batham, S, & Yadav,V. K. et al. (2013). A new video encryption algorithm based on indexed based chaotic sequence. In *Proceedings Fourth International conference Confluence 2013: The Next Generation Information Technology Summit*, September 27–28 (pp. 139–143). Available at IET and IEEE xplorer.
14. Yadav, V. K., & Batham, S. et al. (2013). Hiding large amount of data using a new approach of video steganography. In *Proceedings Fourth International Conference Confluence 2013: The Next Generation Information Technology Summit*, Sept 27–28, pp. 337–343 (Available at IET and IEEE xplorer).

Webliography

15. http://blog.radware.com/security/2013/05/how-much-can-a-DDos-attack-cost-your-business/.
16. http://www.scmp.com/news/hong-kong/article/1534725/cyberattack-threatens-derails-occupycentrals-unofficial-referendum.
17. http://www.digitalattackmap.com/#anim=1&color=2&country=ALL&time=16238&view=map.

Approach to Accurate Circle Detection: Multithreaded Implementation of Modified Circular Hough Transform

Virendra Kumar Yadav, Munesh Chandra Trivedi, Shyam Singh Rajput and Saumya Batham

Abstract Several applications of real-world industries, for example, detection of pellets in pelletization plant, target detection, inspection of manufactured products, ADAS, etc., requires detection of circular features of objects from digital images. Researches have performed in the area of development of algorithms from past few years to detect circular features. Major limitations of these proposed algorithms were either time complexity or accuracy in detection. One powerful concept for circle and ellipse detection is the circular Hough transform and its variants. This paper presents an algorithm which is based on multithreaded concept of modified CHT. Multithreaded approach is used to reduce the time complexity. Simulation results of real industrial images validate the efficiency of proposed algorithm in terms of accuracy in detection and also time efficient.

Keywords Modified circular Hough transform · Accumulators · Multithread · Local maxima · Safe house test

1 Introduction

Every day, we come across several real-world objects that are similar to shape like circle or ellipse. Several researchers were focusing in the area of detection of circular objects or elliptical features from the digital images. Detection of circular

V.K. Yadav (✉) · M.C. Trivedi · S.S. Rajput · S. Batham
CSE Department, ABES Engineering College, Ghaziabad, Uttar Pradesh, India
e-mail: virendrashines@gmail.com

M.C. Trivedi
e-mail: Munesh.trivedi@abes.ac.in

S.S. Rajput
e-mail: ershyamrajput@gmail.com

S. Batham
e-mail: saumyabatham003@gmail.com

© Springer Science+Business Media Singapore 2016
S.C. Satapathy et al. (eds.), *Proceedings of International Conference on ICT for Sustainable Development*, Advances in Intelligent Systems and Computing 408, DOI 10.1007/978-981-10-0129-1_3

objects in digital images is important for image analysis in various computer vision applications such as detection of pellets in pelletization plant, advanced driver assistance systems (ADAS), inspection of manufactured products, target detection, aided vectorisation of drawings, etc. This domain has been researched using various methods in past few years.

With the help of algorithm which was based on swarm intelligence technique (BFO), leads to the automatic detection of circles in the digital image according to authors [1]. The proposed concept of chaotic hybrid algorithm (CHA) according to authors combines the strengths of particle swarm optimization, genetic algorithms, and chaotic dynamics, and involves the standard velocity and position updating rules of PSO with the ideas of GA selection, crossover and mutation [2]. One such scheme of detection was proposed with point triplets possessing right angle property [3]. According to authors, proposed scheme will reduce the enumerations to nC2 from nC3. Authors also presented concept of randomized algorithm (RCD, which is not based on the Hough transform) to overcome the time complexity of CHT [4].

Select four edge pixels in the image and define distance criterion, author named it as RCD concept. If any possibility of circle is detected then, apply an evidence-collecting process to determine whether possible circle is true or not. A two-step circle detection algorithm shows how a pair of two intersecting chords locates the center of the circle [5]. A method that proceeds on two major steps: first, detect N edge points in an $M \times M$ image in both horizontal and vertical scan to construct symmetrical axes using two one-dimensional arrays. The axes cross-point determines the object center [6]. Contribution to this domain still continues. Authors presented an interpretation of the maximum-likelihood estimator (MLE) and the Delogne-Kasa estimator (DKE) for circle center and radius estimator in terms of convolution of an image [7]. In September 2007, SLIDE (subspace-based line detection) method proposed to estimate the center and radius of a single circle [8].

Concept of right triangles inscribed in a circle to detect circle in an image which was basis of semi-random detection (SRD) approach, had also proposed [9]. Sewisy, suggested technique that can overcome the limitation of HT-based algorithm. As HT-based algorithm can detect straight lines in a edge, but the extension of HT to detect circular objects in digital images has been limited by large accumulator storage also the low speed or time consuming. Authors suggested some techniques to overcome the limitation of CHT [10].

An ellipse detection algorithm PRANSAC based on pseudorandom sample consensus was proposed. In addition, the parallel thinning algorithm is employed to eliminate useless feature points, which increases the time efficiency of detection algorithm [11].

Circular Hough transform is used for detection of straight line in an edge. Its variant named as circular Hough transform is most common technique for circle detection in digital images. The Hough transform can be described as transformation to the parameter space from X, Y-plane [12]. Mathematically, equation of circle in x, y-plane is given by

$$r^2 = (x - a)^2 - (y - b)^2 \tag{1}$$

where r is the radius of the circle and a, b are the center of the circle. Equations 2 and 3 are parametric representation of the circle.

$$x = a + r \cos \Theta \tag{2}$$

$$y = b + r \sin \Theta \tag{3}$$

The circle has three parameter r, a, and b, hence the parameter space will belong to R^3.

We applied the CHT algorithm to real industrial images which we obtained from pelletization plant. The image contains the pellets which are circular in nature but not exact circle. Task is to determine the number of pellets and the size of pellets. But the results which we obtained after the implementation of CHT on given industrial images are not encouraging. The result contains counts which is greater than the pellets in the image. This is due to nature of pellets, as they look circular but not exact circles. For detection of pellet in digital images, we extended the concept of local maxima to different accumulators. Concept of safe house is also proposed to obtain more accurate results. We have also proposed the concept that helps in detection also removal of false circles [13]. The proposed concept enhances the accuracy in detection but consumes time [14]. So time is again a factor which motivated us to futher work on this project.

This paper presents an algorithm which works for this type of real images with enhanced accuracy of detection simultaneously consuming less time. The remainder of this paper is organized as follows. This section will followed by the concept of proposed work, proposed algorithm, simulation results, and conclusion, respectively.

2 Proposed Work Concept

Variants of the CHT have been widely implemented, which more commonly reduce its computational complexity or to increase the detection rate of the algorithm. In this section, we have proposed the concept which will increase in the accuracy of detection simultaneously reduction in time of processing.

The main concept of proposed algorithm is discussed in this section. Previously proposed algorithm improves accuracy in detection but in terms of time, the concept is not encouraging [14]. So target is to find out the steps of algorithm which were consuming most of the time. With the help of graph, we are able to locate the steps which is consuming most of the time. As most of the time in previous proposed algorithm is consumed during the calculations involve in making accumulator cell. It means if radius range is large, then numbers of accumulators to be constructed are more; and hence consume more time, making algorithm performance poor. So we modified our proposed algorithm [14]. Here we use the concept of multiple threads.

As construction of accumulators are independent task, divide the task of constructing accumulators among threads. These threads will execute simultaneously hence saving time.

To make concept more clear, suppose a user enters radius range: 22–38. It means numbers of accumulators to be constructed are 38 − 22, i.e., 16. Now suppose we create four threads then each thread makes at least four accumulators. First, assign each thread task of constructing accumulator. So four threads at the same time will construct four accumulators. Similarly, they will do for constructing other accumulators. Now suppose if constructing an accumulator takes one second, then four accumulators will be constructed in one second with the help of four threads thus saving the time of 3 s. The remaining text in this section will give the concept that how we integrated the concept of multiple threads in our previous proposed algorithm.

Find edges using any of the edge detection technique such as Sobel or Canny in the image. In this paper, we have detected edges with the help of canny edge detection technique. The next step is formation of accumulator. The number of accumulators have to be made will depend on the radius range, either given by the user of the application or already defined in program. Suppose radius range defined in program is (R_i, R_j). Then $R_j - R_i$ will give the number of accumulators to be made and searched out in order to carry out desired task of detection. The accumulator is a matrix whose size is same as image size. After finding out the edges in the image, the desired next task is to find out the pixels coordinates that actually lies on the edges. Consider each edge pixel and draw a circle of size having radius equal to R_i (R_i lie within the radius range). The circle is drawn in parameter space (Fig. 1). For parameter space, a-value corresponds to x-axis (Eq. 1) and b-value corresponds to y-axis (Eq. 2), while the z-axis is the radii. Suppose if we consider the radius range 22–38. So $i = 22$ and $j = 38$. Total number of accumulators to be made is equal to $j - i$. Initially, set all the cell values in the accumulator matrix to zero. At the coordinates which lie on the perimeter of the drawn circle in parametric space, increment the value in accumulator matrix. Sweep every edge point in the input image and increment the cell values in the accumulator matrix.

For example, consider radius range from 22 to 38. The numbers of accumulators to be made is 38 − 22 which is equal to 16. Construction of accumulator is an independent task. So if there are five threads, then these five threads can process

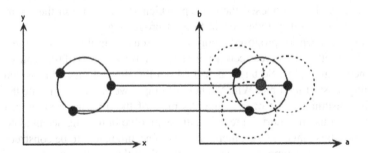

Fig. 1 Parametric space representation (*right*) of a constant radius circle (*left*)

$$A_1 = \begin{matrix}
31 & 22 & 37 & 57 & 46 & 34 & 27 & 20 \\
22 & 67 & 41 & 55 & 34 & 32 & 45 & 26 \\
45 & 24 & 34 & 33 & 21 & 47 & 46 & 64 \\
46 & 34 & 44 & 21 & 29 & 54 & 37 & 19 \\
31 & 34 & 46 & 47 & 32 & 35 & 44 & 13 \\
42 & 19 & 16 & 22 & 37 & 32 & 26 & 21 \\
57 & 51 & 43 & 19 & 23 & 34 & 29 & 31 \\
56 & 23 & 25 & 31 & 43 & 44 & 57 & 49
\end{matrix}$$

Fig. 2 Accumulator matrix sample

five accumulator cells of desired radius in the same time. This enhances the speed of execution.

This process is repeated for all desired radii which belong to radius range. If radius range is (R_i, R_j), then total number of accumulators needed is $R_j - R_i$.

When the preparation of the all the accumulator cells have been completed, then turn attention to the cell values that each accumulator contains. These cell values which an accumulator of particular radius will contain, known as votes or cuts. The cuts or votes simply indicate the number of circles passing through that cell coordinates. First, apply the concept of local maxima within the same accumulator. For each row of accumulator matrix, find out the local maxima. Figure 2 has one sample of accumulator matrix of order 8×8 (Fig. 2). Consider first row of A_1.

$$A_1(1, 1 : 8) = \{31, 22, 37, 57, 46, 34, 27, 20\}$$

Consider window size equals to five. First row of A_1 contains eight elements. Slide this window of size five to entire row to find local maxima.

From the above Figs. 3, 4, 5 and 6, at coordinates (1, 4) local maxima is found. Coordinates (1, 4) means first row and fourth column. Coordinates (1, 4) is temporary considered as center candidate. Coordinates (1, 4) has to pass additional test to confirm his candidature.

Extend the concept of local maxima to the different accumulators. To understand this concept let us consider above example. In the above example considered radius

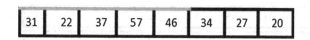

Fig. 3 Sliding window on first row of A_1 (maxima found = 0)

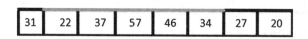

Fig. 4 Sliding window on first row of A_1 (maxima found = 57)

Fig. 5 Sliding window on first row of A_1 (maxima found = 0)

Fig. 6 Sliding window on first row of A_1 (maxima found = 0

range is 22–38, means 16 accumulators to be made. Denote P1 as first accumulator which is for radius equal to 22. Similarly, P2 is for radius equal to 23 and so on. Now elected coordinates, i.e., (1, 4) ϵ P3 (suppose). The coordinate (1, 4) has cell value equals to 57. Now this cell values is compared with the values 41, 18, 27, and 25 which belongs to accumulator P2, P1, P4, and P5, respectively. We are considering five accumulators as our window size is five. We can consider window size according to our application needs. The coordinates (1, 4) again passes the test of local maxima. Now it can be considered for the last test, i.e., safe house test. If the coordinates does not passes this test, then we have to discard the coordinate candidature. Figure 7 will explain this concept more clearly. Extending the concept of finding local maxima not in the same accumulator but also considering some more accumulators depending upon the window size chosen ensures more accuracy in terms of detection. Consideration of one more test is required for these kinds of real industrial images. After applying this concept on some of images may have more accurate results and some images results only reduction in false circles. This is due to nature of the image and source through which we obtained the image.

To remove these false circles obtained during the results, the concept of safe house test comes into role. The things needed are center coordinate and cell value (generally called votes). When we have applied CHT on one of the test images (Fig. 8a), the conducted in order to achieve desired accuracy rate. We store one of the center coordinates and its cell values to some place call it as a safe house. Take another center coordinate and apply distance threshold.

34	20	45	18	26	34	27	20	**P₁**
22	69	39	41	32	20	44	29	**P₂**
31	22	37	57	46	34	27	20	**P₃**
53	34	40	27	23	57	38	17	**P₄**
21	34	46	25	33	37	41	9	**P₅**

Fig. 7 Sliding window on different accumulators

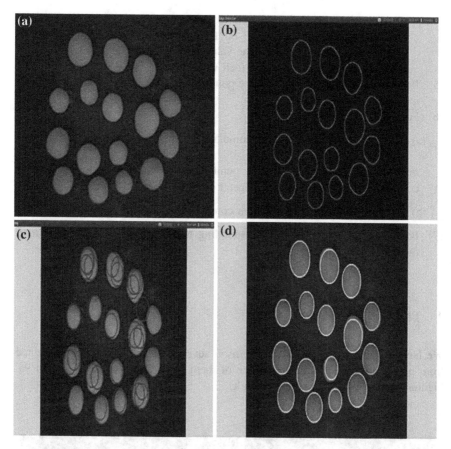

Fig. 8 **a** Original image mulltipellet.jpg. **b** Image after edge detection. **c** Image after applying CHT. **d** Result of proposed concept

3 The Proposed Algorithm

Based on the above concept, proposed algorithm to detect centers and radii is as follows:

(1) Get the edge map of the image using any of the edge detection technique.
    //we have used canny edge detector technique.
(2) Get the radius range from the user.
(3) Create n number of threads and divide the task among these n threads.
    //we have used five threads for our application.
(4) For each edge point

(a) Taking edge point as center with radius r ($r \in$ radius range), draw a circle. At the coordinates which lie on the perimeter of the drawn circle in parametric space, increment the value in accumulator matrix.
//collecting votes in accumulator (cell values)
(5) Repeat step-(3) and (4) for all edge points and all radii defined in radius range in step-2.
(6) For each maxima found

(a) Compare maxima of that accumulator to their consecutive previous and next accumulators.
//confirmation of candidature, the number of accumulators which have to be considered for comparing will depend on the size of considered window.

(b) If step (6) a. is confirmed, then conduct safe house test.

(7) Found parameters (r, a, b) corresponding to local maxima in step-(6), map these to original image.

4   The Proposed Algorithm

We have tested our proposed algorithm on several images which were obtained from pelletization plant industry. Some of them have shown in Figs. 8a and 9a. Figures 8a, and 9a shows positive results.

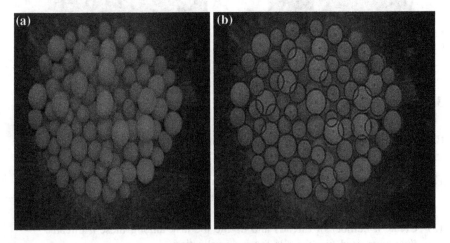

Fig. 9 a Morepellets.jpg. **b** Result of proposed concept

Table 1 Result of proposed concept on some industrial images

S. no.	Image name	Image dimensions	Radius range	CHT	Proposed concept (in terms of detection)		Proposed concept (in terms of time)	
					Using single thread	Using multithread (=5)	Using single thread	Using multithread (=5)
1	multipellet. jpg	640 × 480	24–42	Contains false circles	Accurate result (approx.)	Accurate result (approx.)	14 s	9 s
2	morepellets. jpg	720 × 540	18–40	Contains false circles	Accurate result (approx.)	Accurate result (approx.)	17 s	11

CHT works well for images in detection of circles. But the images which are under consideration contain the pellets. Shapes of pellets are not exact circles but their shape looks similar to circles. Table 1 shown below simplifies the picture of results more clearly.

As we can find from the Table 1 that using multithread improves the algorithm in terms of time.

5 Conclusion

The proposed work extends our previous work [14]. This extension of concept will refine the algorithm in terms of time. This works fine for all images. As previously proposed work is efficient in terms of detection but inefficient in terms of time. This motivates for further extension of the proposed work. Analysis of algorithm shows that most of the time consumed during processing in construction of accumulators. As construction of accumulators is an independent task, multithread concept can be used. The result obtained after implementation of multithread implementation improves the time of detection. It means the algorithm is efficient in terms of time and in detection rate.

Accurate detection of circular objects simultaneously reducing the complexity of algorithm is still the main domain of researches.

Acknowledgments We would like to thanks our professor S. Batham, Varsha, Rahul Paul, and our parents for their valuable suggestions and moral support which they have provided to us during discussion hours.

References

1. Dasgupta, S., et al. (2008). Automatic circle detection on images with an adaptive bacterial foraging algorithm. GECCO'08, July 12–16, 2008, Atlanta, Georgia, USA. ACM 978-1-60558-130-9/08/07.
2. Wu, C.-H., et al. (2010). Chaotic hybrid algorithm and its application in circle detection. *Evo Applicatons' 10: Proceedings of the 2010 International conference on Applications of Evolutionary Computation*—Volume Part I, Volume Part I. Springer-Verlag.
3. Lam, W. C. Y., & Yuen, S. Y. (1996). Efficient technique for circle detection using hypothesis filtering and Hough transform. *IEE Proceedings Visual Image Signal Process, 143*(5) (1996).
4. Chen, T., & Chung, K. L. (2001). An efficient randomized algorithm for detecting circles. *Computer Vision and Image Understanding, 83*, 172–191.
5. Kim, H. S., & Kim, J. H. (2001). A two-step circle detection algorithm from the intersecting chords. *Pattern Recognition Letters, 22.*
6. Goneid, A., et al. (1997). A Method for the hough transform detection of circles and ellipses using a 1-dimensional array. 0-7803-4053-1/97/$10.00@ 997 IEEE.
7. Zelniker, E. E., & Vaughan, I. (2006). Maximum-likehood estimation of circle parameters via convolution. *IEEE Transactions on Image Processing, 15*(4), (2006).
8. Marot, J., & Bourennane, S. (2007). Subspace-based and DIRECT algorithms for distorted circular contour estimation. *IEEE Transactions on Image Processing, 16*(9), (2007).
9. Shang, F., Liu, J., & Zhang, X., & Tian, D. (2009). An improved circle detection method based on right triangles inscribed in a circle. *2009 World Congress on Computer Science and Information Engineering.*
10. Sewisy, A. A. (2000). Detection of circular object with a high speed algorithm. *IEA/AIE '00 Proceedings of the 13th International Conference on Industrial and Engineering Applications of Artificial Intelligence and Expert Systems: Intelligent Problem Solving: Methodologies and Approaches*, pp. 522–533. Springer-Verlag, New York, Inc. Secaucus, NJ, USA ©2000.
11. Song, G., & Wang, H. (2007). A fast and robust ellipse detection algorithm based on pseudo-random sample consensus. *Proceedings of the 12th International Conference on Computer Analysis of Images and Patterns* (pp. 669–676) Berlin, Heidelberg: Springer-Verlag ©2007.
12. Pedersen, S. K. (2007). *Circular hough transform. vision, graphics, and interactive systems.*
13. Yadav, V. K. et al. (2013). False circle detection algorithm based on minimum support percentage and euclidean distance. *Proceedings International Conference on Emerging Trends and Applications in Computer Science (ICETACS-2013)*, September 13–14 (pp. 70–73) (Available at IEEE Xplore).
14. Yadav, V. K. et al. (2014). Approach to accurate circle detection: circular hough transform and local maxima concept. *Proceedings International Conference on Electronics and Communication Systems (ICECS)*, 2014, Feb 13–14, 2014, pp. 28–32 (Available at IEEE Xplorer).

Apache Hadoop Yarn MapReduce Job Classification Based on CPU Utilization and Performance Evaluation on Multi-cluster Heterogeneous Environment

Bhavin J. Mathiya and Vinodkumar L. Desai

Abstract Recently it is observed that Yahoo, Facebook, mobile devices, sensors, scientific instruments, etc., are generating a huge amount of data. It is a challenge to store, manage, process, and analyze this data. Apache Hadoop Yarn is a framework which provides a solution for big data. In this paper, we have evaluated the performance of Apache Hadoop Yarn MapReduce jobs such as Pi, TeraGen, TeraSort, and Wordcount on single cluster node. After evaluating performance; jobs are classified into various classes like low CPU intensive job, high CPU intensive job based on CPU utilization (%). Based on the classification, Apache Hadoop Yarn MapReduce jobs executed on multi-cluster environment and evaluated performance. It is found that execution time has increased for low CPU intensive job and decreased for high CPU intensive job. Also, a total CPU time is decreased for low and high CPU intensive job. In addition, CPU Utilization is decreased for low CPU intensive job and increased for high CPU intensive job when number of nodes increased.

Keywords Hadoop · Apache Hadoop Yarn · HDFS · MapReduce · Pi · TeraGen · TeraSort · Wordcount · Multi-cluster

1 Introduction

In recent years, lot of data is generated everywhere. Big data can be variety, velocity, volume of data. Variety means data can be of various type like structured, unstructured, and semi structured, which is generated by various type of devices

B.J. Mathiya (✉)
C.U. Shah University, Wadhwan City, Gujarat, India
e-mail: bhavinmath@gmail.com

V.L. Desai
Department of Computer Science, Government Science College,
Chikhli, Navsari, Gujarat, India
e-mail: vinodl_desai@yahoo.com

© Springer Science+Business Media Singapore 2016   35
S.C. Satapathy et al. (eds.), *Proceedings of International Conference
on ICT for Sustainable Development*, Advances in Intelligent Systems
and Computing 408, DOI 10.1007/978-981-10-0129-1_4

like sensor network, mobile phones, scientific instruments, websites etc., and it can be of any form like text, image, audio, video, and volumes of data can be Petabyte, Yottabyte generated in a day. Apache Hadoop Yarn is a framework which provides solution for big data. Apache Hadoop Yarn is developed by apache software foundation [1, 2] which has two main components: HDFS (Hadoop Distributed File System) and MapReduce. HDFS is a distributed file system, which stores and processes data in local as well as distributed file system. MapReduce is a programming model, which provides mapper and reducer functions for distributed processing based on key/value pair [3]. Vinod Kumar et al. [1] summarize the design, development, and current state of deployment of the next generation of Hadoop's compute platform: YARN. Apache Hadoop yarn is Hadoop version 2 and also known as a Yarn which means Yet another Resource Negotiator. In this paper, we have evaluated impact of cluster size by increasing node from 1 to 4 linearly and observed impact on performance like execution time, total CPU times, and CPU utilization through evaluating various Apache Hadoop Yarn MapReduce jobs.

This paper is organized as follows. Section 1 gives the introduction, Sect. 2, related work, Sect. 3, Apache Hadoop Yarn architecture with MapReduce and HDFS, classification of Apache Hadoop Yarn MapReduce jobs, Sect. 4, environmental setup, Sect. 5 results and observations of different Hadoop MapReduce jobs in Hadoop Yarn Single node as well as multi-cluster node, and Sect. 6, conclusion and future work.

2 Related Work

Maurya et al. [4] discuss various MapReduce applications such as Pi, Wordcount, Grep, TeraSort, and show experimental results with the finding that the number of nodes increases as the execution time decreases. Joshi [5] executed and evaluated different Hadoop workloads using various hardware and software tuning techniques including BIOS, OS, JVM, and Hadoop configuration parameters tuning. Zheyuan et al. [6] evaluated the performance of Grep, Wordcount, and Sort applications and measured usages of CPU, memory by changes in Hadoop configuration parameters values. Kamal et al. [7] found that the configuration parameter influences the resource allocation using map slot value (MSV) and observed that each Hadoop application has a unique MSV for which it has the best performance. Yao et al. [8] found that a static configuration may lead to low system resource utilizations as well as long completion length and introduced a scheme that uses slot ratio between maps and reduce tasks as a tunable knob. Wang et al. [9] developed predator which gives a guided configuration and evaluated and tuned Hadoop performance through various configuration parameters. Feng et al. [10] conducted an in-depth study of the energy efficiency for MapReduce workloads, evaluated, executed, and evaluated Hadoop performance and found that well-tuned system parameters and adaptive resource configurations improved overall performance and saved energy. Lin et al. [11] proposed a node performance measurement method and evaluated Hadoop performance by running MapReduce programs on a heterogeneous environment. Yamazaki et al. [12] proposed a method that decides the number of task executions in order to use computational resources efficiently based on a load on each computer. Dhok et al. [13] proposed a learning-based scheduler that uses pattern classification for utilization of oriented task assignment in MapReduce.

3 Apache Hadoop Yarn Architecture

In this section, Apache Hadoop Yarn architecture with its components is described [2] (Fig. 1).

Apache Hadoop Yarn architecture contains the following components.

1. Client: Client is user who is submitting the job.
2. Resource Manager: Resource manager is a core part of the Apache Hadoop Yarn architecture. Resource manager is also called a job tracker and is only one per cluster. Resource manager manages job scheduling and execution of Hadoop jobs and allocates resources globally to the respective cluster.
3. Name Node: Name node manages entire namespace for a Hadoop cluster. Multiple names are possible in Hadoop Yarn.
4. Job Scheduler: Resource manager has a pluggable scheduler which is responsible for allocation of resource to the running application. Various job schedulers are available such as FiFo, fair scheduler, capacity scheduler.
5. Node Manager: Node manager is a per machine agent (slave). Node manager manages the lifecycle of task container and provides health report, resource usage to the resource manager.
6. Data Node: Data node is a per machine agent (slave). Data node stores data.
7. Container: Container works as a resource allocator and grants rights to an application to a specific amount of resource such as CPU, memory, etc., on a host.

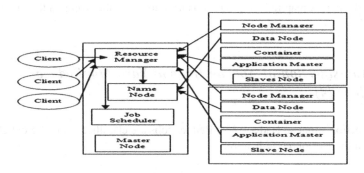

Fig. 1 Apache Hadoop Yarn architecture

8. Application Master: Application master is a per job. The main responsibility of application master is task scheduling, execution work of Hadoop jobs, a task tracker, and local resource allocator.

3.1 MapReduce [2]

MapReduce performs the computational part. It is a programming model based on key value pair. MapReduce provides mapper and reducer function for programing. The mapper function receives blocks of data from HDFS and processes and generates the intermediate key value pair. Reducer function receives intermediate key value pair and processes, shuffles, and generates the final results.

3.2 HDFS [2]

HDFS performs the data part. HDFS has two components. Name node manages name space for whole clusters and Data node stores individual data. HDFS stores blocks of data in file on distributed cluster machines.

3.3 List of Hadoop Yarn MapReduce Job [2]

- Pi: pi is a map reduce program that is used to estimate pi using a quasi-Monte Carlo method.
- TeraGen: TeraGen is used to generate any amount of data.
- TeraSort: TeraSort is used to sort data as per given input. TeraSort takes input from TeraGen and Sort data.
- Wordcount: A map/reduce program is used to count the words in the input files.

3.4 Classification of Apache Hadoop Yarn MapReduce Jobs [4]

In this section, Apache Hadoop MapReduce jobs are classified into different classes based on their resource utilization.

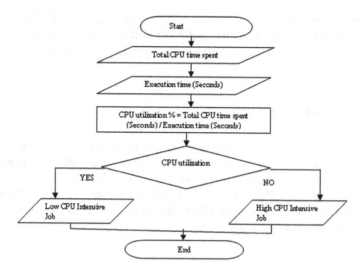

Fig. 2 Flowchart of CPU intensive job classification of job based on CPU

CPU Utilization % = (Total CPU times in Second/Execution time (Seconds))

$$* 100 \tag{1}$$

Classification of job based on CPU utilization

1. Low CPU intensive job
   If CPU utilization is less than 20 %, it is considered as a low CPU intensive job
2. High CPU intensive job
   If CPU utilization is greater than 20 %, it is considered as a high CPU intensive job

Figure 2 explains the flowchart of classification of Hadoop Yarn MapReduce job based on their CPU resource utilization. Here, the total CPU (seconds) and execution time (seconds) spent is taken as input. CPU utilization % is calculated based on the given input. Apache Hadoop Yarn MapReduce job is classified based on the condition that if CPU utilization % is less than 20, it is considered as a low CPU intensive job, otherwise it is considered as a high CPU intensive job.

4 Environmental Setup

The experiments were carried out on Hadoop 2.4 multi-cluster nodes. The Hadoop lab setup contains one master node and three geographically distributed slave nodes. Master node configuration comprises an operating system of Ubuntu 14.04.1 LTS, Intel(R) Core(TM)2 Duo CPU T5870 @ 2.00 GHz, and 3 GB of RAM. Slave nodes configuration comprises an operating system of Ubuntu 12.04.4 LTS, Intel(R) Core(TM) i3-3220 CPU@ 3.30 GHz and 4 GB of RAM. Hadoop 2.4.1, OpenSSH 6.7, Java version "1.7.0_55", etc., softwares were installed in each of the nodes.

5 Results and Observations

In this section, our observations of experiments performed on the Apache Hadoop Yarn MapReduce framework on single cluster node are presented. After evaluating the performance, jobs are classified based on various classes such as low CPU intensive job and high CPU intensive job based on CPU utilization (%). Based on classification Apache Hadoop Yarn MapReduce jobs are executed on multi-cluster environment and evaluated performance.

5.1 Performance Evaluation of Different Apache Hadoop Yarn MapReduce Job on Single Cluster Node

Table 1 and Fig. 3 display different Apache Hadoop Yarn MapReduce job performance results, execution time (seconds), total CPU times (seconds), and CPU utilization (%), which is executed on a single cluster node.

Table 1 Performance evaluation of different Apache Hadoop Yarn MapReduce jobs on single cluster node

Performance evaluation of Hadoop MapReduce jobs	Pi	TeraGen (1 GB data)	TeraSort (1 GB data)	Wordcount (1 GB data)
Execution time (s)	16	63	2144	1285
Total CPU times (s)	1.95	38.61	443.25	257.63
CPU utilization (%)	12.19	61.29	20.67	20.05

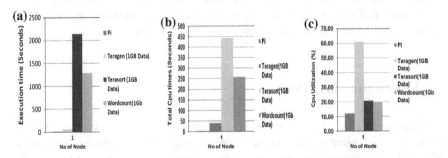

Fig. 3 Performance evaluation of different Apache Hadoop Yarn MapReduce jobs on single cluster node. **a** Execution time (seconds), **b** Total CPU times (seconds), **c** CPU utilization (%)

Table 2 Classification of Apache Hadoop Yarn MapReduce job based on CPU utilization

Classes	Pi	TeraGen (1 GB data)	TeraSort (1 GB data)	Wordcount (1 GB data)
Low CPU intensive job	Y	–	–	–
High CPU intensive job	–	Y	Y	Y

5.2 Classification of Hadoop Job Based on CPU Utilization

It is shown in this section that classification of Apache Hadoop Yarn MapReduce job is based on CPU utilization through evaluating performance on single cluster node. In Table 2, it is observed that Pi is in low CPU intensive job and TeraGen, TeraSort, and Wordcount are high CPU intensive jobs.

5.3 Performance Evaluation of Apache Hadoop Yarn MapReduce Jobs on Multi-cluster Nodes Based on Classification

5.3.1 Evaluating Execution Time (Seconds) of Different Apache Hadoop Yarn MapReduce Job on Multi-cluster

From Table 3 and Fig. 4, it is found that execution time increased for low CPU intensive job and decreased for high CPU intensive job when number of nodes increased.

Table 3 Evaluating execution time (seconds) of different Apache Hadoop Yarn MapReduce job on multi-cluster

Classification of jobs	Low CPU intensive job	High CPU intensive job		
No. of nodes	Pi	TeraGen (1 GB data)	TeraSort (1 GB data)	Wordcount (1 GB data)
1	16	63	2144	1285
2	13	38	409	838
3	21	25	141	208
4	18	26	496	470

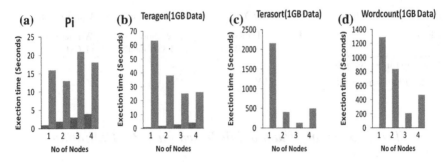

Fig. 4 Execution time (seconds) with increased number of nodes using various Apache Hadoop Yarn MapReduce jobs on multi-cluster nodes. **a** Pi execution time (seconds), **b** TeraGen execution time (seconds), **c** TeraSort execution time (seconds), **d** Wordcount execution time (seconds)

5.3.2 Evaluating Total CPU Times (Seconds) of Different Apache Hadoop Yarn MapReduce Job on Multi-cluster

From Table 4 and Fig. 5, it is found that total CPU time is decreased for low CPU intensive job and high CPU intensive job when the number of nodes increased.

Table 4 Evaluating Total CPU times (seconds) of different Apache Hadoop Yarn MapReduce jobs on multi-cluster

Classification of jobs	Low CPU intensive job	High CPU intensive job		
No. of nodes	Pi	TeraGen (1 GB data)	TeraSort (1 GB data)	Wordcount (1 GB data)
1	1.95	38.61	443.25	443.25
2	1.56	29.94	128.82	128.82
3	1.71	22.09	116.72	116.72
4	1.25	30.6	145.88	145.88

Fig. 5 Evaluating total CPU times (seconds) of different Apache Hadoop Yarn MapReduce jobs on multi-cluster

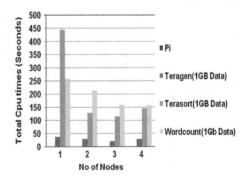

Table 5 Evaluating CPU utilization (%) of different Apache Hadoop Yarn MapReduce job on multi-clusters

Classification of jobs	Low CPU Intensive job	High CPU intensive job		
No. of nodes	Pi	TeraGen (1 GB data)	TeraSort (1 GB data)	Wordcount (1 GB data)
1	12.19	61.29	20.67	20.05
2	12.00	78.79	31.50	25.62
3	8.14	88.36	82.78	76.90
4	6.94	117.69	29.41	33.44

Fig. 6 Evaluating CPU utilization (%) of different Apache Hadoop Yarn MapReduce job on multi-cluster

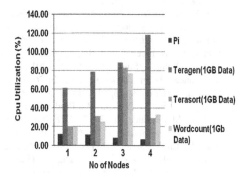

5.3.3 Evaluating CPU Utilization (%) of Different Apache Hadoop Yarn MapReduce Jobs on Multi-cluster

From Table 5 and Fig. 6, it is observed that CPU utilization decreases for low CPU intensive job and increases for high CPU intensive job when the number of nodes increases.

6 Conclusion and Future Work

It is observed that execution time is increased for low CPU intensive job and decreased for high CPU intensive job. Also, total CPU time is decreased for low and high CPU intensive job. In addition, CPU utilization is decreased for low CPU intensive job and is increased for high CPU intensive job when the number of nodes increased. In the future work, Apache Hadoop Yarn MapReduce performance can be tuned by parameter configuration according to job classification.

References

1. Vinod Kumar, V., et al. (2013) Apache Hadoop Yarn: Yet another resource negotiator. In *Proceedings of the 4th annual Symposium on Cloud Computing*, ACM.
2. Apache Hadoop. http://hadoop.apache.org.
3. Dean, J., & Ghemawat, S. (2008). MapReduce: simplified data processing on large clusters. *Communications of the ACM, 51*(1), 107–113.
4. Maurya, M., & Mahajan, S. (2012). Performance analysis of MapReduce programs on Hadoop cluster. In *2012 World Congress on Information and Communication Technologies (WICT)*, IEEE.
5. Joshi, S. B. (2012). Apache Hadoop performance-tuning methodologies and best practices. In *Proceedings of the 3rd ACM/SPEC International Conference on Performance Engineering (ICPE '12)*. ACM, New York, pp. 241–242 doi:10.1145/2188286.2188323 http://doi.acm.org/10.1145/2188286.2188323.
6. Liu, Z., & Mu, D. (2012). Analysis of resource usage profile for MapReduce applications using Hadoop on cloud. In *2012 International Conference on Quality, Reliability, Risk, Maintenance, and Safety Engineering (ICQR2MSE)*, pp. 1500, 1504, 15–18 June 2012.
7. Kamal, Kc. & Freeh, V. W. Tuning Hadoop map slot value using CPU metric.
8. Yao, Y., Wang, J., Sheng B., & Mi, N. (2013). Using a tunable knob for reducing makespan of mapreduce jobs in a hadoop cluster. In *2013 IEEE Sixth International Conference on Cloud Computing (CLOUD)*, pp. 1,8, June 28 2013-July 3 2013.
9. Wang, K., Lin, X., & Tang, W., Predator—an experience guided configuration optimizer for Hadoop MapReduce. In *2012 IEEE 4th International Conference on Cloud Computing Technology and Science (CloudCom)*, pp. 419, 426, 3–6 Dec 2012.
10. Feng, B., Lu, J., Zhou, Y. & Yang, N. (2012). Energy efficiency for MapReduce workloads: an in-depth study. In Zhang, R., Zhang, Y. (Eds.), *Proceedings of the Australasian Database Conference (ADC 2012)*, Melbourne, Australia. CRPIT, vol. 124. ACS, pp. 61–70.
11. Lin, W., & Liu, J. (2013). Performance analysis of MapReduce program in heterogeneous cloud computing. *Journal of Networks, 8*(8), 1734–1741.
12. Kazuki, Y. et al. (2013). Implementation and evaluation of the JobTracker initiative task scheduling on Hadoop. In *2013 First International Symposium on Computing and Networking (CANDAR)*, IEEE.
13. Dhok, J., & Varma, V. (2005). *Using pattern classification for task assignment in mapreduce.* Hyderabad: International Institute of Information Technology.
14. Benslimane, Z., Liu, Q., & Hongming, Z. (2013). Predicting Hadoop Parameters.

Faster Load Flow Analysis

Rahul Saxena, R. Jaya Krishna and D.P. Sharma

Abstract Over the past few decades, load flow algorithms for radial distribution networks have been an area of interest for researches, which has led to improvement in the approach and results for the problem. Different procedures and algorithms have been followed in lieu of performance enhancement in terms of simplicity of implementation, execution time, and memory space requirements. The implementation of load flow algorithm using CUDA parallel programming architecture for a radial distribution network is discussed. The computations involved in serial algorithm for load current, branch impedances, etc., have been parallelized using CUDA programming model. The end result will be an improvement in execution time of the algorithm as compared to the running time of the algorithm over CPU. Finally, a comparison has been drawn between the serial and parallel approaches, where an improvement in execution time has been shown over the functions involved in computations.

Keywords Load flow analysis · Performance improvement · Parallel programming · Branch impedances

1 Introduction

Analysis of electricity distribution networks is performed on the basis of a set of recursive algebraic equations known as load (power) flow equations. Solving these mathematical equations yield us an estimate of parameters like bus voltages, branch

R. Saxena (✉) · R. Jaya Krishna · D.P. Sharma
School of Computing and Information Technology, Manipal University
Jaipur, Jaipur, India
e-mail: rahulsaxena0812@gmail.com

R. Jaya Krishna
e-mail: jayakrishnaa.r@gmail.com

D.P. Sharma
e-mail: dps158@gmail.com

© Springer Science+Business Media Singapore 2016
S.C. Satapathy et al. (eds.), *Proceedings of International Conference on ICT for Sustainable Development*, Advances in Intelligent Systems and Computing 408, DOI 10.1007/978-981-10-0129-1_5

45

currents, and power losses in individual branches of the network. Thus, load flow analysis is a tool to study and analyse various operational and control aspects of distribution network. To perform load flow analysis, various load flow algorithms have been proposed in the past and immense research work has been carried out in the field.

Electricity distribution networks are generally radial in nature and, thus, are called radial distribution networks (RDNs). These RDNs generally suffer from high R/X ratios. Thus, convergence of load flow algorithms is an issue to be addressed. Since RDN can be modeled as a tree-like structure, thus, it is represented in the form of a DAG (Directed Acyclic Graph) as $G(V, E)$.

All the methods [1–5] worked well for transmission systems but have poor convergence performance for RDN due to their high R/X ratio. For this reason, various other types of methods have been proposed. Tripathy et al. [6] proposed a method for solving ill-conditioned power systems, which is similar to Newton–Raphson method. The voltage convergence was good but could not provide efficient optimal power flow calculations. Sharma et al. [7] proposed an efficient load flow algorithm for solution of radial distribution networks using a linear data structure with an algorithmic complexity of $O(n^2)$, where n is the number of nodes. Based upon the idea of paper [7], this paper uses graph theory concept (topological sort) to perform load flow analysis over RDN and further enhancing the efficiency of algorithm over GPU architecture using parallel processing power of CUDA (Compute Unified Device Architecture).

Since RDN can be modeled as a tree-like structure, thus, it is represented in the form of a DAG (Directed Acyclic Graph) as $G(V, E)$. This modeling is explained in Sect. 2.1. Section 2.2 describes the formulation of dependency graph G^T of graph G for evaluation of branch current. Based on leaf and junction nodes, identification is discussed in Sect. 2.3, topological order of nodes is determined of which explanation is done in Sect. 2.4. Section 2.5 describes the evaluation of branch currents and voltage repeatedly in the network until convergence criteria is met and thus evaluating power losses in the network. Next, a parallel implementation using CUDA of above LFA algorithm is discussed in Sect. 3. Here, a parallel version of the functions used in serial computation has been explained. Obtained results are compared in Sect. 4. Finally, Sect. 5 concludes the paper with analysis of results.

2 Design and Analysis of Radial Distribution Network

The large R/X ratio of RDNs results into poor convergence of conventional load flow algorithms, and hence, are dealt with the concept of forward and backward sweeps. For a balanced RDN, the network can be represented as a single line circuit diagram as shown in Fig. 1, where the line shunt capacitances have been neglected as they are very small. Using conventional circuit analysis, the current in the branch $m + 1$ associated with node j is given as

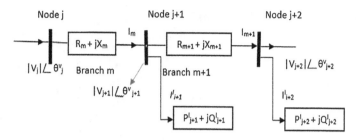

Fig. 1 Electrical equivalent of a section of RDN { V node voltage, I branch current, I^j load current, θ^v node voltage angle, θ^i branch current angle, θ^j load current angle, $P^J(Q^J)$ real (reactive) load}

$$I_m + 1 = I_m + I_j + 1 \tag{1}$$

The branch currents are repeatedly calculated till the substation of RDN is assessed. After the updation of branch currents, the node voltages are given as

$$V_j + 1 = V_j - I_m Z_m \tag{2}$$

where Z_m is the impedance of the branch m given by (R_m = Resistance, X_m = Reactance) and is given as

$$Z_m = R_m + jX_m \tag{3}$$

Once the convergence criterion is met, the real power loss (in Watts) and reactive power loss (in VAR) in a particular branch are computed as [7]:

Real power loss in branch $m = LP_m = |I_m|^2 R_m$

Reactive power loss in branch $m = LQ_m = |I_m|^2 Q_m$

2.1 Modeling of RDN as a Graph

The substation, load buses(nodes) of a RDN network can be treated as vertices V of a graph G, and the branches of RDN can be represented as edges as E of G shown in Fig. 2a).

For a given N-bus (nodes) RDN, we store the relationship between the nodes in the form of a 2D array 'A' of order $N * K$, where K corresponds to maximum degree of branching in the network. The array index position on row corresponds to the sending bus and corresponding to that row, column positions define the receiving nodes to the sending bus. If the node has no more connections, the entry in the

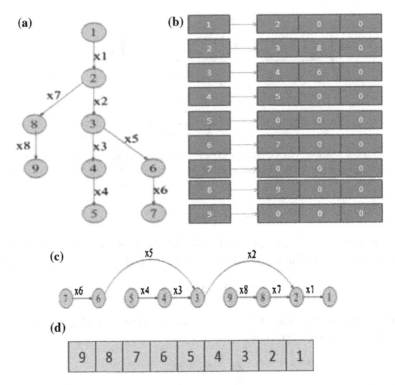

Fig. 2 **a** Graphical representation of RDN. **b** RDN memory representation. **c** Topological order of nodes. **d** TOPO array

column position corresponds to some null value (see Fig. 2b). The track of branches between the nodes is maintained through the data table which contains the relationship between sending node, receiving node, and branch number given by user.

2.2 Dependency Graph Formulation of RDN

Typically, a RDN requires branch current calculations to possess precedence and ordering constraints, i.e., the current in branch X5 (branch between node 3 and 6) depends upon the current in branch X6 (between node 6 and 7). So, the current evaluation in branch X5 can only be evaluated once the branch currents in X6 and

X7 are calculated. The dependencies among branch currents of different branches, thus, can be depicted as a dependency graph G^T of G such that $G^T = (V, E^T)$, i.e., G^T is G with all its edges reversed.

2.3 Identification of Leaf and Junction Nodes

Based upon the modeling of RDN into graph explained in 2.1, leaf nodes correspond to those nodes for which the entry in $A(i, 1)$ corresponds to NULL value (zero here) and the entry $A(i, 2)$ is not equal to NULL implies that the node has more than one receiving node(or outgoing edges). So, as per the graph, leaf nodes are 5, 7, and 9 while junction nodes are 2 and 3.

2.4 Determining Order of Nodes (Topological Sorting)

A topological sort [8] of a graph is a mechanism that gives linear ordering of all the nodes of a graph (Fig. 2c), i.e., if a graph contains a directed edge (u, v), then, this implies that u appears before v in graph. Topological sort of a graph yields the ordering of vertices along a horizontal line so that all directed edges point from left to right. This topologically sorted array is named TOPO as shown in Fig. 2d, stores the nodes of a RDN (graph). Traversing the array left to right gives the order of branch current computation and reverse traversing of the array gives the order of nodes for voltage computation.

2.5 Load Flow Analysis Procedure

(i) Branch impedance and power calculation: The procedure takes an $N * 7$ order input table from user which contains information regarding branch number, sending node, receiving node, resistance and reactance in the branches, and real and reactive power loads at the buses. The procedure evaluates an array 'ZBr' which corresponds to the impedance in the branches of the network and power load 'Pow' at the corresponding receiving buses.

```
Impedance_Pow Cal(B){
     For each branch i and node j of network {
          ZBr(i) = Resultant sum of resistance and reactance of
branch i
          Pow(j) = Resultant sum of real and reactive power load
at node j }}
```

(ii) Branch current calculation: The procedure Branch_current takes array B and array TOPO as input and evaluates branch current in array IBR.

```
Branch_current(B, TOPO)
{
     for each receiving node evaluate load current in
array IL
     for i =1 to length[TOPO] -1        {
        do if(TOPO(i) is not a leaf)    {
               temp=0;
               for j=1 to length[TOPO] -1 {
                       if(TOPO(i) = B(j,2)) {
                             temp=temp + IBR(j); }     }
               IBR(B(TOPO(i), 1)) = temp + IL(TOPO(i));
        } else
        IBR(B(TOPO(i),1)) = IL(TOPO(i));}
```

(iii) Node Voltage calculation: Procedure Voltage for node voltage calculation takes array B, TOPO, ZBr, IBR of graph *G* as input, and evaluates the voltages for the nodes stored in an array *V*.

```
Voltage(B, TOPO, ZBr, IBR){ for
     i = length[TOPO] to 1 {
        do if (TOPO(i) is not leaf){
               for j = 1 to length[TOPO]{
                       if(TOPO(i) = B(j,3))
                       break;}
               V(B(TOPO(i), 3)) = V(TOPO(i)) - IBR(j)*ZBr(j)
} } }
```

(iv) Load Flow Analysis (ZBr, IL, *G*): The procedure of load flow analysis requires three input components—an array of load current IL, array of ZBr, and a graph *G*. The procedure is as follows

1. Prepare dependency graph G of G^T.
2. Calculate array TOPO applying topological sort on G.
3. Evaluate branch currents by calling the procedure Branch_current.
4. Calculate voltages at nodes through Voltage procedure.
5. If the voltages calculated in step 4 converge at each node, then, stop else go to step 3.

3 Parallel Implementation

CUDA Programming Model: For programmers, CUDA [9–12] is a collection of threads running in parallel. A warp is a collection of threads that run simultaneously on a multiprocessor. The number of threads to be executed is decided by the programmer. A collection of threads (called a *block*) run on a multiprocessor at a given time. Multiple blocks may also be assigned to single multiprocessor and they have time-shared execution. Collection of blocks forms a *grid*. Each thread and block is assigned a unique ID based upon which threads access their respective data. Each thread executes a single instruction set called *kernel*. Kernel is the core code to be executed on each thread, i.e., each thread performs the same kernel task on different set of data.

3.1 CUDA Implementation

The paper here presents a parallel implementation of some of the procedures or part of the procedures in load flow analysis. The procedures for branch impedances and power load evaluation at nodes have been made parallel using the block-thread architecture of CUDA as shown in (A). Similarly, for branch current evaluation, corresponding to each node given by TOPO array, we need to map the branch associated with that node to find the branch current for which a parallel search is performed, explained in (B). Load current calculations are made parallel by assigning each node corresponding to each thread shown in (B). For node voltage evaluation, we need to find the receiving bus/es corresponding to each sending bus identified by TOPO array, which can be again done through a parallel search for the receiving end node by generating as much threads as the size of the network, shown in (C). Finally, the branch power loss calculations can also be made parallel as shown in (D).

A.
Branchimpedance_Powercal(B,
Resistance, Reactance, Realpower,
Reactivepower)
{
        id ← get ThreadId for
           each thread

        ZBr[id] ← Resultant of
               resistance and
               reactance at id
               location

        P[id] ← Resultant of
             real and reactive
             power at id location
}

B.
Branch_cur_findbranch(B, A, IBr,
ZBr, P, TOPO)
{
        id ← get ThreadId for each thread

      // Load current evaluation
      corresponding to each node

      IL[id]=(Power load at each id
node) / (Voltage load at each id node)

      // Mapping of branch
corresponding to that
           node

      if (B(id, 3)= node from TOPO
array)
        return branch
}

C.
Voltage_findnode(B, TOPO)
{
      id ← get ThreadId for
         each thread

    if (B(id, 3)=node from TOPO
         array)

      return node
}

D.
Power_loss (IBr, Resistance,
Reactance)
{
      id ← get ThreadId for
         each thread

      Real power loss ← Sum of all

    (IBr(id)*IBr(id)*Resistance(id))

      Reacive power loss ← Sum of all

    (IBr(id)*IBr(id)*Reactance(id))

}

Table 1 33-radial bus system

Function name	Execution time on CPU (μs)	Execution time on GPU (μs)	Number of times kernel execute	Speedup factor (CPU/GPU)
Branch_impedance_power cal()	80.109	7.68	1	10.430
Branch_current()	793.477			1.8232
(i) Load current()		15.2	4	
(ii) Findbranch()		420	128	
Node_Voltage()	690.334			1.4981
(i) Findnode()		460.8	128	
Powerloss()	23.924	4.672	1	5.1207

Table 2 69-radial bus system

Function name	Execution time on CPU (μs)	Execution time on GPU (μs)	Number of times kernel execute	Speedup factor (CPU/GPU)
Branch_impedance_power cal()	148.5164	8.160	1	18.2005
Branch_current()	2794.325			2.68817
(i) Load current()		16.767	4	
(ii) Findbranch()		1022.72	272	
Node_Voltage()	1472.99			1.5927
(i) Findnode()		924.8	272	
Powerloss()	44.141	4.44	1	9.9416

4 Experimental Results

The results are evaluated on NVIDIA Quadro-600 having 16 multiprocessors with 8 processors each with CUDA 6.0 runtime environment on 33-radial bus network is shown in Table 1 and 69-radial bus network is shown in Table 2.

5 Conclusion

A parallel implementation of load flow analysis has been presented over GPU. A function wise improvement using parallel approach has been depicted in Sect. 4. Speedups obtained for 69 radial bus networks are greater than 33 node network which implies better performance as network size gradually increases. Results are expected to improve further if we operate on larger RDNs. Further, other CUDA programming constructs and concepts can be explored to further speedup the process.

References

1. Salama, M. M. A., & Chikhani, A. Y. (1993). A simplified network approach to the VAR control problem for radial distribution systems. *IEEE Transaction on Power Delivery, 8*(3), 1529–1535 (1993).
2. Da Costa, V. M., Martins, N., & Pereira, J. L. R. (1999). Developments in the Newton-Raphson power flow formulation based on current injections. *IEEE Transactions on Power Systems, 14*(4), 1320–1326.
3. Venkatesh, B., Rangan, R., & Gooi, H. B. (2004). Optimal reconfiguration of radial distribution systems to maximize loadability. *IEEE Transactions on Power Systems, 19*(1), 260–266 (2004).
4. Shirmohammadi, D., Hong, H. M., Semlyen, A., & Luo, G. X. (1988). A compensation based power flow for weakly meshed distribution and transmission networks. *IEEE Transactions on Power Systems, 3*, 753–762 (1988).
5. Tinney, W., & Hart, C. (1967). Power flow solution by Newton's method. *IEEE Transactions on Power Apparatus and Systems PAS, 86*(11), 1449–1460 (1967).
6. Tripathy, S., Prasad, G., Malik, O., & Hope, G. (1982). Load-flow solutions for Ill-conditioned power systems by a newton-like method. *IEEE Transactions on Power Apparatus and Systems PAS, 101*(10), 3648–3657 (1982).
7. Sharma, D. P., Chaturvedi, A., Purohit, G. & Shivarudaswamy, R. (2011.) Distributed load flow analysis. *Proceedings International Journal of Electrical Power and Energy Systems Engineering* (pp. 203–206) (2011).
8. Topological Sort. en.wikipedia.org/wiki/Topological_sorting.
9. Kirk, D. B., & Hwu, W. W. (2013). Programming massively parallel processors. ISBN: 978-0-12-415992-1. Published by Elsevier Inc, (2013).
10. Hwu, W. W. (2011). *GPU computing gems* (Emrald Edition). Published by Elsevier Inc (2011). ISBN:978-0-12-384988-5.
11. Study of cuda functions. https://developer.nvidia.com/cuda-zone.
12. Cuda programs profiling. https://developer.nvidia.com/cuda-zone.

Fuzzy Link Based Analysis for Mining Informational Systems

D. Veeraiah and D. Vasumathi

Abstract Web seek clients' dynamically prerequisite watchword and key expression look at interfaces for getting data, and it is regular to extend this model to social data. Expect we are eager about looking at the associations between a few components (or records) included in two distinct stages of the social information source. To this end, in the first stage, a diminished, more reduced, Markov chain containing just the segments of consideration and securing the essential highlights of the preparatory grouping is delivered by stochastic complementation. Co-pertinence is the fundamental presentation of continuously marked information sets connection based examination methodology which does not perform information recovery based question preparation. So in this paper, we propose to create fluffy scan that further upgrades client search for experiences by discovering important arrangements with search for expressions like inquiry search for expressions. A primary computational errand in this model is the high velocity necessity, i.e., every inquiry needs to be reacted to inside of milliseconds to accomplish a quick response and a high question throughput. In the meantime, we likewise require great positioning capacities that consider the region of search for expressions to figure pertinence evaluations.

Keywords Time analysis · Fuzzy instance search records · Relational database approach · Information retrieval

D. Veeraiah (✉)
Jawaharlal Nehru Technological University, Kakinada, Andhra Pradesh, India
e-mail: veeraiahdvc@gmail.com

D. Vasumathi
JNTU College of Engineering, Hyderabad, Telangana, India
e-mail: vasukumar_devara@yahoo.co.in

© Springer Science+Business Media Singapore 2016
S.C. Satapathy et al. (eds.), *Proceedings of International Conference
on ICT for Sustainable Development*, Advances in Intelligent Systems
and Computing 408, DOI 10.1007/978-981-10-0129-1_6

55

1   Introduction

Information extraction from some of the web assets essentially uses machine learning, example revamping, and information mining operations. Web data extraction is a basic issue that has been focused on by a system for distinctive logical mechanical assemblies and in a wide extent of employment [1]. Various approaches to thinking data from the Web are expected to deal with particular issues and work in unrehearsed regions. Distinctive systems, rather, strongly reuse strategies and estimations made in the field of information extraction. This audit goes for giving a composed and comprehensive framework of writing in the field of Web data extraction. The methodology of separating the data from social databases is demonstrated in Figure 1.

A number of diverse exploration regions were created in those relative information extraction systems for getting to the data from different sources and after that perform responsible and other dynamic data recovery techniques in application advancement. A number of procedures were carried out for removing the data from different sources such as information from a percentage of the web handling units. These systems were prepared to create productive and different procedures in an information extraction process. Taking into account the determination of highlights gives productive and other significant results for obliged information in continuing occasions present in the database. As indicated by these components, they are

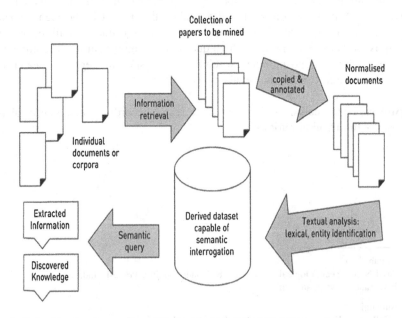

Fig. 1 Data extraction from various data domain registration processes

utilized as part of a request to take in test from information given from delegate stores. Populace-based occasions are obtained in relative and different procedures in the information source. All data are kept up by the impressive information from different sources [2]. This certifiable information originating from different fields is prepared by the held information from database.

An association examination based framework enables one to find affiliations existing between parts of a social data source or, all the more all things considered, a chart. Particularly, because this performance is seen in the perspective of a fascinating go through the data source decoding a Markov chain having the same number of announcements as fragments in the data source [1]. Accept, for occasion, we are amped up for investigating the associations between sections included in two extraordinary tables of a social data source. To this end, a two-stage procedure is arranged. Regardless, a little, reduced, Markov course of action, simply containing the parts of interest—regularly the sections in the two tables—and sparing the key highlights of the preliminary progression, is removed by stochastic complementation. Co-relevance execution of request recuperation logically stamped data sets. Association-based examination may fail in successful data extraction from the web process.

Customers frequently confer typographical lapses in their request. At that point, small PC support on mobile phones, nonappearance of caution, or obliged finding out about the data can realize bumbles. In this circumstance, we cannot discover material answers by finding records with a quest for expressions related to the request unequivocally. This issue can be changed by supporting the indistinguishable quest for expressions, in which we discover plans with a quest for expressions like the request for expr (Fig. 2).

The system discovers a response to the enchantment word and key expression question "teacher weskits" regardless of the way the customer mistyped a prefix of the name "venkata Subramanian". Mixing murky mission with brief quest for expression can offer a far prevalent quest for encounters, especially for versatile customers, who much of the time have the "fat fingers" issue, i.e., each key stroke or tap is drawn out and bungles weakly.

Enter a name, ucinetid, e-mail or phone extension.

prefessorwenkatsu

| Nalni | | Associate | (949) | Computer |
| Venkatasubramanian | nalni | Professor | 824-5898 | Science |

Fig. 2 Query retrieval based on fuzzy instance query retrieval

2 Background Research

Joins with Correspondence Analysis: In this investigation, we proceed to create a Markov chain query recovery model which was produced for proficient question recovery from different gave information bases' significance word development. Emphasis model and other promising thing set the gathering from different information base recovery frameworks regarding operation in late application configurations of question names [1]. Joint examination between diverse essential words in information is introduced in distinctive procedures in the information base vault. Hub examination and other late application procedures may accomplish and perform comparisons on information recovery in data frameworks. When a diminished Markov grouping containing just the hubs of hobby has been obtained, one may need to envision the diagram in a low-dimensional region ensuring as precisely as the region between the hubs could be allowed (Fig. 3).

Preparing an essential dispersion outline, the diminished Markov arrangement is relative to correspondence examination in two exceptional circumstances of interest: a bipartite chart and a star-diagram database. One needs to relate the investigation into diverse promising questions of recovery-based pertinence with the movement of information store support.

3 Problem Statement

In this paper, we consider the accompanying issue: how to join region subtle elements into positioning in moment fluffy search to gauge suitable arrangements productively? The region of related search for expressions in arrangements is a critical metric to focus the significance of the arrangements. Seek questions by and large contain related search for expressions, and answers that have these searches for expressions together are more probable when the client is searching for sought pivotal word. For instance, if the search for inquiry is "Michael Jackson", the client is in all probability searching for the data containing insights about the musical performer Michael Jackson, while the data contains "Subside Jackson". We investigate different distinct options for this critical issue and offer presentation of the thoughts on the tradeoffs of space, time, and reaction top quality. One system is to first discover all the options, gauge the position of every reaction focused on a position work, sort them utilizing the position, and return the top results. Then again, counting every one of these options can be computationally exorbitant when these email location subtle elements are an excess of [3]. This circumstance is more prone to happen in examination to a customary hunt program since inquiry search queries in quick search for are dealt with as prefixes and can have numerous fulfillments. In addition, fluffy search for makes the situation much more confused since there could be numerous search queries with a prefix simply like a question prefix. For instance, the magic word and key expression prefix "cleam" can have

Fig. 3 Document annotation in different query retrievals. The document table contains outcomes of documents while the word table contains outcomes of words

numerous indistinguishable consummations, for example, "clean", "clear", and "cream". As an effect, the mixed bag of choices in moment fluffy pursuit is much greater than that in traditional web index systems.

4   Phrase-Based Indexing and Life-Cycle of a Query

To get over the limitations of the essential strategies, we build up a method relying on expression-based posting. In this segment, we give the points of interest of this methodology.

Expression-Based Indexing: Naturally, the term is a progression of watchwords and words which has high plausibility to show up in the data and concerns. We consider how to actualize the term identified with improved position in this top-k computation structure. We accept that an answer having a related term in the inquiry has a more prominent score than a reaction without such a related term [4, 5]. To have the capacity to still do starting abrogations, we need to open the records containing words first. Case in point, for the inquiry $q =$ hheart, surgeryi, we need to open the data containing the expression "heart surgery" before the data containing "heart" and "surgery" autonomously. Watch that the system sorts the upside down record of an essential word and key expression relying on the importance of its data to the catchphrase and key expression. On the off chance that we buy the upside down record of the pivotal word "surgery" as per the importance to the expression "heart surgery", the best taking care of procurement for another expression, say, "plastic surgery", may be diverse.

For instance, the expression "heart surgery treatment unit" is recorded in the tree in Fig. 4a, besides the catchphrases "heart", "surgery", and "unit". The foliage hubs relating to these conditions are assigned as 5, 3, 11, and 12 individually. The foliage hub for the expression "heart" focuses on its upside down record that contains the data r1, r3, and r4. In incorporation, Fig. 4b uncovers the ahead list, where the decisive word id 3 for the expression "heart" is put something aside for these data.

Fig. 4 Index structure with query processing

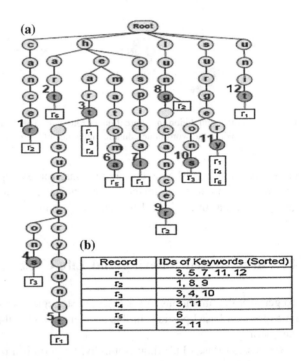

(b)

Record	IDs of Keywords (Sorted)
r_1	3, 5, 7, 11, 12
r_2	1, 8, 9
r_3	3, 4, 10
r_4	3, 11
r_5	6
r_6	2, 11

5 Experiment Evaluation

In this area, we assess the efficiency of the proposed techniques on actual information places. We applied the following methods: (1) Find All ("FA"): We discovered all the answers and came back with the top-k solutions after organizing them as centered on their relevance ranking. (2) Query Segmentation ("QS"): In this approach, we calculated a question strategy depending on legitimate segmentations and ran the segmentations one by one until top-k answers were calculated.

We utilized three data sets as a part of the tests, specifically IMDB, Enron, and Medline. Table 1 uncovers the subtle elements of the data places. We procured the IMDB data set from www.imdb.com/interfaces [6]. We utilized the data as a part of the movies, performing artists, and figures stages, and outlined a table in which every history was a film with a background marked by stars and a background marked by characters. For this data set, we created the concerns from an AOL question log3 and picked those concerns whose went to part was IMDB.com. The Enron data set embodied email data with highlights, for example, timeline, mailer, beneficiary,

Table 1 Data sets for extraction of data

Data sets	IMDB	Enron	Med line
Records	0.8	0.5	10
Distinct keywords	0.76	1.23	4.32
Record length	40	296	152
Data size (in GBs)	248	968	25

subject, and body framework [6]. The Medline data set contains more than 20,000 medical services diaries, and we utilized its subsets of diverse measurements to do tests. In the vague search for tests, we utilized one third as the balanced out altering separation limit. That is, we allowed no typographical oversights if an inquiry prefix is inside of three figures, up to one mix-up if the span is somewhere around 4 and 6, up to two slip-ups if the length of time is somewhere around 7 and 9, etc. All the tests were performed on a Linux program server working an Ie8 10.04 64-bit OS, with two 2.93 GHz Apple Xeon 6-center processor chips, 96 GB capacity, and 2 TB of troublesome commute. To satisfy the high-productivity need of prompt search for, all inventory segments were spared away.

Inquiry Time: We are in correlation of the normal figurings' length of time of FA, QS, and TP as the assortment of search for expressions in question is distinctive. Figure 5 uncovers the results for the three data places. Since TP underpins at most 2-catchphrase concerns, we uncovered the inquiry processing here we are at 1-watchword and 2-magic word attentiveness toward TP. The results for TP uncovered that it did not satisfy the high effectiveness prerequisite of prompt search for. The basic role of its gradualness was posting all the expression puts inside of a predetermined window size; on the ground that it activated to have an excess of fruitions for every inquiry catchphrase and key expression, particularly for brief prefixes. From this exploration, we confirmed that TP is not exceptionally proper for moment search for. FA beat the concerns having more than three search for expressions, while QS outflanked 2-watchword and 3-essential word concerns. Case in point, for the Enron data set the normal question here we are at 2-watchword concerns was 50 ms in FA and 4 ms in QS, while for 4-pivotal word questions it was 7 ms in FA and 18 ms in QS. We talked about the components of these activities from the point of interest in the adaptability assessment of these strategies.

The tests uncovered in Determine 8 utilized stockpiling store for incremental processing. We likewise ran the tests by impairing reserve, yet we sidestepped the results because of the region's limitation indicated in Fig. 5. We watched

Fig. 5 Performance evaluation based on the processing of query in link analysis and Fuzzy search

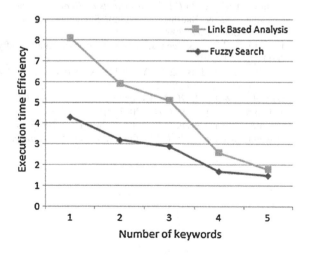

indistinguishable styles as the tests uncovered. On the other hand, without capacity store the response here we are at extensive questions surpassed the 100 ms constraint because of the lack of incremental calculation. We verified that reserving is exceptionally key to the prompt search from the point of view of doing regulated estimations since ensuing concerns in an orderly written work normally shift from one to another by one identity.

6 Conclusion

We examined how to advance current distinct options for fixing this issue, including registering all choices, doing starting retractions, and indexing term sets. We proposed a system to list vital expressions to keep the immense territory cost of posting all word grams. We gave incremental-reckoning criteria for discovering the recorded words in an inquiry adequately, and concentrated on the most proficient method to gauge and position the divisions made up of the ordered words. We in examination our techniques carried out the moment fluffy adjustments of essential strategies. We performed an extremely intensive research by considering region, time, and pertinence tradeoffs of these strategies. Specifically, our tests on genuine data uncovered the productivity of the recommended procedure for 2-watchword and 3-magic word worries that are common in inquiry applications. We established that preparing all the choices for alternate concerns would give the best effectiveness and fulfill the high-proficiency need of quick search for.

References

1. Yen, L., & Saerens, M. (2011). A link analysis extension of correspondence analysis for mining relational databases. In *IEEE Transactions on Knowledge and Data Engineering, 23*(4) (2011).
2. Chaudhuri, S., & Kaushik, R. (2009). Stretching out autocompletion to endure slips. In *SIGMOD Conference*, 2009.
3. Ji, S., Li, G., Li, C., & Feng, J. (2009). Productive intuitive fluffy pivotal word seek. In *WWW*, 2009 (pp. 371–380).
4. Bao, Z., Kimelfeld, B., & Li, Y. (2011). A chart way to deal with spelling remedy in space driven inquiry. In *ACL*, 2011.
5. Bast, H., & Weber, I. (2007). The complete search motor: Interactive, proficient, and towards ir and db reconciliation. In *CIDR*, 2007 (pp. 88–95).
6. Cetindil, I., Esmaelnezhad, J., Kim‡, T., & Li, C. (2014). Proficient instant-fuzzy search with proximity positioning. In *IEEE 30th International Conference on Data Engineering* (2014).

Teaching-Learning-Based Optimization (TLBO) Approach to Truss Structure Subjected to Static and Dynamic Constraints

Ghanshyam Tejani and Vimal Savsani

Abstract Natural frequencies of a structure are of significant importance and their effect becomes dangerous during resonance conditions. During critical dynamic excitation, the effect of frequencies becomes momentous and cannot be ignored. In this paper, teaching-learning-based optimization (TLBO) is proposed for the truss topology optimization (TTO) subjected to static and dynamic constraints. The TTO problems are considered with multiple loading conditions, which append further complexity to the problem due to the multiplication of constraints. All the TTO problems are subjected to natural frequency, element stress, nodal displacement, Euler buckling, and kinematic stability constraints. The TTO is achieved with the deletion of unnecessary elements and nodes from the ground structure. In this method, the troubles arise due to singular solution and unnecessary analysis, hence the finite element analysis (FEA) model is reformed to resolve it. Single stage optimization method is used where size and topology of truss are considered simultaneously. All results reveal that TLBO approaches lighter truss structures compared to the other state-of-art-algorithms.

Keywords Teaching-learning-based optimization (TLBO) · Meta-heuristic · Static and dynamic constraints · Truss topology optimization (TTO)

1 Introduction

Many engineering structures, such as bridges, domes, towers, trusses, frames, etc., are subjected to dynamic excitations and it becomes unsafe when these excitations meet the resonance conditions during service. To protect the structure under such

G. Tejani (✉) · V. Savsani
Pandit Deendayal Petroleum University, Gandhinagar, Gujarat, India
e-mail: p.shyam23@gmail.com

V. Savsani
e-mail: vimal.savsani@gmail.com

© Springer Science+Business Media Singapore 2016
S.C. Satapathy et al. (eds.), *Proceedings of International Conference on ICT for Sustainable Development*, Advances in Intelligent Systems and Computing 408, DOI 10.1007/978-981-10-0129-1_7

unsafe circumstances, natural frequencies need to be controlled by means of convinced restrictions [5]. However, frequency constraints increase the complexity. Static constraints, such as buckling, stress, and displacement, can also have adverse effect and they include additional complicity, which makes the truss topology optimization (TTO) problems further challenging. However, these constraints cannot be neglected in order to assure structure practicality. Most of the TTO problems reported in the literature considered only either static or dynamic constraints. However, few studies are reported considering frequencies and buckling constraints along with stress and displacement constraints simultaneously [5, 14]. Kinematic instable and invalid structures are major problems in the TTO, hence one needs to detect and handle them efficiently to avoid singular structure and unwanted analysis.

Structural optimization can be divided into three categories: size, shape, and topology optimization. Size optimization finds the optimum cross-sectional areas. Shape optimization problems consider the shifting of positions of different nodes. Topology optimization is more challenging because it deals with all the generated different topologies rather than a particular topology to search the finest topology.

In the ground structure method, gradual element removal and addition lead to topology optimization [6]. Due to removal or addition of elements, topology may result in singular topology, which is a challenging issue of the TTO. One way to avoid singularity is to assign tiny values for the removed elements but it adds unnecessary analysis due to the removed element being there in the form of micro-element (element with negligible area). To overcome the above limitations, restructuring of the FEA model is used. In this method, the FEA model is restructured by eliminating the connectivity for the element whenever it is removed. This will avoid unnecessary analysis of removed elements and nodes; also, it will avoid singularity. Most of the investigation neglects the mass at node; however, it is necessary to consider nodal mass because it has a more prominent effect on the overall mass of the TTO problem.

Two different approaches are used for the TTO, depending on the different steps followed for the optimization problems, such as a two stage approach [5, 14] and single stage (simultaneous) approach [2]. In the two stage approach, set of feasible topologies are investigated by considering constant cross-sectional area during first stage and topologies recognized during first stage are optimized for size in second stage. The two stage optimization approach may fail to achieve global optimum solution, if feasible topologies identified in the first stage are not having optimal topology. The single stage optimization approach requires more computational efforts because it deals with simultaneous size and topology optimization.

Teaching-learning-based optimization (TLBO) is a meta-heuristic algorithm developed by Rao et al. [8, 10], which is based on the influence of a teacher on the outcomes of learners. TLBO has proved its capabilities for single and multi-objective optimization problems in terms of achieving global optimum solution, computational cost, and reliability [11]. TLBO has been verified on large size constrained, unconstrained benchmark engineering design problems and proved to be superior performer with other meta-heuristic [9, 13]. Capability of TLBO in the field of structural optimization is not completely addressed so far; however, some researchers have proved its prominence in this field by considering the primary study for the truss optimization [1, 3]. The ability of TLBO has encouraged to formulate the structural optimization problems. In this paper, TLBO algorithm is investigated for the optimization of benchmark structural problems. Two trusses are investigated using TLBO by considering the overall mass as an objective function and stress, displacement buckling, frequency, and kinematic stability as constraints. In this paper, we used the single stage optimization approach, ground structure method, and restructuring of the FEA model for size and topology optimization.

2 Problem Definition

The objective of the TTO is to minimize the mass of truss by finding the best topology and optimum element areas such that it satisfies all its stated constraints. Objective function considers element mass if element exists and a constant mass at node is considered if node exists. Natural frequency, element stress, nodal displacement, Euler buckling, and kinematic stability constraints are considered for this investigation as discussed in the previous sections. Constraint removal technique is adopted to handle the influence of removed element on the constraints. The topology group method is proposed by Xu et al. [14], which reduces the search space due to the removal of worthless topologies. However, this technique is based on two stage optimization method and may not be robust to secure all possible acceptable topologies. Hence, single stage optimization method is adopted in this investigation. The formulation of corresponding optimization problem is as follows:

$$\text{Find, } X = \{A_1, A_2, \ldots, A_m\} \text{ to minimize, } F(X) = \sum_{i=1}^{m} B_i A_i \rho_i L_i + \sum_{j=1}^{n} b_j$$

$$\text{Where, } B_i = \begin{cases} 0 & \text{if } A_i < \text{Critical area} \\ 1 & \text{if } A_i \geq \text{Critical area} \end{cases}$$

Subjected to :

$g_1(X)$: Stress constraints, $|B_i \sigma_i| - |\sigma_i^{max}| \leq 0$

$g_2(X)$: Displacement constraints, $|\delta_j| - |\delta_j^{max}| \leq 0$

$\quad\quad\quad\quad\quad\quad\quad\quad\quad\quad\quad\quad\quad\quad\quad\quad\quad\quad\quad\quad\quad\quad\quad\quad\quad\quad\quad (1)$

$g_3(X)$: Euler buckling constraints, $|B_i \sigma_i^{comp}| - |\sigma_i^{cr}| \leq 0$, where $\sigma_i^{cr} = \dfrac{k_i A_i E_i}{L_i^2}$

$g_4(X) : f_r - f_r^{min} \geq 0$ for some natural frequencies

$g_5(X)$: Size constraints, $A_i^{min} \leq A_i \leq A_i^{max}$

g_6 : Check on kinematic stability

g_7 : Check on validity of structure

where, $i = 1, 2, \ldots, m$ and $j = 1, 2, \ldots, n$

where B_i represents the topological bit, which is '0' for absence and '1' for presence of the element 'i'. A_i, ρ_i, L_i, E_i, σ_i, and σ_i^{cr} are the cross-sectional area, density, length, modules of elasticity, stress, and Euler buckling of the element 'i', respectively. δ_j and b_j are values of nodal deflection and mass of node 'j', respectively. f_r is the rth natural frequency of the structure. The superscript, 'comp' denotes compressive, while 'max' and 'min' denote maximum and minimum allowable limits, respectively. k_i is the Euler buckling coefficient which depends on the cross-sectional geometry of the element.

In this investigation, kinematic stability (g_6) is checked in two steps; the steps are listed as follows [1]: step (i): To validate Grubler's criterion [4] to check the degree of freedom (DOF) and step (ii): To validate the positive definiteness of the structure [12]. Truss structure is called invalid (g_7) if the truss is having absence of loaded node, support node, and undeleted node [7].

TLBO is an unconstrained optimization method; hence, the penalty function approach is adopted to handle stated constraints. For no violation of the constraints, the penalty becomes zero; otherwise, penalty is intended by following criteria [2]:

$$F_{\text{Penalized}} = \begin{cases} 10^9 & \text{if } g_7 \text{ is violated} \\ 10^8 & \text{if } g_6 \text{ is violated in DOF} \\ 10^7 & \text{if } g_6 \text{ is violated with positive definiteness} \\ F(X) + 10^5 * \sum \left(\sum_{p=1}^{4} (|g_p|) \right) & \text{otherwise} \end{cases} \qquad (2)$$

3 Truss Problems and Discussions

This section introduces two benchmark trusses [5, 14] to evaluate effectiveness of the TTO using TLBO. In this study, values for Euler buckling coefficient are considered ($k_i, i = 1, 2, \ldots, m$) 4.0 and nodal mass is considered ($b_j, j = 1, 2, \ldots, n$) 5.0 kg for all problems. The element cross-sectional areas are selected as $A^{\text{lower}} = -A^{\max}$ to offer identical chance for the element existence and the critical area is selected as $A^{\min}$, which is used to eliminate the element. The problems are compared with the published literature results [5, 14]. Xu et al. [14] did not quantify runs for each result and have reported only the best results achieved, while Kaveh and Zolghadr [5] reported the best design over 20 runs. However, in this study, 100 independent runs are used for each problem. The discussion and results are explained as follows:

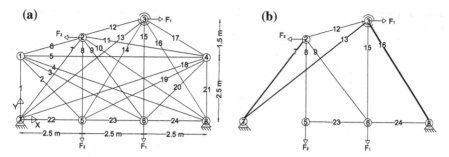

Fig. 1 A 24-bar 2-D truss: **a** ground structure and, **b** optimized by TLBO

3.1 A 24-Bar 2-D Truss

The ground structure of a 24-bar truss is shown in Fig. 1a. This problem was first presented by Xu et al. [14] using 1-D search method and the latter analyzed by Kaveh and Zolghadr [5] using charged system search (CSS) and particle swarm optimization (PSO) algorithms. Multiload conditions are considered as $1 : F_1 = 5 \times 10^4$ N, $F_2 = 0$ and $2 : F_1 = 0$, $F_2 = 5 \times 10^4$ N. The truss is subjected to lumped mass of 500 kg at node 3. The 'min' and 'max' element cross-sectional areas are assumed to be 1 and 40 cm^2, respectively. Natural frequency constraint is $f_1 \geq 30$ Hz, while other parameters are considered as $E = 6.9 \times 10^{10}$ Pa, $\rho = 2740$ kg/m^3, $\sigma_i^{max} = 172.43$ MPa, and $\delta_{5y\&6y}^{max} = 10$ mm.

In this problem, population size and max generations are assumed to be 150 and 200, respectively; hence, it results in 60,000 function evolutions (FE). Table 1 indicates that the lighter truss, with the optimum mass of 118.9573 kg, is obtained by using TLBO. This result is better compared to those of the other trusses reported [5, 14] with no violation of constraints. Optimal topology also satisfies the Euler buckling and stress constraints. The minimum, average, and standard deviation of weight are as 118.957, 137.7608, and 17.2299 kg, respectively, for TLBO. Figure 1b presents the optimal topology obtained by TLBO. Figure 2 shows the evolution graph of an optimal topology using TLBO. It is observed from evolution graph that the result converges nearly within 85 generations.

3.2 A 20-Bar 2-D Truss

The ground structure of a 20-bar truss is depicted in Fig. 3a. No lumped mass is considered for the analysis. Lower and upper bonds of element cross-sectional areas are assumed to be 1 and 100 cm^2, respectively. Multiload conditions are taken as $1 : F_1 = 5 \times 10^5$ N, $F_2 = 0$ and $2 : F_1 = 0$, $F_2 = 5 \times 10^5$ N. The 'min' and 'max' element cross-sectional areas are assumed to be 1 and 100 cm^2, respectively.

Table 1 Optimal design parameters for the 24-bar 2-D truss (mass does not consider lumped mass)

Element no.	Proposed work	Kaveh and Zolghadr [5]		Xu et al. [14]
	TLBO	CSS	PSO	1-D Search
7	19.0222	19.2	20.1	11.0
8	2.9028	3.0	14.8	9.51
9	1.8207	1.4	Removed	15.0
12	4.5278	4.0	2.4	36.5
13	13.7809	14.13	14.9	Removed
14	Removed	Removed	1.2	17.6
15	2.8840	3.3	6.5	13.8
16	23.8843	23.9	23.9	16.0
22	Removed	Removed	Removed	11.02
23	1.0070	1.04	4.7	Removed
24	1.1966	1.4	22.1	14.5
Mass (kg)	**118.9573**	119.75	151.63	167.0
f_1 (Hz)	30.0001	30	30	30
δ_{5y} (mm)	9.0197	8.6	1.2	3.2
δ_{6y} (mm)	9.9019	8.9	8.6	3.0
FE	60,000	–	–	–

Fig. 2 Evolution graph of the 24-bar 2-D truss

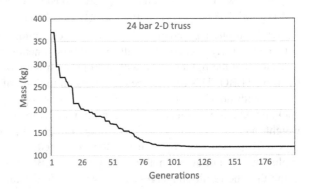

Natural frequency constraints are considered as $f_1 \geq 60$ Hz and $f_2 \geq 100$ Hz. Other design considerations are as $\sigma_i^{max} = 172.43$ MPa, $E = 6.9 \times 10^{10}$ Pa, and $\rho = 2740$ kg/m^3. Optimal topology reported by Xu et al. [14] indicated violation of displacement constraint under load condition 2, which results in 58 mm displacement of node 4 along y-axis [5]. Thus, this problem is solved for two different values for the nodal displacement (case 1: $\delta_{4y} = 10$ mm and case 2: $\delta_{4y} = 60$ mm). Detailed study of both cases is as follows:

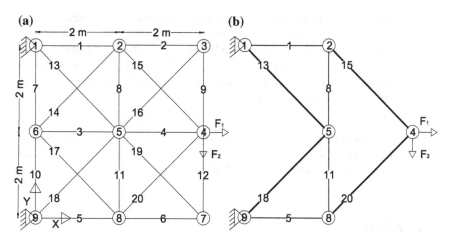

Fig. 3 A 20-bar 2-D truss: **a** ground structure and, **b** optimized by TLBO

Table 2 Optimal design parameters for the 20-bar 2-D truss

Element no.	Case 1			Case 2			
	Proposed work	Kaveh and Zolghadr [5]		Proposed work	Kaveh and Zolghadr [5]		Xu et al. [14]
	TLBO	CSS	PSO	TLBO	CSS	PSO	1-D Search
1	43.6159	39.92	44.05	14.4999	14.74	15.67	53.38
4	Removed	Removed	Removed	Removed	Removed	12.76	Removed
5	43.5199	49.28	42.08	19.0355	19.22	18.41	3.02
8	43.5466	46.41	42.18	19.0481	19.33	22.34	53.38
11	43.4776	41.20	46.75	19.0383	19.03	14.66	3.04
13	61.4395	59.03	63.61	20.5044	20.59	19.19	42.12
15	61.4501	63.15	59.54	20.5045	20.55	26.66	71.97
18	61.4192	64.97	53.46	32.0133	32.40	33.85	53.44
20	61.4673	58.40	71.51	32.0169	32.64	31.17	3.76
Mass (kg)	**315.9135**	317.19	318.23	**150.6529**	151.92	161.88	225.9
f_1 (Hz)	119.4722	120.0	115.1	116.8919	116.6	108.6	60
f_2 (Hz)	191.4311	192.1	186.9	185.7135	185.1	198.6	130.4
δ_{4y} (mm)	10	10	10	24.6051	24	24	58
FE	40,000	–	–	20,000	–	–	–

Case 1: In this case, population size and max generation are assumed to be 100 and 200, respectively. Table 2 indicates that TLBO obtained lighter truss of 315.9135 kg mass in 40,000 FE. The achieved solutions using TLBO are better than other results reported in the literature with no violation of constraints. The minimum, average, and standard deviation of weight are as 315.9135, 368.3899, and 63.1981 kg, respectively, for TLBO.

Case 2: In this case, population size and max generation are assumed to be 50
and 200, respectively. Table 2 reveals that TLBO generated truss with
the optimum weight of 150.6529 kg in 20,000 FE. TLBO reported
lighter trusses than trusses stated in the literature with no violation of
constraints. The minimum, average, and standard deviation of weight are
as 150.6529, 188.0503, and 33.2834 kg, respectively, for TLBO.

Figure 3b illustrates that the optimal topology obtained in all methods is identical for all examples except optimal topology obtained by PSO in Case 2. Optimal
topologies also satisfy Euler buckling and stress constraints in both the cases.

4 Conclusion

In this paper, TLBO algorithm is used to solve the TTO problems subjected to static
and dynamic constraints. All benchmark problems are investigated by considering
natural frequency, element stress, nodal displacement, Euler buckling, and kinematic stability constraints. The simultaneous consideration of the stated constraints
makes the TTO problem complex and challenging, which demands the investigation of effective optimization techniques. The TTO works on the removal of
superfluous elements and nodes, which ends up with great saving in mass and this
advantage is raised up by considering large nodal cost. Consideration of the tiny
value for removed element can avoid singularity; however, this method increases
unnecessary analysis due to presence of removed element and nodes. In this
investigation, the FEA model is reformed, which results in restructuring of stiffness,
mass, and load matrix for each evolution, to avoid singularity and unnecessary
analysis. The effectiveness of TLBO is demonstrated through the benchmark truss
structures using single stage solution approach. Results of TLBO are observed to be
superior as compared to results reported in the literature.

References

1. Camp, C. V., & Farshchin, M. (2014). Design of space trusses using modified teaching
   learning based optimization. *Engineering Structures, 62*, 87–97.
2. Deb, K., & Gulati, S. (2001). Design of truss-structures for minimum weight using genetic
   algorithms. *Finite Elements in Analysis and Design, 37*(5), 447–465.
3. Degertekin, S. O., & Hayalioglu, M. S. (2013). Sizing truss structures using
   teaching-learning-based optimization. *Computers & Structures, 119*, 177–188.
4. Ghosh, A., & Mallik, A. K. (1994). *Theory of mechanisms and machines*. Affiliated East-West
   Press.
5. Kaveh, A., & Zolghadr, A. (2013). Topology optimization of trusses considering static and
   dynamic constraints using the CSS. *Applied Soft Computing, 13*(5), 2727–2734.

6. Kirsch, U. (1989). Optimal topologies of truss structures. *Computer Methods in Applied Mechanics and Engineering, 72*(1), 15–28.
7. Li, L., Liu, F. (2011). *Group search optimization for applications in structural design.* Springer.
8. Rao, R. V., Savsani, V. J., & Vakharia, D. P. (2012). Teaching-learning-based optimization: an optimization method for continuous non-linear large scale problems. *Information Sciences, 183*(1), 1–15.
9. Rao, R. V., Savsani, V. J., & Balic, J. (2012). Teaching-learning-based optimization algorithm for unconstrained and constrained real-parameter optimization problems. *Engineering Optimization, 44*(12), 1447–1462.
10. Rao, R. V., Savsani, V. J., & Vakharia, D. P. (2011). Teaching-learning-based optimization: a novel method for constrained mechanical design optimization problems. *Computer-Aided Design, 43*(3), 303–315.
11. Rao, R. V., & Savsani V. J. (2012). *Mechanical design optimization using advance optimization techniques,* Springer series in advance manufacturing, Springer.
12. Rao, S. S. (2009). *Engineering optimization theory and practice* (4th ed.). Wiley.
13. Satapathy, S. C., & Naik, A. (2014). Modified Teaching-Learning-Based Optimization algorithm for global numerical optimization A comparative study. *Swarm and Evolutionary Computation, 16*, 28–37.
14. Xu, B., Jiang, J., Tong, W., & Wu, K. (2003). Topology group concept for truss topology optimization with frequency constraints. *Journal of Sound and Vibration, 261*(5), 911–925.

Enforcing Indexing Techniques in Berkeley DB Using Implementation of Hilbert Tree Algorithm

Badal K. Kothari, Bhavesh M. Patel and Ashok R. Patel

Abstract The developing enthusiasm inside of the field of information warehousing is transforming more pivotally for decision makers to take a faster and effective decision. The online decision needs a terribly short response time. There are a few calculations and algorithms which are utilized for different indexing strategies and are available for bringing this objective. In multi-dimensional databases, the clammy indexing systems that have pulled in consideration are cube indexing and bitmap indexing. In this research paper, our basic focus is on an approach to enhance the existing cube indexing and storage engine, which will permit us to improve the query resolution and optimization for getting a faster response. For that, we will enforce the Hilbert R-tree algorithm rules with the integration of Berkeley DB and the existing Sidera server cube indexing.

Keywords Data warehouse · OLAP · Multi-dimensional OLAP · Cube indexing · Hilbert R-tree · Berkeley R-tree · Sidera R-tree

1 Introduction

The most important problem in the field of decision support system (DSS) is the querying and accurate and efficient accessing of data in multi-dimensional data stored in the data warehouse. An online analytical processing (OLAP) DBMS, sometimes known as an OLAP storage engine, is responsible for how the data of the data warehouse is stored effectively on a disk and accessed quickly [11, 7, 8].

B.K. Kothari (✉)
Mewar University, Gangrar, Chittorgarh, Rajasthan, India
e-mail: kotbad@gmail.com

B.M. Patel · A.R. Patel
Department of Computer Science, Hemchandracharya North Gujarat University, Patan, Gujarat, India
e-mail: to.patelbhavesh@gmail.com

A.R. Patel
e-mail: arp8265@gmail.com

© Springer Science+Business Media Singapore 2016
S.C. Satapathy et al. (eds.), *Proceedings of International Conference on ICT for Sustainable Development*, Advances in Intelligent Systems and Computing 408, DOI 10.1007/978-981-10-0129-1_8

73

The data cube is a multi-dimensional data model that supports OLAP that allows viewing aggregated data from a number of perspectives. A cube consists of dimensions and measures. For a D-dimensional space, $\{A_1, A_2,...,A_d\}$, we have O (2^d) attribute combinations which are known as views or cuboids or group-bys.

The existing OLAP storage engine like Sidera server employs several components which are used to answer multi-dimensional OLAP queries in a very efficient and effective way. For a D-dimensional space where the fact table is associated with d dimensions, the current Sidera storage engine creates the R-tree indexed data cubes as $2*2_d$ separate standard disk files. Two separate files represent the R-tree indexing for cuboids in a D-dimensional cube [9, 6, 5]. These simple files are not databases in any sense and cannot efficiently support DBMS/OLAP queries.

Berkeley DB provides a high-performance embedded database for key data. Berkeley DB (BDB) stores arbitrary key pairs as byte array and supports multiple data items for a single key. The BDB combines the power of a relational storage engine with simplicity of file system based storage.

In this paper, we will focus on the new approach that integrates the functionality of the Hilbert R-tree for implementing cube indexing and access component of Berkeley DB into the existing Sidera server.

2   Hilbert R-Tree

The performance of the R-trees depends on however how sensible is the algorithmic rule that clusters the data rectangles to a node. Hilbert R-tree curves visit every node during a grid just the ones in noncycle form. For static and dynamic databases, their square measure two differing kinds of Hilbert R-trees is offered. For imposing linear ordering on the data rectangles; Hilbert curves are used by the Hilbert R-tree. Hilbert curves are used to bring the beacon higher ordering within the sense of grouping similar data rectangles along, to attenuate the realm of the ensuring minimum bounding rectangles (MBRs) of multi-dimensional objects within the node [12, 10].

Below Fig. 1 shows the Hilbert curve on a 2 × 2 grid, denoted by H_1. To derive a curve of order i, every vertex of the fundamental curve is replaced by the curve of order two and three. Once the order of the curve tends to time, the ensuring curve may be a form of fractal, with a form dimension of two. The Hilbert curve is often generalized for higher dimensionalities.

Figure 1 shows one such ordering for a 4 × 4 grid. For instance, point (0, 0) on the H_2 curve incorporates a Hilbert value of 0, whereas point (1, 1) incorporates a Hilbert value of 2. The Hilbert value of a rectangle must be outlined.

Hilbert R-tree structures show that form (R, OBJ_ID) contains C_1 as a leaf node, OBJ_ID as a pointer, and R as a real object MBR of the object description records, while form (R, ptr, LHV) contains C_n as a nonleaf node, R as a MBR which contains all children of C_n nodes, ptr as a child node pointer, and LHV as a maximum Hilbert value from all the nodes enclosed by the R [12, 10]. Figure 2 shows rectangles represented in the Hilbert tree.

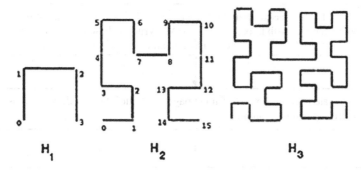

Fig. 1 Hilbert curves of order 1, 2, and 3

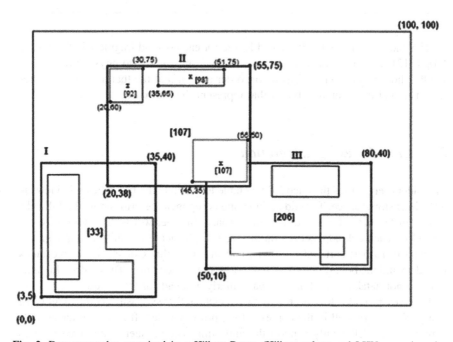

Fig. 2 Data rectangles organized in a Hilbert R-tree (Hilbert values and LHVs are given in brackets)

3 Sidera Server: Hilbert R-Tree Indexed Cube

In the Sidera server, explicit multi-dimensional indexing is provided by parallelized R-trees. The R-tree indexes are unit packed employing a Hilbert space-filling curve. So, that arbitrary k-attribute varies queries which more closely map to the physical ordering of records on disk. For every cuboid fragment on a node, the basic process is as follows:

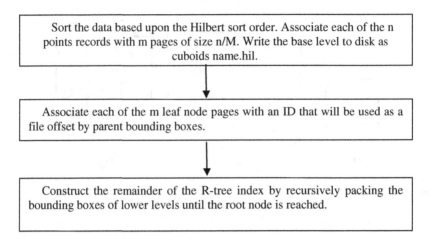

The end result is a Hilbert-packed R-tree for each cuboid fragment in the system [10, 4, 13]. The Hilbert-packed R-tree is stored on disk as two physical files: a .hil file that houses the data in Hilbert sort order and an .ind file that houses the R-tree metadata and the bounding boxes that represent the index tree.

3.1   Cube Indexing Integration

The Sidera server uses the encoded fact table to generate the full cube as 2_d views in a D-dimensional space. The Sidera indexing component describes how the full cube can be materialized for a D-dimensional space. In other words, the indexed R-tree [10, 1] for each cuboid is stored on disk as two physical files: .hil file that contains the data in Hilbert sort order [10, 1], and .ind files that contain the R-tree index metadata and the bounding boxes that represent the index tree. These files (.hil and . ind) are not databases and are not particularly efficient for OLAP queries.

Berkeley provides four access methods such as B-tree, Hash, Recno, and Queue that perform very well in the context of the primary index. It is not sufficient in its current form to efficiently support the multi-dimensional queries that are executed by the Sidera server (Fig. 3).

Berkeley understands the notion of index/data combinations, and it has no mechanism to directly support multi-dimensional R-trees.

The integration process (Berkeley DB and Sidera server) consists of combining the source code for the cube indexing with the code of Berkeley DB. After this integration, Berkeley DB can be used to create a Berkeley DB database with the Hilbert R-tree access method.

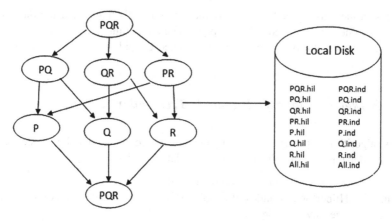

Fig. 3 3-dimensional cube (PQR)

4 Berkeley R-Tree Model

Berkeley supports the storage of many databases, i.e., Berkeley database objects, in one physical file. This physical file contains one master database supported by the B-tree access method that has references to all the databases that are stored in the same file. Keys in this primary B-tree are the database names that are stored in the physical file and the data of the primary B-tree consists of the meta-data page number for each database name [13, 2, 3, 6].

After the integration of the Berkeley and Sidera, instead of $2*2_d$ physical files, it will build the indexed cube as $O(2_d)$. Berkeley database objects for a D-dimensional fact table, one Berkeley database with the R-tree access method for each materialized cuboid.

Berkeley Db physical file contains a master database that has references to Hilbert R-tree indexed cuboids stored as Berkeley database in the same file (Fig. 4).

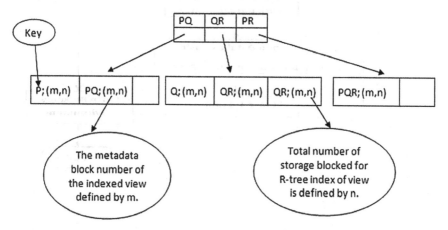

Fig. 4 Berkeley B-tree index that has references to all indexed cuboids stored in one physical file

Keys in the master database are the indexed cuboid's name, and the data contains two values: the indexed cuboid's page meta-data number and the size in terms of the number of blocks required by this indexed cuboid.

4.1 Berkeley Database: Hilbert R-Tree Indexed Cube

Below algorithm represents how the Hilbert R-tree indexed cube is stored as Berkeley database in one physical Berkeley database file.

Input: A set S of cuboids/cube name called C
Output: Hilbert R-tree indexed cuboids stored as Berkeley database objects in
             one physical file
Algorithm:

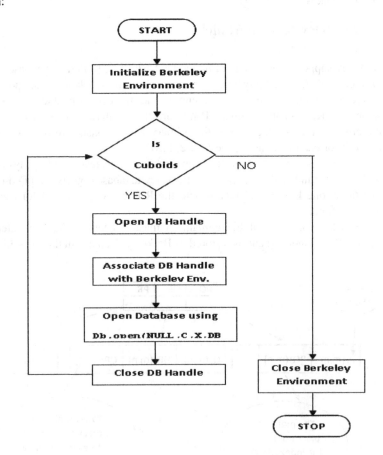

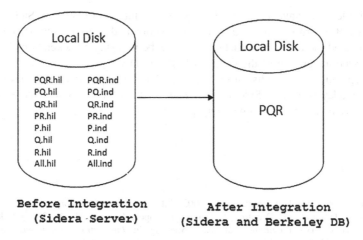

Before Integration
(Sidera Server)

After Integration
(Sidera and Berkeley DB)

Fig. 5 Hilbert R-tree indexing of the three-dimensional cube (PQR) before and after the integration of Berkeley DB and Sidera server

At the beginning, create and open Berkeley database environment handle that encapsulates one or more Berkeley database objects. One for each indexed cuboid in the cube. Then, for each cuboid X in the cube, create Hilbert R-tree indexed view as a Berkeley DB database with the database type DB-RTREE in the open method. Berkeley DB open method with the database type DB-RTREE and a view X means that if the Berkeley database (X) does not exist then, first create it and store in it the Hilbert R-tree index corresponding to cuboids X. In the open method, the name of the physical file that will be used to back one or more Berkeley DB R-tree databases will be the name of the cube (Fig. 5).

When we open a Berkeley DB database handle with a database type equivalent to DB-RTREE, a Hilbert R-tree indexed cuboid is created and stored in one Berkeley DB physical file that has the same name as the cube.

As noted, the indexed cube in Berkeley is represented only in one single physical file; however, it is represented in 2* the number of views in the current Sidera server. It will reduce the index cube's construction time [13, 2, 3]. The primary reason for this reduction is the need for only one physical file to store the index cube in the Berkeley supported Sidera. Storing the indexed cube as $2*2_d$ physical files produces a significant amount of disk thrashing.

5 Conclusions

In this paper, we have described the implementation of Hilbert Algorithm for applying indexing techniques in the integration of Berkeley DB components with the Sidera server. This integration significantly enhances the existing Sidera storage engine. Specifically, Sidera now stores the Hilbert R-tree index in one Berkeley DB physical file. We have conceptually tested the integration of the Sidera with Berkeley DB in terms of the index cube construction and query resolution. We have also compared the index construction for the cube in a single backend node before and after the integration of the codes in Berkeley DB.

Conceptually, our results support the integration approaches that have been taken. Berkeley supported Sidera server is better than the old Sidera server and it significantly boosts the run-time query performance.

References

1. Cwm, common warehouse metamodel. (2003). http://www.cwmforum.org.
2. Eavis, T., Dimitrov, G., Dimitrov, I., Cueva, D., Lopez, A., & Taleb, A. (2007). Sidera: A cluster-based server for online analytical processing. In: *International Conference on Grid Computing, High-performAnce, and Distributed Applications [GADA]*.
3. Giacometti, A., Laurent, D., Marcel, P., & Mouloudi, H. (2004). A new way of optimizing olap queries. *BDA*, 513–534.
4. Jsr-69 javatm olap interface (jolap), jsr-69(jolap) expert group, http://jcp.orgabout Javcommunityprocess/first/jsr069/index.html.
5. Kimball, R., & Ross, M. (2002). *The data warehouse toolkit*. Wiley.
6. Kimball, R., & Caserta, J. (2004). *The data warehouse ETL toolkit*. Wiley.
7. Kothari, B. K., & Patel, A. R. (2014). Cube indexing implementation using integration of Sidera and Berkeley DB. *International Journal of Computer Engineering and Applications (IJCEA), VII*(III) Part I.
8. Microsoft analysis services. http://www.microsoft.com/sqlserver/2008/en/us/analysisservices. aspx.
9. Oracle essbase. http://www.oracle.com/us/solutions/ent-performancebi/business-intelligence/ essbase/index.html.
10. Table, A. (2011). *Query optimization and execution for MOLAP*. 34–44, 78–87.
11. Wikipedia, the free encyclopedia. https://en.wikipedia.org.
12. Xml for analysis specification v1.1, (2002). http://www.xmla.org/index.htm.
13. Zimanyi, E., & Malinowski, E. (2005). Hierarchies in a conceptual model: From conceptual modeling to logical representation. *Data & KNowledge Engineering*.

An Improvement in Performance of Optical Communication System Using Linearly Chirped Apodized Fiber Bragg Grating

Vibha Joshi and Rekha Mehra

Abstract In this paper, a proposal for analyzing the performance of an optical system by using linear chirped apodized fiber Bragg grating has been put forth. Here, we have compared two 10 Gbps systems, one with fiber Bragg grating and the other without fiber Bragg grating. Various parameters used for this analysis are input power, distance, and attenuation coefficient. The performance is analyzed in terms of gain, Q-factor, bit error rate (BER), and eye diagram. It is found that the use of FBG in an optical communication system gives better system performance as compared to a system without FBG.

Keywords Dispersion compensation · Linear chirped apodized fiber bragg grating · Quality factor · Eye diagram · BER

1 Introduction

In fiber optic communication, information is transmitted through an optical fiber in the form of light pulses from the source to the destination through a channel. The light is an electromagnetic carrier wave, which is modulated to carry information. An optical communication system also faces problems, such as dispersion, attenuation, and nonlinear effects that degrade its performance, which are the same as those of any other communication system. Among them, the dispersion effect is dominating and it is difficult to overcome this effect in comparison to other problems. Dispersion can be attributed as a pulse extending in an optical fiber. If the fiber length increases, the dispersion effect also increases. Dispersion causes pulse

V. Joshi (✉) · R. Mehra
Department of Electronics and Communication,
Government Engineering College, Ajmer, India
e-mail: vibha84joshi@gmail.com

R. Mehra
e-mail: mehra_rekha@rediff.com

© Springer Science+Business Media Singapore 2016
S.C. Satapathy et al. (eds.), *Proceedings of International Conference on ICT for Sustainable Development*, Advances in Intelligent Systems and Computing 408, DOI 10.1007/978-981-10-0129-1_9

81

broadening which leads to inter-symbol interference (ISI) in which output pulses of a system overlap. Due to the overlapping of pulses, the output becomes undetectable. Thus, it requires exploring an effective dispersion compensation technique that leads to performance enhancement of the optical system [1, 2].

Several dispersion compensation techniques are available, i.e., dispersion shifted fibers, dispersion-flattened fibers, all-pass filter, raised cosine filter, dispersion compensating fiber, Fiber Bragg Grating, etc. [3].

Thyagarajan, K. Varshney presented a novel design of a dispersion compensating fiber [4]. Knudsen, S.N. Pederson showed the optimization of a dispersion compensating fiber for long haul applications [5]. Luis Miguel Rio de Sousa Ramos and Rui Pedro Martins Alves Ramos presented the characterization of fiber Bragg for dispersion compensation [6]. In this paper, we measure the performance of the system in terms of gain, Q-factor, BER, and eye diagram by varying the system parameters. The network layout is designed and simulated with the help of OptiSystem 7 software.

This paper is divided into five sections: Sect. 2 describes the fiber Bragg grating, Sect. 3 deals with system description including the description of system parameters. Section 4 shows the Results and Discussion, and finally Sect. 5 concludes the paper.

2 Fiber Bragg Grating and Results

Fiber Bragg gratings were introduced in 1980. These gratings are used in an optical fiber as a reflective device in which modulation of the core refractive index takes place at a certain wavelength. Reflection occurs in gratings when the wavelength of signal traveling inside the optical fiber matches to the modulation periodicity of FBG. The reflected wavelength (λ_b) is known as the Bragg wavelength, and it is represented by the following relationship:

$$\lambda_b = 2n_{eff}\Lambda \tag{1}$$

where n_{eff} represents the effective refractive index of the grating in the fiber core and Λ represents the grating period. The operation of fiber Bragg grating depends on parameters such as the reflection of light from grating fringes and on the coupling of the modes [2]. The coupling takes place between the forward and backward propagating fields. According to the coupling process, the two fields show strong coupling if they follow the given Bragg condition:

$$\beta_1 - \beta_2 = 2m\pi/\Lambda \tag{2}$$

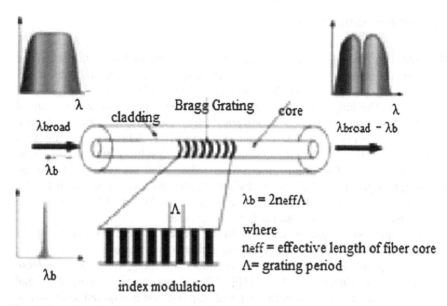

Fig. 1 Operating principle of fiber Bragg grating [7]

where β_1 and β_2 are the phase constants of two coupling modes, Λ depicts the grating period based on the variation of the refractive index (assuming sinusoidal variation), and m is the order of diffraction. For first order, $m = 1$.

Taking two identical inverse propagating modes

$$-\beta_1 = \beta_2 \tag{3}$$

The fiber Bragg diffraction condition becomes

$$2\beta_1 = 2m\pi/\Lambda \tag{4}$$

On substituting the value of grating period Λ from Eqs. (1)–(4), we get [7]

$$\beta_1 = 2\pi n_{\text{eff}}/\lambda_b \tag{5}$$

The principle is as shown below in Fig. 1.

Gratings, which have nonuniform period along their length in FBG, are termed as chirped gratings. The chirp may be linear, may be quadratic, or may even have jumps in the period. A grating could also have a period that varies in any function along its length. The reduction of side-lobe levels in the reflection spectrum is known as apodization [8].

3 System Simulation

For analyzing the system performance (with and without FBG), the schematic experimental setup is shown in Figs. 2 and 3. In this setup, the data transmitter comprises of a continuous wave laser operated at 193.1 THz frequency and power of light 5 dB. The output of pseudorandom bit sequence generator at a bit rate of 10 Gbps is intensity modulated to form pulse trains. This 10 Gbps signal is encoded by a pulse generator and then combined at the Mach–Zehnder modulator with extinction ratio of 30 dB and the modulated signal is then transmitted over a 10 km single mode fiber. The overall link is operated at 1550 nm wavelength. At the receiver, the optical signal is detected by a photodiode with a responsivity of 1 A/W. Then the signal is fed to a BER analyzer and a dual port WDM analyzer to generate an eye diagram and to calculate the Q-factor, gain, etc.

For dispersion mitigation, the fiber Bragg grating is used. The length of the grating, which is used, is 6 mm. Apodization profile is hyperbolic tangent because the tangent profile has a higher reflectivity and a linear chirp is used. Both of these profiles have minimum side lobes [10]. After dispersion mitigation, the signal travels through an optical amplifier. Optical amplification is required to overcome the fiber loss and it also amplifies the signal at the receiver port [11].

Fig. 2 Simulation setup without FBG

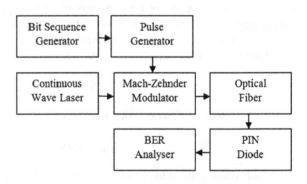

Fig. 3 Simulation setup with FBG

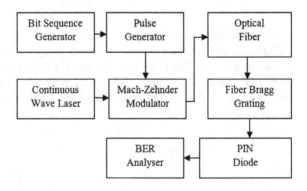

4 Results and Discussion

4.1 By Varying Length of Fiber

For analyzing the effect of different fiber lengths in Figs. 2 and 3, we use the fiber length 5, 10, 15, 20, 25, and 30 km, keeping other parameters constant. The simulation results are tabulated in Table 1. It shows that the length of the fiber is inversely proportional to the value of gain and that with the increase in fiber length the quality factor decreases but it leads to increase in bit error rate. Figure 4a, b represents the eye diagrams of the two systems at 30 km fiber length.

Table 1 Gain, Q-factor, and BER by varying fiber length

Length (km)	Without FBG			With FBG		
	Gain (dB)	Q-factor	BER	Gain (dB)	Q-Factor	BER
5	−4.412508	29.6951	$3.19489e^{-194}$	8.8464356	320.738	0
10	−5.411815	12.5981	$6.84073e^{-037}$	8.8034456	109.315	0
15	−6.41131	11.1091	$6.50641e^{-35}$	8.7629208	61.8682	0
20	−7.411214	8.87484	$2.44851e^{-017}$	8.7192155	32.5762	$4.317462e^{-233}$
25	−8.411501	8.25208	$7.1931e^{-016}$	8.6724654	24.5826	$9.05406e^{-134}$
30	−9.411707	7.84657	$1.71675e^{-015}$	8.6215168	24.217	$7.22408e^{-130}$

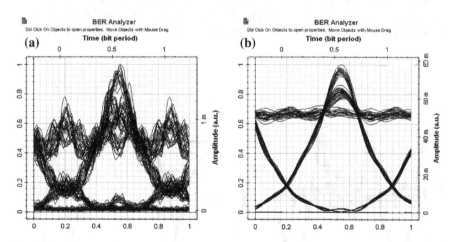

Fig. 4 Eye diagram of system **a** without with FBG at 30 km fiber length and **b** with FBG at 30 km fiber length

Table 2 Gain, Q-factor, and BER by varying input power

Input power (dB)	Without FBG			With FBG		
	Gain (dB)	Q-factor	BER	Gain (dB)	Q-factor	BER
1	−5.412624	12.289	$3.64876e^{-035}$	12.620075	105.496	0
5	−5.411815	12.5981	$6.84073e^{-037}$	8.8034456	106.551	0
10	−5.409251	12.8501	$1.54702e^{-038}$	4.0600618	108.564	0
15	−5.4055	13.6727	$7.19834e^{-043}$	−0.430630	109.315	0
20	−5.449208	14.3871	$2.51205e^{-047}$	−4.289655	122.463	0

4.2 By Varying Input Power

For the analysis of input power variation in the setup of Figs. 2 and 3, we have varied the input power levels by 1, 5, 10, 15, and 20 dB while the other parameters are kept constant. The results of power variations are encapsulated in Table 2. From the simulation, we found that increase in input power reduces the gain due to fiber nonlinearity and increase in quality factor. Figure 5a, b represents the eye diagram of a system without and with FBG at 20 dB input power.

4.3 By Varying Attenuation Coefficient

Analyzing the system performance on the basis of attenuation constant, we change the values of attenuation coefficients by 0.2, 0.5, 1.0, 1.5, 2.0, and 2.5 dB/km. It is found that the value of quality factor without FBG is 12.591 and with FBG is

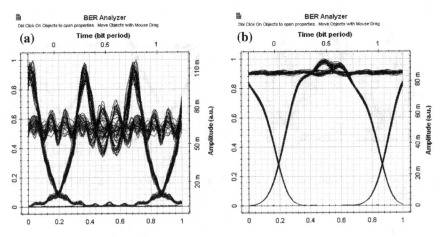

Fig. 5 Eye diagram of system **a** without FBG at 20 dB input power and **b** with FBG at 20 dB input power

Table 3 Gain, Q-factor, and BER by varying attenuation coefficient

Attenuation coefficient (dB/km)	Without FBG			With FBG		
	Gain (dB)	Q-factor	BER	Gain (dB)	Q-factor	BER
0.2	−5.41181	12.5981	$6.84073e^{-037}$	8.803445	109.315	0
0.5	−8.41219	12.3567	$1.595e^{-035}$	8.671395	108.772	0
1.0	−13.4120	10.5515	$2.09366e^{-026}$	8.331189	103.639	0
1.5	−18.4127	6.43687	$5.91741e^{-011}$	7.515784	87.1236	0
2.0	−23.4129	2.58979	0.00477461	5.656756	59.3132	0
2.5	−28.4129	0	1	2.415039	38.2659	$6.5859e^{-321}$

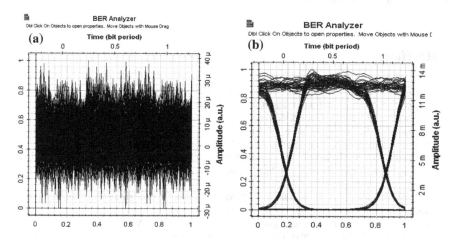

Fig. 6 Eye diagram of system **a** without FBG at 2.5 dB/km attenuation coefficient and **b** with FBG at 2.5 dB/km attenuation coefficient

109.315, which is also depicted in Table 3. This parameter variation also shows that with the increase in the value of attenuation coefficient, the gain and quality factor reduce. Figure 6a, b represents the eye diagram of system without and with FBG at 2.5 dB/m attenuation coefficient.

On comparing the performance of the above two systems by varying length, input power, and attenuation coefficient, it is clear that the system having fiber Bragg grating provides better results because it has simple structure, low insertion loss, high wavelength selectivity, polarization insensitivity, and compatibility with optical fiber.

5 Conclusion

By the simulation results, it can be concluded that the system having fiber Bragg grating provides better results. Optical fiber length and attenuation coefficient are directly proportional to the BER while the gain is reducing with the increase in length. On the other hand, increase in fiber length decreases the quality factor. Also, increase in input power reduces the gain. But the overall performance of the system with fiber Bragg grating is better than the system without fiber Bragg grating. We can extend this work by changing the position of fiber Bragg grating and there are a lot of parameters whose effects gain and Q factor can be taken. More simulation can be conducted in future taking more parameters into consideration and this may provide more perfect results.

References

1. Keiser, G. (2011). *Optical fiber communications*. India: Wiley.
2. Keiser, G. (2000). *Signal degradation in optical fiber* (pp. 113–115). McGraw-Hill.
3. Agrawal, G. P. (2002). *Fiber-optic communication systems* (3rd ed.). India: Wiley.
4. Thyagarajan, K., Varshney, R. K., Palai, P., Ghatak, A. K., & Goyal, I. C. (1996). A novel design of a dispersion compensating fiber. *IEEE Photonics Technology Letter, 8*, 1510.
5. Knudesen, S. N., Pedersen, M. O., & Nielsen, G. (2000). Optimization of dispersion compensating fibers for cable long haul applications. *IEEE Electronics Letter, 36*(25), 2067–2068.
6. de Sousa Ramos, L. M. R., & Ramos, R. P. M. A. (2004). Characterization of fiber Bragg for dispersion compensation. *Thesis of postgraduate* (pp. 4–18).
7. Kashyap, R. (1999). *Fiber Bragg gratings* (3rd ed.). Academic Press.
8. Hill, K. O., & Meltz, G. (1997). *Fiber Bragg grating technology fundamentals and overview*.
9. Liu, W., Shu-qin, Chang, L., Lei, M., & Sun, F. (2010). *The research on 10 Gbps optical communication dispersion compensation systems without electric regenerator*. IEEE, 3rd International Conference on Image & Signal Processing (CISP).
10. Li, P., Ning, T. G., Li, T. J., Dong, X. W., & Jian, S. S. (2005). Studies on the dispersion compensation of fiber Bragg grating in high-speed optical communication system. *Acta Physica Sinica, 54*, 1630–1635.
11. Optisystem design. *Optiwave Corporation 7 Capella Court Ottawa*, Ontario, Canada.

Enhancement of Minimum Spanning Tree

Nimesh Patel and K.M. Patel

Abstract Importance of minimum spanning tree is used to find the smallest path which includes all the nodes in the network. Minimum spanning tree can be obtained using classical algorithms such as, Boruvka's, Prim's, and Kruskal. This research paper contains a survey on the classical and recent algorithms which used different techniques to find minimum spanning tree. This research paper includes a new method to generate a Minimum Spanning Tree. It also contains comparisons of Minimum Spanning Tree algorithms with new proposed algorithms and its simulation results.

Keywords Graph · LC-MST · MST · Tree · DG · UG

1 Introduction

A Minimum Spanning Tree is obtained from the graph. It is a tree of weighted graph in which the total weight of all its edges is a minimum of all such possible spanning tree of the graph. Minimum spanning Tree must be found from the Graph. A Graph is formulated with the help of edges and vertices. In which each edge is interlock with a pair of vertices [1–4].

Mostly, Graphs are divided into two different categories which are directed graph (DG) and undirected graph (UG). In directed graph, set of vertices are connected together, where all the edges are directed from one vertex to another. A DG is also called as digraph or a directed network. In contrast, a graph where the edges are bidirectional is called an UG. In the directed graph, edges have a direction associated with them. An undirected graph edges have no orientation, i.e., edge (a, b) is identical to the edge (b, a). The maximum number of edges in an undirected graph without a self-loop is (n* (n − 1)/2) [5] (Fig. 1).

N. Patel (✉) · K.M. Patel
Department of Computer Engineering, R K University, Rajkot, India
e-mail: 011neem011@gmail.com

K.M. Patel
e-mail: kamlesh.patel@rku.ac.in

© Springer Science+Business Media Singapore 2016
S.C. Satapathy et al. (eds.), *Proceedings of International Conference on ICT for Sustainable Development*, Advances in Intelligent Systems and Computing 408, DOI 10.1007/978-981-10-0129-1_10

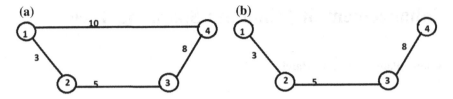

Fig. 1 Graph and its MST. **a** Graph of four nodes and four vertices. **b** Minimum spanning tree of the (**a**) graph

Applications of MST are: (i) It offers a method of solution to other problems such as, clustering and classification problems, network reliability. (ii) It is used in the telephone networks, utility circuit printing, links road network, obtaining an independent set of circuit equations for an electrical network, design of computer and communication networks, electrical circuits, islands connection, pipeline network, etc. (iii) It is also used to find the approximation solution for the NP hard problems.

Objectives of MST are: (i) To minimize cost of the spanning tree for both directed and undirected graph. (ii) To minimize load on the network. (iii) To eliminate the cycle from the graph from the MST. (iv) To improve the complexity of the MST.

2 Literature Survey

There are various classical algorithms available which are described below.

Kruskal's, Prim's, and Boruvka's algorithm is used to find a MST for a connected weighted undirected graph. It means, when the total weight of all the edges is minimized in the tree at that time it finds a subset of the edges which forms a tree which includes every vertex [6–8].

Kruskal algorithm: Using simple data structure Kruskal algorithm complexity is O(e log e) [9] time, or equivalently, O(e log v) time. Where v is the number of vertices and e is the number of edges in the graph and [1, 2].

Prim's algorithm: Using a simple binary heap data structure Prim's complexity is O(|e| log |v|), where |v| is the number of vertices and |e| is the number of edges. Using Fibonacci heap dense graph complexity is O(|e| + |v| log |v|), which is asymptotically faster [2, 10].

Boruvka's algorithm: Boruvka's algorithm has taken O(log v) iterations of the outer loop until it terminates. Therefore, it takes O(e log v) [15] time to run. Where v is the number of vertices in graph and e is the number of edges.

Visit, Mark, and Construct MST algorithm [5]:

In this algorithm, adjacency matrix is used which helps to reduce the step at the time of constructing MST. This method is purely used for the undirected graph. So the weight of the 1 to 2 vertices is same for the 2 to 1 vertices. See the Fig. 2a, in which edges 1 to 2 contain 52 weights and the weight for the edges 2 to 1 is also 52. So, it is same for undirected graph.

(a)   (b)   (c)

$$\begin{bmatrix} 00 & 52 & 10 & 16 \\ 52 & 00 & 09 & 05 \\ 10 & 09 & 00 & 02 \\ 16 & 05 & 02 & 00 \end{bmatrix} \begin{bmatrix} 00 & 52 & 10 & 16 \\ 52 & 00 & 09 & 05 \\ 10 & 09 & 00 & 02 \\ 16 & 05 & 02 & 00 \end{bmatrix} \begin{bmatrix} 52 & 10 & 16 \\ 00 & 09 & 05 \\ 09 & 00 & 02 \end{bmatrix}$$

Fig. 2 n*n weighted matrix reduce into m*m weighted matrix where (m = n − 1)

Here, first row and last column are never used during the implementation because edges 1 to 2 has the same 2 to 1 and edges 1 to 1 is always 0 or automatically eliminated because of generating cycle. See the Fig. 2b, c. Thus, (m^2) steps will be performed. So, complexity is O (m^2), where m is $(n − 1)$ [11].

3 Proposed Algorithms and Their Results

Proposed algorithm-I: It uses the above algorithm techniques to construct a MST. So, using this technique gives a better output at certain situation [5].

Algorithm:
Input: For the undirected weighted graph G weight matrix M = [wij] m × m. Where m=n-1 and [wij] = weight of the edges in the graph.
Output: Minimum Spanning Tree T of G.
          Remove edges as a Red.
          Total possible edges in the graph declare as a Ted.
          Vertices declare as a V.

Step 1: Start
Step 2: Set element of M[i,j] as 0 or Infinite which has no edges in the existing graph G and start increasing Red.
Step 3: Repeat Step 4 to Step 5 until Red =Ted-V+1 elements of upper triangular matrix of M where, all the nonzero elements are marked.
Step 4: Search the upper triangular weigh matrix M either column-wise or row-wise to find the unmarked nonzero maximum element M[i,j] which, is the weight of the corresponding edge eij in M.
Step 5: If the corresponding edge eij of selected M[i,j] forms cycle with the existing matrix in the graph G then
Set M[i,j] = 0
          Else
Mark M[i,j]
Step 6: Construct the MST T including only the marked elements from the upper triangular weight matrix M which shall be the desired MST T of G.
Step 7:Finish

Practical result of this proposed algorithm-I is given in Figs. 3 and 4.
Proposed algorithm-II: Union find data structure gives a better output during the implementation of Kruskal algorithm. Proposed Idea-I algorithm technique is not

Proposed Idea-1 --> For Simple Data Structure with Insertion Sort		
Vertices	With All possible Edges weight	With Edges < 2*vertices -1
10	870	758
20	2441	1282
30	19440	1809
40	48292	4040

Fig. 3 Proposed algorithm-I with different Sorting techniques

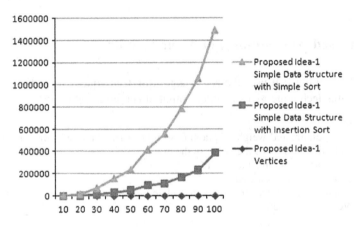

Fig. 4 Proposed algorithm-I chart for various outputs

work effectively because this algorithm sort the edges in decreasing order. Where union find data structure work with the increasing order only. For the better result, proposed idea-II algorithm is discovered with a new method to construct MST which works as quickly as Union find data Structure (Figs. 5 and 6).

Our method is based on Adjacency matrix. Here, our algorithms are divided into two phase which is Adjacency vertices and Adjacency Edges. (i) Adjacency vertices are neighbor vertices of current vertices. (ii) Adjacency Edges are neighbor edges of current edges. Now, during the construction of MST our algorithm sometimes may perform only phase if adjacency vertices phase is not able to find

Proposed Idea-2 -->Adjacency Based		
Vertices	With All possible Edges weight	With Edges < 2*vertices -1
10	869	821
20	1710	1358
30	3464	1947
40	7426	4129

Fig. 5 Proposed algorithm-II in different situation of the same number of vertices

the MST in that situation. Adjacency Edges phase will perform only with those graphs whose value is last modified by Adjacency Vertices phase. In short, Algorithm can work with first phase or first and second or both phases. See the algorithm step which is given below.

Algorithm:
Input: For the undirected weighted graph G weight matrix M = [wij] m × m. Where m=n-1 and [wij] = weight of the edges in the graph.
Output: Minimum Spanning Tree T of G.

> Remove edges as a Red.
> Total possible edges in the graph declare as a Ted.
> Vertices declare as a V.

Step 1: Start
Step 2: Set element of M[i,j] as 0 or Infinite which has no edges in the existing graph G and start increasing Red.
Step 3: If Red =Ted-V+1
> Then Go to Step16
> Else
> Go to Step 4

Step 4: Search the upper triangular weigh matrix M either column-wise or row-wise to find the unmarked nonzero maximum element M[i,j] which, is the weight of the corresponding edge eij in M.
Step 5: Take a adjacency vertices of the corresponding edge eij of selected M[i,j] and create a Subset of those edges vertices for adjacency vertices.
Step 6: If any one vertices are match with value of another subset excluding corresponding edges vertices then cycle is detected so eliminate those edges from the current graph and set it 0 and those graph becomes a current graph for the next maximum edges.
Step 7: Step 5 and Step 6 will repeat till Red =Ted-V+1or all corresponding edges value finds as zero (eij=0).
Step 8: If Red =Ted-V+1
> Go to Step 16
> Else
> Go to step 9

Step 9: Now, once eij=0 then use the last modified graph and consider as a current graph.
Step10: Select the maximum edges weight value from the current graph.
Step 11: Make a Set of adjacency edges of the corresponding edges.
Step12: Based on Set value create a Subset of adjacency edges and compare the subset value of one edge with another edges.
Step 13: If edges are match then cycle is detected. So, eliminate those edges from the graph and modify eij =0 in to the current adjacency matrix.
Step 14: Repeat Step10 to Step 13 till Red =Ted-V+1.
Step 15: If Red =Ted-V+1
> Go to step 16

Step 16: Stop
Step 17: Construct the MST T including the weight of the graph.
Step 18: Finish

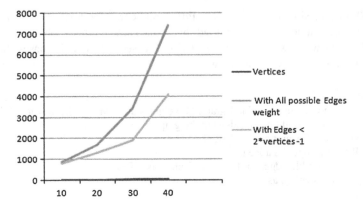

Fig. 6 Proposed algorithm-II chart in different situation of the same number of vertices

4 Conclusions

This paper presents classical algorithms and advance MST algorithm and its practical result. From the practical result it is observed that complexity is very high, because of cycle in the graph and the edges with the same weight. So, it could be improved by using different data structure to detect cycle. It is also observed that time taken by the algorithm is improved with the help of Proposed Idea-I, where complexity remain as it is O(e log v). Complexity is improved for at the time of sorting just because of first row and last column form the adjacency matrices which is never used during implementation. So, for that it takes $O(m^2)$ instead of $O(n^2)$ where $m = n - 1$. Here, from the Proposed Idea-II we can clearly see that time taken by this algorithm is comparatively less than Method-6. This will happen because in this algorithm cycle from adjacency matrix. So, in feature with little bit modification adjacency based cycle detection technique helps to find MST for directed graph also.

References

1. Cormen, T. H., Leiserson, C. E., Rivest, R. L., & Stein, C. (2009). *Introduction to algorithms* (3rd ed., pp. 624–642). PHI Learning Private Ltd.
2. Mehlhorn, K., & Peter, S. (2008). *Algorithms and data structures: The basic toolbox.* Springer.
3. Wu, B. Y., & Chao, K.-M. (2004). *Spanning trees and optimization problems.* CRC Press.
4. Motwani, R., & Raghavan, P. (2010). *Randomized algorithms.* Chapman & Hall/CRC.
5. Nesetril, J., Nesetrilova, H., & Milkovs, E. (2001). Otakar Boruvka on minimum spanning tree problem translation of both the 1926 papers, comments, history. *Discrete Mathematics, 233* (1), 3–36.
6. Hassan, M. R. (2012). An efficient method to solve least-cost minimum spanning tree (LC-MST) problem. *Journal of King Saud University—Computer and Information Sciences, 24,* 101–105.

7. Dagar, S. (2012). Modified prim's algorithm. *IJCIT, 03*(02). ISSN:2218-5224.
8. Kruskal, J. B. (1956). On the shortest spanning subtree of a graph and traveling salesman problem. *Proceedings of the American Mathematical Society, 7,* 48–50.
9. Bazlamacci, F., Hindi, K. S., & Cuneyt. (2001). Minimum-weight spanning tree algorithms a survey and empirical study. *Computers & Operations Research, 28*(8), 767–785.
10. Eisner, J. (1997). *State of the art algorithms for minimum spanning trees a tutorial discussion.*
11. Mandal, A., Dutta, J., & Pal, S. C. (2012). A new efficient technique to construct a minimum spanning tree. *International Journal, 2*(10).

Feasibility Study of Proposed Architecture for Automatic Assessment of Use-Case Diagram

Vinay Vachharajani, Sandeep Vasant and Jyoti Pareek

Abstract Manual assessment is especially a tedious and time-taking process when the number of students enrolled in a subject is large. Since decade, researchers are constantly working on innovative and automatic way to improve the assessment process. MCQ Online exam is one of the successful examples where the whole evaluation process is fully automatic. However, to evaluate diagrammatic answer is still a challenge. Diagrammatic answer of given Use-Case scenario is highly subjective from student to student and still it can be correct. In such situation, if there is a tool which can automatically assess the student answer and generate a quality feedback will be useful for teachers and students. Objective of this research is to know the feasibility of such tool from the appropriate stakeholders. To find the feasibility of proposed architecture for Automatic Assessment of Use-Case Diagram, online survey has been conducted and detailed statistical analysis of the responses is discussed in this paper.

Keywords Automatic assessment · Use-case diagram · Feasibility study · Computer assisted assessment · E-learning

1 Introduction

In the era of high powered computing, teachers across the world tend to use various automated tools for assessing the student's work. Assessment may involve different kinds of activities and highly varies from person to person. Traditional evaluation

V. Vachharajani (✉) · S. Vasant
School of Computer Studies, Ahmedabad University, Gujarat, India
e-mail: vinay.vachharajani@ahduni.edu.in

S. Vasant
e-mail: sandeep.vasant@ahduni.edu.in

J. Pareek
Department of Computer Science, Gujarat University, Gujarat, India
e-mail: drjyotipareek@yahoo.com

© Springer Science+Business Media Singapore 2016
S.C. Satapathy et al. (eds.), *Proceedings of International Conference
on ICT for Sustainable Development*, Advances in Intelligent Systems
and Computing 408, DOI 10.1007/978-981-10-0129-1_11

process consists of checking short notes, essay writing, multiple choice questions, diagram evaluation, and many more. Manual evaluation of answer sheet or grading is been a challenge for Universities, Institutes, and evaluators especially when a number of students appearing in exam is large.

In case of multiple choice questions, system can be fed with desired result and assessment of such type of questions is very easy and straightforward. Short notes, Essay writing, and other text-based answers can be assessed automatically through CASE tools by identifying keywords or semantic analysis with the help of Artificial Intelligence techniques. Numerous researchers have explored the area of automatic evaluation of text-based answers in the past few years [1–3]. It is observed that students can express their subject knowledge better when answering text-based response in their exam. Though, when it comes to explaining through diagrams, student may draw vague or imprecise diagrams and it becomes a challenge for a teacher to evaluate it [4]. Since last decade, researchers around the world are exploring the area of automated assessment for diagrams such as ER-Diagrams, UML Diagrams, Biological Flow Diagrams, Chemical Structure Diagrams, etc. [5–10]. Area of automated assessment of ER-Diagrams has been explored by many researchers and results for the same are very encouraging [11–14]. UML diagrams mainly consist of Use-Case, Sequence, Activity, Class, and Collaboration diagrams. From all these diagrams, Use-Case is having its own significance because of its ability to explain overall functionality of the system to be developed. Area of automated assessment of Use-Case Diagrams has been explored little or not at all [15].

Hence, our focus in this work is to find the feasibility of creating an automated tool which will compare and evaluate the Use-Case diagram drawn by students with model diagram drawn by a subject expert.

2 Literature Review

Exhaustive review of 25 research papers related to automatic assessment of Use-Case diagram is carried out in this research paper and it reveals that very less work has been done.

However, little work is done on automated assessment of E-R diagrams but the scope of this work is very limited. Wu reported in his research [16] about assessing distance learning student's performance using natural language processing approach to analyze class discussion messages. As per Vasant, use of ICT enabled tools in E-learning domain helps to enhance their knowledge in their related domain and improves the overall teaching-learning experience [17].

In [18], authors have focused on the need for automated evaluation of Java programming assignment and exam solutions submitted by students for effective use of ICT in E-learning domain. Manual evaluation of Java programs submitted by large number of students is subject to manual errors and variations. Since decade, various researchers are working in the area of automatic assessment of diagrammatic answers. They have addressed the challenges and issues while comparing the diagrams drawn by the students with the model solution drawn by an expert [4–6]. In [19], author has discussed an automated assessment of class diagrams by rule-based approach. He has considered UML class diagrams as graphs. Graph query language was applied to check the rules defined to evaluate the class diagram. In his results, he has achieved the higher grade of accuracy except few rules, which were not checked by an automated tool due to the missing roles, names, and direction of association shown by the student as compared to manual checking.

Although, few issues such as different spellings of the same term and correct direction of association based on their names can be resolved by string analysis and natural language processing. In [9], Thomas et al. have designed a system for automatic assessment of free form diagrams. They have developed an e-assessment system, known as OpenMark, which can draw two-dimensional objects like rectangles, circles and lines for joining them by integrating it with Moodle—an LMS. OpenMark has a facility that students can attempt multiple questions and examine the corresponding model answers. Marks are given on the basis of marking scheme used to compare a student's answer with the model solution.

Authors in [5] have explained five-staged framework to explain an approach to the automatic interpretation of graph-based diagrams. The framework for evaluating diagrams is described in the domain of ER diagram and grading scheme is prepared to assess the performance of students. This is done by comparing students diagram with specimen solutions representing correct solutions to an assessment question. Natural language processing techniques have been used to determine synonyms, match names, and calculate the similarity measures [10].

Experiments have been carried out to check the accuracy of the tool and quantitative as well as qualitative feedbacks have been provided [20]. The area of evaluation of precise diagrams such as diagrams in mathematical proofs as per [21] and visual query interface to GISs in [22, 23] have also been explored. Noraida, Zarina, and Sufian have used the techniques of notations extraction and automatic assessment to evaluate the class diagrams. On the basis of evaluation of the system, feedback is generated and communicated to the students. The system gives the feedback in the form of comments on the students' diagram in a text format [7].

In [24], authors have mainly focused on effective label matching of model Use-Case diagram, drawn by an expert and students by providing them an Editor. Authors have described an architecture of application which provides label matching process. In this research, they have tried to match the labels by considering the problems of misspelling words, abbreviating words, synonyms, etc. The results of experiments have been analyzed with this editor.

3 Proposed Architecture for Automatic Assessment of Use-Case Diagram

Goal of our research is to automate the process of assessment of diagrammatic answers of students mainly Use-Case diagrams. This can be achieved by

- Accepting and storing the Use-case data, prepared by an expert.
- Accepting and storing the Use-case data, prepared by students, in database in the format appropriate for its assessment.
- Automating the assessment of the Use-case diagrams thus stored in the database.

The proposed architecture for the same is given in Fig. 1 [15].

For accepting and storing data there are two approaches. In the first approach, third party tool can be used for creating and accepting Use-case diagram drawn by students [25].

In the second approach, one interface can be developed for drawing Use-case diagram and comparing it with the model diagram drawn by an expert. In this approach, every component of Use-case diagram, drawn by students and an expert, is extracted and stored in the database for assessment. During assessment process not only quantitative but also qualitative feedback is provided by the system.

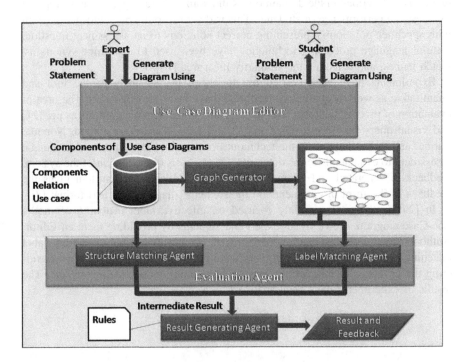

Fig. 1 Proposed architecture for automatic assessment of usecase diagram

There are three major components of the proposed architecture: Graph generator, Evaluation agent, and Result generating agent.

Graph Generator: In this stage of assessment, every basic components related to use-cases like actor, use-case, relationship like include, extend, association, generalization drawn by students and an expert, extracted and stored in the data-base, from which graph will be generated automatically using graph generator.

Evaluation Agent: This agent will work simultaneously on label matching agent and structure matching agent. Label matching agent will match the labels of Use-Case diagram drawn by students and an expert by applying various algorithms [24]. Structure matching agent will match the structures of graphs of Use-Case diagrams drawn by students and an expert. The structure matching agent and label matching agent process simultaneously while assessing the Use-Case diagrams drawn by students and an expert.

Result generating agent: This agent will take input from evaluation agent. It will generate quantitative and qualitative results. Different rules will be applied to generate these feedbacks.

4 Results and Discussion of Online Survey of Teachers and Students

The online survey of 55 teacher and 313 students from different states have been carried out to analyze the responses for feasibility study of the proposed architecture of automatic assessment of Use-Case diagram. In this survey, the average experience of teacher who is teaching the subject is more than 5 years, which is very substantial to understand the importance of the requirement of the proposed tool. Equitable responses of the students of MCA, ICT, CE, IT, M.Sc. (Computer Science), M Sc. IT have been recorded. MS VIsio and IBM Rational Rose are being used widely by the teachers and the students across the states. However, few of them are using other tools. When asked, about knowledge of automatic assessment tools, 98 % of respondents have denied knowing such tools and 2 % have end up listing tools which only draws the Use-Case rather than assessing it. All the teachers feel that it is highly essential to have such tools to reduce the labor work in assessing Use-Case diagrams. 90 % students believe that it will be useful to get prompt, accurate, and unbiased feedback for their answers. Marks, Annotation, Model Diagram, and Color Scheme for correct and incorrect association, processes, relationship, and actors would be the much needed features suggested by the students. Active list of users and user wise marks are the required features for teachers.

To verify the responses received for the feasibility study of automatic assessment of Use-Case diagram, Chi-Square test analysis is applied on requirement of the e-assessment tool regardless of teacher or student. From the results shown in

Table 1 Chi-square test analysis of dependency of requirement of the proposed tool for teacher/student

Analysis of dependancy of requirement of the automatic assesment of Use Case diagram tool and person category

Observed Frequencies						Calculations		
	Column variable					fo-fe		
Category of person	Requirement of tool	Non requirement of tool	No opinion	Total				
Faculty	50	2	3	55		-2.30978	0.804348	1.505435
Student	300	6	7	313		2.309783	-0.80435	-1.50543
Total	350	8	10	368				

Expected Frequencies								
	Column variable					(fo-fe)^2/fe		
Category of person	Requirement of tool	Non requirement of tool	No opinion	Total				
Faculty	52.30978261	1.195652174	1.49456622	55		0.10199	0.541107	1.516383
Student	297.6902174	6.804347826	8.50543478	313		0.017922	0.095083	0.266457
Total	350	8	10	368				

Data	
Level of Significance	0.05
Number of Rows	2
Number of Columns	3
Degrees of Freedom	2

Results	
Critical Value	5.991464547
Chi-Square Test Statistic	2.538941953
p-Value	0.280980228
Do not reject the null hypothesis	

Expected frequency assumption is met.

Table 1, null hypothesis is not rejected at 5 % significant level. To find the correlation between ratings given by teachers/students and need of the automatic assessment tool, Karl Pearson's correlation coefficient is found out and it is 0.98.

5 Conclusion

At last, it is concluded from the responses received from teachers and students that features like Marks obtained, Annotation, Model Diagram, Color Scheme for correct and incorrect association, processes, relationship and actors, Active list of users and user wise obtained marks are vital for the proposed tool. Strong evidences derived from statistical analysis of responses advocates the need of automatic assessment tool for Use-Case diagram which will be path breaking in the area of e-learning especially in assessment of diagram-based answers.

References

1. Burstein, J., Claudia, L., & Richard, S. (2001). Automated evaluation of essays and short answers. In *Proceedings Fifth International Computer Assisted Assessment Conference, Loughborough University, UK, Learning & Teaching Development* (pp. 41–45). Loughborough University.
2. Shermis, M. D., & Burstein, J. C. (2003). *Automated essay scoring: a cross-disciplinary approach*. Mahwah, NJ, USA: Lawrence Erlbaum Associates.
3. Foltz, P. W., Gilliam, S., & Kendall, S. (2000). Supporting content-based feedback in on-line writing evaluation with LSA. *Interactive Learning Environments, 8*(2), 111–127.
4. Hoggarth, G., & Lockyer, M. (1998). An automated student diagram assessment system. *ACM SIGCSE Bulletin, 30*(3), 122–124.
5. Thomas, P. G., Waugh, K., & Smith, N. (2007). Learning and automatically assessing graph-based diagrams. Beyond control: learning technology for the social network generation. *Research Proceedings of the 14th Association for Learning Technology Conference (ALT-C)*, 4–6 September, Nottingham, UK (2007).
6. Thomas, P., Neil, S., & Kevin, W. (2008). Automatic assessment of sequence diagrams.
7. Ali, N. H., Shukur, Z., & Idris, S. (2007). Assessment system for UML class diagram using notations extraction. *International Journal on Computer Science Network Security, 7*, 181–187 (2007).
8. Thomas, P., Smith, N., & Waugh, K. (2009). Automatically assessing diagrams. *Proceedings IADIS International Conference e–Learning* (Vol. 2009).
9. Thomas, P., Waugh, K., & Smith, N. Automatically assessing free-form diagrams in e-assessment systems. *1st HEA Aiming for Excellence in STEM Learning and Teaching Annual Conference, Imperial College London* 2012.
10. Thomas, P, Smith, N., & Waugh, K. G. (2007). Computer assisted assessment of diagrams. *ACM SIGCSE Bulletin, 39*(3). ACM.
11. Thomas, P., Waugh, K., & Smith, N. (2006). Using patterns in the automatic marking of ER-diagrams. *ACM SIGCSE Bulletin, 38*(3). ACM.
12. Thomas, P., Waugh, K., & Smith, N. (2005). Experiments in the automatic marking of ER-diagrams. *ACM SIGCSE Bulletin, 37*(3). ACM.
13. Thomas, P. G. (2003). *Grading Diagrams Automatically*. Technical Report of the Computing Department, Open University, UK, TR2004/01, 2003.
14. Thomas, P., Waugh, K., & Smith, N (2005). Experiments in the automatic marking of ER-diagrams. *ACM SIGCSE Bulletin, 37*(3). ACM.
15. Vachharajani, V., & Pareek, J. (2014). A proposed architecture for automated assessment of use case diagrams. *International Journal of Computer Applications, 108*(4), 35–40.
16. Wu, Y. B., & Xin, C. (2004). Assessing distance learning student's performance: A natural language processing approach to analyzing class discussion messages. *Proceedings. ITCC 2004. International Conference on Information Technology: Coding and Computing, 2004* (Vol. 1.) IEEE.
17. Vasant, S., & Bipin, M. (2015). A case study: embedding ICT for effective classroom teaching & learning. *Emerging ICT for Bridging the Future-Proceedings of the 49th Annual Convention of the Computer Society of India (CSI)* (Vol. 1). Springer International Publishing.
18. Patel, A., Dhaval, P., & Manan, S. (2015). Towards improving automated evaluation of Java program. *Emerging ICT for Bridging the Future-Proceedings of the 49th Annual Convention of the Computer Society of India (CSI)* (Vol. 1). Springer International Publishing.
19. Striewe, M., & Michael, G. (2011). Automated checks on UML diagrams. *Proceedings of the 16th Annual Joint Conference on Innovation and Technology In Computer Science Education.* ACM.
20. Waugh, K., Thomas, P., & Smith, N. (2007). Teaching and learning applications related to the automated interpretation of ERDs. *24th British National Conference on Databases, 2007. BNCOD'07.* IEEE.

21. Jamnik, M. (1998). Automating diagrammatic proofs of arithmetic arguments. Ph.D. thesis, University of Edinburgh.
22. Anderson, M., & McCartney, R. (2003). Diagram processing: Computing with diagrams. *Artificial Intelligence, 145*(1), 181–226.
23. Donlon, J. J., & Forbus, K. D. (1999). Using a geographic information system for qualitative spatial reasoning about trafficability. *Proceedings of the Qualitative Reasoning Workshop.*
24. Vachharajani, V., Jyoti, P., & Sunil, G. (2012). Effective label matching for automatic evaluation of use–case diagrams. *2012 IEEE Fourth International Conference on Technology for Education (T4E),* IEEE.
25. Vachharajani, V., & Jyoti, P. (2012). Use case extractor: XML parser for automated extraction and storage of use-Case diagram. *2012 IEEE International Conference on Engineering Education: Innovative Practices and Future Trends (AICERA).* IEEE.

Extraction of Web Content Based on Content Type

Manish Kumar Verma, Sarowar Kumar, Kumar Abhishek
and M.P. Singh

Abstract Today, World Wide Web has become an integral part of our life. We have entered into a digital era where everything we need is available online. For every task or information we think of, there exists a website for it. With so many websites running over the internet the amount of useless scripts, images, ads, videos, link have increased exponentially. These irrelevant information is making the sites heavy and taking a lot of resources to load properly. If these types of contents are removed from the site or at least restrict them from loading, then the surfing speed will improve a lot and a more precise and concise site will be loaded which will be easier to view and accurate. This paper proposes a method to load the contents of a website like links, images, videos, etc., as per user requirement and demand. Runtime tests have been performed on different types of websites such as educational sites, blogs, personal websites, e-commerce sites, etc. Results from these tests have been included in this paper which emphasize the fact that a concise and on demand loading of heavy web contents make web surfing easier and efficient.

Keywords PHP · Data mining · Absolute and relative URLs · CSS · JavaScript · DOM document · XAMPP

M.K. Verma (✉) · S. Kumar · K. Abhishek · M.P. Singh
Department of CSE, NIT, Patna, India
e-mail: manish123977@nitp.ac.in

S. Kumar
e-mail: sarowarkumar@gmail.com

K. Abhishek
e-mail: kumar.abhishek@nitp.ac.in

M.P. Singh
e-mail: mps@nitp.ac.in

© Springer Science+Business Media Singapore 2016
S.C. Satapathy et al. (eds.), *Proceedings of International Conference on ICT for Sustainable Development*, Advances in Intelligent Systems and Computing 408, DOI 10.1007/978-981-10-0129-1_12

1 Introduction

When WWW started the website content was static and contained only texts, but as the web technology progressed the contents on the web became dynamic and included more and more multimedia contents like images, videos, animations, ads, etc. At first it was a success, then later on excessive ads and multimedia increased overhead on network traffic and deviated the user from the specific info he/she is looking up on a website. Nowadays on most of the website we see a lot of ads, pop ups, and useless animation and the relevant content is very less and scattered. The paper focuses on how to filter out relevant content from a website on user demand. One way to implement this is making changes on the server side so that the data a user gets on the client side on HTTP request is already organized and concise. But this cannot be done as a normal user does not have access to content on a server. An alternative approach is that we create a third party which will do the task of filtering for the user and display the on demand content on the user's web browser.

This paper proposes a method in which by using certain PHP string functions and string parser we can filter out the content of web pages using only the URL of the web page. Although the paper is inspired by previous works, the method used here is quite simple and uses only a few lines of code.

2 Related Works

There have been very few researches [1–6] in this domain of filtering web content on the basis of their types whether images or links or videos. Analysis of some research papers helped in improving and making this paper more efficient.

Gunasundari and Karthikeyan [7] have proposed a method of extracting web content on the basis of link text density, link amount, link amount density, node text length, and HTML tree. Using these parameters an algorithm was designed which was used to extract results from the nodes which meet the conditions of precise content extractions.

Suhit et al. [8] proposed a method of parsing web content using DOM. DOM means Document Object Model which is parsing tree-like structure of HTML tags very easily and effectively.

Kuppusamy and Aghila [9] developed a technique of filtering web page content by building a model which divides the web page into segments. They also incorporated personalization in the model to enhance filtering and make it specific.

Azad et al. [10] improved it further and treated ads, banners, etc., as Noise and developed a method to extract the content using DOM tree along with filters and jsoup.

3   Proposed Methodology

The approach used in this paper is influenced by previous works and uses methods and techniques which eliminate the disadvantages and improves the good points present in them. Now in order to filter out the contents of web page we gather the source of web page by using the URL entered and stored it in a variable. Now this string variable is parsed completely and look for any matching string defined by regular expression for script tag and if any match found it's removed.

Now as per the first choice for links only web page we filter all tags from the variable containing source code and allow only the link tag and a few basic tags. Then the result is "echo" on browser. Only the necessary tags are passed on to browser. This option is quite helpful when we need only the text data from a web page.

For choice of text and images only img tag and a few basic tags are allowed and store it in a DOM Document object [11]. Then the "src" attribute of those images are reset whose URL are relative to that particular web page and make them static URL.

For choice of text and videos we allow only web page and the tags which are used to embed videos in a web page.

Thus, by using a few PHP built-in functions [12, 13], the source code of web page can be parsed and filter content as per the users demand effectively and efficiently at very high speed.

4   System Implementation

4.1   The Algorithm for Designing User Interaction Form

Step 1. Open form tag
Step 2. set action attribute = # and method = post
Step 3. Create input text field for URL input
Step 4. If URL set then value = URL
Step 5. else value = blank
Step 6. Create radio buttons for choice Link, Image, Video
Step 7. If any choice set then make it checked
Step 8. Else set link as checked by default
Step 9. Create submit button
Step 10. close form tag

4.2   The Algorithm for Filtering Content Using URL

Step 11. Store the **input URL** from form in a variable
Step 12. store the choice from form in a variable
Step 13. if URL is null
Step 14. print error message and retry
Step 15. else
Step 16. get source code using URL
Step 17. store the output in a variable as **pagedata**
Step 18. match R.E. for script tag and replace with blank in pagedata
Step 19. If **choice = link**
Step 20. strip all tags except <a> and some other basic tags
Step 21. echo pagedata in browser
Step 22. If **choice = video**
Step 23. strip all tags except <video>, <embed> etc. And some other basic tags
Step 24. echo pagedata in browser
Step 25. If **choice = image**
Step 26. strip all tags except <img> and few basic tags
Step 27. create DOM Document object and loadHTML data to it
Step 28. search for <img> tags and store it in array
Step 29. **Loop start**
Step 30. Get attribute "src" value for each <img> tag
Step 31. if value not absolute
Step 32. change its value to absolute
Step 33. **Loop end**
Step 34. save HTML data into pagedata
Step 35. echo pagedata into browser
Step 36. End

The result obtained after each test run may vary for different web pages and URLs. The code runs well for web pages where multimedia content have an absolute source URL.

5   Experiments and Code Validation

The PHP code is uploaded and run on local XAMPP Apache plus PHP server [14]. A lot of websites have been tested and out of these the results of some of famous websites are shown below. The URL is taken as user input and using this URL further processing is done.

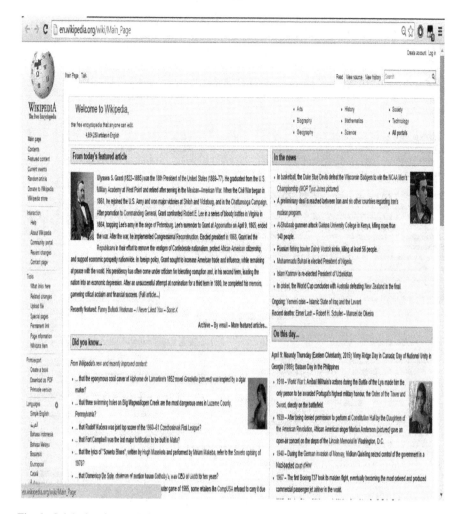

Fig. 1 Original web page [15]

With respect to hardware configuration a Core i5 processor with 2.5 GHz system speed and 8 GB RAM is used. The speed of internet connection used in this setup is 7 Mbps (Figs. 1, 2, 3 and 4).

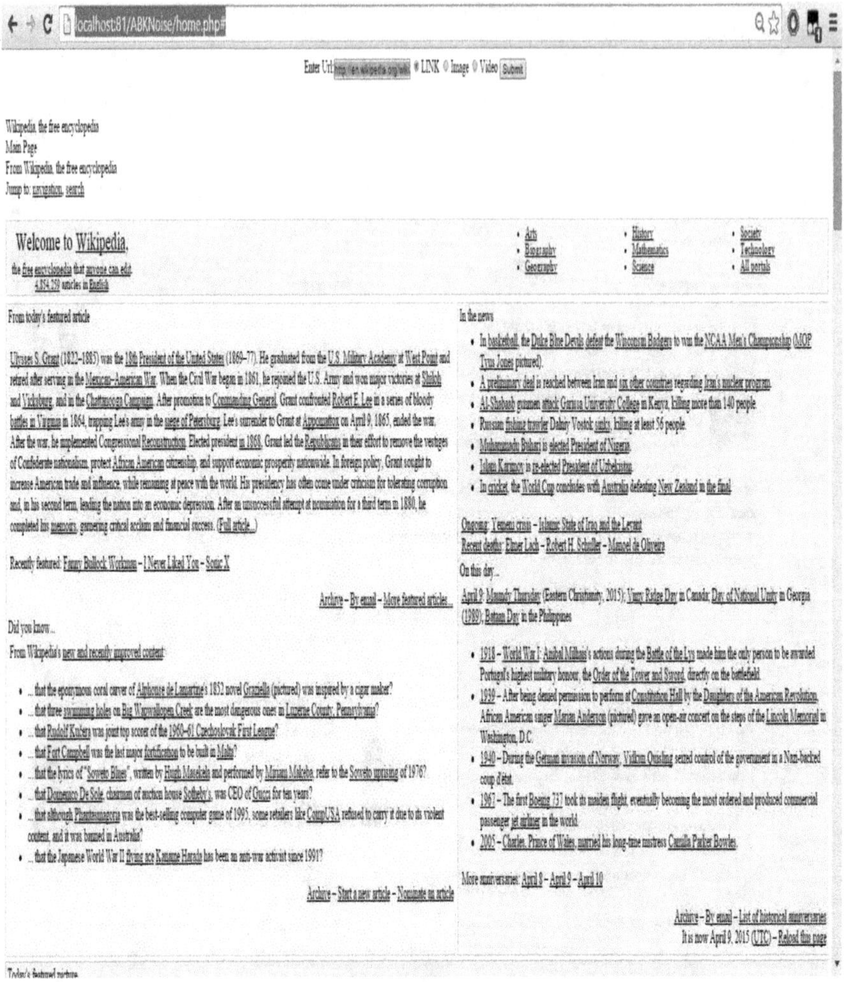

Fig. 2 Filtered with only links allowed [15]

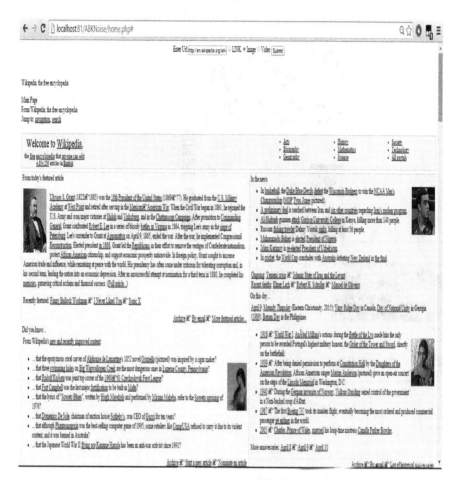

Fig. 3 Filtered with images allowed [15]

Fig. 4 Original and filtered web page with video allowed [15]

6 Conclusion

Most of the web pages contain different types of multimedia elements like images, video, animation, ads, etc., and it is really hard for a common user to extract data from a web page as per his/her requirement. Many researches have been done in this field, but they are not efficient enough. The method proposed in this paper is new, simple, and different from the rest as it uses minimum resources and gives the best result. The input form helps to get input from the user in an interactive and easy way. It uses both the built-in PHP functions and DOM Document object to filter out the required web content and display to the user.

7 Future Work

The current implementation to filter out web page content is not good with styling or designing. Further research might help to extract the content as per need without hampering the style/design of web pages.

Also, the method is not very effective for some pages following the HTTPs protocol only. An effort is being made to extract content from such pages.

Some web pages have a very heavy design theme and take a lot of time to load. Finally, the technique used in this paper can be improved so that we can display the content of such web pages in a lightweight theme keeping the rest contents same.

References

1. Hu, G., & Zhao, Q. (2010). Study to eliminating noisy information in web pages based on data mining. *Sixth International Conference on Natural Computation (ICNC 2010)* (pp. 660–663).
2. Nithya, P., & Sumathi, P. (2012). Novel pre-processing technique for web log mining by removing global noise and web robots. *National Conference on Computing and Communication Systems (NCCCS)* (pp. 1–5). doi:10.1109/NCCCS.2012.6412976.
3. Lan, Y., Bing, L., & Xiaoli, L. X. (2003). Eliminating noisy information from web pages for data mining. In *Proceedings of the ninth ACM SIGKDD International Conference on Knowledge Discovery and Data Mining* (pp. 296–305).
4. Android Development Tools. http://www.mkyong.com/java/jsoup-html-parser-hello-world-examples.
5. Suhit, G., Gail, K., David, N., & Peter, G. (2003). DOM-based content extraction of HTML documents. In *Proceeding WWW '03 Proceedings of the 12th International Conference on World Wide Web* (pp. 207–214).
6. Zhu, F., Gong, C., Yao, H., & Dong, W. (2009). Enhancing site style tree construction algorithm. *Wuhan University Journal of Natural Sciences, 14*(2), 129–133.
7. Gunasundari, R., & Karthikeyan, S. (2012). A study of content extraction from web pages based on links. *International Journal of Data Mining & Knowledge Management Process (IJDKP), 2*(3).
8. Gupta, S., Kaiser, G., Neistadt, D., & Grimm, P. (2003). DOM-based content extraction of HTML documents. *Proceedings of the 12th International Conference on World Wide Web*.
9. Kuppusamy, K. S., & Aghila, G. (2012). A personalized web page content filtering model based on segmentation. *International Journal of Information Sciences and Techniques (IJIST), 2*(1).
10. Azad, H. K., Raj, R., Kumar, R., Ranjan, H., Abhishek, K., & Singh, M. P. (2014). Removal of noisy information in web pages. *ICTCS '14*, November 14–16 2014. Udaipur, Rajasthan, India.
11. PHP Manual Page for DOM Document. http://php.net/manual/en/class.domdocument.php.
12. PHP String manual. http://php.net/strings.
13. PHP Manual for string functions. https://php.net/ref.strings.
14. XAMPP Package. https://www.apachefriends.org/index.html.
15. VideoSift. http://videosift.com/.

User Based Collaborative Filtering Using Bloom Filter with MapReduce

Anita Shinde and Ila Savant

Abstract Recommender systems help to solve excess information problem. Collaborative filtering is the most extensively used methods for recommendation. CF produces high quality recommendations based on likings of society of similar users. Collaborative filtering is based on assumption that people with same tastes choose the same products. Collaborative filtering does not perform well for large systems and it also suffers from sparse data. This paper proposes a novel approach where user based CF uses Bloom filter to filter out redundant intermediate results and helps to get better output. The bloom filter is implemented in the MapReduce phase.

Keywords Collaborative filtering · Mapreduce · Hadoop · Recommender system · Scalability · Bloom filter

1 Introduction

Recommender systems are mainly developed due to excess use of Web and e-commerce. Recommender systems are used to determine whether a client will satisfy with a specific item and to generate a catalog of N items that will be liked by client [1]. There are various applications, where Recommender systems have been used, such as recommending items customer will like to purchase, i.e., TV programs, movies, song a customer will find pleasant, discovering web pages. There are different methods developed for designing the recommender systems that make

A. Shinde (✉) · I. Savant
Marathwada Mitra Mandal's College of Engineering, Pune, India
e-mail: anitashinde04@yahoo.co.in

I. Savant
e-mail: ilanaresh@rediffmail.com

© Springer Science+Business Media Singapore 2016
S.C. Satapathy et al. (eds.), *Proceedings of International Conference
on ICT for Sustainable Development*, Advances in Intelligent Systems
and Computing 408, DOI 10.1007/978-981-10-0129-1_13

use of historical information. Among them, collaborative filtering is mainly doing well and extensively used method. In this method [2], the recommendations for each user are generated by discovering a region of analogous users and then suggesting products of their interest.

Recommender systems are broadly categorized into three types dependent on procedure used for making recommendations [3]: In content-based method, recommendations are made based on items user purchased in earlier period. In collaborative recommendation methods, the data about people with similar tastes are used to generate recommended products. In hybrid recommendation method, join the previous two approaches.

We have studied and presented a comparison of different recommendation strategies in paper [4]. Along with this, comparison of various collaborative filtering algorithms is discussed in details.

The remaining of this paper is structured as follows. Section 2 explains related work of collaborative filtering with MapReduce, Bloom filter, and Mapreduce with Bloom filter. Section 3 introduces system overview which explains steps of algorithm and system architecture diagram. Section 4 describes the evaluation metric, i.e., speedup. Section 5 shows experimental results which include datasets description and various graphs. Finally, in Sect. 6 we conclude the paper and discuss future work.

2 Background

2.1 Collaborative Filtering with MapReduce

Though Collaborative filtering procedures are mostly used in many recommender systems, its computational difficulty is high due to which it cannot be used in bulky size systems [5]. This paper has implemented user based collaborative filtering procedure on Hadoop to reduce the scalability issue of it. Hadoop platform executes on MapReduce framework. With the help of Hadoop platform, we can have parallel execution of program and larger program can be divided into many small subproblems. Collaborative filtering algorithm is based on three assumptions. Customer has same liking and interests which are constant. With the help of these two, we can forecast his choice. In collaborative filtering, we compare user's behavior with other user's behavior, to identify the active users nearest neighbor. Using his neighbor's preferences, we can guess his preferences.

MapReduce is a distributed working representation developed by Google com. It has two parts: Map part and Reduce part. Map section accepts input key and its value pairs and generates output key and its value pairs. Reduce section takes intermediate key and all its corresponding values, generates smaller set of values.

Collaborative filtering algorithm with MapReduce on Hadoop platform is described in three stages: Data fragmentation stage, Map stage, and Reduce stage.

2.2 Bloom Filters

A Bloom filter [6] is helpful to check whether given data is available in collection or not. It has an m-bits group and k self-sufficient hash functions. Every bit in the group is initialized to 0. When the data is included into the group, using k hash functions the data is hashed k times, and the place in the group equivalent to the hash values is made to 1. While checking whether data is available in group or not, every bit of its k hash positions of the group are examined. If present values in all bits are 1, we can say that the data is present in collection.

The subsets of Universe U are represented using a Bloom filter [7]. The paper describes how to obtain Bloom filters that are parallel, rapid, precise, memory-efficient, scalable, and flexible.

In [8] distance aware bloom filters are explained. BF-based schemes have the problem of false positive as many resources are scattered in the network. To tackle this problem, the hint information will be divided and the fragments will be disseminated in a bandwidth-efficient manner and the fragment information will be summed up and quick search decisions will be made for query.

The paper [9] describes the survey of various network applications where Bloom filters can be used. There are various examples where we would like to use a list in a network. Especially when space is concern, a Bloom filter may be an excellent alternative to keeping an explicit list.

In [10] the bloom filters are designed in a scattered manner which mainly helps to reduce intermediate records which in turn helps to improve system output.

2.3 MapReduce with Bloom Filter

MapReduce is programming framework widely used for bulky size systems statistics study. In data analysis, join operation is the fundamental operations. But join operation is not carried out capably by MapReduce because all the records in the datasets are considered when only small number of records are applicable to join operation. This problem is reduced by using bloom filter in MapReduce which helps to enhance the join performance. Bloom filters are designed in disseminated manner and mainly used to take out unnecessary transitional results. The paper [10] describes how the number of intermediate results are reduced which in turn improves the join performance.

Bloom filter construction has two steps as local filter construction and global filter construction. Local filters are constructed in Map phase. Its output is given to job tractor which designs global filter. The global filter is applied to datasets. The filter helps to reduce the number of intermediate key/value pairs which in turn finishes the task early. This enhances the performance of join operation.

3   System Overview

3.1   Algorithm

CF was implemented and studied using three different approaches as sequential, Hadoop with MapReduce, and Hadoop with MapReduce using Bloom filter. In sequential approach, the system is running on standalone machine with no parallel processing. The user based collaborative filtering algorithm is executed sequentially. In Hadoop with MapReduce approach, the algorithm uses parallel processing by dividing data among available number of nodes to achieve speedup. In Hadoop with MapReduce using Bloom filter, local and global filter are constructed to reduce the number of intermediate results which in turn enhances the performance of system.

The local filters are constructed here using random function. Random function selects users randomly and similarity is calculated for selected users only. The similarity between user under consideration and other selected users is calculated using cosine similarity measure given by Eq. 1.

$$\text{sim}(x, y) = \frac{\sum_{s \in S_{xy}} r_{x,s} r_{y,s}}{\sqrt{\sum_{s \in S_{xy}} r_{x,s}^2} \sqrt{\sum_{s \in S_{xy}} r_{y,s}^2}} \tag{1}$$

The rating is calculated using selected users given by Eq. 2.

$$r_{x,s} = \bar{r}_x + \frac{\sum_{y \in S_{xy}} (r_{y,s} - \bar{r}_x) \text{sim}(x, y)}{\sum_{y \in S_{xy}} \text{sim}(x, y)} \tag{2}$$

The global filter is constructed using accuracy function. Accuracy function compares the calculated rating with existing rating in dataset and finds accuracy of each rating. The accuracy values are sorted, top values are selected and other values are ignored. Selected ratings are reduced by reducers and final ratings are shown to user.

3.2   System Architecture Diagram

See Fig. 1.

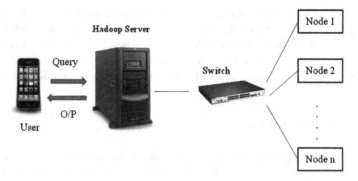

Fig. 1 System architecture

4 Evaluation Metric

The runtime of standalone machine and Hadoop platform is required to be compared. Speedup is taken as significant criterion to measure our algorithm's power. Speedup is calculated using following equation:

$$Speedup = \frac{T_a}{T_s} \tag{3}$$

where T_a is average running time with Hadoop platform, T_s is standalone's running time.

5 Experimental Results

5.1 Dataset

In the experiment, NetFlix dataset is used. The NetFlix Datasets consists of around 17,770 movies and 4,80,189 users with more than 100 million ratings. The ratings of each user are not necessary to be same. It is widely used for Movie Recommender system. Ratings are in range between 1 and 5 stars for each movie.

5.2 Result

We have implemented User Based Collaborative algorithm on Hadoop platform. For experiments, cluster of five nodes is constructed in which one is name-node and others are Datanodes. The softwares used for the experiments are Java JDK 1.6 and

Hadoop MapReduce framework. The dataset NetFlix is used in experiments. The main goal is to analyze the system speedup on a standalone machine, with Hadoop platform, with Hadoop using bloom filter. Each dataset is given to 1 machine, 2 machines, 3 machines, 4 machines, and 5 machines, respectively. The dataset is divided into 100 clients, 200 clients, 300 clients, 400 clients, and 500 clients.

5.2.1 Comparison Graphs

Figure 2 shows Comparison of Sequential, Hadoop, and Hadoop with Bloom Filter for response time on Standalone Machine. It has been observed that response time of Hadoop is better than sequential algorithm and response time for Hadoop with Bloom Filter is better than Hadoop. Hadoop with Bloom filter has local and global filters which reduces intermediate results which help for faster execution of query (Tables 1 and 2).

Figure 3 shows Comparison of Sequential, Hadoop, and Hadoop with Bloom Filter for Load Completion on Standalone Machine. It has been observed that load

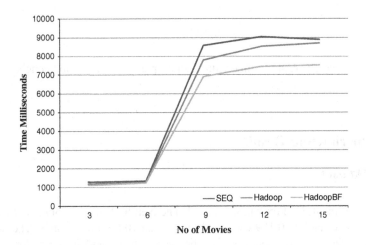

Fig. 2 Comparison of SEQ, Hadoop, and Hadoop BF

Table 1 Response time for sequential, Hadoop, and Hadoop BF

S. no.	No. of movies	Sequential (ms)	Hadoop (ms)	Hadoop with bloom filter (ms)
1	3	1304	1200.70	1139.99
2	6	1353	1283.48	1263.28
3	9	8586	7788.74	6930.14
4	12	9056	8532.80	7458.47
5	15	8897	8707.72	7541.16

Table 2 Load completion by sequential, Hadoop, and Hadoop with bloom filter

S. no.	No. of movies	Sequential	Hadoop	Hadoop with bloom filter
1	3	0.21	0.65	0.77
2	6	0.10	0.58	0.75
3	9	0.06	0.42	0.50
4	12	0.04	0.43	0.60
5	15	0.03	0.46	0.61

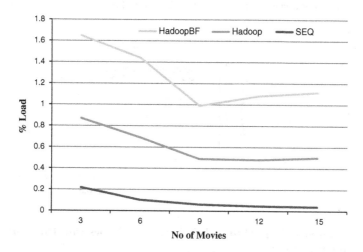

Fig. 3 % Task completed for SEQ, Hadoop, and Hadoop BF

completed by Hadoop is greater than sequential and load completed by Hadoop with Bloom Filter is greater than Hadoop.

Table 3 shows Comparison of response time with 2, 3, 4, and 5 Nodes with and without Bloom Filter. It has been observed that response time of query is reduced in Hadoop with Bloom filter compared to Hadoop. With Bloom filter, the number of intermediate results is reduced as it helps to generate the recommendation list with less response time (Fig. 4).

From the above diagram it has been observed that time required for distributed hadoop is greater than time required for distributed hadoop with bloom filter.

Table 3 Average response time for Hadoop and Hadoop with bloom filter on 2, 3, 4 and, 5 nodes	S. no.	No. of nodes	DHadoop	DHadoop with bloom filter
	1	2	1439.4	1308.08
	2	3	1277.22	1121.44
	3	4	1167.72	1038.79
	4	5	1071.87	914.71

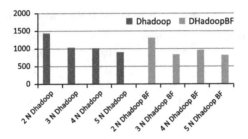

Fig. 4 Comparison of average response time with 2, 3, 4, and 5 nodes with and without bloom filter

Fig. 5 Speedup with 5 nodes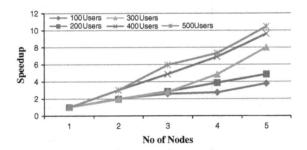

5.2.2 Comparison of Speedup with 2, 3, 4, and 5 Nodes with Different Number of Users

Here the recommendation process is dependent on the splitting up of clients. When recommendations for several users are made on Hadoop platform, it allots calculation to N machines. Theoretically, N will be the speedup where N is number of machines.

Figure 5 illustrates comparison of Hadoop with 1, 2, 3, 4, and 5 Datanodes using 100 clients, 200 clients, 300 clients, 400 clients, and 500 clients. As we can see from the graphs, if we raise the number of nodes, the speedups boosts linear. Dataset should be larger then only superiority of Hadoop can be achieved.

6 Conclusion

The recommendation system generates recommendations for users which will help them in the decision-making process. We have compared three methods as Sequential CF, CF with Hadoop, and CF with Hadoop and Bloom Filter. It is observed that use of bloom filter in MapReduce has a greater speedup, even when the number of nodes goes on increasing. We have tried for 5 nodes.

Through the experimental study, we can see that our algorithm enables Collaborative Filtering on Hadoop platform to give excellent output. There is 10 % improvement in response time using Bloom filter compared to Hadoop without bloom filter. The algorithm gives linear speedup for 5 datanodes. Future work would be to see if the same linear speedup can be achieved as the number of nodes increases.

References

1. Deshpande, M., & Karypis, G. (2004). Item-based top-N recommendation algorithms. *ACM Transactions on Information Systems, 22*(1), 143–177. doi:10.1145/963770.963776.
2. Pagare, R., & Shinde, A. (2013, November). Recommendation system using bloom filter in mapreduce. *International Journal of Data Mining & Knowledge Management Process (IJDKP), 3*(6). doi:10.5121/ijdkp.2013.3608.
3. Adomavicius, G., & Tuzhilin, A. (2005, June) Toward the next generation of recommender systems: A survey of the state-of-the-art and possible extensions. *IEEE Transactions on Knowledge and Data Engineering, 17*(6), 734–749. doi:10.1109/TKDE.2005.99. http://ieeexplore.ieee.org/stamp/stamp.jsp?tp=&arnumber=1423975&isnumber=30743.
4. Pagare, R., & Shinde, A. (2012, June) Article: A study of recommender system techniques. *International Journal of Computer Applications, 47*(16), 1–4. New York, USA: Foundation of Computer Science.
5. Zhao, Z.-D., & Shang, M.-S. (2010, Jan). User-based collaborative-filtering recommendation algorithms on Hadoop. *Knowledge Discovery and Data Mining.* In *WKDD '10. Third International Conference on* (pp. 478–481), 9–10 Jan 2010 doi:10.1109/WKDD.2010.54. http://ieeexplore.ieee.org/stamp/stamp.jsp?tp=&arnumber=5432528&isnumber=5432448.
6. Bloom, B. H. (1970, July). Space/time trade-offs in hash coding with allowable errors. *Communications of the ACM (CACM), 13*(7), 422–426. doi:10.1145/362686.362692.
7. Dillinger, P. C., & Manolios, P. (2004). Bloom filters in probabilistic verification (pp. 367–381). Springer Berlin Heidelberg. ISBN:978-3-540-23738-9. doi:10.1007/978-3-540-30494-4_26.
8. Zhang, Y., & Liu, L. (2013, August). Distance-aware bloom filters: Enabling collaborative search for efficient resource discovery. *Future Generation Computer Systems (2012), 29*(6), 1621–1630. doi:10.1016/j.future.2012.08.007.
9. Broder, A., & Mitzenmacher, M. (2002). *Network applications of bloom filters: A survey.* Allerton Conference. http://www.eecs.harvard.edu/michaelm/NEWWORK/papers.html.
10. Lee, T., Kim, K., & Kim, H.-J. (2012). Join processing using bloom filter. In *MapReduce RACS'12* October 23–26, 2012, San Antonio, TX, USA. 2012 ACM 978-1-4503-1492-3/12/10. doi:10.1145/2401603.2401626.

ICT-Based Facilities Management Tools for Buildings

Ashaprava Mohanta and Sutapa Das

Abstract The advancement in architectural, engineering, and construction industries leads to optimization of resources, where facilities management (FM) is a great concern. During the entire lifecycle of a building, the operation and maintenance (O&M) phase is the longest and revenue-generating phase. FM tools are used for O&M phase which reduces the operational cost and ensures organizational goal. Among innumerable FM tools, major ones are CAFM, CMMS, EAM, and IWMS. This paper presents a thorough literature research on major categories of ICT (information and communication technology)-based solutions available in the market which suit the changing trends of FM for the current decade.

Keywords Facilities management · ICT · Operation and maintenance

1 Introduction

The global trends in architectural, engineering, and construction (AEC) industries in past few decades, which ignited facilities management (FM) are increase in construction demand, advancement in technologies, changes in users' requirements, and efficient use of spaces to increase productivity [1]. The efficiency is mainly gauged in the longest period of the life cycle, i.e., the operation and maintenance (O&M) phase of a building, which have evolved into the concept of FM from the mere necessity of breakdown maintenance. The reduction of the operation cost was the main goal. Gradually it was implemented in real estate, project management, and lease management as well as in sustainable measures with advancement in

A. Mohanta (✉) · S. Das
Department of Architecture & Regional Planning, Indian Institute
of Technology Kharagpur, Kharagpur 721302, India
e-mail: ashaprava.m@gmail.com

S. Das
e-mail: sutapa@arp.iitkgp.ernet.in

© Springer Science+Business Media Singapore 2016
S.C. Satapathy et al. (eds.), *Proceedings of International Conference
on ICT for Sustainable Development*, Advances in Intelligent Systems
and Computing 408, DOI 10.1007/978-981-10-0129-1_14

technologies [2]. Thus, the effective management of a built facility ultimately determines the overall performance of a building [3]. The factors responsible for driving the high growth of FM in the market are high growth in infrastructure development and enhanced return on investments (ROI) capabilities among others. Over the forecast period 2014–2019, FM tools are expected to experience high adoption in Asia-Pacific and the Middle East an Africa regions [4]. This paper focuses on the brief description about functions under FM, their sourcing strategies, and available FM tools.

1.1 Facilities Management

FM is originated at developed countries of West, the concept of FM application is gradually becoming popular in the developing countries [5]. FM is defined by International Facility Management Association (IFMA) as "integration of processes within an organization to maintain and develop the agreed services which support and improve the effectiveness of its primary activities." There are some other definitions provided by IFMA as "the practice of coordinating the physical work-place with the people and work of the organization; integrates the principles of business administration, architecture and the behavioral and engineering sciences" and "a profession that encompasses multiple disciplines to ensure functionality of the built environment by integrating people, place, process and technology" [6]. Hence, it can be concluded that FM is mainly focused on the activities concerning 'building services,' 'space services,' and 'services to people' [5].

FM is a multifaceted profession. There are 11 core competencies of FM defined by a global job task analysis (GJTA) based on global survey and analysis based on responses from facility managers from 62 countries [7]. These tasks are performed by different sourcing strategies as mentioned below, but mostly dominated by outsourcing [5]:

- *In-house*: The facilities are assured by appointing employee directly by the organization.
- *Outsourcing*: The facilities are commissioned by an external organization through contracts.
- *Public Private Partnership (PPP)*: the service provider and the organization shares both the responsibilities as well the benefits to deliver the required facilities.
- *Total Facilities Management (TFM)*: the overall facilities are provided by a single external organization. It ensures the facilities to be delivered, monitored, controlled, and performed according to the objectives of the organization.

1.2 Outsourcing Strategies

The outsourcing of facilities in terms of contracts is commonly adopted though it has a mixed response. The misunderstanding about the type and level of facilities in the contract preparation stage is the common cause of problems in the failure of outsourcing [3]. However, the reasons for adopting outsourcing are (1) cost saving; (2) adjust to fluctuation of work; (3) access to better high quality skills; (4) allows the organization to concentrate on the core business; and finally (5) the level of facilities can be ensured, which cannot be achieved in case of in-house sourcing [8, 9]. About 70 % of the organization uses outsourcing strategies for FM [10].

For ensuring the fulfillment of the objective of an organization 'outsourcing' of technological resources is required to improve efficiency, reduce costs, and streamline all operations. There are eight practical guidelines for evaluating FM software tool. These are [11]:

- *Coverage of facilities*: The proper understanding of the coverage area of FM tool and the level of facilities to be provided based on contract. It should be cost efficient and upgrades the IT as per the up-gradation in technologies without disruption of the services. It should provide a periodic backup of the information.
- *Core Functionality*: The core functionality of FM tools covers a wide range including space management, move management, maintenance management, asset management, lease management, room reservation, and environmental sustainability.
- *Usability*: It should be simple, easy, and functional for all users.
- *Successful implementation/integration*: It should be able to satisfy all the goals of the organization through easy and economic interfaces with the existing tool.
- *Ease of configuration*: It should provide a solution to the current situation as well as have the ability to facilitate the future needs with less hassle.
- *IT structure*: The IT structure of the tools must fit the requirements of an organization and can be either web-based or internally hosted.
- *Mobility*: It is based on software which is accessible in the form of mobile applications. It should be capable of all now and future mobility needs of the organization.
- *Security*: It should have secure authentication processes to access the service level. It helps the administration to have access control and to maintain the privacy as per the demand.
- *Adoptability*: The success of any FM tool is based on how fast it adopts with the latest technology as well as adjustment with the existing tool [12].

Overall, the software should support the organizations' objectives without adding additional features which will lead to unnecessary investment. There should be mutual understanding between the software vendor and the user to provide additional tools or to upgrade the tools for growing demand of the organization.

2 ICT in Facilities Management

From the previous discussion, it is apparent that FM is a complex process which needs a systematic approach. Hence, the application of information and communication technology (ICT) supports facility managers to play their role efficiently. The advancement in ICT also caters to the better level of FM. However, this ICT software consist of basic information handling capacities, such as to capture, store, process, and distribution of information in order to reduce the complexities of the FM [9].

FM software is chosen according to the requirements of the organization to the cater the facilities based on the information about real estate, spaces, assets, services, processes, costs, and customer requirements. As soon as the software is finalized, the software monitor the facilities performance facilitated by the service provider and also helps to implement changes [13]. Figure 1 represents the facility lifecycle management framework. Corresponding to the five major phases of a building lifecycle, i.e., plan, design, build, operate, and decommission, the FM system consist of following five major supporting technologies to fulfill the business process based on associated competencies [14] as (1) Capital planning and management system; (2) Cost estimating and project management system; (3) Computer-aided facility management system (CAFM); (4) Building automation system; (5) Computerized maintenance management systems (CMMS).

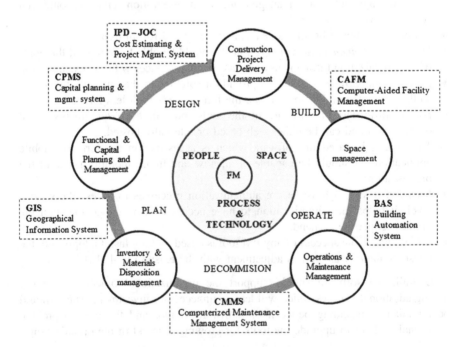

Fig. 1 Facility lifecycle management framework [14]

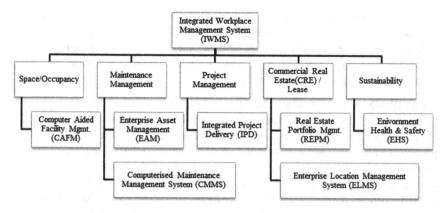

Fig. 2 Different FM software according to facility management purposes [17, 23]

Figure 2 suggests the appropriate technology as per the facilities required [15]. The FM tools are beneficial all along the lifecycle of a building. In this paper, the FM tools only for O&M phase of buildings are considered. The most commonly used options are integrated workplace management system (IWMS), CAFM, CMMS, and enterprise asset management (EAM). These management systems are discussed briefly with their features and applications.

2.1 IWMS

The IWMS is described as "enterprise solutions that support the management of facilities, physical assets, real estate, and the associated services in support of workplace management." [16, 17] IWMS provides overall FM solution [18–20].

"It addresses all operational areas of the five major functional areas, namely,

1. *Space Management* supports the optimization of space usage. This module, with features including CAD/BIM integration, strategic facilities planning and reservation scheduling, manages occupancy and tracks the movement of people and assets within the portfolio.
2. *Operations and Maintenance* allow for the tracking and operational efficiency of physical assets that enable a facility to operate. This module also typically lets users be predictive and/or manage asset condition assessments so they know the right protocols to use.
3. *Real Estate Management* controls the transactions that drive selection and delivery of portfolio assets in a timely and appropriate manner and includes Lease Administration.
4. *Capital Project Management* allows management of large capital budgets against multiple projects. It helps to under the return on investment (ROI) for

each alternative project and helps to decide the best. It also allows users to actually manage the tasks, budgets and schedules of facility projects.

5. *Sustainability and Energy Management* are for the increasingly "green" environment facility requirements and include integration with Building Management Systems (BMS)" [21].

The IWMS is chosen when the vast facilities are required to address in an organization. The individual technologies can be chosen when the facilities are used for definite purposes such as for space management, computer-aided facility management technology; for maintenance management, computerized maintenance management system, etc., is used. However, the success of this software is based on accuracy and consistent updating of information [21]. "Gartner published an analysis on IWMS savings stating a 10–15 % cost reduction in space by effective and efficient management, 5–8 % costs savings by process improvements and better contracting, and 5–8 % savings in lease costs by professional lease administration" [22].

2.2   CAFM

The CAFM is "collection of tools used for organizing and managing various activities within the facilities such as; client contract whereby the client is aware of the equipment, locations and services catered for; material, stock, purchases and equipment replaced for repairs; procurement, the subcontractors service and management service; the services rendered in accordance to service level agreement and other reactive maintenance; work history carried out on equipment; and the strategy used to manage the assets with the engineering instructions to do so at a schedule." [18, 19, 23] Overall the activities are related to space, occupants/users, and the service providers. The CAFM tools can integrate the information from computer-aided drawings (CAD) regarding space usage type and the type of facilities to be catered [8].

2.3   CMMS

The CMMS are comprehensive software applications for a variety of equipment and materials. It can generate work order request for a piece of equipment when the maintenance conditions are not satisfied. It also provides preventive maintenance of the service [18, 19, 24]. Overall it helps to monitor the operational level of any equipment and also provides updates for any kind of causality in the system [25].

2.4 EAM

The EAM advanced CMMS with "a more detailed infrastructure; manages finances and human resources, and other administrative departments through an entire enterprise. EAM's goal is to optimize asset productivity and prolong its lifespan while minimizing the cost of ownership. It is used for large organizations" [18, 19, 24].

3 Discussion

Based on the various definition of FM, it can be concluded that FM is integrating people, process and technology, and space. FM is concerned about three types of facilities such as services to people, building services, and space services. Outsourcing is the best sourcing strategies for any kind of facilities. The web-based IT structure is the hassle-free technology, which avoids discontinuity of services during upgradation of technologies. Based on the literature about FM tools, IWMS provides an overall solution for FM, but cannot provide specific features of other FM tools such as CAFM, CMMS, etc. Figure 3 shows the schematic overlapping of solutions.

The appropriate FM tools for O&M phase can ensure the reduction in operational cost for a lifespan of building. CAFM tools solely deals with space, its users and users' activities of an organization; CMMS tools support the users' activities by maintaining the quality and level of the facilities provided, and EAM tools considers the financial aspects of the organization for future growth as well as to maintain the current facilities. These tools are specialized in the core functions for which it is developed, whereas IWMS is like "jack of all trade and master of none." It provides all solutions for an organization. EAM and IWMS are useful for large

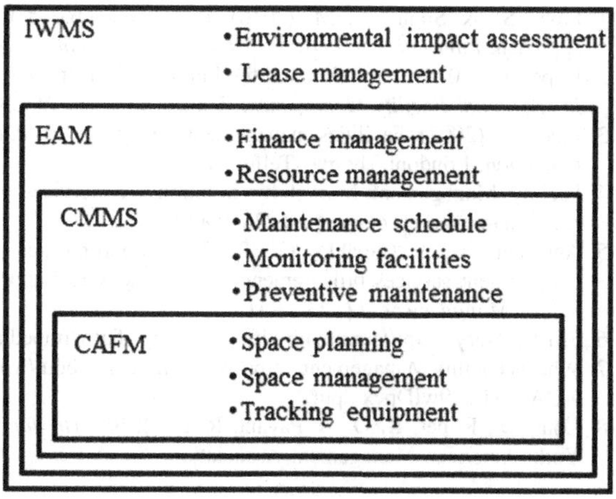

Fig. 3 FM tools based on the level of features

IWMS	• Environmental impact assessment • Lease management
EAM	• Finance management • Resource management
CMMS	• Maintenance schedule • Monitoring facilities • Preventive maintenance
CAFM	• Space planning • Space management • Tracking equipment

organizations with multiple projects but may be idealistic for small and medium organizations. Using the idealistic solution as per organization's goals and requirement can optimize the operational cost and ensure the organizational growth.

The application of BIM in AEC industries has improvised the FM tasks by providing accurate data; and other features like energy analysis, a time schedule for the project, etc. The integrated efforts of all stakeholders can bring more successful and efficient projects considering the environmental aspects from the preconstruction phase of the building. FM is adopted by the developing countries in this decade for the development of infrastructure and optimizing the resource usages. The successful implementation of FM tools can ensure the success of projects. So, the appropriate tool should be chosen.

4 Conclusion

The advancement of ICT for FM tools is supporting the growth in AEC industries. The outsourcing strategy of facilitating services is quite common. The performance of the facilities is based on various indicators; still the fulfillment of the organizational goals is the primary indicator. During O&M phase of the building, different FM tools are used. IWMS tools are used for overall FM functions, whereas other FM tools such as CAFM, CMMS, and EAM have their own specialties. The use of BIM in AEC industry is providing the FM tools accurate information which helps facility manager to execute their role in a better way. Thus, adopting FM tools in AEC industries is very vital for developing countries for efficient management of scarce resources.

References

1. Lavy, S., & Shohet, I. M. (2010). Performance-based facility management-an integrated approach. *International Journal of Facility Management, 1*(1), 1–14.
2. Roper, K. (2014). Evolution and the Future of FM. In *Innovation in the Built Environment—International Facility Management*. United kingdom: Wiley Blackwell.
3. Finch, E. (2007). Facilities management. In *Intelligent Buildings: Design, Management and Operation*. London: Thomas Telford Ltd.
4. Facility Management Market by Solutions & Services—Worldwide Market. http://www.researchandmarkets.com/reports/2849609/.
5. Ancarani, A., & Capaldo, G. (2005). Supporting decision-making process in facilities management services procurement: A methodological approach. *Journal of Purchasing and Supply Management, 11*, 232–241.
6. FM Glossary. http://community.ifma.org/fmpedia/w/fmpedia/facilities-management.aspx.
7. What is Facility Management. http://www.ifma.org/about/what-is-facility-management#sthash.arDAnzXL.K9wIDpcx.dpuf.
8. Cotts, D., Roper, K. O., & Payant, R. P. (2010). *The facility management handbook*. New York: American Management Association.

9. Barrett, P., & Baldry, D. (2003). *Facilities management towards best practice*. UK: Blackwell Publishing.
10. Interserve: White Paper: The future of facilities management. http://pages.planonsoftware. com/WhitePaperInterserveUS_1Registration.html.
11. iOffice: White Paper: 8 Practical Guidelines to evaluating facilities management softwares. http://www.iofficecorp.com/ebook/8-practical-guidelines-to-evaluating-facilities-management-software.
12. iOffice: White Paper: Productivity Handbook for Busy Facilities Managers. http://www. iofficecorp.com/ebook/productivity-handbook-for-busy-facilities-managers.
13. Facility Management Softwares. http://planonsoftware.com/us/glossary/facility-management-software/.
14. Peter Cholaki: iwms and eam are buzzwords. https://buildinginformationmanagement. wordpress.com/2013/07/18/iwms-and-eam-are-buzzwords/.
15. Hanks, S. why-eam-cmms-cafm-fmis-cifm-iwms-and-other-acronyms-are-confusing. http:// www.iwmsnews.com/2009/12/why-eam-cmms-cafm-fmis-cifm-iwms-and-other-acronyms-are-confusing/.
16. Śliwiński, B., & Gabryelczyk, R. (2009). Development of process modelling for facility management. In D. Hans Robert Hansen (ed.), *1st CEE Symposium on Business Informatics* (pp. 101–110). Vienna.
17. Bell, M. Magic quadrant for IWMS in North America. http://lib-resources.unimelb.edu.au/ gartner/research/123700/123789/123789.html.
18. Halligan, A. A plain english guide to facility & asset management software. http://blog. softwareadvice.com/articles/cafm/plain-english-guide-to-facility-asset-management-software-1030812/.
19. Spence, D. CAFM CMMS EAM and IWMS deciphering the jargon. http://www. officespacesoftware.com/blog/bid/294528/CAFM-CMMS-EAM-and-IWMS-Deciphering-the-Jargon.
20. Rowley, K. Why do leading organizations implement Integrated Workplace Management Systems (IWMS)? European Facilities Management Conference. Tririgia Research. http:// informationmanagement.solutionsdaily.com/mediaFiles/tririga_why%20implement%20iwms. pdf.
21. Hanks, S. Driving the strategic decision: Why more companies are moving to IWMS. http:// www.iwmsnews.com/2014/12/driving-strategic-decision-companies-moving-iwms/.
22. White Paper-What is IWMS? http://planonsoftware.com/us/register/what-is-iwms/.
23. Elmualim, A., & Pelumi-Johnson, A · Computer-aided facility management (CAFM) of intelligent buildings: concept and opportunities. In *CIB W070 Conference in FM* (pp. 87–94). Edinburgh.
24. Bertolini, D., & Lif, A. White Paper—CMMS vs EAM: What is the difference. http:// download.ifsworld.com/home/if1/page_888/cmms_vs_eam_whats_the_difference.html.
25. Sinopoli, J. (2010). Facility management systems. In *Smart Building Systems for Architects, Owners, and Builders*. United States: Elsevier.

Study the Effect of Packet Drop Attack in AODV Routing and MANET and Detection of Such Node in MANET

Kajal S. Patel and J.S. Shah

Abstract MANET is a network which is created on temporary basis. No preexisting infrastructure required for setting up MANET. Nodes of this network are used on mobile so topology is not fixed. All participant nodes are coordinated with each other for setting up MANET. This network uses wireless media and for forwarding packets from source to destination, they actually broadcast the packets. Before sending data from source to destination node, the source node searches for route from source to destination. As no special routers available in such network each node acts as a router and cooperates in routing. Special routing protocols are available for MANET like DSDV, DSR, AODV, etc. Due to attacker nodes and selfish nodes these protocols are vulnerable to many types of malicious activities like packet drop, packet delay, packet modification, etc. In this paper, we have implemented a packet drop attack and study its effect on route discovery time of AODV and throughput. We modified AODV routing to detect such nodes and shows improvement in studied parameters.

Keywords MANET · AODV · Packet drop · Selfish node · Modified AODV

1 Introduction

Mobile ad-hoc network (MANET) is a network which will always be created on temporary basis. No pre existing and pre installed infrastructure required for setting up such network. Nodes of this network are mobile so topology dynamically changes with time. All participant nodes are coordinated with each other for setting

K.S. Patel (✉)
Computer Department, Government Engineering College, Rajkot, India
e-mail: kspldce@gmail.com

J.S. Shah
Computer Engineering, Gujarat Technological University, Ahmedabad, India
e-mail: jssld@yahoo.com

© Springer Science+Business Media Singapore 2016
S.C. Satapathy et al. (eds.), *Proceedings of International Conference on ICT for Sustainable Development*, Advances in Intelligent Systems and Computing 408, DOI 10.1007/978-981-10-0129-1_15

up MANET. This network uses wireless media and for forwarding packets from source to destination, they actually broadcast the packets. Also, before sending data from source to destination node the source node searches for route from source to destination. As no special routers available in such network each node acts as a router and cooperates in routing process. Special routing protocols are available for MANET like DSDV, DSR, AODV, etc. These protocols works good in normal environment but due to attacker nodes and selfish intermediate nodes they are vulnerable to many types of malicious activities like traffic analysis, packet drop, packet delay, packet modification, etc. [1].

Due to some characteristics of ad-hoc Network routing process in ad-hoc network is totally different from the wired networks [2]. They are as follows:

- More chances of error in sent data due to much interference in transmission
- Due to battery operated nodes and to conserve less energy low transmission range is used.
- Nodes in this network are mobile, so topology frequently changes also link frequent breakage.
- The link between two nodes may be unidirectional sometimes. Also if node is not part of any active route it may enter into sleep mode.
- Environmental conditions also affect the operation of ad-hoc network.

2 AODV Routing in MANET

The routing protocols of MANET can be of two types reactive or proactive. The DSDV routing is an example of proactive routing. It maintains a route table and each node periodically receive route update information and update route table with latest route. If more mobile nodes, more routing traffic is generated due to frequent route updates. So other option is reactive protocol. DSR is an example of reactive routing in MANET. In this protocol, source node search for route when it wants to send data to any destination. Dynamically, then needed route is established and maintained.

No route table is maintained on each node. DSR works better with high mobility network but not good for network in which nodes are almost stationary. The solution is the AODV routing. This protocol uses some features of DSDV and some from DSR.

It is ad-hoc on demand distance vector routing protocol. It uses RREQ, RREP, and RERR packets for performing various routing operation. These messages are received on port no 654 and UDP is used [3, 4].

The followings are the various values which store in route table of AODV routing [3, 4]:

Destination IP Address, Destination Sequence Number, Interface, Hop Count (number of hops needed to reach destination), Last Hop Count, Next Hop, List of Precursors, Lifetime (expiration or deletion time of the route), Routing Flags.

In AODV routing when source node want to send some data to destination node, first it search for any active route in its own routing table toward destination. If such route found, it uses that route for sending data packet otherwise it broadcasts RREQ packets to its neighbors. On receiving RREQ packet, each node will search for route entry toward destination stored in its routing table. If route is available in routing table it will send (unicast) route reply to source on the same path from where RREQ came. If not route found, the neighbors also broadcast the same RREQ packets to their neighbor. This continues until destination nodes receive the RREQ or any node which has route toward destination. In any case, an RREP is unicast to source node on the same path from where RREQ came. On receiving RREP, source node uses that route for sending data packets. While sending data packet if any intermediate node of active route changes its position route may break. In such case, neighbor of that node which is in active route detects link break and tries to do route repair. If it is not succeeded in repairing route it will send RERR packet to source to inform it about route break. After receiving RERR source node it will initiate route searching process again by broadcasting RREQ packet.

I have chosen AODV protocol for my research because it works efficiently with high mobile environment as well as environment with low mobility. If topology of network changes frequently due to high mobility of nodes AODV works like DSR and if all nodes are static or less number of nodes are mobile it works like DSDV as route table is maintained at each node. Also new implementation of AODV stores multiple routes from source to destination. This is due to multiple route reply received by source for each path from source to destination.

In this paper we have used OPNET network simulator for implementation. We have implemented a malicious node model which periodically drops the packets. Then, we studied the effect of such malicious node model on route discovery time and throughput of the network. Also we have modified the existing AODV routing which detects node with such malicious behavior and avoid them in route. This will improve the route discovery time and throughput of network compare to AODV.

3 Experiment Setup

We use MANET model in OPNET to simulate AODV network. Malicious node is a node which is a part of active route in MANET and periodically drops data and control packets (AODV routing). I have implemented malicious node in such a way that it updates the packets at IP layer such that they are dropped by higher layer and thus it drops the network throughput and increase route discovery time [5, 6].

Fig. 1 Without malicious
node

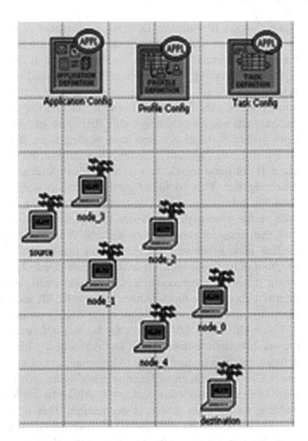

After creating the malicious node model, I have compared the performance of AODV routing protocol and Wireless LAN by creating two scenarios one is without the malicious node (reliable_aodv) and other is with malicious node (unreliable_aodv). Both the scenarios (reliable_aodv and unreliable_aodv) are shown in Figs. 1 and 2. In Fig 2 encircled nodes are malicious. The traffic used for simulation is UDP traffic which flows from source node to destination node. The simulation runs for 60 min. All the nodes in the wireless LAN are stationary nodes [7].

4   Result Showing Effect of Malicious Node

The results for simulations are given below. The Fig. 3 compares the throughput of Wireless LAN which shows that the throughput of wireless LAN is less when malicious nodes are added in the network. As per Fig. 4 the introduction of unreliable node in wireless LAN causes high route discovery time in network while AODV routing.

Fig. 2 With malicious node

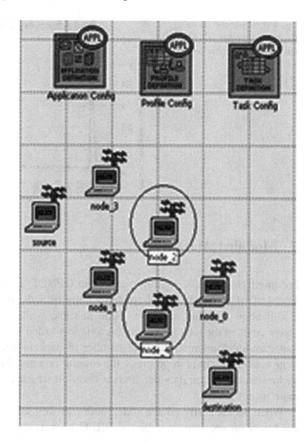

Fig. 3 Comparison of
throughput for reliable and
unreliable scenarios

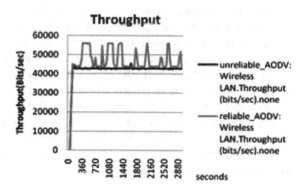

Fig. 4 Comparison of route
discovery time for reliable
and unreliable scenarios

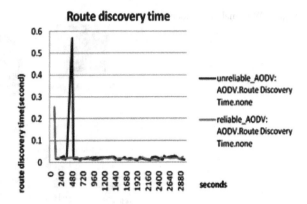

5 Modified AODV

For modifying AODV routing protocol in OPNET, I have to modify the process
model of AODV(aodv_rte). I created a copy of aodv_rte as my_aodv_rte and did
modification to avoid malicious node in route while route is being established.
I have counted the total number of packets forwarded by the node which works as a
router and also counted the total number of packets dropped by the node. While
route establishment is in progress, the routing process checks the value of dropped
packets and based on its value it takes decision whether to include that node in the
route or not [5, 6].

When packet is arrived at a node, routing process checks whether to forward it or
not based on sequence number of RREQ packet, source address and destination
address, and the diameter of the network. If source address and destination address
of the packet are same, node decided to drop the packet and it increments drop-
ped_pkts counter by one and drops the packet. This logic is implemented in
function static void my_aodv_rte_pkt_arrival_handle (void). Route establishment
between source and destination is initiated by destination by sending RREP packet
to node which sends the first RREQ packet. The receiver of RREP again forward
the RREP packet to next predecessor until source reached. While using route
establishment I check value of counters and consider node in route only if counters
have specific values (here dropped_pkts<incoming_pkts*0.20).

After modifying AODV as My_AODV I have created third scenario shown in
Fig. 5 with modified AODV as routing protocol with malicious nodes (encircled) in
previous scenario and compare the results.

Fig. 5 unreliable_Mvaodv

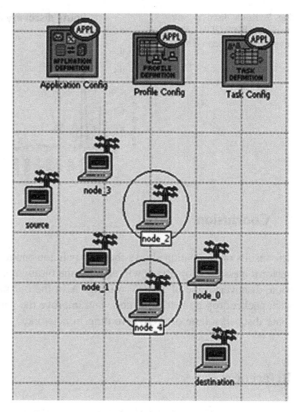

Fig. 6 Throughput for
unreliable AODV and
unreliable modified AODV

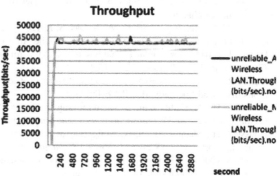

6 Improvement in Results

The results of simulation are shown in the following figures. From Fig. 6 we can
say that with modified AODV the throughput of the wireless LAN is improved. The
Fig. 7 shows improvement in the route discovery time by using modified AODV
routing protocol.

Fig. 7 Route discovery time
for unreliable AODV and
unreliable modified AODV

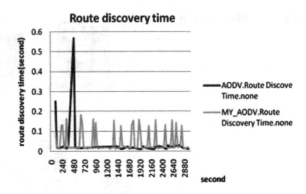

7 Conclusion

The results of simulation shows that the malicious node in wireless network which
perform message drop attack will degrade the throughput of network and take more
route discovery time of AODV routing. Also the our modified AODV will detect
such packet drop by malicious node and improve the wireless LAN throughput and
route discovery time of AODV routing in network.

References

1. Macker, J., & Corson, S. (1997). Mobile Ad hoc Networks (MANET). http://www.ietf.org/
   charters/manet-charter.html(IETF.
2. Perkins, C. E., Belding-Royer, E. M., & Das, S. R. (2002). Ad hoc on-demand distance vector
   (AODV) routing draft-ietf-manet-aodv-10.txt Mobile Ad hoc Networking Group.
3. Hosek, J. (2011). Performance analysis of MANET routing protocols OLSR and AODV.
   Electrorevue, 2, 22–27 ISSN:1213–1539.
4. Das, S. R., Perkins, C., & Royer, E. (2000). Performance comparison of two on-demand routing
   protocols for ad hoc networks. IEEE INFOCOM.
5. Houssein, H., & Shahrestani, S. A. Improving the reliability of ad-hoc on demand distance
   vector protocol. *WSEAS Transactions on Communications*.
6. Houssein, H., Shahrestani, S. A. Mitigation of the effects of selfish and malicious nodes in
   ad-hoc networks. *WSEAS Transactions on Communications*.
7. Suresh, A. (2004). Performance analysis of ad hoc on-demand distance vector routing (AODV)
   using OPNET Simulator. University of Bremen.

UDP Flooding Attack Detection Using Information Metric Measure

Debojit Boro, Himant Basumatary, Tribeni Goswami
and Dhruba K. Bhattacharyya

Abstract UDP flooding is one of the most pursued DDoS attack among the attackers. Extensive research solutions focused on other DDoS flooding attacks could be found, but little work to deal UDP flooding attack traffic exists. Due to the stateless nature of UDP, the detection of the attack is very difficult and can effectively throttle the victim with unwanted traffic. In this paper, we present a solution to detect UDP flooding attack based on generalized entropy information metric and also determine the malicious source IP (SIP) addresses by carrying out the attack. We conduct our experiment on some captured attack traffic and the results demonstrate that the proposed solution can effectively detect UDP flooding attack along with the malicious SIP addresses.

Keywords DDoS · UDP · Information metric · Renyi's entropy · Shannon's entropy

1 Introduction

DDoS flooding attack has emerged as one of the popular attack tool among the attackers. The attack is orchestrated by coordinating large number of compromised computers in the Internet to send unsolicited traffic to a victim, until it cripples to serve any of its legitimate users. Due to the Internet vulnerability, the attackers also

D. Boro (✉) · H. Basumatary · T. Goswami · D.K. Bhattacharyya
Department of Computer Science and Engineering, Tezpur University,
Tezpur 784028, India
e-mail: deb0001@tezu.ernet.in

H. Basumatary
e-mail: basumatary.himant@gmail.com

T. Goswami
e-mail: tribenigoswami16@gmail.com

D.K. Bhattacharyya
e-mail: dkb@tezu.ernet.in

© Springer Science+Business Media Singapore 2016
S.C. Satapathy et al. (eds.), *Proceedings of International Conference
on ICT for Sustainable Development*, Advances in Intelligent Systems
and Computing 408, DOI 10.1007/978-981-10-0129-1_16

143

spoof the SIP addresses of the attack packets making it more elusive and hard to detect. Recent reports present alarming situation, where the attack intensity has been continually increasing and has shot up to 400 GBPS in 2014 [1], 300 GBPS in 2013, and 75 GBPS in 2012 [2]. Among the many protocols used for flooding, UDP is one of the most prominent protocol used for flooding nowadays. Recent reports on UDP flooding launched by anonymous hacker group could be found in [3, 4]. Though several methodologies for curbing DDoS flooding attacks have been proposed [5–7], but little work for exactly handling UDP flooding could be found. Or, even if the methodologies proposed also applies to UDP, it is still skeptical that whether the key assumptions made in these works will also hold for UDP traffic.

In this paper, we propose a solution for UDP flooding attack detection at the victim-end since this attack is mainly bound at a victim server using its destination IP (DIP) address and port number within the packets. At the same time, it can also overwhelm the network bandwidth and resources with irrelevant UDP traffic until it shuts down. Our solution considers the UDP flooding targeted to a particular destination in two different scenarios: (i) Fix SIP address with random change of destination port number, and (ii) Randomly spoofed SIP address.

The main contribution of this paper is as follows:

1. We propose a solution based on generalized entropy information metric to detect UDP flooding attack based on non-spoofed scenario with large variations of destination port number and spoofed based scenario with large SIP address variation. We could also extract the malicious SIP addresses by launching the attack.
2. We establish the effectiveness of our solution by testing on various captured network attack datasets. The result demonstrates that our method can detect UDP flooding attack in both the scenarios.

The rest of the paper is organized as follows. Sections 2 and 3 discusses the background and the related work. Section 4 describes our proposed solution. In Sect. 5 we discuss the experimental results to establish the effectiveness of our solution. Finally, in Sect. 6 we draw our conclusion and future work.

2 Background

UDP is a connectionless transport-layer protocol that is basically used when higher preference is put on faster delivery of message or datagram than quality or reliability. Unlike TCP, UDP offers little transport services such as data checksumming and port number multiplexing. The UDP flooding attack is launched by sending innumerable number of UDP packets to random destination port of a victim. If the victim finds no application to be running at that port upon checking, it replies with ICMP destination unreachable packets. This eventually leads to network bandwidth clogging. Alternatively, the attacker also sends randomized spoofed SIP address packets to the victim until all of its networks resources and available bandwidth cripples.

In [8], notable increase in UDP flows is reported in the Internet traffic traces due to usage of many P2P applications such as BitTorrent, edonkey, etc., and ephemeral ports above 1024. With such manifold applications and port numbers, launching flooding attacks using UDP is on the rise and being a stateless protocol, prevention strategies adopted is based on connection state such as TCP SYN cookies which cannot be employed. Therefore, it is imperative to have an effective measure to handle UDP flooding attacks for both spoofed and non-spoofed scenarios discussed above, and not let the users from being deprived of those services. Also at the same time, safeguarding cost and reputation of those organizations.

3 Related Work

In this section, we present some recent works done on detection of UDP flooding attacks. In [9], only change of arrival rate of new IP addresses coupled with CUSUM algorithm was used as a detection feature for fast detection of high rate flooding attack. The method reported faster detection and considered UDP flooding attack under both constant source and varied source IP addresses. A high rate flooding attack detection in backbone networks using two-dimensional (2D) Sketch data structure was proposed in [10]. The method employed Least Mean Square (LMS) filter and Pearson Chi-square divergence on random aggregation of rows in the data structure with high detection accuracy and low false alarm rate. [11] adopted proportional packet rate assumption for UDP traffic, which is based on the fact that under normal operations the packet rate is proportional in both directions. However, the method does not consider UDP flooding attack under spoofed SIP addresses, and under such circumstances it cannot find the malicious SIP addresses. A novel approach based on finding the entropies of Conditional Random Fields (CRF) such as SIP and DIP, source, and destination port of each IP header packets was proposed in [12]. The method was based on fusion of signature and anomaly based methods for DDoS attack detection and uses L-BFGS algorithm for generation of detection model before which, all the IP flow entropies are normalized to a single entropy metric. The method can also detect new attacks. A flooding attack detection method based on collaborative approach using multiple autonomous agents was proposed in [13]. Regulation of bandwidth allocation for attackers and legitimate users using fuzzy inference system was adopted. However, the method assumes the agents to be residing within the victim machine that may not be feasible during heavy flooding attack and do not consider the UDP flooding attack using spoofed SIP addresses.

As observed, all the existing methods for DDoS flooding attack also handle UDP flooding, but few also address the scenario when flooding is done using spoofed SIP addresses. Most of them assume UDP flooding with constant source. But none of the work addresses UDP flooding with random destination port variation. In this paper, we address both the UDP flooding attack scenarios, i.e., (i) fix SIP address with random destination port number and (ii) randomly spoofed SIP address.

4 Proposed Method

In Fig. 1 we depict the detection architecture with our victim-end solution known as *Flooding Attack Detector (FAD)* for UDP flooding attack detection. FAD is a traffic monitoring system with high processing capacity and large memory, that passively monitors the forwarded incoming traffic to sense any deviation from the normal traffic behavior using generalized information metric measure. The victim server is connected to the Internet through a L3 switch, firewall, and an edge router. We consider only the incoming traffic for attack detection in this paper. During normal conditions, incoming traffic from the router/firewall are directly forwarded to the victim server through the switch. The switch also forwards a copy of the traffic seen at one of its port to the FAD through port mirroring while forwarding the traffic to the victim server.

For a particular destination, our method considers two detection features for UDP flooding attack: (i) total *destination port change (d_port_change_count)* for non-spoofed SIP addresses with random destination port and (ii) *SIP address change (SIP_change_count)* for randomly spoofed SIP addresses. Since our method is a victim-end solution, therefore either of non-spoofed or spoofed UDP packets will only have the DIP address of the victim. For each duplicate forwarded packets to FAD by the switch, the FAD access the UDP packet header and generates a unique ID (UID) from the SIP address. A node with six components, i.e., *UID, SIP, d_port_change_count, d_port_change_count_entropy, SIP_change_count, SIP_change_count_entropy* is created, and a KD-Tree with UID as a single key as shown in Fig. 2 is created for each time interval *t*. We choose KD-Tree to make our solution scalable for randomly spoofed SIP address UDP flooding attack, where large number of SIP addresses may be generated. In that case, usage of any fixed size data structure may limit the storage of SIP address information in computer's memory.

While reading the switch forwarded packets in each interval *t* by the FAD, the FAD increments the *d_port_change_count* component in a tree node for every change of destination port by a SIP. Similarly, the component *SIP_change_count* is incremented for every change of SIP *X* to SIP *Y*. After the complete KD-Tree is built for an interval, we find the randomness of the *d_port_change_count* and

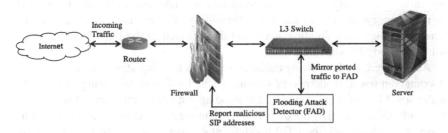

Fig. 1 Detection architecture

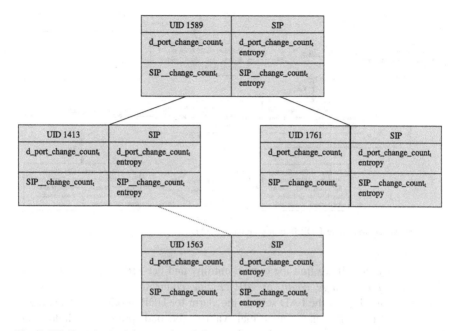

Fig. 2 KD-Tree for storing entropies of *d_port_change_count* and *SIP_change_count* in FAD

SIP_change_count in each node. The tree is deleted and recreated in each interval. To find the randomness we use the information metric measure known as Renyi's entropy [14] of order α, which is a mathematical generalization of Shannon's entropy [15]. The Renyi's entropy is defined as follows.

Definition Given a discrete probability distribution P, where $P = p_1, p_2, \ldots, p_n$ such that $\sum_{i=1}^{n} p_i = 1$, $p_i \geq 0$, and $i = 1, 2, 3, \ldots, n$. Then the Renyi's entropy of order α is defined as

$$H_\alpha(\chi) = \frac{1}{1-\alpha} \log_2 \left(\sum_{i=1}^{n} p_i^\alpha \right) \tag{1}$$

where $\alpha \geq 0$ and $\alpha \neq 1$
when $\alpha \to 1$, $H_\alpha(\chi)$ converges to Shannon's entropy.

$$H_1(\chi) = -\log_2 \sum_{i=1}^{n} p_i^2 \tag{2}$$

In Eq. 1 for probability distribution P in each interval, i represents the SIP address and p_i is either the normalized *d_port_change_count* (c) (i.e., $p_i = c_i / \sum c_i$) or *SIP_change_count* (c) (i.e., $p_i = c_i \sum c_i$) where $i = 1, 2, 3 \ldots N$. In each time interval t, we perform pre-order traversal of the KD-Tree to calculate the overall entropy and per SIP address (UID node) entropy for both *d_port_change_count* and

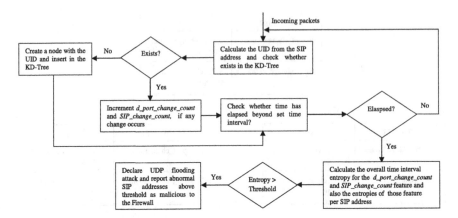

Fig. 3 Flowchart to handle UDP flooding attack by FAD

SIP_change_count. If we find the overall entropy and per SIP address entropy for either of *d_port_change_count* or *SIP_change_count* to be beyond a preset threshold (using Eq. 1), the FAD sets on the alarm for UDP flooding attack and we record the SIP's that have abnormal values (beyond preset threshold) for *d_port_change_count_entropy* or *SIP_change_count_entropy* as the malicious SIP addresses. These malicious SIP addresses are reported by the FAD to the firewall to be configured and to be blocked until the attack is over. A flowchart representation of our method in FAD is shown in Fig. 3.

4.1 Complexity Analysis

The method by the FAD takes $O(n\log(n-1)) + O(n)$ time in each interval t to detect UDP flooding attack for each individual order α of information metric, where n is the number of UDP packets within a time interval t. The additional $O(n)$ time is because of the pre-order traversal of the KD-Tree.

5 Experimental Results

We evaluated the effectiveness of our method on UCLA [16], NUST [17], and CAIDA 2013 [18] datasets. UCLA dataset contains UDP packet traces collected at the edge router of Computer Science Department in 2001. We used both the passive and attack traces of UCLA dataset. NUST traffic dataset contains only the attack traces that have large variation of SIP addresses. CAIDA 2013 contains approximately 10 min (i.e. 600 s) of anonymized passive traffic traces from CAIDA's *equinix-chicago* and *equinix-sanjose* monitors on high-speed Internet backbone

Table 1 UDP packet rate of the datasets (per 5 s)

Dataset	Normal	Attack
UCLA	120	9274
NUST	NA	1100
CAIDA 2013	NA	146,751

NA Not Available

links. In CAIDA 2013 dataset, large variations in SIP address could be seen and we expect those variations to be detected by our method.

The choice of the duration of time interval is pivotal to the problem of attack detection. Small time interval leads to more computation per interval. Whereas, large time interval leads to accumulation of more packets that may be throttling for computation itself. Therefore, for all the datasets we choose to sample them into 5 s interval and the total observation is done for 300 s based on the normal and attack packet rate of the various datasets as shown in Table 1. As discussed already, the KD-Tree is created and deleted for each interval, during which the overall entropy is calculated for both UDP *d_port_change_count* and *SIP_change_count* features and the respective SIP's. We apply both the Renyi's entropy of order α using Eq. 1 (where α is varied within a range of 0–12) and the Shannon's entropy using Eq. 2 (shown in all the figures).

Figure 4 shows the entropies for the destination port number change over varied order α in UCLA dataset, when UDP flooding attack traces were injected within normal traces. The figure shows the spikes in the regions where the attack traces were injected and the normal entropy fluctuates within the range of 0–6. We consider this range as the threshold for destination port number change and for UDP flooding attack detection. NUST dataset do not have normal traces, therefore we show the graphs for both the attack detection features in Figs. 5 and 6 for attack traces only. Much variation in destination port change ranging between 0 and 12

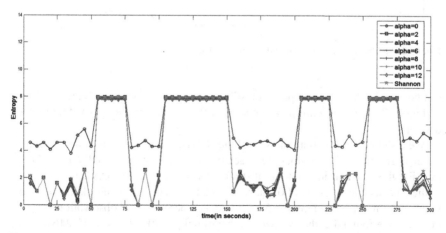

Fig. 4 Destination port number change during attack in UCLA dataset

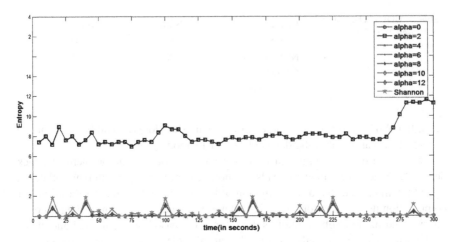

Fig. 5 Destination port number change during attack in NUST dataset

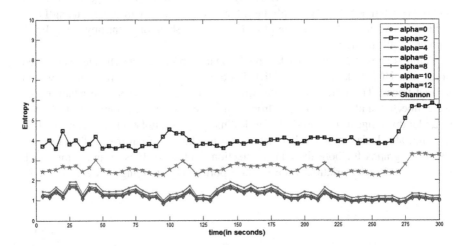

Fig. 6 SIP address variation during attack in NUST dataset

can be seen in Fig. 5 as compared to SIP variation in Fig. 6 ranging between 1 and 6. Figure 7 show the entropies for huge SIP address variation in CAIDA 2013 dataset.

We also take into account the size of the KD-Tree when FAD is under randomly spoofed SIP address UDP flooding attack that can generate many UID nodes in the tree. For this, we consider a high rate traffic dataset such as CAIDA 2013 that have traffic rate as much as 50,000 packets/s. In this dataset, we check the average number of unique SIP address in 5 s time interval across a total observation window of 300 s. We found that the average number of unique SIP address is 12,246 and the

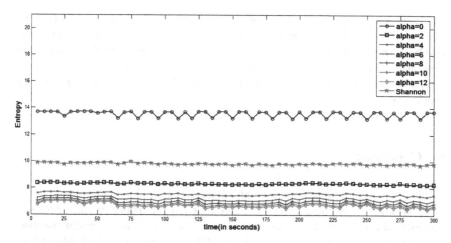

Fig. 7 SIP address variation in CAIDA 2013 dataset

KD-tree with 12,246 nodes can easily accommodate such large number of SIP address without any concern for the size.

We also evaluated our method based on the computational time. All the datasets have varying UDP packet rate as shown in Table 1 and we report the average computation time taken by our method on different datasets in Table 2. Since NUST dataset do not have normal traces, we report only the computation time for attack traces. Similarly for CAIDA 2013 dataset, we report the computation time under attack only, as it is not completely attack-free traffic. In the table, both Shannon's

Table 2 Average computation time for UDP flooding attack detection on the datasets

Attack Scenario	Information Metric Measure	Dataset	Normal (per 5s)							Attack (per 5s)						
Non-spoofed attack (dest_port_change)	Shannon's Entropy	UCLA	0.0002							0.0092						
		NUST	NA							0.0004						
		CAIDA (2013)	NA							0.1483						
	Renyi's Entropy		$\alpha=0$	$\alpha=2$	$\alpha=4$	$\alpha=6$	$\alpha=8$	$\alpha=10$	$\alpha=12$	$\alpha=0$	$\alpha=2$	$\alpha=4$	$\alpha=6$	$\alpha=8$	$\alpha=10$	$\alpha=12$
		UCLA	0.0002	0.0002	0.0002	0.0002	0.0002	0.0001	0.0002	0.0091	0.0091	0.0091	0.0091	0.0092	0.0092	0.0091
		NUST	NA	NA	NA	NA	NA	NA	NA	0.0005	0.0044	0.0041	0.0042	0.0046	0.0046	0.0049
		CAIDA (2013)	NA	NA	NA	NA	NA	NA	NA	0.1498	0.1488	0.1477	0.1478	0.1477	0.1475	0.1490
Spoofed attack (SIP_change)	Shannon's Entropy	UCLA	0.0001							0.0087						
		NUST	NA							0.0004						
		CAIDA (2013)	NA							0.1422						
	Renyi's Entropy		$\alpha=0$	$\alpha=2$	$\alpha=4$	$\alpha=6$	$\alpha=8$	$\alpha=10$	$\alpha=12$	$\alpha=0$	$\alpha=2$	$\alpha=4$	$\alpha=6$	$\alpha=8$	$\alpha=10$	$\alpha=12$
		UCLA	0.0001	0.0002	0.0002	0.0002	0.0001	0.0002	0.0002	0.0088	0.0089	0.0090	0.0089	0.0089	0.0090	0.0090
		NUST	NA	NA	NA	NA	NA	NA	NA	0.0004	0.0004	0.0004	0.0004	0.0005	0.0004	0.0004
		CAIDA (2013)	NA	NA	NA	NA	NA	NA	NA	0.1443	0.1441	0.1456	0.1457	0.1454	0.1463	0.1456

**NA: Not Available

and Renyi's entropy takes same amount of time in normal conditions. But CAIDA 2013 takes the highest amount of time for both the entropy method in either of the attack scenarios as it has highest UDP packet rates. But comparatively, computation time taken is very minimal for traffic of 5 s interval. Similarly, our method takes less computation in other datasets as well. Hence, our method is fast in UDP flooding attack detection.

6 Conclusion

In this paper, we proposed a victim-end solution for UDP flooding attack detection using information metric measure that can filter out the malicious SIP addresses. The method can detect UDP flooding attack for both the abnormal destination port change that leads to generation of numerous ICMP destination unreachable packets and spoofed SIP address attack. The method has been found scalable in context of CAIDA 2013 dataset. High packet rate did not indicate significant degradation of computation performance. Further, the use of KD-Tree has also found to be very effective in storing the packet informations, rather than storing in fixed size data structure. The tree can grow without any perturbations of the limitation of the size of data structure during spoofed SIP flooding attack. As a future work, we plan to propose a solution for UDP fragmentation attack and extend the solution to handle non-UDP traffic such as ICMP, TCP, and HTTP alongwith IP traceback.

Acknowledgments This work is supported by Ministry of Human Resource and Development (MHRD), Government of India, under Frontier Areas of Science and Technology (FAST).

References

1. Arbor Networks: Worldwide Infrastructure Security Report. http://www.techworld.com/news/security/worlds-largest-ddos-attack-reached-400gbps-says-arbor-networks-3595715/.
2. The Availability Digests: Surviving DNS DDoS Attack? http://www.secure64.com.
3. The Business of Technology (Bits): Hackers Step Up Attacks After Megaupload Shutdown. http://bits.blogs.nytimes.com/2012/01/24/.
4. Anonymous DDoS Activity. https://www.us-cert.gov/ncas/alerts/TA12-024A.
5. Chen, Y., Hwang, K., & Ku, W. S. (2007). Collaborative detection of DDoS attacks over multiple network domains. *IEEE Transactions on Parallel and Distributed Systems, 18,* 1649–1662.
6. Chou, J., Lin, B., Sen, S., & Spatscheck, O. (2009). Proactive surge protection: A defense mechanism for bandwidth-based attacks. *IEEE/ACM Transactions on Networking, 17,* 1711–1723.
7. Keshariya, A., & Foukia, N. (2010). DDoS defense mechanisms: A new taxonomy. In J. G. Alfaro, G. N. Arribas, N. C. Boulahia, & Y. Roudier (Eds.), *Data privacy management and autonomous spontaneous security* (Vol. 5939, pp. 222–236)., LNCS Heidelberg: Springer.

8. Zhang, M., Dusi, M., John, W., & Chen, C. (2009). Analysis of UDP traffic usage on internet backbone links. In *9th Annual International Symposium on Applications and the Internet (SAINT 2009)* (pp. 280–281). Seattle: IEEE.

9. Ahmed, E., Mohay, G., Tickle, A., & Bhatia, S. (2010). Use of IP addresses for high rate flooding attack detection. In K. Rannenberg, V. Varadharajan, & W. Christian (Eds.), *Security and privacy—silver linings in the cloud* (Vol. 330, pp. 124–135)., IFIP Advances in Information and Communication Technology Heidelberg: Springer.

10. Salem, O., Makke, A., Tajer, J., & Mehaoua, A. (2011). Flooding attacks detection in traffic of backbone networks. In *36th IEEE Conference on Local Computer Networks* (pp. 441–449). Bonn: IEEE.

11. Bardas, A. G., Zomlot, L., Sundaramurthy, S. C., Ou, X., Rajagopalan, S. R., & Eisenbarth, M. R. (2012). Classification of UDP traffic for DDoS detection. In *5th USENIX Conference on Large-Scale Exploits and Emergent Threats* (pp. 7–7). Berkeley: USENIX Association.

12. Chen, S. W., Wu, J. X., Ye, X. L., & Guo, T. (2013). Distributed denial of service attacks detection method based on conditional random fields. *Journal of Networks., 8,* 858–865.

13. Preetha, G., Devi, B. S. K., & Shalinie, S. M. (2014). Autonomous agent for DDoS attack detection and defense in an experimental testbed. *International Journal of Fuzzy Systems, 16,* 520–528.

14. Renyi, A. (1961). On measures of entropy and information. In *4th Berkeley Symposium on Mathematical Statistics and Probability* (pp. 547–561). University of California Press.

15. Shannon, C. E. (1948). A mathematical theory of communication. *The Bell System Technical Journal, 27,* 379–423.

16. UCLA CSD Packet Traces. http://www.lasr.cs.ucla.edu/ddos/traces/.

17. NUST. http://wisnet.seecs.nust.edu.pk/projects/nes/implementation.html.

18. CAIDA 2013. http://www.caida.org/data/passive/passive_2013_dataset.xml.

Generalized MCDM-Based Decision Support System for Personnel Prioritization

Aarushi and Sanjay Kumar Malik

Abstract Human resource is the utmost important ingredient in producing the best in an organization. It is the talent and creativity of the working individuals, which acts a key factor in achieving the targeted success. Deeper analysis reveals that, application of a particular talent at the correct place is immensely important for the growth of an organization and the refinement of the skills of the person. This paper presents an algorithm for precise mapping of an appropriate candidate(s) for a particular job. The algorithm is designed using MCDM methods namely—AHP and TOPSIS. In order to optimize the problem, twelve generic criteria have been chosen, which enable the prioritization of the appearing candidates with respect to a profession. The algorithm discussed in this paper, adequately deals with the problem of personnel selection considering all the uncertainties associated, including both the qualitative and quantitative measures required to prioritize the candidates.

Keywords Multiple criteria decision making (MCDM) · Analytic hierarchy process (AHP) · Technique for ordered preference by similarity to ideal solution (TOPSIS) · Decision maker (DM)

1 Introduction

The choice of a qualified individual for a particular post of a professional is an art. Here, the term 'qualified' is broader in nature that encapsulates human behavior, cognitive abilities, academic qualification, and also physical and psychological constructs. On the other hand, experts involved in their task should also need to have a multidimensional mindset. They have to choose their best from several

Aarushi (✉) · S.K. Malik
Department of Computer Science and Engineering,
Hindu College of Engineering, Sonepat, Haryana, India
e-mail: aarushi175@gmail.com

© Springer Science+Business Media Singapore 2016
S.C. Satapathy et al. (eds.), *Proceedings of International Conference
on ICT for Sustainable Development*, Advances in Intelligent Systems
and Computing 408, DOI 10.1007/978-981-10-0129-1_17

closely similar individuals. Hence, such a choice is a kind of an art to explore out the best among them, in all the spheres of the job. With the passage of time, the mindset of the youth is drifting from simply getting a job. Rather, they want to work in a domain that explores out the talent/potentials they possess. Thus, this area of human resource has gained strong interest of the researchers in the twentieth century.

Personal selection is a two-way mapping process, where the selection is done by matching the features of the job with the features of the candidates appearing for the selection. So, this process is more or less like an art form where subtle matching is done precisely. The work force of an organization is an important driving force leading it to unmatchable heights. Hence, such a selection urge for a planned methodology where the results are not just accurate but also dynamically consistent.

The technique discussed here serves as among the fine solutions of the problem of Personal Selection. The technique covers both academic and non-academic factors for making a selection. Further, the mathematical basis provides the fitness value for each of the appearing candidate, upon which unvarying choice can be made easily. This techniques works above the traditional ways by applying some mathematical formulations, and, gives the fitness value which is less subjected to selector's behavior. In this paper, we have tried to bridge this gap by an MCDM-based methodology.

2 Literature Review

2.1 Personnel Selection and MCDM

The term 'Personnel Selection' is dynamic in nature. Earlier, when the 'requirement' of the man power was higher than the available population, the selection was not a complex task and neither getting a job was difficult. Hence, the process was simple. With time, as the situation gradually became contrast, the task of selection had to be scrutinized. Due to the dynamic nature, this area has received a lot of importance from researcher's perspective. Late nineteenth century has marked many noticeable researches in this area. Though the process of selection is as old as human race, the first documented selection test was Chinese Civil Servant Exam, in 605 A.D. The history of Personnel Selection is interesting as it depicts the vital twists and turns involved, based on the current trends and requirements. The handbook of Oxford-A history of Personnel Selection and Assessment covers all such evident trends [1]. It owes too much to psychometric theory and art of integrating selection systems with the psychological and behavioral perspective of candidate and requirements of the job [2].

Personnel Selection is the process of mapping a set of individual with unique features to a particular domain of the job. Hence, more will be the association, better is the choice. It has now become a boundless term, determining the input quality of the personnel and determining its associations with the requirement of the task or job [3]. This paper has tried to cover the selection process from a multidimensional view, i.e., other than psychological and academic fulfillment; behavioral patterns, etc., have also been discussed subtly [4]. This paper has highlighted the solution of the problem using Multiple Criteria Decision Making. MCDM is an old method which can be traced back to the years 1706–1790 with the simple technique of Benjamin Franklin [5]. In 1975, first organization as Special Interest Group on MCDM was developed and later it was named as International Society on MCDM [6].

MCDM today is a network of so many techniques, having a wide applicability in scientific and non-scientific area [7]. The techniques of MCDM have their own corns and pros, summarized by Velasquez and Hester [8]. The task of personnel selection when done in conjunction with MCDM, avoids inconsistent selections avoiding voluntary judgmental notions. The task is very complex and has immense vagueness involved [9].

2.2 MCDM Method

MCDM is the composition of criteria, set of alternatives, and their comparison in some manner. This comparison is among the alternatives and criteria act as the parameters for comparison. Different techniques under MCDM do the same task following different principles [10, 11].

The methodology discussed in this paper is a combined approach of two techniques, bringing out the better of the two available methods under one shade. Both AHP and TOPSIS, individually serve as the best solutions to many of the selection problems [12–14].

Analytical Hierarchy Process (AHP) is one of the strongest technique developed during the 1970 s and finally in 1980, Salty [15], presented the technique comprehensively. AHP serves as the best situation where the criteria are huge and there is uncertainty associated with the ground conditions. This methodology breaks the problem into a hierarchical structure. It is a comprehensive and rational methodology, encompassing the following steps sequentially:

Step 1: Formation of a Hierarchical Decision Matrix of the Problem: This matrix is composition of alternatives (row wise) and criteria (column wise). The value (X_{ij}) in the matrix, is the weight of an alternative with respect to a criterion as given in the following matrix as example.

	C_1	C_2	C_m
A_1	X_{11}	X_{12}	X_{1m}
A_2	...	...	...
A_n	...	...	X_{nm}

Step 2: Comparative Judgment: The process of comparative judgment is a two-level process. Initially, a matrix of type $A_{N \times N}$ is formed in which the relative comparison of alternate with another alternate is done, thus here N is the number of alternatives. Likewise M metrics of the form $A_{M \times M}$ are generated, where M is the number of criteria under consideration. (In this paper, we are basically interested in criteria comparison only i.e. latter one) So, this level of comparison consist of M metrics (one with respect to each criteria) containing the mutual comparison of the criteria among themselves. The entries in these matrices, e_{ij} is done according to AHP original measurement scale from one to nine, as given in the following Table 1.

Step 3: Normalization: After the development of the comparison matrices both for the criteria comparison and the alternative comparison, it is the time to construct the normalized matrix out of these. A normalized element r_{ij} of the normalized matrix is obtained using Eq. (1) The equation written here is basically for the normalization of the elements of the criteria comparison matrix (M is the number of criteria).

$$r_{ij} = \frac{e_{ij}}{\sum_{i=1}^{M} e_{ij}} \tag{1}$$

where each element e_{ij} of matrix $A_{M \times M}$ is divided by the sum of the elements of its respective column.

Step 4: Formation of the Eigen Vector: Eigen Vector is the average of the summated value across each row in the normalized decision matrix. After the normalized value for each criterion has been obtained, the weight vector is to be calculated as follows, using Eq. (2). Likewise, each criterion is assigned as 'W', which determines the weight of the criterion. The weight vector associated with each entity in the comparison (criteria), generated using Eq. (2), is the value which decides the priority of each criteria.

$$W = \frac{\sum_{i=1}^{M} r_{ij}}{M} \tag{2}$$

Table 1 Scale of the AHP [15]

Values	Definition
1	Equal importance
3	Moderate importance
5	Strong importance
7	Very strong importance
9	Extremely important
2, 4, 6, 8	For the intermediate values
Reciprocals	For vice versa comparison, if i to j is 3, then j to i is 1/3

Step 5: Consistency Check: The validation step involved in AHP is a consistency check, done using Eq. (3) where CI is the Consistency Index and RI is the Random Index. The application consistency check on the matrix verifies whether the comparison matrix is consistent.

$$CR = \frac{CI}{RI} \tag{3}$$

The value of CI can be calculated using Eq. (4) and RI using the standard Table 2, one can find CR. A perfectly consistent value of CR corresponds to 0, which is basically an ideal case and that is not practically feasible. A value of CR which is equal or less than 0.1 indicates acceptable level of consistency in the pair wise matrix, hence acceptable. However, if the value of CR exceeds 0.1, the comparison values are inconsistent and evaluators need to review their judgments.

$$CI = \frac{\lambda\max - m}{m - 1} \tag{4}$$

Step 6: Synthesizing comparisons across various levels to obtain the final weight of alternatives. This is a recursive step that involves the application of the above discussed steps on different matrices of the alternatives with respect to each of the criterion. This particular step will not be a part of the proposed methodology in this paper as it is not required.

Despite wide applicability, AHP narrows due to inefficiency in providing ideal consistency. The probability of inconsistency increases when number of alternatives and criteria are large.

TOPSIS It is yet another technique of MCDM, developed by Hwang and Yoon in 1980s, using the same constituents as any other MCDM methodology, i.e., set of alternative, a set of criteria, and the aim is an optimal selection. The principle of TOPSIS is that the optimal choice should have minimum distance from the ideal solution and maximum distance from the negatively ideal solution.

The major mathematical steps involved in this technique are:

Step 1: Initially it begins up with the matrix, which consists of alternatives, Criteria, weights associated with the criteria, and the scores of the alternatives with respect to each of the criteria. N is the number of alternative $(A_1, A_2 \ldots \ldots A_n)$ and M is the number of criteria $(C_1, C_2 \ldots \ldots C_m)$. X_{ij} is the score of ith alternative with respect to the jth criteria and W_i is the weight of jth criteria. The weight has been calculated earlier using AHP.

Step 2: In this step, the decision matrix is normalized by squaring X_{ij} and summating each column. Calculate the square root of this resulting column wise summation and then divide each column of decision matrix. This step can be performed using Eq. (5).

Table 2 Values of RI with respect to the number of criteria

1	2	3	4	5	6	7	8	9	10	11	12	13	14	15
0.00	0.00	0.58	0.90	1.12	1.24	1.32	1.41	1.45	1.49	1.51	1.48	1.56	1.57	1.59

$$Z_{ij} = \frac{X_{ij}}{\sqrt{\sum_{i=1}^{m} (X_{ij})^2}} \tag{5}$$

Step 3: In this step, the normalized values are multiplied with the weight associated with its corresponding criteria.

Step 4: Identification of the ideal and the negatively ideal case with respect to each criterion. Here, we differentiate between the maximizing criteria and the minimizing criteria. Maximizing criteria are those for which higher the value more is the efficiency. For example, in 'past experience', more is the experience, better will be the candidate. On the other hand, minimizing criteria are those for whom lesser the value better is the choice. For example, in cost, lesser is the cost of commodity (other factors remaining constant), better will be the product. So, highlight both the ideal and negatively ideal value in each column by taking into concern whether the criteria is maximizing or minimizing. T^* is the ideal set of criteria and T' is set of negatively ideal values.

Step-5: Separation from the Ideal and Negatively Ideal Solution: Now we will determine the separation from the ideal solution and separation from the negatively ideal solution by subtracting I^* and I' individually from just previous matrix and squaring these values.

Step 6: Now, sum up the values row wise, for all the criteria for each individual alternative and then take the square roots of the summated values given as T^* and T' for ideal and negatively ideal solution respectively.

Step 7: The final score (S_i) of each alternative can be calculated using Eq. (6),

$$S_i = \frac{T'}{T^* - T'} \tag{6}$$

Criteria/Attributes of Personnel Selection Problem In order to differentiate among the candidates appearing for a job selection, some parameters are required. These parameters are a kind of constraints that help to eliminate the unwanted features and scrutinize the overall process. The criteria are based on the context in which selection is being done. In addition to the psychological perspective involved in the selection, the organizational features are equally necessary to be considered [16]. Before, we proceed, the criteria on the basis of which candidates will be prioritized are necessary to be elicited and these criteria are based on job and organizational features [2, 4]. This paper has elicited twelve generic criteria, irrespective of a particular context in which selection is being done. Moreover, the criteria have been chosen by taking into consideration the personality traits [17]. For, the convenience, they have been divided into two categories, namely Objective Criteria and Subjective Criteria [18].

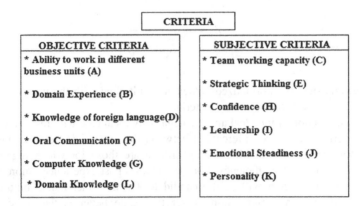

Fig. 1 Criteria chosen for personnel selection (objective and subjective criteria)

Objective Criteria These are the quantified measures whose values can be predicted or calculated approximately, using some evident proofs. There exists means to provide values to these criteria.

Subjective Criteria Subjective criteria are the internal measures for which a fixed value is difficult to be assigned. These are manipulative in nature. Some ways like questionnaire, group activities, situations, etc., can help in knowing these attributes of the person to some extent.

The concept of generalization has been acutely applied on criteria selection. Specialization can be further applied on these, depending on the context of selection problem. Figure 1, has given a lucid depiction of the criteria chosen in this paper. The coherent portrayal, defines the two categories implicitly. Table 3, has enlisted the criteria selected in the paper. It is a summarized view of each one of them. It relates that how and to what extent each criterion has the influence on the resulting comparison.

3 Proposed Methodology

The proposed method begins with the problem analysis, in order to identify all its features which will be used as the parameters to find the optimal selection, at the later stages. In the current study, the paper analyses the generic process of selection, to elicit the eligible work force. This analysis will result in the features which are the 'criteria' against which the candidates will be examined. Following this step we had twelve criteria as discussed above.

The algorithm is described using flowchart given in Fig. 2. The techniques which are used here have been elaborated in the previous sections of the paper. After the criteria elicitation, the next step involves the role of AHP. Here, a mutual comparison among the criteria will be made, in order to generate the Eigen vector.

Table 3 Criteria and their description

S. no.	Criteria	Description
1	Ability to work in different business units	It refers to the ability of an individual to adjust in the different working environments or working locations. Any organization seeks for it, to make the employees work in the area where they are best suited. Easily adjustable nature of a person is more suited to the organization
2	Domain experience	Domain experience is the key factors that hone the knowledge and talents possessed by a candidate. More is the experience; more is the potential of a person in handling sudden situation of their domain
3	Team working capacity	It reflects ones flexibility in adjusting with their peers and collogues. A perfect balance of professional terms and coordination among the team members, work as a catalyst in achieving the desired goals
4	Knowledge of foreign language	Employees of the organization are timely deported to various locations in the world, depending on the needs. So, knowledge of foreign languages is always a benefit for the candidate as well as for the organization
5	Strategic thinking	This factor is very critical to be examined as it tells how well the person is apt in facing difficult and unexpected situations and getting through them. The other means of strategic thinking is the basic aptitude of a person in taking at the spot decisions
6	Oral communication	The clarity of representing one's thought is an art. Oral communication skills add positivity to the overall personality and in turn the candidate gets opportunities to represent the organization at large
7	Computer knowledge	It is the need of the hour, to be apt by using computers. In the era of automation, the basic knowledge is a necessary and unavoidable fact. Therefore, person with some basic knowledge of computers is always preferred
8	Confidence	It is an indirect measure, which can be calculated from few of the above-discussed attributes. Confident personnel have positivity in his/her behavior, which is very important for the growth of the organization
9	Leadership	Leadership quality makes a person to shine in the crowd. A leader being the part of the team, leads them, assigns the right job to the candidate. People with this quality, prove to be good team leaders and they are always at the time of hiring
10	Emotional steadiness	The emotional steadiness is a key of establishment in professional life. It covers all the phases of human behavior, which should be handled patiently in the work place. An emotionally stable person proves to be the winner in the team game
11	Personality	Personality is composed of one's looks, confidence, communication skills, habits, and behavior. A well

(continued)

Table 3 (continued)

S. no.	Criteria	Description
		groomed personality has a better and easy establishment, and generally experts seek for this attribute
12	Domain knowledge	A good domain knowledge, facilitates even a fresher to be selected among the experienced ones. Hence, this feature holds much importance

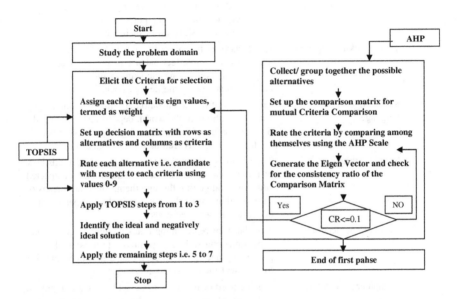

Fig. 2 Proposed methodology of integrated MCDM-based algorithm

Eigen vector is a set of values whose composition yields 'unity'. Each value in the set of the vector corresponds to each criterion compared. The value tells that to what extent the criterion has the influence on the results of the selection process. Table 4, is the comparison matrix of the criteria, on which the selected steps of AHP is shown using Eqs. (1–4). Figure 3, displays the comparison of the calculated Eigenvalues.

3.1 Decision Support System

The selected generic criteria with their Eigenvector will serve as the decision support system for rest of the real-time selection problems. This paper has precisely presented the values which can be used directly in many problems. So, here we generate a decision matrix with columns as Criteria (along with their respective

Table 4 Criteria comparison using AHP

	A	B	C	D	E	F	G	H	I	J	K	L
A	1	2	2	4	3	2	3	3	2	2	3	3
B	1/2	1	1	3	2	1	2	1	2	2	3	1
C	1/2	1	1	3	2	1	2	2	1/3	1	2	1/3
D	1/4	1/3	1/3	1	1/2	1/3	2	1	1/3	1/3	1	1/4
E	1/3	1/2	1/2	2	1	1/2	1	1	3	2	2	1/3
F	1/2	1	1	3	2	1	2	2	2	3	2	2
G	1/3	1/2	1/2	1/2	1	1/2	1	1/2	1/2	4	1/2	1/4
H	1/3	1	1/2	1	1	1/2	2	1	2	3	3	1/4
I	1/2	1/2	3	3	1/3	1/2	2	1/2	1	1	2	1/3
J	1/2	1/2	1	3	1/2	1/3	1/4	1/3	1	1	1	1/3
K	1/3	1/3	1/2	1	1/2	1/2	2	1/3	1/2	1	1	1/4
L	1/3	1	3	4	3	1/2	4	2	3	1/3	4	1

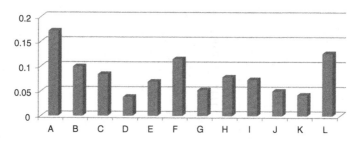

Fig. 3 Eigenvalues of the criteria and their mutual comparison

Eigen value, termed as 'weight') and Alternatives as rows. The values in this matrix are the score of each alternative with respect to each criterion. Here, the alternatives are the five candidates, so the scores will be the examination score of the candidate with respect to each criterion. Table 5, is the summarized view of this step. After this, the usual steps of TOPSIS will be applied on this matrix using Eq. (5) and step 3 (as given in the TOPSIS method details). From the resulting matrix, now we will find the ideal and the negatively ideal solution, both of which are the exceptional cases. The aim is to find a value between them. The value which is between these but more nearer to the ideal case will be the best solution. So, ideal and negatively ideal solutions are the bounds between which our solution will lie. The subsequent steps in the procedure are composed of step 5, 6, and 7, using Eq. (6). Figure 4, gives the final values attained by each candidate, and the comparison among those. The alternative with the highest value is the best option, followed by 2nd value and so on.

Table 5 Decision matrix (P1, P2.... are alternatives i.e. candidates)

	A	B	C	D	E	F	G	H	I	J	K	L
	0.1717	0.0996	0.0842	0.038	0.069	0.1147	0.0523	0.0782	0.073	0.05	0.042	0.126
P1	4	7	3	2	2	2	2	2	4	4	2	5
P2	4	4	6	4	4	3	7	4	4	5	4	4
P3	7	6	4	2	5	5	3	5	5	6	5	6
P4	3	2	5	3	3	2	5	3	3	3	3	3
P5	4	2	2	5	5	3	6	4	3	3	4	4

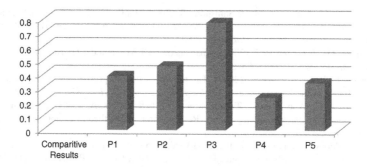

Fig. 4 Comparison of the values obtained by 5 candidates (*X*-axis denote the five candidate and *Y*-axis denote their calculated values)

4   Conclusions

Personnel Selection has been a keen area of research and there is still a huge scope of more, due to the inclusion of human psychology [18, 19]. The application areas include, educational institutes, corporate houses, retail and marketing, and even daily life cases [20, 21]. The proposed methodology is a blend of both the widely accepted MCDM techniques which aim for consistent prioritization of the candidate and not merely the selection of the best. Such an approach, ultimately enables the comparative analysis of the alternatives, thus giving a wide area of choice. Further, the decision support system given in the paper will enable easy selection in other hiring processes as all the DM left to do is to score each of the candidate, rest of the algorithm will solve itself and provide the experts directly with the results. The paper has given a generic mathematical model of selection which can be applied in many real-time selection issues, other than personnel selection (of course the criteria will be chosen accordingly).

References

1. Andrew, J. V., & Laura, L. K. B. (2012). A history of personnel selection and assessments. *The Oxford handbook of personnel selection and assessments* (Chapter 2, pp. 9–25).
2. Robertson, I. T., & Smith, M. (2012). Personnel selection. *Journal of Occupational and Organizational Psychology, 74*, 48–67.
3. Dursun, M., & Ertugrul Karsak, E. (2010). A fuzzy MCDM approach for personnel selection. *Expert System with Application, 37*, 4324–4330.
4. Murphy, K. R. (2012). Individual differences. *The Oxford handbook of personnel selection and assessments* (Chapter 3, pp. 31–47).
5. Dyer, J. S., Fishburn, P. C., Stener, R. E., Wallenius, J., & Zionts, S. (1990). Multiple criteria decision making, multiattribute utility theory: The next ten years. *Management Science, 38*(5), 645–654. JAY-STOR.
6. Stener, R. E., & Zionts, S. *Facts about MCDM, international society on MCDM,* www.mcdmsociety/facts.html.

7. Alias, M. A., Hashim, S. Z. M., & Samsudin, S. (2008). Multi criteria decision making and its applications: A literature review. *Asia-Pacific Journal of Information Technology and Multimedia, Jurnal Teknologi Maklumat, 20*(2).
8. Velazquez, M., & Hester, P. T. (2013). Analysis of MCDM methods. *International Journal of Operational Research, 10*(2), 56–66.
9. Roberson, C. L. K., & Perry, E. (2007). The multiple category problems: Category activation and inhibition in the hiring process. *Academy Management Review, 32*(2), 529–548.
10. Karmperis, A. C., Aravossis, K., Tatsiopoulos, I. P., & Sotirchos, A. (2013). Decision support models for solid waste management: Review and game-theoretic approaches. *Waste Management, 3*, 3–5.
11. Martin Aruldoss, T., Laxmi, M., & Prasanna Venkatesan, V. (2013). A survey on multi criteria decision making methods and applications. *Science and Education Publishing, 1*(1), 31–43.
12. Sun, X., & Li, Y. (2010). An intelligent multi-criteria decision support system for systems design. In *10th AIAA, Aviation Technology, Integration & Operation Conference*, (Chapter: 10.2514/6, AIAA).
13. Bhutia, P.W., & Phipon, R. (2012). Application of AHP and TOPSIS method for supplier selection problem. *IOSR Journal of Engineering, 2*(10), 43–50.
14. Agarwal, P., Sahai, M., Mishra, V., Bag, M., & Singh, V. (2011). A review of multi-criteria decision making techniques for supplier evaluation and selection. *International Journal of Industrial Engineering Computations 2*, 801–810.
15. Saaty, T. L. (1980). *The analytic hierarchy process*. New York: McGraw-Hill.
16. Polyhart, E.R., & Schneider, B., (2012). The social and organizational context of personnel selection. *The Oxford handbook of personnel selection and assessments*, (Chapter 4, pp. 48–67).
17. Rothstein, M. G., & Goffin, R. D. (2006). The use of personality measures in personnel selection: What does current research support? *Human Resource Management Review, 16*, 155–180.
18. Hough, L. M., & Oswald, F. L. (2000). Personnel selection: Looking towards the future and remembering the past, annual reviews. *Psychology, 51*, 631–664.
19. Balezentis, A., Balezentis, T., & Bravers, W. K. M. (2012). Personnel selection based on computing with words and Fuzzy MULTIMOORA. *Expert System with Application, 39*, 7961–7967.
20. Norddin, N. I., Ibrahim, K., & Aziz, A. H. (2012). Selecting new lecturers using analytical hierarchy process, statistics in science, business and publishing. *IEEE Explore*, 1–7.
21. Celik, M., Kandakoglu, A., & Deha, I. Er. (2009). Structuring fuzzy integrated multistage evaluation model on academic personnel recruitment in MET institutions. *Expert System with Application, 36*, 6918–6927.

A Proactive Dynamic Rate Control Scheme for AIMD-Based Reactive TCP Variants

Hardik K. Molia

Abstract TCP—Transmission Control Protocol is a transport layer protocol. TCP's default interpretation of any loss as a cause of congestion works well with the wired networks but wireless networks suffer from other losses like channel loss, route failure loss which are not distinguishable by TCP. Several TCP variants have been proposed for various types of networks to differentiate various losses and act accordingly. Reactive variants focus on congestion detection and recovery approach while proactive variants focus on congestion avoidance approach. This paper modifies conventional AIMD—Additive Increase Multiplicative Decrease scheme of reactive TCP variants to add a proactive factor for dynamic rate control. Rate control factor is measured from the network stability predicted by analyzing past few RTT—Round Trip Time values. Simulation has been performed in NS 2.35 and tested over scenario-based MANETs.

Keywords TCP · Slow start · Congestion avoidance · Fast retransmission · Rate control · MANET

1 Introduction

TCP—Transmission Control Protocol is a Transport Layer protocol which provides process to process ordered delivery of bytes. TCP also performs flow control, error control, and congestion control. Traditional TCP was initially used with the wired networks. Wired networks often suffer from network congestion. Congestion may cause packet drop at intermediate nodes. TCP considers any loss as a cause of congestion only. TCP tries to detect congestion and slows down the transmission rate. This concept works well with the wired networks where in majority of the cases packet loss is due to network congestion. Wireless networks have their own

H.K. Molia (✉)
Government Engineering College, Rajkot, Gujarat, India
e-mail: hardik.molia@gmail.com

© Springer Science+Business Media Singapore 2016
S.C. Satapathy et al. (eds.), *Proceedings of International Conference on ICT for Sustainable Development*, Advances in Intelligent Systems and Computing 408, DOI 10.1007/978-981-10-0129-1_18

vulnerabilities and so many other losses like route failure losses, channel losses corrupt or drop packets. In wireless networks, TCP's default interpretation of any loss as a congestion loss degrades performance significantly. Ideally, TCP should be able to slow down the transmission rate on congestion loss, retransmit the lost or corrupted packets on channel loss, and wait for re-establishment of new route on route failure loss. Several TCP variants have been proposed to differentiate various causes of packet loss and make TCP act accordingly [1, 2].

TCP variants can be broadly classified into two categories. Reactive TCP variants (TCP Reno, TCP NewReno, TCP Cubic) focus on congestion detection and recovery while Proactive TCP variants (TCP Vegas) focus on congestion avoidance. Most of the Reactive TCP variants are based on AIMD—Additive Increase Multiplicative Decrease Scheme-based congestion control. This paper introduces an enhanced AIMD scheme where a proactive effort is added to introduce dynamic rate control. The main motive of this scheme is to enhance TCP for wireless networks which may not be the best at all the time but on an average it should be stable [2, 3].

2 TCP Congestion Control

2.1 Congestion

A computer network uses intermediate devices like routers and switches to forward packets from source to the destination. These forwarding devices have limited amount of buffers to queue packets. If the packet arrival rate is higher than packet processing rate then input queue overflows in a while. If the packet departure rate is slower than packet processing rate then output queue overflows in a while. Both of these cases cause discard of some packets when input and/or output queues have no capacity to store packets. This situation is called congestion. If not handled properly, the situation may become worst and may cause the network collapse. TCP tries to detect such situation and tries to slow down the transmission rate to reduce congestion. As a part of error control, it retransmits discarded packets [2].

2.2 Sliding Window

Sender TCP keeps track of sent and acknowledged bytes, sent but not yet acknowledged bytes, can be sent bytes, and cannot be sent bytes using a sender sliding window. Receiver TCP keeps track of bytes which are received and delivered to the process, bytes which are received but not yet delivered to the process, bytes can be received, and bytes cannot be received using a receiver sliding window. Size of sliding window decides the current transmission rate which is the maximum number of bytes which the sender can send without waiting for an acknowledgment in middle of transmission. Sender TCP increases or decreases the size of sliding window as per current network congestion and receiver's capability. The size of sender sliding window wnd is minimum for two values [2].

$$wnd = MINIMUM\,(rwnd, cwnd)$$

TCP Receiver informs sender about the maximum rate at which it is capable of receiving bytes through TCP Header's Window Size field known as rwnd—Receiver's advertised Window. TCP Sender approximates amount of network congestion using AIMD known as cwnd—Congestion Window [2].

3 AIMD—Additive Increase Multiplicative Decrease

3.1 Introduction

Standard TCP's congestion control has three phases known together as AIMD scheme. These phases are slow start (exponential increase), congestion avoidance (additive increase), and congestion detection (multiplicative decrease). As a reliable protocol, TCP confirms ordered delivery by receiving ACKs—acknowledgments. Standard TCP is purely ACK clocking where the speed at which ACKs are receiving takes part in deciding the transmission rate. AIMD scheme is based on increasing or decreasing transmission rate by modifying cwnd size as per the arrival of ACKs. RTT—Round Trip Time is the time interval between a sender who sends something and receives its acknowledgment sent by the destination. TCP's AIMD is explained with the reference of TCP Reno Variant. Later on TCP NewReno is explained and used for the proposed scheme [3].

3.2 Slow Start-Exponential Increase

Initially TCP Sender should start with a very slow rate of transmission rather than with a higher rate which may overwhelm the network. Initially, cwnd is set to 1 or 2 MSS—Maximum Segment Size. TCP is a byte oriented protocol and so every unit of measurement is in terms of bytes. For simplicity, we will refer the packet term as a transmission unit. Receiver sends an ACK on receiving expected packets in order. Sender increments cwnd by 1 on receiving an expected ACK. So cwnd gets doubled after a completion of a round. It can be also said that slow start phase doubles cwnd per RTT so it is said to be an exponential increase. A threshold value called ssthresh equal to the half of the maximum window size is set by the system. TCP remains in the slow start phase until cwnd reaches the ssthresh. Once cwnd becomes greater than ssthresh, TCP enters into congestion avoidance phase [3, 4].

3.3 Congestion Avoidance-Additive Increase

Crossing ssthresh informs TCP sender to increase transmission rate linearly rather than exponentially. In congestion avoidance phase, TCP increments cwnd by 1 not with every expected ACK but on completion of every round only. It can be also said that congestion avoidance phase increments cwnd by every 1 RTT so it is said to be linear increase. TCP stays in this phase until it predicates a loss either by RTO —Retransmission Time Out or receiving three Duplicate Acknowledgments [3, 4].

3.4 Congestion Detection-Multiplicative Decrease

TCP should decrease transmission rate by decreasing size of cwnd on detection of congestion. TCP uses two events to predicate congestion and acts accordingly. RTO event is considered as strong possibility of congestion. TCP sets ssthresh to half of the current sender window size. TCP sets cwnd to initial size and switches to slow start phase. Receiving 3 Duplicate ACKs is a weaker possibility of congestion. TCP sets ssthresh to half of the current sender window size. TCP sets cwnd to the value of ssthresh and switches to congestion avoidance phase. In both the cases TCP retransmits the oldest unacknowledged packet [3, 4]. Slow start and Congestion Avoidance phases are shown in Fig. 1 [2] and AIMD scheme is represented graphically in Fig. 2 [2].

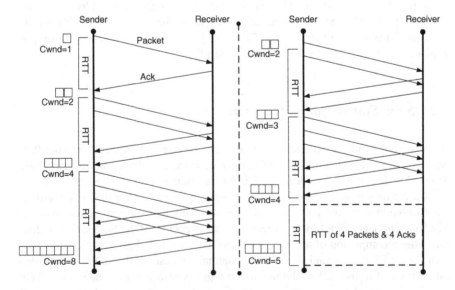

Fig. 1 TCP's slow start and congestion avoidance phase

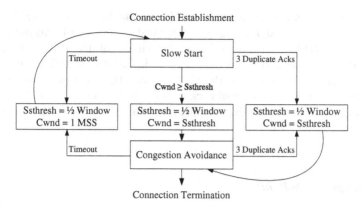

Fig. 2 TCP's AIMD scheme

3.5 Congestion Detection-Multiplicative Decrease

Three Duplicate ACKs event indicates there are still packets getting received but the one expected was missing. At that time, rather than completely slowing down the transmission rate, TCP switches to the congestion avoidance phase. TCP retransmits the oldest unacknowledged packet—which was asked by the three duplicate ACKs. This retransmission is earlier than of the RTO event and so called Fast Retransmission. Retransmitting the oldest unacknowledged packet does not smooth out the situation. TCP should ensure that there must not be any other packet loss from the already sent packets. After fast retransmission, TCP enters into a fast recovery phase which acts as an intermediate stage between fast retransmission and congestion avoidance. TCP stays in fast recovery until it detects whether receiver has started receiving packets by getting an ACK [3, 4].

TCP Reno performs well when there is one packet loss in a window of outstanding packets. Reno comes out of the fast recovery phase once it receives a fresh ACK. Reno does not care whether the new ACK, acknowledges all the outstanding packets or not. So if there are multiple packet losses in a single window, Reno does not perform well. NewReno has enhanced fast recovery phase. NewReno stays in the fast recovery phase until all the outstanding packets are acknowledged successfully [5, 6].

4 Proactive Dynamic Rate Control

4.1 Introduction

Sender TCP calculates size of cwnd as per the AIMD scheme. Sender's transmission rate is limited by the size of cwnd and rwnd. In most of the cases, size of cwnd is smaller than of rwnd and so sender's transmission rate is limited by cwnd only. This rate control is dynamic as value of cwnd is continuously changing as per the various AIMD-based events. It has been noticed that due to lack of loss differentiation capabilities, TCP's performance is degraded in MANETs even when AIMD scheme has worked properly. Conventional TCP does not transmit more than (cwnd/packet_size) number of packets per unit time. Or in other words cwnd number of bits per unit time. This measure is amount of data TCP sender transmits before expecting an ACK [7].

4.2 Proposed Scheme

In a proactive effort, a further restriction over the increase in transmission rate can be introduced. The proposed scheme restricts the transmission rate by a factor β, where $0 \leq \beta \leq 1$. In proposed scheme, the transmission rate is dynamically set to $\beta*$ cwnd bits per unit of time. In other words, the transmission rate is $\beta*$ (cwnd/packet_size) packets per unit of time. Value of β decides the amount of reduction in the existing size of cwnd. When β is 1, there is no reduction in the size of cwnd. When β is 0, cwnd is set to the initial size.

This scheme tries to approximate instability of a network by analyzing past few RTTs. The choice of number of RTTs to process is a difficult task. The reasonable choice would be equal to the number of RTTs which have been calculated since last RTO expired. RSD—Relative Standard Deviation provides approximate variation over the data in the form of percentage. After analyzing past few RTTs, if RSD is high, it indicates high instability of the network. If RSD is low, it indicates high stability of the network. Dynamic Rate Control will be activated on both of these events (Tables 1 and 2).

Dynamic_Rate_Control() process is called on occurrence of following events. Dynamic Rate Control will be activated when Stability $\leq$ Variation_Threshold for AIMD Mode and when stability $\geq$ Variation_Threshold for RTO Mode.

Table 1 Parameters

Variable name	Purpose
β	$0 \leq \beta \leq 1$. Rate control factor
Num_RTT_Samples	Total number of most recent RTT values to process
Variation_Threshold	Variation beyond which the rate control is activated
Mode	"AIMD" mode when procedure is called between switching from one phase to another phase "RTO" mode when procedure is called on expiration of RTO—retransmission time out timer

Table 2 Events

Event	Mode	Variation_Threshold (%)
Whenever TCP switches from slow start phase to congestion avoidance phase	AIMD	35
Whenever TCP switches from fast retransmission to congestion avoidance phase	AIMD	35
Whenever RTO occurs. If dynamic rate control has reduced rate then TCP will not slow down the rate completely even if it switches to slow start phase	RTO	65

Pseudocode

Dynamic_Rate_Control(Num_RTT_Samples, Variation_Threshold, mode)

```
1. Collect Most Recent Num_RTT_Samples of RTT
2. Find Mean.
3. Find Variance.
4. Find Standard Deviation.
5. Find RSD (Relative Standard Deviation)    // if RSD > 100 then RSD = 100
6. temp = 100 – RSD                          // indicates stability
7. If (temp ≥ Variation_Threshold and mode == "RTO") or
      (temp ≤ Variation_Threshold and mode == "AIMD")
   Then
         a.  β = temp / 100                   // Stability's upper bound is 100 %
         b.  Cwnd = β * Cwnd                  // Changing size of Congestion Window
   End If
End
```

5 Simulation

The scheme has been implemented in NS 2.35 and tested over various scenario-based MANETs with a TCP connection carrying continuous FTP traffic. Results are shown in Figs. 3 and 4 of two different scenarios with Simulation time is of 50 s and mobility redefine interval is of 5 s.

6 Conclusion

The main purpose of this scheme is to add dynamic congestion window reduction based on the approximated network stability. This scheme is an additional layer over the traditional AIMD scheme. In most of the cases, the scheme has shown

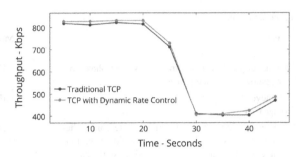

Fig. 3 MANET with 5 nodes and mobility speed 25 s

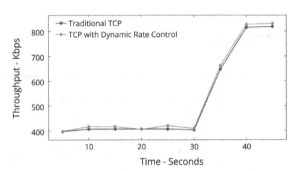

Fig. 4 MANET with 10 nodes and mobility speed 30 s

improvements. Future work can be carried out by introducing dynamic values for thresholds for both the modes of rate control. New modes and their corresponding threshold values can be introduced.

References

1. Holland, G., & Vaidya, N. (1999). Analysis of TCP performance over mobile ad hoc networks. In *Proceedings of the 5th Annual ACM/IEEE International Conference on Mobile Computing and Networking* (pp. 219–230). Seattle, Washington, United States, August 15–19 1999.
2. Forouzan, B. TCP/IP protocol suite (4/e). McGraw-Hill.
3. Richard Stevens, W. TCP/IP illustrated, the protocols (Vol. 1). Addison-Wesley.
4. Abed, G. A., Ismail, M., & Jumari, K. (2012). Exploration and evaluation of traditional TCP congestion control techniques. *Journal of King Saud University—Computer and Information Sciences, 24*, 145–155.
5. Qureshi, B., Othman, M., & Hamid, N. A. W. (2009). Progress in various TCP variants. *Computer, Control and Communication*. IC4.
6. Mathis, M., & Mahdav, J. Forward acknowledgment refining TCP congestion control. Pittsburgh Supercomputing Center—ACM SIGCOMM.
7. Mast, N., & Owens, T.J. (2011). A survey of performance enhancement of transmission control protocol (TCP) in wireless adhoc networks. *EURASIP Journal on Wireless Communications and Networking, 2011*, 96.

Investigating Varying Time Black Hole Attack on QoS over MANET

Sourabh Singh Verma, R.B. Patel and S.K. Lenka

Abstract Mobile Ad hoc Networks are vulnerable to various attacks due to its nature, as it is a multihop process between sender and receiver. One of these attacks is black hole attack, which will simply delete any packet instead of forwarding it. Sometimes these kinds of malicious nodes may behave as true node giving impression that it is not malicious, while rest of the time they are dropping all packets, so affecting the QoS. Discovering such kind of malicious nodes are very difficult as isolating them is not easy because their malicious behavior is in on and off fashion. In our paper, we had inspected such nodes which are behaving like a black hole for variable time intervals on different parameters. Further for deeper study, we have considered different number of such malicious nodes to give more insight on comparisons.

Keywords Black hole · QoS · Varying time attack · Malicious node · AODV attacks

1 Introduction

Mobile Ad hoc Network is decentralized and infrastructure less network. It is ad hoc, which means it is a temporary network that does not require any centralized control, so it does not depend on any fixed infrastructure [1, 2]. Nodes in Manets

S.S. Verma (✉) · S.K. Lenka
CSE, IT Department, FET, Mody University of Science and Technology, Lachmangarh 332311, India
e-mail: ssverma.fet@modyuniversity.ac.in

S.K. Lenka
e-mail: sklenka.fet@modyuniversity.ac.in

R.B. Patel
CSE Department, CCET, Chandigarh, Punjab, India
e-mail: rbpatel@ccet.ac.in

© Springer Science+Business Media Singapore 2016
S.C. Satapathy et al. (eds.), *Proceedings of International Conference on ICT for Sustainable Development*, Advances in Intelligent Systems and Computing 408, DOI 10.1007/978-981-10-0129-1_19

requires multihop transmission to send any data, which makes MANET vulnerable to many attacks such as black hole and flooding attack [3]. These attacks can be explained as:

Flooding Attacks: Flooding nodes will flood Route Request (RREQ) packet to unknown destination that does not exist, all nodes on the way will broadcast this RREQ packet that results in flooding. The resources in the network is consumed by such packages which will ultimately result in denial of services [4–6].

Black Hole Attack: Black Hole [7, 8] nodes behaves as normal in route discovery process but after route is established through it, a black hole will drop all the packets it received by any other nodes. That means it transmits all the routing packet but drop all the data packet which makes it very complicated to handle.

Many types of attacks can be encountered in MANET like sinkhole, wormhole, selfish nodes, etc. Any node with such a malicious behavior can cause various impairment to networks [9]. In this paper, we are evaluating black hole attack using AODV as routing protocol.

The rest of the paper is organized as; Sect. 2 covers brief notes on background and literature survey. Section 3 contains performance evaluation of AODV routing protocol under black hole attack with variable malicious active percentage time and with different numbers of malicious nodes, results are evaluated for different parameters. At last, concluding remarks in Sect. 4.

2 Background and Literature Survey

Various research works have been done to know malicious behavior outcomes on MANET and some of them are discussed here. In [10] fixed number of malicious node are used to evaluate different attacks on MANET with different mobility. Attacks [11, 12] are evaluated with different parameters while considering number of malicious nodes as a baseline. In [13] author studied different attacks including black hole while specially considering with number of attackers and their positions.

Some of the work has been done to detect the malicious node by reading its behavior [14, 15], in behavior based malicious nodes detection the nodes in neighbor of such nodes will keep track of the package sent to it. While behavior detection or trust management will force overhead to the nodes' behavior so that nodes can bypass such malicious node for path maintenance. The attack will affect the QOS [10, 11], while our basic QoS models such as IntServ [16] and Differv [17] are helpless in any such attacks, that is why behavior detection or trust management is required to bypass such nodes.

3 Performance Evaluation of AODV Under Varying Rate

For evaluation, we had considered three different {3, 7, 8} numbers of malicious (on/off black hole) nodes. While considering that black hole malicious nature of these nodes are activated for {100, 75, 60, 50, 20, 10, 0} % time of the overall simulation.

Performance Consideration Bases:

(a) Packet Delivery Fraction (PDF): PDF versus different numbers of malicious nodes with variable interval malicious nature.
(b) Bandwidth consumption of network versus different numbers of malicious nodes with variable interval % of malicious nature.
(c) Number of Packet Drop in malicious environment with numbers of malicious nodes.
(d) Throughput of network versus different numbers of malicious nodes with variable interval % of malicious nature.

Bandwidth consumed by Flood RREQ versus varying time flooding (under 4, 3 and 2 malicious nodes).

3.1 Simulation Setup

For simulation environment we have used ns2 [18]. All above performances are evaluated in AODV routing protocol.

NS2 Parameter. Simulation setup for various evaluations

Simulation parameter	Value
Simulation time	100 s
No. of nodes	50
Area	500 × 500 m
Traffic	CBR
CBR rate	0.25
Motion	Random
Routing protocol	AODV
Black hole varying duration % in simulation	{0, 10, 20, 50, 60, 75, 100}
No. of malicious nodes	3, 7, 8
Transport layer	UDP
Node maximum speed	10 m/s
Maximum connection	40
Pause time	2

3.2   Simulation Results

A. Packet Delivery Fraction (PDF): PDF versus different numbers of malicious nodes with variable interval malicious nature.

   In Fig. 1 it can be seen that how varying active time of black hole attack effect the network's PDF. From this evaluation it can be seen that if malicious node active % time is high and number of such nodes are high then number of packet delivery fraction is low. It can also be noticed if active time of such nodes are 0 % then PDF will be same as network will behave as normal in all the cases, while on other side if all such nodes are active for all the time then PDF will be least.

B. Bandwidth consumption of network versus different numbers of malicious nodes with variable interval % of malicious nature.

   From Fig. 2 we had shown bandwidth occupied by data packet, it can be concluded from the figure that bandwidth is least consumed when active time % of malicious nodes are highest, this is because packets are dropped by such nodes and so bandwidth is wasted while on the other hand bandwidth is utilized by packets as they are not dropped.

C. Number of Packet Drop in malicious environment with numbers of malicious nodes.

   We can see from Fig. 3 that higher numbers of packet are dropped if numbers of malicious nodes are high and/or varying active time of such nodes is high. In our simulation we had considered fixed queue size, i.e., 50.

D. Throughput of network versus different numbers of malicious nodes with variable interval % of malicious nature Fig. 4.

   Throughput is the number of successful packet received over network in a unit of time. In our case size of each packet is 512 bytes and throughput is evaluated in kbps. Throughput 100 % malicious behavior is least because at such scenario number of packet dropped will be high.

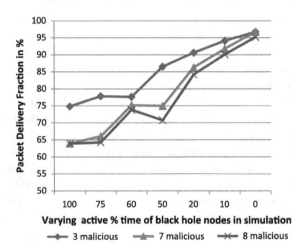

Fig. 1 PDF versus different numbers of malicious nodes with variable interval malicious nature

Fig. 2 Bandwidth versus
malicious nodes with variable
interval % of malicious nature

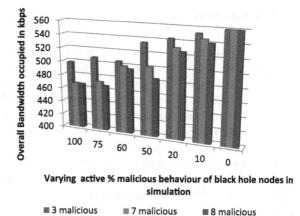

Fig. 3 Number of packet
drop in malicious
environment with numbers of
malicious nodes

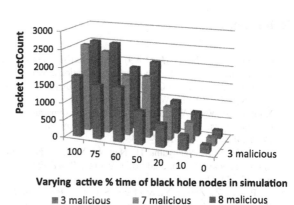

Fig. 4 Throughput versus
malicious nodes with variable
interval % of malicious nature

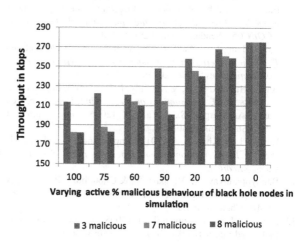

4 Conclusions

As we saw that if nodes' active % malicious behavior is high than their negative impacts on the network is high. Similarly, if number of malicious nodes are high then also it will impact the network negatively. In over experiment we have shown that how it affects different parameters of the network like throughput, packet deliver fraction, bandwidth used, and numbers of packet dropped. However, detecting malicious nodes which behave like normal nodes sometime and malicious for rest of the time, is very hard because its behavior is mixed which you cannot trust or distrust it completely. While on the other side a node that is behaving malicious for all the time can be easily identified. In future, we will try to solve to bypass varying active black hole attacks by detecting it and by updating entries in route table.

References

1. Corson, S., & Macker, J. (1999). Mobile ad hoc networking (MANET): Routing protocol performance issues and evaluation considerations. *RFC* 2501, January 1999.
2. Basangi, S., Conti, M., Giordano, S., & Stojmenovic, I. (2004). *Mobile ad hoc networking* (pp. 282). IEEE Press, Wiley-Interscience.
3. Xing, F., & Weng, W. (2006). Understanding dynamic denial of service attacks in mobile ad hoc networks. *MILCOM'06 Proceedings of the 2006 IEEE Conference on Military Communications* (pp. 1047–1053).
4. Kannhavong, B., Nakayama, H., Nemoto, Y., Kato, N., & Jamalipour, A. (2007). A survey of routing attacks in mobile ad hoc networks. *Proceedings of Wireless Communications, IEEE* (pp. 85–91), Oct 2007, Issue 5.
5. Yi, P., Dai, Z., Zhong, Y., & Zhang, S. (2005). Resisting flooding attacks in ad hoc networks. *Proceedings of the International Conference on Information Technology: Coding and Computing (ITCC'05)* (pp. 657–662), April 2005.
6. Eu, Z., & Seah, W. (2006). Mitigating route request flooding attacks in mobile ad hoc networks. *Proceedings of the International Conference on Information Networking (ICOIN'06)*, Sendai, Japan, January 2006.
7. Tamilselvan, L., & Sankaranarayanan, V. (2007). Prevention of blackhole attack in MANET. *The 2nd International Conference on Wireless Broadband and Ultra Wideband Communications, IEEE*, 2007.
8. Aad, I. Hubaux, K. P., & Knightly, E. W. (2007). Impact of denial of service attacks on ad hoc networks. *IEEE Transactions on Networking*, Reference: LCA-ARTICLE-2007-011.
9. Barbir, A., Murphy, S., & Yang, Y. (2006). Generic threats to routing protocols. *IETF RFC* 4593. (Status Informational).
10. Ehsan, H., & Khan, F. A. (2012). Implementation and analysis of routing attacks in MANETs. *IEEE 11th International Conference on Trust, Security and Privacy in Computing and Communications*.
11. Bandyopadhyay, A., Vuppala, S., & Choudhury, P. (2011). A simulation analysis of flooding attack in MANET using NS-3, February 28. *IEEE 2nd International Conference on Wireless VITAE*, Chennai, India.
12. Nguyen, H. L., & Nguyen, U. T. (2008). A study of different types of attacks on multicast in mobile ad hoc networks. *Ad Hoc Networks*, 6(4), 32–46.

13. Verma, S. S. & Patel, R. B., & Lenka S. K. (in press). Analyzing varying rate flood attack on real flow in MANET and solution proposal "Real Flow Dynamic Queue (RFDQ)". *International Journal of Information and Communication Technology, Inderscience, 7.*

14. Patel, M., & Sharma S. (2013). Detection of malicious attack in MANET a behavioral approach. *IEEE 3rd International Advance Computing Conference (IACC)*, 22–23 February 2013.

15. Sui, A. F., Guo, D. F., & Zhao, D. S. (2011). An effective method to mitigate route query floods in MANETs. *IEEE 13th International Conference on Communication Technology (ICCT)* (pp. 25–28), September 2011.

16. Braden, R., Clark, D., & Shenker, S. (1994). Integrated services in the Internet architecture: An overview. Technical Report 1633.

17. Black, D. (2000). Differentiated services. *RFC* 2475.

18. Fall, K., & Varadhan, K. *Ns manual*. The VINT Project.

Hybrid Approach to Reduce Time Complexity of String Matching Algorithm Using Hashing with Chaining

Shivendra Kumar Pandey, Hari Krishna Tiwari
and Priyanka Tripathi

Abstract String matching is an integral part of various important areas of computer science viz search engines, DNA matching, information retrieval, security, and biological systems. String matching algorithms are used in finding a pattern in a string. In this paper, the authors have given a novel algorithm to reduce the time complexity of string matching. The proposed algorithm is based on the concept of hashing with chaining. Further, the authors have found reduced time complexity in most of the cases and equal in few.

Keywords String matching · Hashing · Chaining · Time complexity

1 Introduction

Purpose of string matching algorithm is to find out the given pattern p ($p = p1, p2, p3....pm$) in the string t ($t = t1, t2, t3, t4....tn$) [1]. If more than one pattern is matched in the string simultaneously, then it is known as multiple patterns matching algorithm otherwise it is known as a single pattern matching algorithm [2–4]. The single pattern matching techniques is generally used in computer network security viz. intrusion detection, viruses, and dirty network information [1]. There are many different existing algorithms that reduce the time and space complexity of the string matching algorithms. Most popular are Rabin Karp, Boyer Moore, Brute force, Knuth Morris Pratt, String Matching with Finite Automata, Parameterized String Matching Algorithm, String Matching with Suffix Tree [1, 4–6].

S.K. Pandey (✉) · H.K. Tiwari · P. Tripathi
Department of Computer Application and Research, National Institute
of Technical Teachers Training and Research, Bhopal, India
e-mail: shivendrapandey786@gmail.com

H.K. Tiwari
e-mail: csehktiwari@gmail.com

P. Tripathi
e-mail: ptripathi@nitttrbpl.ac.in

© Springer Science+Business Media Singapore 2016
S.C. Satapathy et al. (eds.), *Proceedings of International Conference
on ICT for Sustainable Development*, Advances in Intelligent Systems
and Computing 408, DOI 10.1007/978-981-10-0129-1_20

185

The remainder of the paper is as follows. Section 2 gives complete literature survey of the string matching algorithms. In Sect. 3, authors have proposed the algorithm with an example and in Sect. 4 result and time complexity analysis of algorithm has been discussed. Further, in Sect. 5 time complexity comparison with the other existing algorithms. Last Sect. 6, conclusion of this work.

2 Literature Survey

The String matching algorithms have two approach to minimize the search complexity first one is by applying the preprocessing on the pattern and the second one is by applying the preprocessing on a string [7–10]. Some of the algorithms are discussed in the brief.

2.1 *Brute Force Matching*

This technique is also known as a Naïve String Matching algorithm. Brute Force technique matches the pattern with the string character by character and slides one character right at a time [3]. Brute Force string matching algorithm does not require any preprocessing time and the number of comparisons in the Brute Force algorithm are nearly $2n$ [3]. The Brute Force technique is the basic and simplest method to search a pattern in a given string [3] as described in Fig. 1.

For example:
Let Σ (set of input) = $\{a, b, c....z\}$
String $T = aabacabbbcb$
Pattern $P = bbcb$

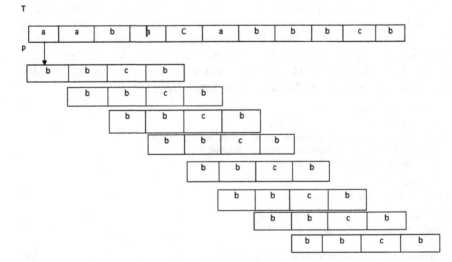

Fig. 1 Working process of Brute Force algorithm

The main drawback of Brute Force technique in the worst case is it takes $O(n*m)$ time to search a pattern in the given string.

2.2 Rabin Karp String Matching Algorithm

Rabin Karp string matching algorithm takes two phases that are preprocessing time and matching time, this algorithm uses hashing technique to find out the pattern in the string [4]. This algorithm is used to find the numerical pattern in the give text. In this algorithm, first it divides pattern by a predefined prime number 'q' and calculates remainder of the pattern. In next step, it calculates the remainder of first 'm' character of text 'T' [4]. If the remainder of first 'm' characters and the remainder of the pattern 'P' are equal, then only it performs matching. Otherwise, no need of comparison [4, 5]. The main concepts of Rabin Karp string matching algorithm uses hashing to find the pattern in the given string. The running time analysis is described in Table 1.

2.3 Knuth Morris Pratt

Knuth Morris Pratt developed an algorithm for pattern matching which takes linear time $O(n)$ to find a particular pattern [4, 11, 12]. KMP algorithm will also take two phases to find a pattern [11]. That is prepreposing time and the searching time where it makes the prefixes and matches respectively at each phase [4]. That is carried out by the prefix function and the KMP matcher respectively [11].

- Preprocessing phase: $O(m)$ space and time complexity
- Searching phase: $O(n + m)$ time complexity (independent from the alphabet size)
- At most $2n - 1$ character comparisons during the text scan

2.4 Boyer Moore Algorithm

The Boyer Moore Algorithm achieves expected sublinear runtime using three ideas [13, 14]. The right to left scan, the bad character shift rule, and the good suffix shift rule [6, 13]. The Boyer Moore Algorithm is especially suitable if the string size is large (Example: natural language) and the pattern is large [13, 14]. The complexity of Boyer Moore Algorithm can be shown to linear-time behavior, only if 'P' does not occur in 'T'. Otherwise the worst-case complexity of Boyer Moore Algorithm is still $\Theta(nm)$ [6].

3 Proposed Algorithm

Proposed algorithm has two working phases, in the first preprocessing phase authors have divided the string 'S' in a substring each of size, pattern 'P' and we assigned an integer value to all the substring, then with the help of a hash function $H(n)$ store them in a specific location [4]. For each substring integer unique value repeat the same process (make a unique value, and then store in the hash table with the help of a hash function). To store the substrings integral value hashing technique has been used, during the process of storing data into the hash table with the help of hash function, i.e., (H(integer value) = [integer value%$(n - m + 1)$]). If we get the same modulus (%) for more than one integer value, then to resolve the collision hashing chaining is used. This is the most suitable and efficient way to store the substrings integer values, and authors have used dynamic memory allocation to store substring integer values to use only as much memory as needed. In the searching phase, to search a pattern in the hash table, first find the unique value of pattern and then use hash function, that maps it to the key (modules for the pattern) and to access the pattern match link list integer value with the pattern integer value up to the last node.

String $S = a\ b\ a\ b\ c\ a\ b\ d\ a\ b\ c\ d\ a\ b\ c\ d$

Pattern $P = a\ b\ c\ d$

3.1 Phase: 1 Preprocessing

Step 1.a

First assign the unique value to the pattern (Fig. 2).

Step 1.b

Convert it to unique value (Fig. 3).

Step 1.c

Then make a substring for a string and assign a unique code for the each substring (Fig. 4).

Step 1.d

Store all the substring into the hash table with the help of hash function—H(integer value) = [integer value% $(n - m + 1)$](Fig. 5).

a	b	c	d
0	1	2	3

Fig. 2 Assign code for substring

Pattern	0123

Fig. 3 Unique value of pattern

Fig. 4 Concatenations of values for every substring

0	0101
1	1012
2	0120
3	1201
4	2013
5	0130
6	1301
7	3012
8	0123
9	1230
10	2301
11	3012
12	0123

Fig. 5 Hash table representation of integral values

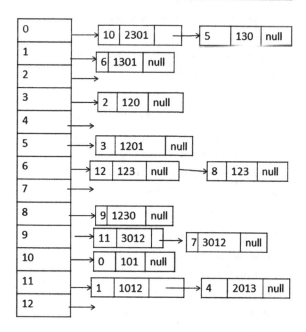

3.2   Phase: 2 Searching

In phase 2, proposed algorithm is used to find out searched pattern in the hash table. If pattern is found then return the location of string otherwise no pattern found in the string.

```
Algorithms used in Pre-procesing phase:

// algo.1:-make the unique value for searched pattern

m=size of patter,n-m=number of sub string.

For( int j=0 to m)
{
n=(int)arr[i]-97;
num=num*10+n;
j++;
}

// algo.2:-make the unique value for sub stings of a string

for(int i=0 to (n-m))
{
for(int j=0 to m)
{
temp = (int)string[stringindex]-97;
num = num*10 + temp;
stringindex++;
j++;
}
insertToHash(num, i);
}

// algo.3:-to insert the values in the hash

table insertToHash(int key, int index)
{
hashIndex = key % eleCount;
struct node *newnode createNode(key,index);
if (!hashTable[hashIndex].head)
{
hashTable[hashIndex].head = newnode;
hashTable[hashIndex].count = 1;
return;
}
newnode->next = (hashTable[hashIndex].head);
hashTable[hashIndex].head = newnode;
hashTable[hashIndex].count++;
return;
}

Algorithms used in searching phase:


// algo.4:- to search the key pattern in hash table

searchInHash(int num)
{
hashIndex = num % eleCount, flag = 0;
struct node *myNode;
myNode = hashTable[hashIndex].head;
if (!myNode)
{
return NULL;
}
while (myNode != NULL)
{
if (myNode->key ==num)
{
myNode->index;
flag = 1;
break;
}
myNode = myNode->next;
}
if (!flag)
return NULL
}
```

4 Result and Analysis

Proposed algorithm is based on the process of dividing the string into the substrings and converting into integer values, and then searching for a particular pattern in those substrings, so the whole process in divided into two following phases.

Phase I: Preprocessing
Phase II: Searching

Preprocessing time is the time, taken to make substrings of the string, each of size 'm' that is the size of pattern, then convert those substrings into integer values with the help of the ASCII code of the characters, and store them into a particular space, then apply hash function to store the value in the hash table. Authors have used hashing with chaining technique to resolve the conflict of more than one integer values having same modulus that results in collision. Same process is repeated for each substring.

Searching time is a time to find a pattern in the string with the help of hash function in the hash table. So in case if there is a link list, then follow the link list till the time list is not pointing to a null value. Else report that pattern is not found in the string.

4.1 Preprocessing Time

Total number of substrings of a pattern size 'm' is $(n - m)$. To assign integer value to one substring, takes 'm' times then to make '$(n - m)$' integer value of substrings, each of size 'm' it takes m $*(n - m)$. So that

1- Time taken to assign a unique integer value for substrings is $O(m*(n - m))$
2- Time taken to assign a unique integer value for pattern $O(m)$

Total time in preprocessing = [time to assign a unique value to a substring + time taken to assign a hash table + time to assign a unique value to the pattern]

$$\text{Total pre-processing time} = O(m * (n - m) + m)$$

4.2 Searching Time

Searching technique is based on the hash function where the authors have used hashing technique to reduce the access time complexity of an algorithm.

If we are searching a pattern for which there is null value or only one integer value stored in the hash table.

Table 1 Time complexity comparison

Algorithm name	Preprocess time complexity	Search time complexity
Boyer Moore	$O(m + n)$	$O(mn)$
Brute Force	No preprocessing	$O(mn)$
KMP	$O(m)$	$O(m + n)$
Quick Search	$O(m + n)$	$O(mn)$
Rabin Karp	$O(m)$	$O(mn)$
Proposed Algorithm	$O(m(n - m) + m)$	$O(1)$ and $O((n - m) + 1)$

$$\text{Best case } O \tag{1}$$

If we get same modulus for each integer value from hash function then it works as a linear link list so in this case proposed algorithm takes maximum time.

$$\text{Worst case } O((n - m) + 1)$$

5 Comparisons with the Existing Algorithms

See Table 1.

6 Conclusions

In this paper, authors have presented string matching algorithm. This is a substring based algorithm that uses the ASCII value of a character to make the integer value and compare with the integer value of pattern. This is very fast algorithm as authors have presented in the comparison table (Table 1. Time complexity comparison). The number of comparison is very less as compare to other algorithms. The limitation of proposed algorithm is, it takes $O(n - m)$ extra space to store the substrings in the hash table. As preprocessing is required only once when we start the searching if the size of the pattern to be searched is same as the previous pattern. In most of the cases proposed algorithm takes constant time to search a pattern in the string or, to find if the pattern is present or not present in the string.

References

1. Norton, M., & Roelker, D. (2003). The new Snort. *Computer Security Journal, 19*(3), 37–47.
2. Boyer, R., & Moore, J. (1977). A fast string searching algorithm. *Communications of the ACM, 20*(10), 762–772.

3. Pandey, S. K. et al. (2014). A study on string matching methodologies, (IJCSIT). *International Journal of Computer Science and Information Technologies, 5,* 4732–4735.
4. Karp, R. & Rabin, M. (1987). Efficient matching algorithms. *IBM Journal* of *Research* and *Development, 31*(2), 249–260.
5. Takáč, Ľuboš. (2013). *Fast exact string pattern-matching algorithm for fixed length patterns.* Žilina, SlovakRepublic, TRANSCOM: University of Žilina.
6. Chang, C., & Wang, H. (2012). Comparison of two-dimensional string matching algorithms. *International Conference on Computer Science and Electronics Engineering, 3,* 608–611. doi: 10.1109/ICCSEE.2012.29.
7. Yuan, L. (2011). An improved algorithm for Boyer-Moore string matching. In *2011 International Conference on Chinese Information Processing, Computer Science and Service System (CSSS)*, pp. 182–184. IEEE. doi:10.1109/CSSS.2011.5974722. Print ISBN: 978-1-4244-9762-1.
8. Alshahrani, A. M. (2013). Exact and like string matching algorithm for web and network security. *2013 World Congress on Computer and Information Technology (WCCIT)*, pp. 1–4. Print ISBN:978-1-4799-0460-0INSPEC. Accession Number:1382632. doi:10.1109/WCCIT. 2013.6618726.
9. Sheik, S. S., Aggarwal, S. K., Poddar, A., Balakrishnan, N., & Sekar, K. (2004). A FAST pattern matching algorithm. *Journal* of *Chemical Information* and *Computer Sciences, 44,* 1251–1256.
10. Exact String Matching Algorithm Animation Java. http://www-igm.univ-mlv.fr/ ~ lecroq/ string/node1.html.
11. Fuyao, Z. (2009, December 19–20). A string matching algorithm based on efficient hash function. *IEEE Information Engineering and Computer Science*, pp. 1–4. doi:10.1109/ ICIECS.2009.5363191. Print ISBN:978-1-4244-4994-1.
12. Knuth, D., Morris, J., & Pratt, V. (1977). Fast pattern matching in strings. *SIAM Journal on Computing, 6*(2), 323–350.
13. Boyer-Moore (1995). Parameterized pattern matching by type algorithms. In *Proceedings of the 6th ACMSIAM*, pp. 541–550. San Francisco, CA.
14. Notes on Boyer Moore String Mathing Algorithm. http://www.cs.ucdavis.edu/ ~ gusfield/ cs224f11/bnotes.pdf.
15. Franeka, F., Jennings, C. G., Smyth, W. F. (2007). A simple fast hybrid pattern-matching algorithm. *Journal of Discrete Algorithms, 5*(4), 682–695. December 2007.

INDTime: Temporal Tagger—First Step Toward Temporal Information Retrieval

Parul Patel and S.V. Patel

Abstract In today's era, a large amount of digitized information is generated every day. Retrieving information about a specific time period is one of the demands of information retrieval system. Such time-related information can be helpful to add time dimensions in existing information processing and information retrieval systems. In this paper, we represent INDTime, a rule-based system that is aimed to recognize and normalize temporal expressions present within the text document. It is designed to automatically annotate temporal expressions as per TIMEX3 annotation standard. The system has been evaluated on three datasets, one of them is WIKIWAR dataset and the other two are manually annotated datasets. We achieved an average precision of 94.15 %, recall of 92.55 %, and f-measure of 91.37 % on three datasets for recognizing temporal expressions and average precision of 90.95 %, recall of 92.55 %, and f-measure of 91.37 % for normalization.

Keywords Temporal expression · Temporal tagger · Rule-based approach

1 Introduction

Temporal information processing has received great attention in the natural language processing research community over the past few years. Temporal information present in the document can be used for applications such as question answering, document summarization, exploring search result in timeline, etc. For implementing such applications, the first step is to identify and normalize temporal information present in the document for further processing. Processing of temporal

P. Patel (✉)
M.Sc (I.T) Programme, VNSGU, Surat, Gujarat, India
e-mail: parul.patelns@gmail.com

S.V. Patel
Department of Computer Science, VNSGU, Surat, Gujarat, India
e-mail: patelsv@gmail.com

© Springer Science+Business Media Singapore 2016
S.C. Satapathy et al. (eds.), *Proceedings of International Conference on ICT for Sustainable Development*, Advances in Intelligent Systems and Computing 408, DOI 10.1007/978-981-10-0129-1_21

Table 1 Examples of temporal expression in document

While members of delegation from Andhra and Rayaseema regions argued **on Monday** that a partition would ruin the state, votaries of Telangana were gathering here to make their own pitch. The delegation is set to meet Shinde and other senior leaders **on Tuesday** amid indications that they bluntly tell the Centre that the region would not brook a repeat of **December 2009** when the Centre raised hopes of Telangana only to smother it.

After India gained its independence **in 1947**, the city became the capital of Madras State, which was renamed as Tamil Nadu **in 1969**. The violent agitations **of 1965** against the compulsory imposition of Hindi in the state marked a major shift in the political dynamics of the city and eventually it had a big impact on the whole state. **On 26 December 2004**, an Indian Ocean tsunami lashed the shores of Chennai, killing 206 people in Chennai and permanently altering the coastline.

AHMEDABAD: One of the world's biggest kite festivals, the French Dieppe International Kite Festival, will see participation by four Indians **this year**. The festival will be held **between September 6 to September 14.** is the first time that a Gujarati has been chosen for the festival since its launch **in 1989**

expressions involve both recognizing time-denoting expressions and finding their meaning. It is known as TERN (Temporal Expression Recognition and Normalization). The task is challenging because some temporal expressions (implicit) are straightforward to recognize and normalize (e.g., 24/3/2010), whereas some are context-dependent (e.g., next Friday, Last Christmas, etc.). So finding an appropriate value for temporal expressions requires analysis of surrounding tasks (Table 1).

From the above examples, it is clear that news or any other document that contains temporal expressions implicitly or explicitly denote past or future events, which can be very useful in the search. For example, in the search engine if the user fires a query like "Election from 2012 to 2014 in India," it means user is interested in documents that contain information about election from the year 2012 to 2014. The current search engine lacks in interpreting the temporal meaning of the expression present in the user query. Recognition of such temporal expression from free text and presenting them with semantics in a running text is a challenging task. It must have the capability to identify temporal expressions from input text and convert them into canonical form to store in database for more accurate comparison, sorting, etc.

In order to address the issue, we have developed a rule-based system that not only recognizes but normalizes temporal expression present in the document. It supports all Indian festivals with different semantics and all popular events that happened in India. Moreover, the system is generic to accept new festivals and events that can be added in the future without any change in rule.

The paper is organized as follows. Section 2 covers the literature review. In Sect. 3, research methodology is described. Section 4 includes the evaluation of our system on various datasets. Limitations of the system are discussed in Sect. 4. Section 5 includes conclusion and future work.

2   Literature Review

The Message Understanding Conferences (MUCs) in 1996 and 1998 played a significant role wherein evaluations were covered during conference covering recognition of TEs to inspire the research community, while novel contributions toward the normalization of TEs were made in 2000 [1]. KUL is a machine learning-based system for recognition and normalization of temporal expression with 0.85 % precision and 0.84 % with recall [2]. Chronos is a rule-based system which involves tokenization, statistical part-of-speech tagging, and multiwords recognition based on a list of 5000 entries retrieved from WordNet. The text is processed by a set of approximately 1000 basic rules that recognize temporal constructions and gather information about them that is expected to be useful in the process of normalization. This is followed by the application of composition rules, which resolve ambiguities when multiple tag placements are possible. The results in terms of F-measure on ACE 2004 data are 92.6, 83.9, 87.2 % for detection, recognition, and determining the VAL attribute value, respectively [3]. Heideltime is high quality rule-based tagger for temporal expression recognition and normalization with 0.90 % precision and 0.82 % recall [4]. Jelena developed a system for temporal information extraction and interpretation for Serbian language with precision of 0.93 %, recall of 0.96 %, and F-score of 0.94 % [5]. SUtime is the library for recognizing and normalizing temporal expressions developed by Stanford University. It is a rule-based system developed in Java [6].

We have observed that most of the above systems do not support festivals as temporal expressions and some of them support only festivals like Christmas, which fall on fixed days. Most of them do not support compound temporal expressions (e.g., from 2 December, 2001 to 3 March, 2002.) They consider them as two separate expressions. These limitations make the above tools inappropriate for use or extend easily if we want to use them on Indian news documents because if we analyse Indian news documents, we find that most Indian festivals do not have fixed date (e.g., Diwali, ID, etc.).

We have developed a rule-based temporal expression recognition and normalization system which eliminates the above limitations. Further, it is generic enough for further extension for future change. We have used library of SUTime and an improved version of temporal tagger, which overcomes the limitations of SUtime and supports the dynamic behavior of Indian festivals and range queries.

3 Research Methodology

3.1 Document Preprocessing

The document preprocessing phase is important for any IE application that transforms document into useful format. First, we extracted individual sentences from the text document. These sentences were tokenized and disambiguated into their lexical category by means of statistical POS tagger. Multiword selection is required at this stage because most of the temporal expressions are multiword (e.g., last Thursday, next Monday, next Diwali, etc.). Such multiword expressions are recognized using WordNet [7].

3.2 Recognition

Extracted token or multiwords are checked individually with rules stored in a rule file. Rule file is a combination of regular expressions. It contains more than 1000 rules. Selection of a rule is a time-consuming process so that it is divided into two steps: First, it tries to match with simple rules which contain names of weekdays, names of months, names of seasons, names of festivals, dates in various formats (e.g., 21 December 2014, May 15, 2014/12/12, 12/4/2014, etc.), some popular days, etc. Complex rules then are used to extract complex temporal expressions such as 'day before tomorrow,' or relative temporal expressions such as before, after, ago, recently, earlier, later, yearly, monthly, since, etc. Our basic rules recognize temporal expressions as well as gather surrounding information around the temporal expressions for annotation. For e.g., more than, few, before, later, last. This information is also required for normalization phase. Finally, all recognized temporal expressions are analysed again and ambiguous temporal expressions not likely to be temporal expressions are removed. For e.g., we have a rule indicating that if a temporal expression is a single word 'may' and the POS tag is not a noun then it is not likely to be a season of autumn but a verb used with sentences ('I may come there'). Such expressions are removed before going to normalization phase (Fig. 1).

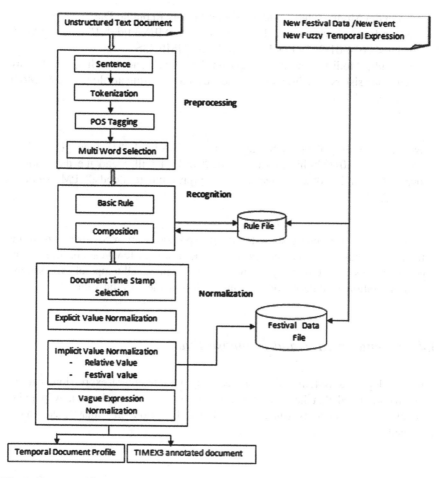

Fig. 1 System architecture

3.3 Normalization

During this phase, all temporal expressions are assigned their possible absolute value. First, all explicit temporal expressions are normalized to their absolute values by selecting their appropriate function. Second, all relative temporal expressions are assigned their absolute value using document timestamp. Third, all relative temporal expressions that belong to some popular day, event, or festival are normalized to their respective absolute values by extracting year from timestamp and searching respective year from database. For e.g., consider the following example.

(1) Narendra Modi visited Gandhi Ashram on this Republic Day.
    Narendra Modi visited Gandhi Ashram on <TIMEX3 tid="t1" type="DATE"
    value="2014-01-26"> this Republic Day</TIMEX3>.
(2) Narendra Modi will be in Srinagar on this Diwali and "I will spend the day
    with our sisters and brothers affected by the unfortunate floods," PM Modi
    tweeted.

Narendra Modi will be in Srinagar on <TIMEX3 tid="t1" type="DATE"
value="2014-10-23">this Diwali</TIMEX3>& "I will spend the day with
our sisters and brothers affected by the unfortunate floods," PM Modi
tweeted.

During recognition phase some temporal expressions are recognized which are
difficult to normalize for specific absolute value. For e.g., few years back, in early
days, few weeks later. Our recognition module recognizes all temporal expressions,
but normalization module is not normalizing it to some specific value.

3.4 Extending Rule and Database File

After developing a system with good accuracy, new temporal expressions can be
easily incorporated that have been missed out or new festivals or new popular
incidents, events such as tsunami, attack on Taj, riots in Gujarat, etc. have
happened.

3.5 TIMEX3 Annotated File

After normalization phase, it generates annotated file in which all temporal
expressions are annotated as per TIMEX3 standard. It includes Timex Id, Type, and
Value attribute of TIMEX3. Timex Id uniquely identifies temporal expressions
within the document. Type attribute is assigned one of the four values: DATE,
TIME, DURATION, and SET. DATE describes calendar time. For
e.g.,'Friday,''October 1, 1996,' 'the second of November,' 'yesterday,', 'in October
of 1953.' TIME expression defines time of the day. For e.g., '5 min ago,' '2 min
later,' '10 min to 5.' DURATION expression describes explicit durations like

Table 2 Plaintext as an input to the system

> *Speed News » 13 Feb 2014,22:10 IST   Those who indulge in hooliganism and eve teasing on Valentine day, beware. Ludhiana police claims to go very hard against such elements and have put in fool proof arrangement to deal with such people. Doctors from the oncology department of Government Medical College and Hospital (GMCH) and volunteers of NGO CanKids will be celebrating Valentine's Day by organizing a blood donation camp. It will be held at the promenade of Futala lake. Nearly 150 supporters in saffron were on roads in a 'Chetna rally' to convey the message. In a prelude to discourage western culture in state, supporters of Bajrang Dal- youth wing of the Vishwa Hindu Parishad- have on Wednesday warned people against celebrating Valentine's day or indulging in any indecent act on February 14.*

Table 3 TIMEX3 annotated output

> *speed News<< <TIMEX3 tid="t1" type="TIME" value="1985-02-13T22:10">13 Feb 2014,22:10</TIMEX3> IST Those who indulge in hooliganism and eve teasing on <TIMEX3 tid="t2" type="DATE" value="1986-02-14">Valentine day</TIMEX3> , beware. Ludhiana police claims to go very hard against such elements and have put in fool proof arrangement to deal with such people.Doctors from the oncology department of Government Medical College and Hospital (GMCH) and volunteers of NGO CanKids will be celebrating <TIMEX3 tid="t3" type="DATE" value="1986-02-14">Valentine's Day</TIMEX3> by organizing a blood donation camp. It will be held at the promenade of Futala lake.Nearly 150 supporters in saffron were on roads in a 'Chetna rally' to convey the message. In a prelude to discourage western culture in state, supporters of Bajrang Dal- youth wing of the Vishwa Hindu Parishad- have on <TIMEX3 tid="t7" type="DATE" value="1985-09-18">Wednesday</TIMEX3>   warned people against celebrating <TIMEX3 tid="t8" type="DATE" value="1986-02-14">Valentine's day</TIMEX3>   or indulging in any indecent act on <TIMEX3 tid="t9" type="DATE" value="1986-02-14">February 14</TIMEX3>.*

'2 months,' '5 years,' '2 days before.' SET expression defines set of times, for e.g., 'twice a week,' '4 days a week,' 'every day,' etc. The following example looks at annotation file and temporal document profile (Tables 2 and 3).

3.6   Temporal Document Profile

After normalization phase, extracted temporal expressions with their normalized value are stored into vector called temporal document profile, which can be used further in applications such as time line generation, question answering, and document summarization. Temporal document profile contains all temporal expressions present in the document and their corresponding normalized value. It contains the starting position of the sentence in the document, where temporal expression is present and the ending position of the sentence. These positions can be used to extract sentences where temporal expressions occur in their document. Table 4 describes temporal profile generated for input text of Table 2.

Table 4 Temporal document profile

Temporal expression	Normalized value	Start position	Date	Month	Year
13 Feb 2014, 22:10	2014-02-13T22:10	15	13	2	2014
Valentine day	2014-02-14	86	14	2	2014
Valentine's day	2014-02-14	280	14	2	2014
Wednesday	2014-02-12	450	12	2	2014
Valentine's day	2014-02-14	480	14	2	2014
February 14	2014-02-14	525	14	2	2014
Valentine day	2014-02-14	596	14	2	2014

4   Evaluation

For evaluation, we have selected three datasets. Each of them is rich in temporal expressions. DataSet 1 is Wikiwar corpus, which contains approximately 2671 temporal Expressions. DataSet 2 is a collection of biographies of well-known persons of India. Dataset 3 is a collection of 100 news articles of different periods. These data were collected from the Times of India. Figure 2 shows analysis of different types of temporal expressions present in the different datasets (Table 5).

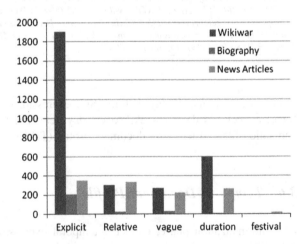

Fig. 2 Comparing number of temporal expression in different datasets

Table 5 Result of evaluation on different datasets

	Precision (%)		Recall (%)		F-measure (%)	
DataSet 1	92.84	90.21	99.63	94.37	96.11	92.24
DataSet 2	90.72	89.32	94.34	93.58	92.49	90.40
DataSet 3	98.90	93.34	97.24	89.72	98.06	91.49
Average	94.15	90.95	97.07	92.55	95.53	91.37

5 Limitations

Resolving relative temporal expressions is challenging. We have considered document creation date as reference date for normalization of relative temporal expressions. But in some cases, it is not appropriate to use DCT as reference date. For e.g., 'He concluded the 2012 annual general meeting by saying "The next year will be an important year for them."' In this example, next year should be normalized using previous sentence date instead of DCT.

6 Conclusion and Future Work

We have developed INDTime system for extraction and normalization of temporal expression from English text documents. The system has been evaluated on three datasets including WikiWar corpus. We have achieved good results such that INDTime can be used in good applications like exploring search results on timeline, question answering, document summarization, etc. In future, heuristics around the text and previous sentences can be used to select dynamic reference date.

References

1. Mani, I., & Wilson, G. (2000). Robust temporal processing of news. In *Proceedings of the 38th Annual Meeting on Association for Computational Linguistics* (pp. 69–76). Hong Kong.
2. Pustejovsky, J., Ingria, B., Sauri, R., Castano, J., Littman, J., Gaizauskas, R., et al. (2004). The Specification Language TimeML. In I. Mani, J. Pustejovsky, & R. Gaizauskas (Eds.), *The language of time: A reader*. Oxford: Oxford University Press.
3. Negri, M., & Marseglia, L. (2005). Recognition and Normalization of Time Expressions: ITC-irst at TERN 2004. Technical Report WP3.7, Information Society Technologies, February 2005.
4. Strötgen, J., & Gertz, M. (2010). HeidelTime: High quality rule-based extraction and normalization of temporal expressions. In *Proceedings of the 5th International Workshop on Semantic Evaluation, ACL 2010* (pp. 321–324). Uppsala, Sweden, 15–16 July 2010.
5. Jaćimović, J. (2012). Recognition and normalization of temporal expressions in Serbian texts. *BCI'12*, September 16–20, 2012, Novi Sad, Serbia.
6. Chang, A. X., & Manning, C. D.: SUTime: A library for recognizing and normalizing time expressions. In *Eighth International Conference on Language Resources and Evaluation (LREC 2012)*.
7. Manning, C. D., & Fellbaum, C. (1998). *WordNet: An electronic lexical database*. Cambridge: The MIT Press.

A Secure Text Communication Scheme Based on Combination of Compression, Cryptography, and Steganography

Kirti Saneja, Mukesh Kumar and Pallavi Sharma

Abstract Nowadays security plays a crucial role in distinct fields of science and technology. Vast quantity of information security and information hiding algorithms are available for textual data shield. This research mainly focuses on amalgamation of the advantages of compression, encryption, and steganography schemes in order to enhance the notion of security. The offered systems provide lossy techniques for compression and encryption, i.e., jpeg, mpeg, and block cipher methods. These techniques are secure but have high computational costs. In this paper, a Secure Text Communication Scheme is proposed with a combination of compression using Text Converter Technique (TCT), encryption using bits to Latin1 Unicode function, and steganography done using NT–DCT. These newly defined algorithms have advantages of high encryption speed, reduced processing time, and provide more reliable private data. The main goal is to hide a huge amount of text behind a single image without causing any distortion in its original image quality.

Keywords NT-DCT · TCT · Steganography · Encryption · Compression

1 Introduction

In today's scenario, due to the quick evolution of the Internet and network technology, the protection of data becomes significant. The main aim is to reduce computational expenditure and stimulate security of data as well as reduce

K. Saneja (✉) · M. Kumar · P. Sharma
Department of Computer Engineering, The Technological Institute
of Textile and Sciences, Bhiwani, India
e-mail: kirtisaneja92@gmail.com

M. Kumar
e-mail: drmukeshji@gmail.com

P. Sharma
e-mail: pallavisharma.cp@gmail.com

© Springer Science+Business Media Singapore 2016
S.C. Satapathy et al. (eds.), *Proceedings of International Conference
on ICT for Sustainable Development*, Advances in Intelligent Systems
and Computing 408, DOI 10.1007/978-981-10-0129-1_22

redundancy of data in order to store or send data in an efficient form. So, now Secure Steganography becomes an important aspect of study for researchers.

1.1 Compression

It is an operation where information can be reduced without difficulty in order to diminish the space and time for transmission. This information can be an image, text, audio, fax, video, etc. There are two types of compression techniques which are as follows (Fig. 1).

(i) Lossless compression—It reduces the quantity of source data transmitted in such a way that when compressed data is decompressed, no alteration will occur. The lossless technique can be further classified in the following categories:

    i. Run length encoding—It is an effective algorithm that is used to send one item from a large number of repeating items.

    ii. Huffman encoding—It is used to represent the compressed binary stream that is generated from a large number of symbols and tokens.

    iii. Lempel-Ziv—It is based on table-based lookup algorithm that compresses large files into smaller files.

(ii) Lossy compression—It does not redevelop an accurate copy of information after decompression. Here some information gets lost after decompression. Lossy technique can be further classified into the following categories:

    i. JPEG is a well-known standard that is used to compress the image files.

    ii. MPEG is used for encoding and compression video images.

    iii. MP3 is used to compress continuous flow of sound into small files.

Fig. 1 Types of compression

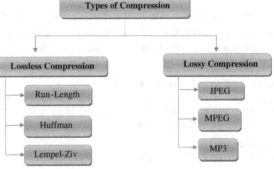

1.2 Cryptography

It is the technique to accomplish security by encoding the original data into non-readable format. It has two elementary functions, encryption and decryption. Encryption is the process of altering plain text data into cipher text. Decryption is the reverse process of encryption. It is the process of retrieval of original data from encoded data. The basic block diagram of cryptography is shown in Fig. 2.

The various ways of classifying cryptographic algorithms are:

- Secret key cryptography
- Public key cryptography
- Hash functions
- Secret key cryptography (SKC): It uses a single key for both encryption and decryption (Fig. 3).
- Public key cryptography (PKC): It uses one key for encryption and another for decryption (Fig. 4).
- Hash functions: It uses a mathematical alteration to irreversibly "encrypt" information (Fig. 5).

Fig. 2 Block diagram of cryptography

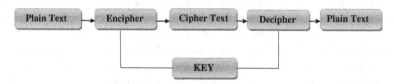

Fig. 3 Secret key cryptography

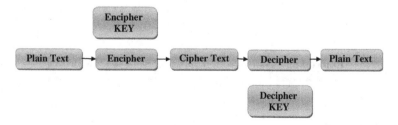

Fig. 4 Public key cryptography

Fig. 5 Hash functions

1.3 Steganography

It is basically a powerful protection scheme used to mask the private information behind conventional information to establish an unseen communication. This conventional information can be about any digital object and it produces a stego medium. It provides better scope in such a way that one cannot know the existence of hidden data and an unsanctioned user cannot access the original message.

There are different types of steganography, as described below:

- Text steganography
- Image steganography
- Audio/video steganography
- Text steganography: It includes a small amount of data redundancy, therefore it is frequently used.
- Image steganography: It is generally used for hiding data. It provides a simple process to transfer information over the Internet.
- Audio/video steganography: Both are used to create secure hidden data in the form of audio and video.

The basic block diagram of steganography is shown in Fig. 6.

The message is transformed into encrypted form with the help of an encryption key, which is known only by the sender and receiver. The message cannot be retrieved by anyone without using the encryption key. Steganography is an approach that has been developed to handle the imperfection of cryptographic techniques. Thus, steganography hides the existence of data so that none can detect its presence. For the improvement of network security, both steganography and cryptography concepts are combined with compression approach.

According to the Secure Text Communication Scheme, it is divided into various different phases. In the first phase Text Converter Technique (TCT) is used to compress the text into bits by selecting the characters one by one. If the scanned character belongs to the same category as the next character, then find its location value. If the scanned character belongs to a different category as the next character,

Fig. 6 Steganography

then find its location value as well as its category value. Based on this priority, generate the code words. In the second phase encryption is used to encrypt data by calculating the length of bits and converts them into Latin1 Unicode. In the third phase convert the text into an image. Finally, in the last phase NT-DCT approach is used to hide the data. In this phase, the cover image is first separated into RGB components and these components are transformed separately to frequency domain using New Technique Discrete Cosine Transform (NT-DCT). By using this new technique, data or text is hidden in the fourth pixel to hide a greater amount of data in the same image.

2 Literature Survey

In the paper "A Method for Binary Image Thinning using Gradient and Watershed Algorithm" by Kamaljeet Kaur, Mukesh Sharma, thinning is basically reducing a 'thick'digital object to 'thin'skeleton. Thinning is one of the most frequently used methods to know the geometrical feature of objects. Bedwal and Kumar [1] suggested a new algorithm to hide an RGB image in another RGB image based on LSB insertion technique. Bharti and Soni [2] recommended a technique of dividing the image into 4 * 4 level sub blocks and hiding the data in diagonal pixels of these sub blocks. Gupta and Sharma [3] suggested a signature hiding standard. The pixel of RGB-based 24-bit image using LSB technique is where 8 bits are used to show red color, 8 bits show green color, and the last 8 bits blue. The first 2 bits of 8 of them show the color of the next pixel using RGB color combination, then the third bit will be ignored. The next two bits define the difference between the current and the next pixel. Again, the 6th bit will be ignored and next two bits, that is, 7th and 8th define how much bits of signature will be hidden. In [4] the authors propose an algorithm that first enciphers the text using an extended hill cipher algorithm and then the encrypted text is masked behind the 6th, 7th, and 8th pixels of the brightest (gray value from 224 to 255) and darkest pixels (gray value from 0 to 31) that are scattered randomly throughout the image. Sharma [5], "Image Compression Using Enhanced Haar Wavelet Technique," describes data compression which can be lossy or lossless and is required to decrease the storage requirement and improve data transfer rate. One of the best image compression techniques is using wavelet transform. It is comparatively new and has many advantages over others. In [6] Kamaljeet Kaur, Mukesh sharma. "A Method for Binary Image Thinning using Gradient and Watershed Algorithm", in this paper, thinning is basically reducing a 'thick'digital object to 'thin'skeleton. Thinning is one of the most frequently used methods to know the geometrical feature of objects. In [8] Thomas Leontin Philjon and Venkateshvara Rao. Metamorphic Cryptography—A Paradox between Cryptography and Steganography Using Dynamic Encryption. In [10] B.Schneier, Description of a new Variable-Length Key, 64-bit Block Cipher (Blowfish) Fast Software Encryption. In [11] Mukesh Sharma "Image Compression Using Enhanced Haar Wavelet Technique", it describes data compression which can be lossy or lossless is required to decrease the storage requirement and better data transfer rate. One of the best image compression techniques is using wavelet transform. It is comparatively new and has many advantages over others. In [12] Ratinder Kaur, V.K. Banga "Image Security using Encryption based Algorithm".

The remainder of the paper is organized as follows: Sect. 3 describes the Proposed Methodology (A Secure Text Communication Scheme). Section 4 analyzes the results. Section 5 concludes the paper.

3 Proposed Methodology

The proposed method consists of four stages. In the first stage, Text Converter Technique is applied (TCT) to decrease transmission time. It is used to cover more data behind an image. In the second stage, encryption is applied that provides higher security and in the next stage convert the text into an image. Finally, in the end stage data hiding is accomplished by hiding the data using New Technique of DCT (NT-DCT). The main intention of this work is to enhance the security and lossless recovery of data. The algorithms in each phase are discussed as follows.

The process of embedding the data is shown below.

Step 1   Enter the Text File.
Step 2   Compress the Text File using TCT.
Step 3   Encrypt the bits into Latin1 Unicode.
Step 4   Convert the text into an image.
Step 5   Hide the data using the New Technique of DCT.
Step 6   Generate the Stego image that hides the text file (Fig. 7).

Fig. 7 Embedding process

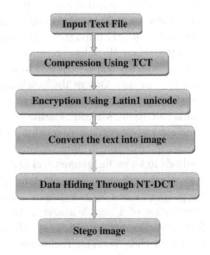

Embedding Algorithm

1. Input text file, cover image
2. For compression apply TCT in Steps 3–7.
3. Scan the text file to be compressed.
4. If the scanned character belongs to the same category as the next character, then find out its location value.
5. If the scanned character belongs to a different category than the next character, then find out its location value as well as its category value.
6. Based on this priority, generate the code words.
7. Encrypt the data by calculating the length of bits and convert them into Latin1 unicode.
8. After that convert the text into an image.
9. Then separate RGB component of CI as Rimg, Bimg, Gimg.
10. Calculate the NT–DCT on each component using

    Rimg=dct2 (Rimg)
    Gimg=dct2 (Gimg)
    Bimg=dct2 (Bimg)

11. Embed the ECTEXT into the Bimg

    For i=1: length (ECTEXT)
    Bimg (i, 4)=Bimg (i, 4) +ECTEXT (i)
    end

12. Concatenate Rimg, Gimg, and Bimg to get StegoImg.
13. Find Inverse NT-DCT of StegoImg

    StegoImg=idct2 (StegoImg)

The output of the embedding algorithm is the Stego image that has the text hidden behind it.

Extracting: The process of extraction of original data from the Stego image is shown below.

Step 1   Read the Stego image.
Step 2   Retrieve the data using NT-DCT.
Step 3   Convert the image into text
Step 4   Decrypt the Latin1 Unicode data into text.
Step 5   Decompress the decrypted data using reverse of TCT.
Step 6   Retrieve the original file (Fig. 8).

Extracting Algorithm: The extracting algorithm includes the reverse steps of the embedding algorithm which are given below.

1. Input cover image CI and stego image SI.
2. Separate the RGB components of both CI and SI. CIRimg, CIGimg, CIBimg and SIRimg, SIGimg, SIBimg are RGB components of CI and SI respectively.
3. Take NT-DCT of each component separately.

Fig. 8 Extracting process

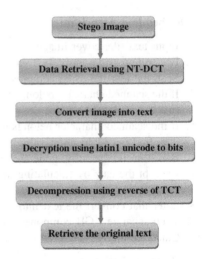

CIRimg=dct2 (CRimg)
CIGimg=dct2 (CGimg)
CIBimg=dct2 (CBimg)
SIRimg=dct2 (SRimg)
SIGimg=dct2 (SGimg)
SIBimg=dct2 (SBimg)

4. Extract text by subtracting the DCT coefficient of cover image from stego image.

For i=1: length (TEXT)
TEXT (i)=SBimg (i, 4)-CBimg (i, 4)
End

5. Convert the image into the text.
6. Decrypt the data by calculating the Latin1 Unicode and convert them into bits.
7. Decompress the data using reverse of TCT.
8. Original text file found.

4 Result

The chief purpose of steganography is to identify defects in the embedded message. So, the analyzed consequences are based on Peak Signal to Noise Ratio (PSNR) value and Mean Square Error (MSE). A Secure Text Communication Scheme was implemented and executed using MATLAB R2011b. This technique is applied over various cover images with different values of plain text and keys. It includes compression with Text Converter Technique and cryptography with bits to Latin1

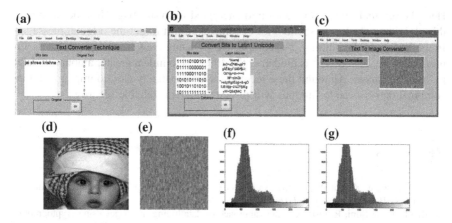

Fig. 9 **a** Text compression using text converter technique. **b** Convert bits into Latin1 unicode. **c** Convert text into image. **d** Shows the cover image. **e** Stego image. **f** Histograms of cover image. **g** Histograms of stego image

Unicode function. The overall time of compression, encryption, decryption, and decompression are compared and presented here. The proposed algorithm is more efficient than the previous algorithms in terms of compression ratio as well as time taken to carry out the whole mechanism.

Figure 9 shows:

a. Text compression using text converter technique.
b. Convert bits into Latin1 unicode.
c. Convert text into image.
d. Shows the cover image.
e. Shows the stego image.
f. Shows histograms of cover image.
g. Shows the histograms of stego image.

Obtained values of the parameters are:
PSNR = + 59.28 dB
MSE = 0.1538

5 Conclusion

In this paper some stages (compression, cryptography and steganography) are used for security enhancement that makes it difficult to detect the presence of hidden message. But in various cases if the attacker has attacked the carrier of message, then he will not be able to get the original message because here the message is in encrypted form. For compression, a new technique (TCT) is used to compress the data and cryptography is used, which is much better than DES and Triple DES.

Here the original text is converted into encrypted form of image. In order to break this algorithm one has to spend a lot of time and effort to get the original message. In the last stage data hiding is accomplished by using New Technique of DCT. In this technique, data or text is hidden in the 4th pixel to hide a greater amount of data in the same image. Although these techniques are separately easy to implement, the combination of these techniques will provide efficient, reliable security.

References

1. Bedwal, T., & Kumar, M. (2013). An enhanced and secure image steganographic technique using RGB-Box mapping. *IET Digital Library*.
2. Bharti, P., & Soni, R. (2012). A new approach of data hiding in images using cryptography and steganography. *International Journal of Computer Applications, 58*(18), 1–4.
3. Gupta, K., & Sharma, M. (2014). Signature hiding standard: Hiding binary image into RGB based image. In *2014 International Conference on Information and Communication Technology*.
4. AL-Abiachi, A. M., Ahmad, F., & Ruhana, K. (2011). A competitive study of cryptography techniques over block cipher. In *UKSim 13th IEEE International Conference on Modelling and Simulation 2011*.
5. Gaikwad, D. P. Prof., Jagdale, T., Dhanokar, S., Moghe, A., & Pathak, A. Hiding the text and image message of variable size using encryption and compression algorithm in vedio steganography. *International Journal of Engineering Research and Application(IJERA), 1*(2), 102–108.
6. Kaur, K., & Sharma, M. (2013). A method for binary image thinning using gradient and watershed algorithm. *International Journal of Advanced Research in Computer Science and Software Engineering*.
7. Thakur, J., & Kumar, N. (2011).DES, AES and Blowfish: Symmetric key cryptography algorithms simulation based performance analysis. *International Journal of Emerging Technology and Advanced Engineering, 1*(2), December 2011.
8. Philjon, T. L., & Rao, V. (2011). Metamorphic cryptography—A paradox between cryptography and steganography using dynamic encryption. In *IEEE-International Conference on Recent Trends in Information Technology, ICRTIT 2011*.
9. Sharma, M. (2012). Smiley Gandhi compression and encryption: an integrated approach. *International Journal of Engineering Research and Technology* .
10. Schneier, B. (1994). Description of a new variable-length key, 64-bit block cipher (Blowfish). In *Fast Software Encryption. Cambridge Security Workshop Proceedings* (pp. 191–204), December 1993. Springer-Verlag.
11. Sharma, M. (2012). Image compression using enhanced haar wavelet technique. In *International Conference 2012*.
12. Kaur, R., & Banga, V. K. (2012). Image security using encryption based algorithm. In *International Conference on Trends in Electrical, Electronics and Power Engineering (ICTEEP 2012)*, July 15–16, 2012 Singapore.

A Comparative Analysis of Feature Selection Methods and Associated Machine Learning Algorithms on Wisconsin Breast Cancer Dataset (WBCD)

Nileshkumar Modi and Kaushar Ghanchi

Abstract The sector that affects all the citizens as a common factor is the health sector. It has huge data and important information about patients and their health. Now it is the need of the hour to utilize those enormous data and to extract knowledge, which will be useful to all the stake holders related to electronic health records. In this paper, we have focused on feature selection techniques to gain high quality attributes to enhance the mining process. Feature selection techniques touch all sectors, which require knowledge discovery. In this study, we have made a comparison between different feature selection methods and associated machine learning algorithms on Wisconsin Breast Cancer dataset, Wisconsin Diagnostic Breast Cancer (WDBC), and Wisconsin Prognostic Breast Cancer (WPBC) of UCI Repository. The study found that fusion of classification algorithms perform better on WBCD. It also revealed that IG performs better on WBCD, whereas IG and CFS give good results on WPBC, and CA gives best results on WDBC.

Keywords Feature selection methods · Classification · Breast cancer analytics

1 Introduction

"Data mining means extraction of knowledge from large databases" [1], [2]. Data mining process gives the knowledge required for further decision. Data mining extracts knowledge by applying rules and patterns generated by machine learning

N. Modi (✉)
Narsinhbhai Institute of Computer Studies & Management, Kadi, India
e-mail: drnileshkumarmodi@gmail.com

K. Ghanchi
Navgujarat College of Computer Applications, Ahmedabad, India
e-mail: kaush18@gmail.com

© Springer Science+Business Media Singapore 2016
S.C. Satapathy et al. (eds.), *Proceedings of International Conference on ICT for Sustainable Development*, Advances in Intelligent Systems and Computing 408, DOI 10.1007/978-981-10-0129-1_23

algorithms. As we know, in the recent era, we are flooded with data. Datasets generated by the systems are highly dimensional and gigantic in size. "To apply data mining algorithm on these kinds of datasets causes many complexities and errors too. So feature selection is the solution to dilute such complexities" [3]

Feature selection is the process of identifying relevant attributes for use in applying machine learning algorithm. Attribute selection methods provide a ranks for each attribute. The attribute having the best rank for the measure is chosen. Basically, there are two main categories of feature selection methods such as filter and wrapper.

In the current work, we have focused on breast cancer disease. In breast cancer, malignant (cancer) cells form in the breast tissue. The damaged cells can invade surrounding tissues, but with early detection and treatment, most people continue a normal life [4]. According to the World Health Organization, breast cancer is the most common cancer among women worldwide. "Out of eight women, one woman is diagnosed with breast cancer [4]." It is second leading cause of death in women. It is also estimated that in 2013, 39,620 women died due to breast cancer [5], [6]. In Sect. 2, we will discuss the literature review and then in Sect. 3, will explain different feature selection methods in data mining. In the next Sects. 4and 5 we are focusing on our experiment and results of experiment.

2 Related Work

Many authors have worked on feature selection methods and its impact on different datasets. Ashraf et al. [7] presented their study of feature selection techniques on three datasets on thyroid, hepatitis, and breast cancer of UCI repository. The results conclude that attribute feature selection methods can improve the performance of learning algorithms. However, no single feature selection method best satisfies all datasets and learning algorithms. Hall and Holmes [8] have presented a benchmark comparison of six attribute selection techniques. Ashraf et al. [9] have presented their study on information gain and adaptive neuro-fuzzy inference system for breast cancer diagnoses. In this paper, author presented a new approach for breast cancer diagnosis using the combination of an adaptive network-based fuzzy inference system (ANFIS) and the information gain method. Results show that the accuracy of the proposed algorithms reaches to 98.23 %, which can vary on other datasets. Azhagusundari and Thanamani [10] have presented their study on feature selection based on information gain. In this paper, author discusses about an algorithm based on discernibility matrix and information gain to reduce attributes. Naseriparsa et al. [3] have presented their study on hybrid feature selection methods, which takes advantages of wrapper subset evaluation with a lower cost and improves the performance of a group of classifiers.

3 Feature Selection Methods

Feature selection is also known as attribute selection or variable selection in data mining. "Feature selection can significantly improve the clarity of the resulting classifier models and often build a model [11], [12]." The main objective of feature selection methods is to improve the understanding of underlying business. Also to improve efficiency of data to be mined, and to improve the prediction performance of the predictors in the model [13]. Attribute selection techniques have two types such as "filter" and "wrapper." Filters are methods, which rank variables as per its usefulness. It is used as a unprocessing step [14]. These methods can be independent of target variables. While wrappers evaluate attributes using accuracy estimates provided by the actual target learning algorithm.

3.1 Information Gain (IG)

Information gain is widely used for selecting features or attributes in preprocessing of datasets. It measures the numbers of bits of information obtained for categorizing the prediction by knowing the presence or absence of a term in a document. If x is an attribute and c is the class, the following equation gives the entropy of the class before observing the attribute:

$$H(s4x) = -\sum xP(x)\log 2P(x)$$

Here $P(c)$ is the probability function of variable c. The conditional entropy of (c) given (x) (post entropy) is given by:

$$H(c|x) = -\sum xP(x)\sum cP(c|x)\log 2P(c|x)$$

Now the difference between prior entropy and post entropy will be known as information gain. And calculated as:

$$H(c,x) = H(c) - H(c|x)$$

3.2 Principal Component Analysis

PCA is a very good method to decrease the high dimensionality of big datasets to fewer dimensions, which are easier for humans to understand and visualize. Each principal component is a linear combination of the observed variables. PCA is an unsupervised method, meaning that no information about groups is used in the dimension reduction.

3.3   Relief

Relief was proposed by Kira and Rendell in 1994. The success of the algorithm is
due to the fact that it is fast, easy to understand and implement, and accurate even
with dependent features and noisy data [15]. The algorithm is based on a simple
principle.

The Relief algorithm basically consists of three important parts:

1. Calculate the nearest miss and nearest hit;
2. Calculate the weight of a feature;
3. Return a ranked list of features or the top k features according to a given
   threshold.

4   Experiment Methodology

To perform the experiment, here we have used Weka tool. It is a very powerful
open source machine learning tool, which is Java based.

For the experiment, we have used

- Wisconsin Breast Cancer dataset (WBCD),
- Wisconsin Diagnostic Breast Cancer (WDBC)
- Wisconsin Prognostic Breast Cancer (WPBC)

These datasets were created by William Wolberg, W. Nick Street, and Olvi L.
Mangasarian. WBCD contains 699 (as of 15 July, 1992) records. In this dataset,
there are ten attributes and a class attribute. The class attribute has two possible
values such as "benign" and "malignant." Class distributions in dataset are as
follows:

Class value	Instances
Benign	458
Malignant	241
Total	699

Attributes of datasets are as follows:

S. no	Attribute name	Range
1	Sample code number	Id Number
2	Clump thickness	10–Jan
4	Uniformity of cell size	10–Jan
5	Uniformity of cell shape	10–Jan

(continued)

(continued)

S. no	Attribute name	Range
6	Marginal adhesion	10–Jan
7	Single epithelial cell size	10–Jan
8	Bare nuclei	10–Jan
9	Bland chromatin	10–Jan
10	Normal nucleoli	10–Jan
11	Mitoses	10–Jan
12	Class	2–benign 4–malignant

Wisconsin Diagnostic Breast Cancer (WDBC) contains 569 instances and 32 attributes such as ID, diagnosis, and 30 real-valued input features. Diagnosis attribute has two classes such as M = malignant, B = benign.

Here, we have omitted the first attribute's sample code number as it is a unique value. Attributes from serial number 2 to 11 are used and the last attribute class.

Ten real-valued features are computed for each cell nucleus:

1. radius (mean of distances from center to points on the perimeter)
2. texture (standard deviation of gray scale values)
3. perimeter
4. area
5. smoothness (local variation in radius lengths)
6. compactness (perimeter^2/area − 1.0)
7. concavity (severity of concave portions of the contour)
8. concave points (number of concave portions of the contour)
9. symmetry
10. fractal dimension ("coastline approximation" − 1)

Class distribution in WDBC dataset is as follows:

Class value	Instances
Benign	357
Malignant	212
Total	569

Wisconsin Prognostic Breast Cancer (WPBC) dataset contains 198 instances and there are 34 attributes such as ID, outcome, and 32 real-valued input features. Here outcome attributes have two possible values such as R = recur, N = nonrecur. Ten real-valued features are computed for each cell nucleus:

1. radius (mean of distances from center to points on the perimeter)
2. texture (standard deviation of gray scale values)
3. perimeter

4. area
5. smoothness (local variation in radius lengths)
6. compactness (perimeter^2/area − 1.0)
7. concavity (severity of concave portions of the contour)
8. concave points (number of concave portions of the contour)
9. symmetry
10. fractal dimension ("coastline approximation" − 1)

Class distribution in WDBC dataset is as follows:

Class value	Instances
Recur	151
Nonrecur	047
Total	198

To gain a reasonable finding from feature selection methods, we have considered four machine learning algorithms; two from Bayes category and two from Trees category of learning methods.

From bayes category, we have chosen Bayes Network classifier and simple naïve Bayes classifier, and from tree category we have chosen random forest classifier and J48.

We have used the following feature selection methods for attribute selection.

Information gain	(IG)
Relief	(R)
CFS	(CFS)
Correlation attributes	(CA)

5 Experiment Results

Performance of classification algorithms on WBCD, WDBC, and WPBC without applying feature selection method is as the following table:

FS method	WBCD	WDBC	WPBC
Simple NB	95.99	92.61	67.17
Bayes Net	97.14	95.25	74.75
RF	96.85	95.96	60.81
J48	94.56	92.97	73.74

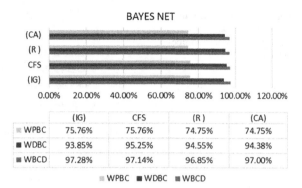

Fig. 1 Results for feature selection methods with Bayes network classifier

	(IG)	CFS	(R)	(CA)
WPBC	75.76%	75.76%	74.75%	74.75%
WDBC	93.85%	95.25%	94.55%	94.38%
WBCD	97.28%	97.14%	96.85%	97.00%

Now we have applied four feature selection methods on three breast cancer dataset and analysed the performance of classification algorithm. Bayes Net classification accuracy on WBCD, without applying any classification is 97.14, which turns to 97.28 after applying information gain (IG), increases by 0.14. In case of CFS, there is no change while R and CA give little results. The same process on WDBC, gives a better result by applying CFS and in case of WPBC, without applying FS its accuracy is 74.75, and IG and CFS give best results as 75.76 and shows an increase by around 1.00. Figure 1 shows these result analyses.

Classification accuracy of simple naïve Bayes on WBCD, is 95.99, which turns in 96.42 after applying IG, R, and CA. It shows an increase by 0.43. The same process on WDBC, gives a very good result by applying CFS with an increase by 1.94 % and in case of WPBC, without applying FS its accuracy is 67.17, and IG and CFS give best results as 74.24 and 75.25, respectively. Figure 2 shows these result analyses.

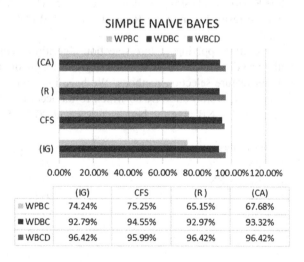

Fig. 2 Results for feature selection methods with simple naïve Bayes classifier

	(IG)	CFS	(R)	(CA)
WPBC	74.24%	75.25%	65.15%	67.68%
WDBC	92.79%	94.55%	92.97%	93.32%
WBCD	96.42%	95.99%	96.42%	96.42%

Fig. 3 Results for feature
selection methods with
random forest decision tree
classifier

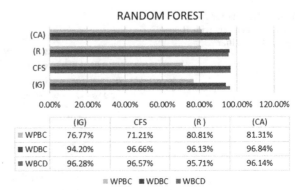

	(IG)	CFS	(R)	(CA)
WPBC	76.77%	71.21%	80.81%	81.31%
WDBC	94.20%	96.66%	96.13%	96.84%
WBCD	96.28%	96.57%	95.71%	96.14%

Fig. 4 Results for feature
selection methods with
decision tree J48

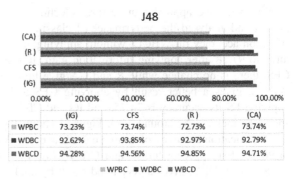

	(IG)	CFS	(R)	(CA)
WPBC	73.23%	73.74%	72.73%	73.74%
WDBC	92.62%	93.85%	92.97%	92.79%
WBCD	94.28%	94.56%	94.85%	94.71%

Classification accuracy of random forest on WBCD, is 96.85, which gives a decreasing result r applying all FS methods such as IG, CFS, R, and CA (Fig. 3).

Classification accuracy of J48 on WBCD is 94.56. CFS gives the same accuracy, while R performs best by an accuracy of 94.85. The same process on WDBC, gives the same result by applying R and in case of WPBC, CFS gives the same result (Fig. 4).

Here after applying feature selection algorithms, classification does not result with great difference. So we decided to go for fusion of algorithms on three datasets without feature selection methods. We have implemented various fusions of classification algorithm on WBCD, which result according to the following table.

Here J48 + NB give the best accuracy with **97.00 %**.

Fusion classifier	Accuracy
J48 + BN	94.85
J48 + KNN + BN	95.31
J48 + RF	95.72
J48 + NB	97.00
J48	94.28

FUSION Classifier	ACCURACY
J48 + BN	94.85
J48+KNN+BN	95.31
J48+RF	95.72
J48+NB	97.00
J48	94.28

ACCURACY

6 Conclusion

Here we have applied FS and classification algorithm on three breast cancer datasets such as WBCD, WDBC, and WPBC of UCI Repository. According to the result, it seems that no single classification algorithm performs best in terms of classification accuracy. While applying fusion classifiers, intermingling of J48 and naïve Bayes gives the best result of 97.00 (without applying feature selection). Simple naïve Bayes and j48 alone give different results. While applying feature selection on three datasets, giving different results on all three datasets. It seems that IG performs best on WBCD while IG and CFS perform better on WPBC. CA performs well on WDBC. Here, we cannot ascertain any single feature selection algorithm for best performance as it depends on the type and attributes of the dataset. Future work should identify best fusion on other two datasets and applying feature selection and fusion classifier on breast cancer datasets.

Acknowledgments Wisconsin Breast Cancer dataset (WBCD), Wisconsin Diagnostic Breast Cancer (WDBC), and Wisconsin Prognostic Breast Cancer (WPBC) databases were obtained from the University of Wisconsin Hospitals, Madison from Dr. William H. Wolberg. We have used these datasets for further research and we are thankful to them for providing datasets.

References

1. Han, J., K. M, *Data mining concepts and techniques*, 3rd edn.
2. Fayyad U., Piatetsky-Shapiro, G., & Smyth, P. (1996). From data mining a knowledge discovery in databases, american association for artificial intelligence.
3. Naseriparsa, M., Bidgoli, A.-M., & Varaee, T. (2013). A hybrid feature selection method to improve performance of a group of classification algorithms. *International Journal of Computer Applications, 69*(17), 0975–8887.
4. National Breast Cancer Society. http://www.nationalbreastcancer.org/breast-cancer-facts.
5. Breast Cancer Facts and Figures 2013–2014. American Cancer Society.
6. Cancer Information. http://www.cancer.org/research/cancerfactsstatistics/breast-cancer-facts-figures.
7. Ashraf, M., Chetty, G., & Tran, D. Feature selection techniques on thyroid, hepatitis, and breast cancer datasets.
8. Hall, M. A., & Holmes, G. (2003). Benchmarking attribute selection techniques for discrete class data mining. *IEEE Transactions on Knowledge and Data Engineering, 15*.
9. M. Ashraf, Le, K., & Huang, X. (2010). Information gain and adaptive neuro-fuzzy inference system for breast cancer diagnoses. Presented At the International Conference on Computer Sciences and Convergence Information Technology (ICCIT), Seoul.
10. Azhagusundari, B. & Thanamani, A. S. (2013). Feature selection based on information gain. *International Journal of Innovative Technology and Exploring Engineering (IJITEE), 2*(2), ISSN: 2278–3075.
11. Aloraini, A. (2012). Different machine learning algorithm for breast cancer dataset. *International Journal of Artificial Intelligence and Applications (IJAIA), 3*(6).
12. Kim, Y. S., Nick Street, W., & Menczer, F. In *Feature Selection in Data Mining*. University of Iowa, USA.
13. Ladha, L., & Deepa, T. (2011). Feature selection methods and algorithms. *International Journal on Computer Science and Engineering, 3*, 1787–1790.
14. Kohavi, R., & John, G.H. (1997). Wrappers for feature subset selection. *Artificial Intelligence 97*, 273–324.
15. Kononenko, I. (1994). Estimating attributes: analysis and extensions of RELIEF. In *Machine Learning ECML-94*.

ACE (Advanced Compression Encryption) Scheme for Image Authentication

Bhavna Sangwan and Sakshi Bhatia

Abstract In ancient times, there was a need to protect data. Due to the pervasive range in the field of security, several technologies came into existence. But the whole idea of securing data from attacks took a quick turn in the world of Internet. Here, data accuracy is the major issue. As in the previous existing methods, compression of data becomes an important research field that provides an efficient way for compressed data transmission over the Internet. The technique proposed here publicizes the concept of reliability and transmission mechanism. This is a faster and simple method of compression, and describes different lossless image compression algorithms for high performance. The newly adapted method produces better image quality without any loss of data. The suggested approach highlights security of data and provides a practical approach for compression of an image. The proposed ACE scheme works on binary concepts of compression of the image and encryption by reducing number of bits per pixel.

Keywords ACE scheme · Compression · Encryption · Image

1 Introduction

Image compression plays an important role in data compression. The main focus is to reduce the redundancy of the image data in order to store or transmit data in an effective form. So due to these factors image compression becomes an important issue of study for researchers. If the coalition between one pixel and its neighbor pixels is high, then it can compress an image easily or the values of one pixel and its next pixels are similar. It reduces the total number of bytes of an original image

B. Sangwan (✉) · S. Bhatia
Department of Computer Engineering, The Technological Institute
of Textile & Sciences, Bhiwani, India
e-mail: erbhavna26@gmail.com

S. Bhatia
e-mail: sakshi.bhatia@gmail.com

© Springer Science+Business Media Singapore 2016
S.C. Satapathy et al. (eds.), *Proceedings of International Conference on ICT for Sustainable Development*, Advances in Intelligent Systems and Computing 408, DOI 10.1007/978-981-10-0129-1_24

without any distortion in its quality; it will take less time and less hard disk space to send data from one place to another. Already existing standards for image compression and decompression are also beneficial to achieve goals like PNG and GIF. The most extensively used form of lossy image compression is JPEG. JPEG which is developed by the Joint Photographic Experts Group is compatible with both RGB and grayscale image format (GIF). Now, the whole frame is first divided into number of blocks of 16 × 16 pixels each because the compression algorithm runs on a block size of 16 × 16 pixels.

Compression is divided into two categories

1. Lossy Image Compression
2. Lossless Image Compression

In lossy compression, some data are discarded during compression and cannot be recovered. Lossy technique can be further classified in the following categories

  i. Predictive coding
 ii. Transform coding (FT/DCT/Wavelets).

Lossy technique has much greater compression ratio than lossless technique. But it also distorts the image quality. In lossy compression schemes, quantization is the main process behind compression. It acts as a control knob for trading off image quality for bit rate. It is the primary way for data reduction in images. This step of quantization can be reconstructed but its contradiction is not possible. Methods like JPEG are usually an acceptable standard and have a good quality outcome (Fig. 1).

Lossless compression technique is one of the most suitable image compression techniques, where no information regarding the image is lost. There are more extensively used areas in compression like Medicative analysis, military, remote observer, multimedia, radar, etc. This lossless technique is further classified into following categories

   i. Run length encoding
  ii. Huffman encoding
 iii. Arithmetic encoding
  iv. Entropy coding

Fig. 1 Lossless and lossy compression

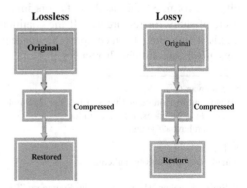

Fig. 2 encryption and
decryption

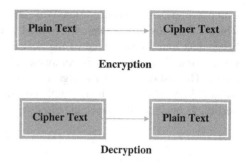

The demonstrated ACE approach works on the basis of lossless compression and decompression algorithms. Main focus of this paper is to compress images by reducing number of bits per pixel required to represent it and to decrease the transmission time for transmission of images and then reconstructing back.

Encryption

It is the process in which clear text gets transformed into cipher text for security purposes. Decryption is just the reverse of encryption process which in turn provides the original text (Fig. 2).

According to ACE image compression scheme, it compresses the image by dividing it into various categories. Each category contains uniform number of pixel values. Its supreme benefit is that for every scanned pixel of the same category, there is no need to represent category value again and again. Only pixel value is enough to define the position of the image pixel. In this way, total number of bits available in the image gets compressed and well-preserved.

2 Literature Survey

Krishan Gupta, Mukesh Sharma, Pallavi Sharma [1] suggested a lossless compression-based Kmp technique. The need for an efficient technique for compression of images is ever increasing because the original images need large amounts of disk space, which seems to be a big disadvantage during transmission and storage. Gupta et al. [2] suggested three different KG versions for image compression. Even though there are so many compression techniques already present, a better technique which is faster, memory efficient, and simple which surely suits the requirements of the user is needed. Gupta and Sharma [3] suggested a RGB-based KMB image compression technique. The technique used here, is more helpful in reducing the bandwidth of an image and to speedup of its availability, reliability, and transmission rates. Gupta et al. [4] suggested a new lossless KMK technique for image compression. Due to the increasing demands of various compression techniques, there is also a huge need of compression of an image. Compression of an image in data storage and transformation is very important and valuable, for this purpose researchers are always looking for different techniques to resolve this issue. Sharma [5, 6], "Image

Compression Using Enhanced Haar Wavelet Technique," describes data compression which can be lossy or lossless is required to decrease the storage requirement and better data transfer rate. One of the best image compression techniques is using wavelet transform [7, 8]. It is comparatively new and has many advantages over others. The rest of the paper is organized as follows: Sect. 3 describes the proposed methodology (ACE for image authentication). Section 4 analyses the results. Section 5 concludes the paper and discusses future work, respectively [9, 10].

3 Proposed Methodology

According to ACE image compression scheme, first divide the image into categories of equal size. Each category of image contains pixels. And each category size and category number will depend upon the number of pixels in which we want to represent. As pixels of an image lie in the ranges 0–255. So in our technique 16 different categories are made and each category contains 16 pixels like in first category first 16 pixels are located, i.e., 0–15 and so on. Thus, to represent each category value and pixel value five bits are used. First five bits are used for category representation and next five bits are used for pixel representation. But every time there is no need of category value for identifying the category number as necessary; one time means the same category pixel needs only one pixel value.

For example, first of all, category values come, then its all pixels, every time the pixel value is needed but if and only if the pixel of different category comes then there is need become to take category value. Hence, ACE technique takes first of all category value then always pixel value to continue for all pixels of same category whenever next category pixels not come. When next category pixels come, then again take category value than always pixel value to continue for all pixels of the same group whenever next category pixel does not come.

ACE Compression Algorithm

1. First scan the image for compression.
2. Then divide the whole image into different number of categories, whose range lies between 0 and 15 in binary format.
3. Each category contains different number of pixels, whose range lies between 16 and 31. The first pixel value of first category is the same as the first pixel value of all other categories of an image.
4. Similarly, all pixel values of one category are having the same pixel values of other category at the same index value.
5. If the scanned pixel value is of the same category then calculate only its pixel value.
6. Else, the scanned pixel value belongs to different category then calculate its both categories as well as pixel value.
7. Traverse all the pixels of an image until it reaches the end of the pixels of every category.
8. Get the compressed image (Fig. 3).

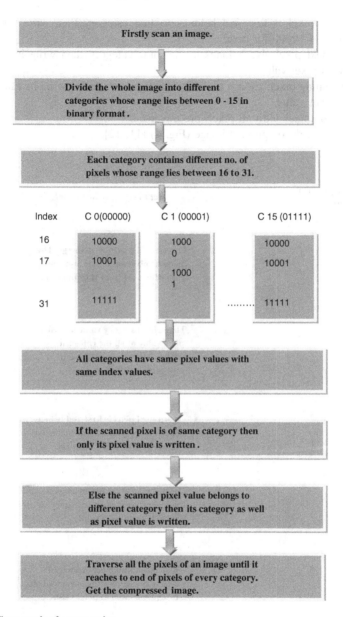

Fig. 3 Flow graph of compression

ACE Decompression Algorithm

1. Select the compressed image.
2. Traverse all pixels of an image. Then, check its pixel value and category value.
3. Find category value as well as pixel value of all the categories whether they belong to the same category or different.

4. If scanned pixel is first and belongs to first category then write 00000 for category and 10000 for pixel.
5. Then, if next pixel also belongs to the same category then write its category value with pixel value.
6. Else upcoming pixel belongs to different category then also write its category as well as pixel value.
7. Process continues until all pixels are traversed.
8. Then retrieve decompressed image (Fig. 4) [11, 12].

Fig. 4 Flow graph of decompression

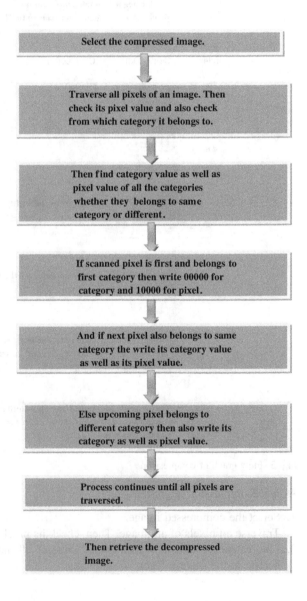

Table 1 Flow graph of decompression

No. of different images	Simple required no. of bits (bytes)	ACE scheme required no. of bits (byte)
640 kB	655,360	307,200
168 kB	172,032	51,348
2.25 MB	2,097,402	33,792
3.38 kB	3452	1864
244 kB	249,856	174,080
150 kB	153,600	112,640

4 Result

According to ACE (advanced compression encryption) cheme for image authentication, the compression is totally based on category values and pixel values. Here, we need not represent category values of pixels again and again. So, this approach compresses the enormous amount of data using an efficient and innovative method. It reduces the number of bits after transforming the pixel and category values into least binary values. So it can also be applicable to perform encryption by converting pixels into binary and reduction of bits. This approach is easily understandable using clustering mechanism. It follows the divide and conquer strategy. Hence, using ACE technique, whole image will be compressed, mostly from 40 to 70 % easily. After analysing this algorithm, average image compression is found 50 % in this case for 10 different images. This scheme is implemented on MATLAB 2011b (Table 1).

5 Conclusion

ACE scheme is a diminutive algorithm that is used to represent data in less number of bits as compared to other predefined methods available. The proposed technique sends the data at faster rate and also it consumes less space in memory. It is cost efficient and provides better authenticity. This approach is equivalent to pixel values and category values. After analysis, we have come to a conclusion that this scheme provides a powerful way to represent each pixel in five bits. There is no need to represent its category value if it belongs to the same category. Each category consists of 16 pixels. The capability of the presented algorithm is excellent for large-sized images. These images take more time to compress but show more compression ratio without causing any distortion as compared to small images. The described scheme provides better encoding without any loss of data. It will not only reduce the storage requirements but also overall execution time and is more beneficial for super computers too. As for the future studies, it might be an interesting topic to consider this method in lossless image compression algorithms.

References

1. Gupta, K., Sharma, M., & Sharma, P. (2014). Lossless compression based Kmp technique. In *Optimization, Reliabilty, and Information Technology (ICROIT), 2014 International Conference.*
2. Gupta, K., Sharma, M., & Baweja, N. (2014). Three different KG version for image compression. In *Issues and Challenges in Intelligent Computing Techniques (ICICT), 2014 International Conference.*
3. Gupta, K., & Sharma, M. (2014). RGB based KMB image compression technique. In *Optimization, Reliabilty, and Information Technology (ICROIT), 2014 International Conference.*
4. Gupta, K., Sharma, M., & Saneja, K. (2014). A new lossless KMK technique for image compression. In *Optimization, Reliabilty, and Information Technology (ICROIT), 2014 International Conference.*
5. Sharma, M. (2012). Image compression using enhanced haar wavelet technique. In *International Conference 2012.*
6. Kaur, R., & Banga, V. K. (2012). Image security using encryption based algorithm. In *International Conference on Trends in Electrical, Electronics and Power Engineering (ICTEEP 2012)*, Singapore, July 15–16, 2012.
7. Netravali, A. N., & Haskell, B. G. (1995). *Digital pictures—representation, compression, and standards*, 2nd ed. London: Plenum Press.
8. Bilgin, A., & Marcellin, B. W. (2003). JPEG2000: Highly scalable image compression. In K. Sayood, (Ed.) *Lossless Compression Handbook*, San Diego, CA: Elsevier, 2003, Chap. 18, pp. 351–369.
9. Netravali, N., & Haskell, B. G. (1995). Digital pictures—representation, compression, and standards, 2nd ed. New York: Plenum Press.
10. Olyaei, A., & Genov, R. (2005). *Mixed-signal haar wavelet compression image architecture. MIWSCAS '05.* Ohio: Cincinnati.
11. Gonzalez, R. C., & Woods, R. E. *Digital Image Processing*, 2nd ed. Pearson Education.
12. Gaikwad, D. P., Jagdale, T., Dhanokar, S., Moghe, A., & Pathak, A. Hiding the text and image message of variable size using encryption and compression algorithm in video steganography. *International Journal of Engineering Research and Application (IJERA) 1*(2), 102–108.

An Efficient and Interactive Approach for Association Rules Generation by Integrating Ontology and Filtering Technique

Rahul Divakar Jadhav and Arti Deshpande

Abstract Association rule mining identifies the correlation among the set of items provided in the database. Although Apriori, frequent pattern mining, and other algorithms are proposed in the literature for association rule generation, these are statistical methods. In such cases, mining is completely uncontrolled because once data is supplied to algorithm; it produces results according to the predetermined methodology. Many times generated rules lack user's expectations and hence need arises for methodologies with traditional algorithms. To overcome the aforesaid drawback, we propose the usage of ontology and filters along with frequent pattern tree mining algorithm for getting the desired results. Graphical structures are used for generation of ontologies. This paper thus proves and indicates the use of ontology and filters, and their proper implementations to obtain optimum and desired results through utilization of the above mentioned improved technique for data mining.

Keywords Datamining · Association rule · Ontology · Graph

1 Introduction

An association rule is defined over the implication $G \rightarrow H$. Item set G and item set H is the antecedent and the consequent of the rule, respectively, here intersection of G and H is zero [1].

R.D. Jadhav (✉)
Fr. C. Rodrigues Institute of Technology, Vashi, Navi Mumbai 400703, India
e-mail: rahulj212@gmail.com

A. Deshpande
G. H. Raisoni College of Engineering, Nagpur, India
e-mail: artideshpande75@gmail.com

© Springer Science+Business Media Singapore 2016
S.C. Satapathy et al. (eds.), *Proceedings of International Conference on ICT for Sustainable Development*, Advances in Intelligent Systems and Computing 408, DOI 10.1007/978-981-10-0129-1_25

Support Database has huge number of data items. The support on rule G to H is found as the number of times the items G and H have occurred to the total number of items in database [2].

$$\text{support } (G \rightarrow H) = \frac{\#\text{tuples\_containing\_both\_G\_and\_H}}{\#\text{total\_no\_of\_tuples}} \qquad (1)$$

Confidence The confidence on the association rule of "item G" over "item H" can be found on the basis of support as the ratio of "support of rule $G \rightarrow H$" to the "support of G".

$$\text{confidence } (G \rightarrow H) = \frac{\#\text{tuples\_containing\_both\_G\_and\_H}}{\#\text{tuples\_containing\_G}} \qquad (2)$$

Lift The ratio of "confidence of rule $G \rightarrow H$" to the "support of consequent G" is the lift value of given rule G over H.

Ontology Ontology is an inventory of things defined over a particular domain, which uses a specific language to describe the concepts of domain under consideration. Ontology is a set of concept, relationships over concepts, instance, and a Graph G. Graph is based on relations of concepts which is directed and does not have a cycle [3].

2 Description

The proposed framework is basically objective based, in the beginning only the user has to feed the knowledge and goals for which he is looking for to the system. An interactive interface is proposed to change the objectives or parameters, if the user is not satisfied with the generated results Fig. 1.

After applying FP-Growth (frequent pattern growth) algorithm on large datasets, a number of association rules get generated. Ontology is formed for the items available in database, so based on that ontology user selects the items of interest. Due to this, the other rules get filtered and only the rules containing interested items remain for further processing. Finally, pruning operator and MICF (minimum improvement constraint filter) gives the actionable rules as per the user's expectations. The experimental results given in Sect. 5 show the reduction in large number of unwanted rules generated by classical rule mining algorithm. The entire system is divided into four main phases as below:

1. Association rule mining
2. Ontologies
3. Filtering step
4. Interactive approach for association rule mining

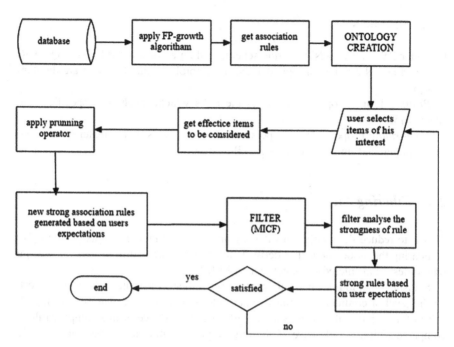

Fig. 1 Flowchart of the proposed system

2.1 Association Rule Mining Based on Frequent Pattern Tree Algorithm

Most important feature of FP-growth algorithm is that, it does not generate candidate item set unlike Apriori algorithm. A huge list of frequent data items is compressed in the form of frequent pattern tree, which preserves the association information of item set. Compressed database on each fragment of pattern forms a conditional data base, which then separately mines this database [4].

Input: the database and a threshold value of minimum support
Output: the frequent pattern tree of database.

2.2 Creating Ontology

Expressiveness of ontology is done by rule schemas RS (m1, m2 ... ($\rightarrow$) n1, n2 ...)
It combines abstraction and pruning constraints. Rule schema is defined over concept C of prepared ontology O. Mi and Ni are subsets of concept C.

2.2.1 Operators

Consider $RS_1 : (\langle M \rightarrow N \rangle)$, as rule schema and an association rule AR_1: $Q \rightarrow R$, where M and N are the concepts defined from ontology, and Q and R are the item sets [3].

Pruning Pruning operator removes the rules which match with specified rule schema [5].

Conforming Property of this operator is to retain rules which match with given rule schema, and unmatched rules are filtered out.

2.3 Filtering Step

This is to reduce the number of rules, by applying certain constraints or filters. By comparing the association rules derived and the rules derived from user knowledge ontologies, final filtered rules get generated [6].

Statistical methods such as minimum improvement constraint filter (MICF) work on the level of confidence of rules. When rule is consisting of many items on both sides of association then it is complex rule. The confidence of this complex rule is compared with the confidence of all of its simplification. Filter deletes complex rule if it does not enhance its simplification [7].

E.g. Notebook, pencil $\rightarrow$ pen (confidence = 85 %)
Notebook $\rightarrow$ Pen (confidence = 90 %)
Pencil $\rightarrow$ Pen (confidence = 83 %)
The rule Notebook, Pencil $\rightarrow$ Pen is removed

2.4 Interactive Approach for Association Rule Mining

Throughout these phases, interactive approach inculcates task analysis so that at each stage objective of association is retained. Ontology phase provides understanding and visualization of arrangement of given items and concepts. If user is not satisfied with the selection then new selection can be proposed through the process of new item selection [3].

3 Implementation

For experimental purpose we are using a data set of transactions from "supermarket. csv" generated by data generator tool [8]. This file has 4627 transactions each representing individual customer's basket, i.e., things he/she buys from supermarket in a single visit in one day. There are 30 items available for customers to buy.

3.1 Generation of Association Rules

We used Apache Mahout Framework for implementation of frequent pattern mining algorithm [9]. Mahout provides distributed computing environment to create Frequent Pattern tree. Totally, 1539 rules are generated for a given example of supermarket.

3.2 Generation of Ontology

We consider a same database of super market with 30 items and divided them into various categories as shown in Fig. 2. The "Neo4j" graph structure is used to generate relationships among various categories which show relationships with

Ontology Filter selected **Health and Body, Grocery -> Grocery, Furniture**

Set Filter

Fig. 2 Ontology of supermarket

various leaf nodes representing as product names in given database. We have used "cypher" text as proposed in Neo4j documentation for creating the graph structure. This "cypher" first creates categories and act as nodes which connect with other nodes and then the relationships are defined [10].

```
CREATE (grocery:CATEGORY {name:'Grocery'}) /*category
definition*/
```

Here "CREATE," "CATEGORY," names are keywords. With this syntax we are creating categories "grocery," "bakery," "dairy," "frozen," "meat_seafood," and there represented names are "Grocery," "Bakery," "Dairy," "Frozen," "Meat and Seafood," respectively. The items which are already present in the input database need to be defined as "ITEM" in cypher.

```
CREATE (milk:ITEM {name:'milk'})/*item declaration*/
```

The relationships between various categories are defined as,

```
(grocery)-[:FURTHER_CATEGORIED_INTO] → (bakery),
(grocery)-[:FURTHER_CATEGORIED_INTO] → (dairy),
(grocery)-[:FURTHER_CATEGORIED_INTO] → (frozen),
```

3.3 Minimum Improvement Constraint Filter (MICF)

The filtering criteria described in MICF filter definition applied on supermarket scenario shows that rules like menssprey, creames → milk, has (69.5 %) confidence whereas, menssprey → milk, has (69.7 %) and creames → milk, has (68.3 %). Then menssprey, creames → milk is filtered out.

4 Observation

4.1 It Has Been Observed from the Experiments Carried Out for Association Rule Generation Phase

- With the decrease in the support counts, the association rules generated increase, however, the final set of strong rules depend on the confidence level.
- If support level is very high then very small range of rules are available to qualify the minimum support level. It does not mean that all are optimized rules. The number of rules is very less because the range we selected is short.
- If the frequencies of items vary in a large scale, then the following problems can appear

- Rules involving rare items will not be found if minimum support level is too high
- Combinatorial explosion happens when minimum support level is too low because too many ways are available for frequent items to get associated with each other

4.2 It Has Been Observed from the Experiments Carried Out for the Ontology Generation Phase

- According to the number of items in the dataset, the ontology graph generated varies.
- According to the constraints of the various users, different users will input different rule schemas
- If user selects many items then enormous number of rules gets generated.
- When user selects root or parent item in ontology then all child items also get selected and are considered as user's selection.
- When user selects root node on both left and right sides of the rule then all items in ontology are considered as qualified and no item can be pruned for output

  - In this case output of the first module is the same as output of the second module.
  - Only filtering can be used to cut down generated rules.

4.3 It Has Been Observed from the Experiments Carried Out for the Filtering Phase

- Final output of the method will be those rules which pass through filter.
- If no rules get qualified for filtering then all rules pass through it and the rules which generated through first module will be shown.
- However, this may not always be the case, since such results also depend on many other factors such as dataset chosen, level of confidence chosen, and so forth.

5 Results

Experiment is performed on a sample dataset of supermarket. Association rules generated only with FP tree algorithm are 1539 for given data items of supermarket. If filtering technique is used then resultant rules may get filtered out but only 10 rules, i.e., 1 % reduction occurred. Graphical representation of this result is shown in Fig. 3. As per our proposed method where we have combined filtering technique with ontology, then user can select items of his interest and prune the rules which are not of user's interest.

(a)   **(b)**

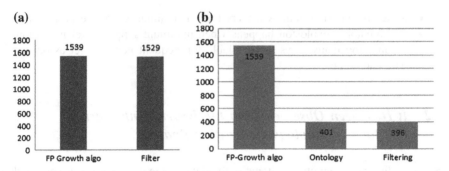

Fig. 3 Graphical results comparing number of rules generated through statistical and proposed methods. **a** Rules generated using statistical method. **b** Rules generated using proposed method

6 Conclusion

We have discussed the problem of using only statistical methods for association rule generation. Unlike statistical methods objective based method is more efficient as it filters out unwanted rules and gives the actionable rules as per the interest of user. In our framework, user has provision to monitor his goals of choosing items of his own interest through ontology phase. We have used ontology concept to integrate user objectives and filtering phase to filter out the redundant rules. So our proposed framework has combined objective-based and statistical approaches together to generate more precise and actionable rules based on user's expectations. Through experimental results, we have proved that our approach improves the performance of market basket analysis.

References

1. Imielinski, T., Swami, A., Agrawal, R. (1993). Mining Association rules between sets of items in large databases. in *ACM SIGMOD*, pp. 207–216.
2. Sulthana A. R., & Murugeswari, B. (2011). ARIPSO: Association rule interactive postmining using schemas and ontologies). In *International Conference on Emerging Trends in Electrical and Computer Technology (ICETECT)*, March 2011, pp. 941–946.
3. Guillet, F., Marinica, C. (2010). Knowledge-based interactive post mining of association rules using ontologies. In *IEEE Transactions*, June 2010, pp. 784–797.
4. Pei, J., Yin, Y., & Han, J. (2004). Mining frequent patterns without candidate generation: A frequent-pattern tree approach, 1st edn, Vol. 8. In H. Mannila (Ed.), Kluwer Academic Publishers.
5. Rahul J. (2015). Interactive approach for generation of association rules by using ontology. In *International Conference on Nascent Technologies in the Engineering*, pp. 215–217.
6. Antunes, C., & Jacinto, C. User-driven ontology learning from structured data. In *Computer and Information Science (ICIS), 2012 IEEE/ACIS 11th International Conference*, 2012, May, pp. 184–189.

7. Bayardo, R. J., Agrawal, R., Constraint-based rule mining in large, dense databases. In *Dimitrios Gunopulos Research Report, IBM Research Division*, California, p. 5.
8. Red gate SQL data generator. http://www.red-gate.com/products/sql-development/sql-data-generator/.
9. Apache mahout project. http://mahout.apache.org/.
10. Neo4j graph. http://neo4j.com.

Detection of Wormhole Attack in Wireless Sensor Networks

Jyoti Yadav and Mukesh Kumar

Abstract Although Secure routing is the demand of the hour in the application of wireless sensor networks (WSN), secure routing is a challenge in itself due to the inherently limited capacities of sensor nodes. Routing attacks are a major challenge in the process of designing of effective and robust security mechanisms for WSNs. This main focus of this paper is on wormhole routing attack. The measures which have been proposed till now to counter the routing attacks suffer from flaws that essentially prove the fact that they are not good enough for use in large-scale WSN deployments. We need light weight and hard-bearing security mechanisms due to inherent limitations found in WSNs. In this paper, we discuss a wormhole detection method: WGDD (wormhole geographic distributed detection).

Keywords Ad hoc · WSN · Wormhole

1 Introduction

Mobile ad hoc and wireless networks are now considered to be most prevalent when we talk about modern days communications.

Various types of malicious attacks on wireless sensor networks have been described very well in earlier books; they are classified as laptop class and mote class attacks, outsider and insider attacks, active and passive attacks. The secure location, network routing, also data aggregation and clustering protocols are affected by wormhole. To deal with this attack is very difficult task because of the fact that the attacker does not need to make any changes in any legitimate nodes or have access to any cryptographer keys [1, 2].

An introduction of easy and effective ways to detect and locate wormhole is given in this paper.

J. Yadav (✉) · M. Kumar
The Technological Institute of Textile and Sciences, Bhiwani, India
e-mail: Jyotiyadav1302@gmail.com

M. Kumar
e-mail: drmukeshji@titsbhiwani.ac.in

© Springer Science+Business Media Singapore 2016
S.C. Satapathy et al. (eds.), *Proceedings of International Conference on ICT for Sustainable Development*, Advances in Intelligent Systems and Computing 408, DOI 10.1007/978-981-10-0129-1_26

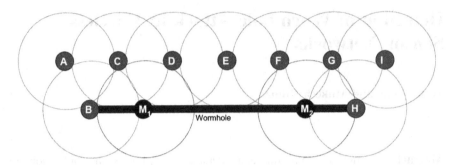

Fig. 1 Two or more malicious nodes collaborate in setting up a shortcut link between each other

2 Wormhole Attack

In wormhole attacks, a low-latency link is created between two points in the network by the attackers. This achievement can be attained by either forcing two or more sensor nodes of the network to compromise or add n entirely new set of naughty nodes to the network. At the one end of the link, the attacker collects data packets using low-latency link are sent and are replayed at the other end at the establishment of the link [3]. The base station is deceived as wormhole attack changes the network data flow. It is not easy to detect wormhole as the replayed information is usually valid (Fig. 1).

When and if the wormhole attackers get hold of link, then it can be dangerous in several ways for the network. This attack can be launched even when the cryptographic keys are absent and even giving up hold of any legal node which is a part of the network. The wormhole attack cannot be defended by cryptographical means as the wormhole attackers actually do not create influenced packets separately; simple re-transmission of packets is done, by which all storing and transmitting checks are bypassed. So wormhole attacks should be defended effectively [4, 5].

3 Related Work

1. Certain efforts have been put to defend against wormhole attacks, into signal processing techniques and hardware design. Data bits are resistant to closed wormholes, if being transferred in a special modulating way which is only the neighbor nodes know.
2. Second potential solution to wormhole detection problem is the integration of prevention methods with intrusion detection systems. Because the packets which are sent from the wormhole and the fine nodes are identical hence to isolate the attacker is difficult using the software-only approach.

3. The packets are prevented against moving farther than the transmission range for radio with the help of a technique called "packet leases." Not able to be changed and free physical metric, like-time delay or geographical location can be brought to use so as to detect wormhole attack. In order to overcome wormhole attacks, it restricts the maximum distance of transmission, with the help of either tight time synchronization or existence information.

Drawbacks

- Unfortunately, highly synchronized clocks are needed for this purpose and this is a drawback.

4. In order to be sure that the packet recipient is at a certain distance from the sender *Geographical lease* is there. The location of sending node and its time of sending are included in sending packet. The distance upper bound between the sender and node is measured by the receiving node, when the packet reaches the receiving node. To verify neighbor relation, loosely synchronized clocks and location information are used.

Drawbacks

- Since every node must be knowing its own location and every node should have loosely synchronized clocks, this is a drawback.
- Because clock synchronization demands resources; therefore, this limits the applicability of packet leases in WSNs.

5. Beacon nodes are also known as Anchor nodes. The in line idea is to take help of the discovered locations of beacon nodes which at the starting were used in discovery of location in WSNs.
   For each beacon node to be aware of its distance from another beacon nodes and also their coordinates, we use a hope counting technique. Although the straight line distances obtained by the coordinates are not affected abiding by the law of mathematics, few hop distances can be significantly lessened with a wormhole link.

6. If every node has access to nodes current location information, this results in another method, and depends on loosely synchronized clocks. Each node, this method attaches time and its location to the packet forwarded by it, and this information is secured with the help of an authentication code. The destination node of the packet is then verified by the coordinates of the nodes (i.e., it verifies if the coordinates reported are within the range of communication) and speeds. Similarly Hu proposed geographical packet leashes, and this approach is likely to work fine in cases where GPS coordinates are appropriate [6].

Drawbacks

- All the verification is done by the end node and this in itself is a disadvantage.

7. Another wormhole prevention method, somehow alike temporal packet leashes, finds roots in the time of flight of individual packets.

8. Another method called SECTOR makes uses of hardware which is specialized to enable faster transmission of one-bit challenge messages and that too with no role of CPU, so that processing delays are minimized. A distance-bounding algorithm is used in order to calculate the distance in two interacting nodes by sector.

9. Hu and Evans et al. [7] had given a solution for wormhole attacks for networks that states that every node should be having directional antennas. Nodes should be using specific sectors of their antennas while directional antennas are to be used so as to interact with one another. Hence, node which receives a packet from its neighboring node definitely has information regarding neighbor's location that in turn has knowledge of the inclination of the neighboring node in relation to itself. Information about the extra bit boosts the detection of wormhole even easier as compared to networks which have exclusive Omnidirectional antennas.

Drawbacks

- The use of directional antenna is not possible for sensor networks, thus this is a drawback.

10. For the graphical visualization of the wormhole occurrence, Wang et al. [8] proposed a method in sensor networks that are static by reconstruction of the sensor layout with the help of multidimensional scaling. Multidimensional scaling is used by MDS-VOW so as to again construct the network and the detection by visualization of the oddity introduced by the wormhole, on behalf of neighbors space from a server center.

Drawbacks

- A central controller is required by this approach, and hence this approach is not suitable for decentralized networks.

11. On the basis of the use of 'Guard' nodes that are location aware (LAGNs) Lazes et al. [9] proposed other method to prevent against wormhole attacks. The guard node is inherited by them so as to know the direction of flow of messages between nodes. A wormhole attack may be detected by a node during the fractional key distribution with the use of the property of single guard and the communication range property. It is considered that an identical message is received multiple times because an illegal entity is replaying the message due to effects of multipath.

Drawbacks

- The weakness of this system is that to know the location the guard nodes are required.

- Lazos's method is although good but, it appears to suit more to dense stationary sensor networks.

12. Another detection technique was proposed by N. Song et al. It is an easy statistical analysis (SAM)-based scheme. Main concern is the comparative frequency of every link appearing in the group of the assumed routes. The difference is measured between the most frequently appearing link and the second most frequently appearing link from the set of the obtained routes. Values are much higher in the presence of wormhole as compared to wormhole less network [10].

4 Proposed Work

This geographically distributed wormhole detection algorithm takes help of counting the hop method as a detection procedure. After we are finished with running detection method, all nodes in the network make a set of hop counts of the neighboring nodes within a range of 1 or k hops from itself. In the very next step, any suitable algorithm is run for the calculation of the shortest path for all node pairs, and a map using multidimensional scaling is obtained. At last diameter is brought to use so as to detect wormholes by being able to point out differences in the maps.

Probe procedure In order to identify wormhole, a probe procedure is used which fills the network with bootstrap node messages so as to enable every node in the network to measure the hop value in between them and the bootstrap node.

Bootstrap Node a probe message is created by this node x with $(i = idx)$ to completely fill the network. In the second step, all probe messages whose origin is node are dropped. The coordinate of the hop is

$$\text{hop\_}x = 0 \text{ and offset\_}x = 0.$$

Other Nodes hop value calculated by node a, the neighboring node is given name b; min number of hops to reach bootstrap node from a is termed as hop_a and it is initialized as MAXINT. The hop coordinate for node a is a joining of hop_a and offset_a. The node set that can be reached from a in single hops is Na, and no. of nodes in $N_a = |N_a|$.

$$\text{offset}_a = \frac{\Sigma_{b \in N_a}(\text{hop}_b - (\text{hop}_a - 1)) + 1}{2(N_a + 1)}$$

Local map computation procedure a map is obtained by all nodes for their neighbors on the basis of hop coordinates calculated earlier. When the hop coordinates have been obtained, every node makes a request to its neighboring nodes that are in the range of 1 or k hops so as to get the hop coordinates. Shortest paths between the node pairs are computed by each node that are 1 or k hops distant from each other using any algorithm like Dijkstra's algorithm. After this MDS on the matrix is applied to get the first two or more large eigenvalues and eigenvectors so as a 2D (or 3D) local map can be obtained. A computational and a memory cost is attached with all nodes.

Detection Procedure Use of the map is in the wormhole detection. when the calculations of the nodes are over for the local circle nodes, it gives the local map.

On behalf of earlier observation, diameter is used as a detection factor of wormholes. we calculate diameter d for a node a as

$$\text{distance} = \text{maximum } (\text{distance}(b, c))/2,$$

where $b, c \in Na$
N_a set of neighboring nodes of a,

   In 2D,

$$\text{distance}(a, b) = \text{sqrt}((x - x_0)2 + (y - y_0)2)$$

where (x, y) and (x_0, y_0) = node coordinates of a and b, respectively.

we in an experiment used 1000 nodes in a uniform placement with placement error $\pm 0.5r$, where width of a small square is $r = 2$ m. since wormhole creates shortcut, the computed map for node's neighboring area is distorted; hence the diameter of resulting map shall be greater comparatively, $2d = 49$ m.

So as to make sure if diameter feature is convincing in the detection of wormhole in entire network, diameter is calculated for every node in the very same 1000-node network in the absence and presence of wormhole. Computed diameter for every node is shown in Fig. 2. The red circles represent the end locations of these two wormholes. It is visible that inspite of the closeness of two wormholes, the diameter peaks still appear in the nodes that are close to the wormhole ends, from calculations, four peak values are 22.2, 22.6, 24.8, and 25.2 m.

Hence, by computing the d, algorithm can run in isolation for every nodes; in addition, local map for the neighboring area is also obtained.

Results: So as to check for the correctness of attack detection against multiple values, the two proposed measures are:

FDR (False Detection Rate): The frequency of false recognition of the characteristics which are identical to be treated as different ones by the detection system, thus resulting in failure is termed as FDR, a normal localization error is an example of FDR.

Null FDR value signifies no-false alarms in case of wormhole detection.

FTR (False Toleration Rate): The frequency of false recognition of the characteristics which are different to be treated as identical ones by the detection system, thus resulting in failure.

Null value of FTR signifies success in detection.

For the above-mentioned experimental data,

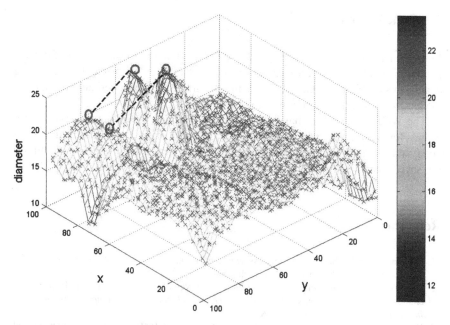

Fig. 2 Diameter measurement with two wormholes

In case of a long wormhole like >3 hops,

$$\text{detection rate} = 100\%$$

In case of shorter wormholes where hops <3,

$$\text{detection rate} = 80\%$$

5   Comparative Analysis as Compared to Other Wormhole Detection Methods

This detection algorithm has advantages over other wormhole detection algorithms because

1. Does not require anchor nodes.
2. Does not require special hardware.
3. Does not require the manual setup of networks.
4. Quickly finds the wormhole locations.
5. Since the algorithm is of distributed type, every node can surely identify the alterations made by a wormhole, which enhances the probability of wormhole detection.

6 Conclusions

In particular, we gave a look over the wormhole attack up to some level of details. Although number of various countermeasures had been proposed for defending against this attack, almost all of these suffer from flaws which essentially leave them ineffective in large-scale deployments of WSN.

The WGDD algorithm makes use of a hop counting technique as a probe procedure for wormholes, with the help of MDS local maps are constructed at each node, and for distortion detection diameter is used. Hence the proposed method has got several advancements over previous techniques.

References

1. Rehana, J. (2009). Security of wireless sensor network. *Seminar on inter networking* (TKK T-110.5190). Helsinki University of Technology.
2. Padmavathi, G., & Shanmugapriya, D. (2009). A survey of attacks, security mechanisms and challenges in wireless sensor networks. *International Journal of Computer Science and Information Security (IJCSIS), 4*(1 & 2).
3. Hanapi, Z. M., Ismail, M., Jumari, K., & Mahdavi, M. (2009). Dynamic window secured implicit geographic forwarding routing for wireless sensor network. *World Academy of Science, Engineering and Technology, 3*.
4. Deng, H., Li, W., & Agrawal, D. P. (2002). Routing security in wireless ad hoc networks. *IEEE Communications Magazine*, 70–75.
5. Karlof, C., & Wagner, D. (2003). Secure routing in wireless sensor networks: attacks and countermeasures. *Ad Hoc Networks Journal: Special Issue on Sensor Network Applications and Protocols, 1*, 293–315 (Elsevier Publications).
6. Kavitha, T., & Sridharan, D. (2010). Security vulnerabilities in wireless sensor networks: a survey. *Journal of Information Assurance and Security, 5*, 31–44 (Lukman Sharif and Munir Ahmed 183).
7. Naeem, T., & Loo, K. K. (2009). Common security issues and challenges in wireless sensor networks and IEEE 802.11 wireless mesh networks. *International Journal of Digital Content Technology and its Applications, 3*(1), 89–90.
8. Lee, J. C., Leung, V. C. M., Wong, K. H., Cao, J., & Chan, H. C. B. (2007). Key management issues in wireless sensor networks: Current proposals and future developments. *IEEE Wireless Communications*, 76–84.
9. Bojkovic, Z. S., Bakmaz, B. M., & Bakmaz, M. R. (2008). Security issues in wireless sensor networks. *International Journal of Communications, 2*(1).
10. Raj, P. N. & Swadas, P. B. (2009). DPRAODV: A dynamic learning system against blackhole attack in AODV based MANET. *International Journal of Computer Science Issues (IJCSI), 2*.

Comparative Analysis of MCML Compressor with and Without Concept of Sleep Transistor

Keerti Vyas, Ginni Jain, Vijendra K. Maurya and Anu Mehra

Abstract In this paper, an analysis of different compressors is done with and without the concept of sleep transistor. In VLSI design, compressor is an important part of multiplier as these are used to reduce the space equipped by partial products as the maximum space in a multiplication process is taken by partial products. The compressor architecture given are exclusively based on combination of three-level MCML gate with the concept of sleep transistor as these are generally suitable for improvement in speed, power consumption, and area the concept of sleep transistor is useful in reducing the leakage current. 16 nm, CMOS technology were used for designing using Tanner EDA 14.1 version of these compressor this results in up to 40 % reduction of power consumption.

Keywords MCML · Sleep transistor · Compressor · Leakage current

1 Introduction

In many digital signal processors and many microprocessors, multiplication is a key operation and there are three steps basically. Using Booth encoding partial products are generated first and then by using Wallace tree is formed using compressor

K. Vyas (✉) · G. Jain · V.K. Maurya
ECE Department, Geetanjali Institute of Technical Studies, Udaipur, Rajasthan, India
e-mail: kkvyas18@gmail.com

G. Jain
e-mail: jain24.ginni@gmail.com

V.K. Maurya
e-mail: maurya.vijendra@gmail.com

A. Mehra
Amity University, Noida, Uttar Pradesh, India
e-mail: amehra@amity.edu

© Springer Science+Business Media Singapore 2016
S.C. Satapathy et al. (eds.), *Proceedings of International Conference on ICT for Sustainable Development*, Advances in Intelligent Systems and Computing 408, DOI 10.1007/978-981-10-0129-1_27

251

circuits number of partial products is reduced. At the end for summation of two rows is done using two input adder.

Most critical operation in multiplication is partial product reduction as the area and performance of the multiplier is mainly influenced by it. As the three rows of partial product is converted into two by compressor circuit name 3 to 2 compressor is given. Alternatively, the concept of sleep transistor can be understood as a PMOS or NMOS transistor with elevated threshold voltage that connects a permanent power supply to "virtual power supply." Transistor switching is controlled by power management unit [1]. Here several high-order compressors with sleep transistor concept, specifically 4 to 2, 5 to 2, and 7 to 2 compressors, have been designed in an attempt to reduce leakage current and speed up compression operation.

In a previous work, different architectures up to 7 to 2 compressor was done in MCML [2]. The CMOS complimentary implementation of compressor architecture was basically done then these designs were compared with compressor architectures that are designed using three-level MCML gate. Results of this study show that compressor designed using these three input MCML gates designed are better than other implementation in terms of speed, power consumption, and silicon area. Figures 1, 2 shows the earlier proposed three input XOR gate and carry generate circuit using those the architecture was designed.

Also as power reduction is an important part of any designing, a MCML technology was presented earlier that reduces the power consumption. Sleep mode and active mode are two operating modes of that MCML technology. In series with supply voltage (VDD), a sleep transistor is inserted in the sleep mode; the transistor is disconnected from VDD to reduce the power consumption and in active mode regular operation is performed as the transistor is connected to power supply. This will result in minimization of leakage current [3] (Fig. 3).

Fig. 1 Three input XOR gate without sleep transistor [2]

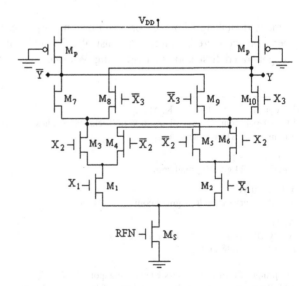

Fig. 2 Carry generate circuit without sleep transistor [2]

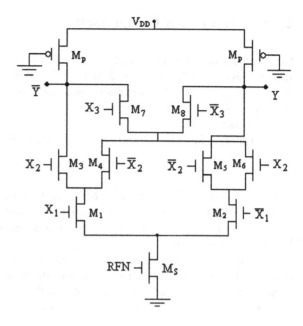

Fig. 3 Earlier proposed MCML circuit with sleep transistor [3]

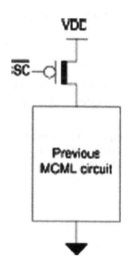

Here we have combined the concept of sleep transistor with this compressor architecture, proposed early as a result along with increased speed and reduced silicon area, power consumption, and leakage current is also minimized. In Sect. 2, the compressor architectures used are described, results of simulation are presented in Sect. 3 and conclusion is given in Sect. 4.

2 Architectures of Compressors

Same architecture of compressor as proposed in [2] is designed with and without
sleep transistor. 3 to 2, 4 to 2, 5 to 2, and 7 to 2 compressor architectures are shown
in this paper and power consumption results are also compared for all in case of
with and without sleep transistor.

3 to 2 compressor:

It is simply a full adder. Equation used for designing is

$$X_1 + X_2 + X_3 = \text{Sum} + 2 \text{ carry.} \tag{1}$$

Realization of 3 to 2 compressor in MCML can be done using just two three input
circuits which are XOR and circuit for carry generation is given in Figs. 4 and 5,
respectively. A sleep transistor is also added in design which leads to decrease in
leakage current (Fig. 6).

In comparison to earlier presented compressor architectures in [2], this architecture is better in power consumption and delay as in that the best possible
architectures of all architectures was presented. We just combined them with sleep
transistor and improved in terms of performance.

4 to 2 compressors

Equation used for designing is as shown (Fig. 7):

$$X_1 + X_2 + X_3 + X_4 + \text{Cin} = \text{Sum} + 2(\text{Carry} + \text{Cout}) \tag{2}$$

Four inputs X_1, X_2, X_3, X_4 and sum-weighted same, while C_{OUT} and carry are
higher by one bit order. Another architecture of compressor which is without sleep
transistor is also simulated.

Fig. 4 Three input XOR gate
used for compressor
designing

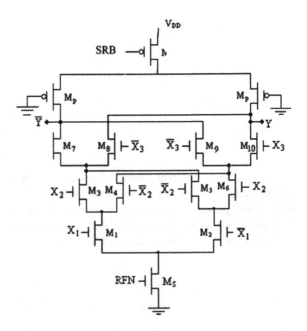

Fig. 5 MCML carry
generation circuit (CGEN)
used for compressor
designing

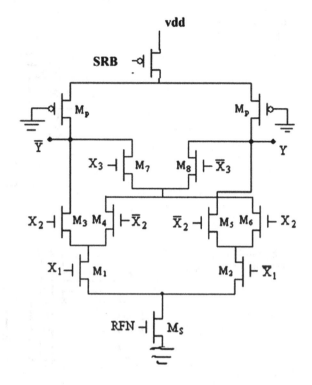

Fig. 6 Architecture of 3 to 2
compressor

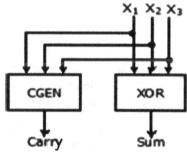

5 to 2 Compressor:

There are five inputs X_1, X_2, X_3, X_4, and X_5 and carry and sum are outputs. Additionally, there are two more inputs these are carry bits Cin_1 and Cin_2 are two carry bits output $Cout_1$ and $Cout_2$. The equation used for designing:

$$X_5 + X_4 + X_3 + X_2 + X_1 + Cin_2 + Cin_1 = Sum + 2(Carry + Cout_1 + Cout_2) \quad (3)$$

For this we have implemented two designs. One is designed with both sleep transistor and sleep transistor is shown in figure. The earlier design is described in [2] (Fig. 8).

Fig. 7 Architecture of 4 to 2
compressor

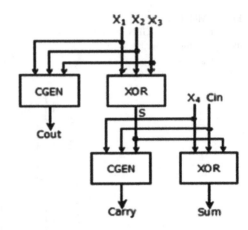

Fig. 8 5 to 2 compressor
architecture 7 to 2 compressor

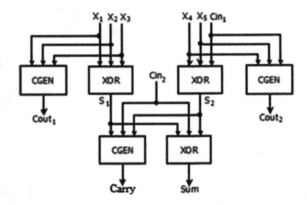

Seven inputs are there X_1, X_2, X_3, X_4, X_5, X_6, and X_7 outputs are carry and sum. Additionally, two input carry bits are there Cin_1 and Cin_2 and two carry bits at output $Cout_1$ and $Cout_2$. Expression used for designing is (Fig. 9):

$$X_4 + X_3 + X_1 + X_2 + X_5 + X_6 + X_7 + Cin_1 + Cin_2$$
$$= Sum + 2x(Carry + Cout_1) + 4Cout_2 \tag{4}$$

In this also the $Cout_1$ and carry are higher by one bit order in weight than X inputs and sum, while $Cout_2$ is 2 bit orders higher. Here we also compare the same architecture proposed in [4] and with sleep transistor as given in this paper.

3 Results of Simulation

In MCML gates, transistors were sized such that compressor is simulated in Tanner EDA 14.1 version using CMOS 16 nm technology. Tanner EDA belongs to Tanner Research. It provides tools for designing of schematic and simulation of

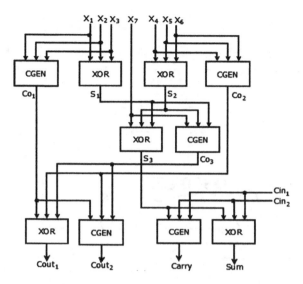

Fig. 9 7 to 2 compressor architecture

circuits, waveform analysis and layout editing, extraction of net list, and design rule
check verification. The tables and waveforms display the changes obtained after
insertion of sleep transistor (Waveforms 1, 2).

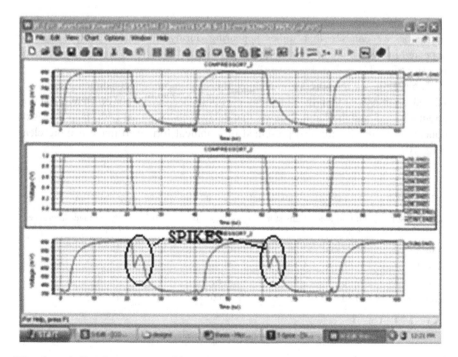

Waveform 1 7 to 2 compressor without sleep transistor (with spikes)

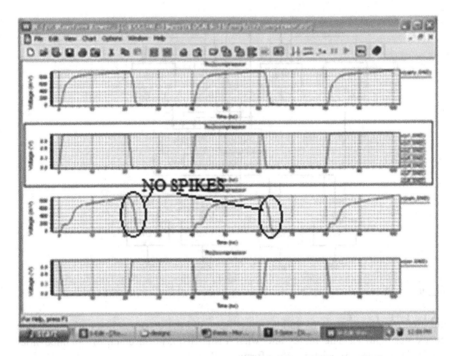

Waveform 2 7 to 2 compressor architecture with sleep transistor (without spikes)

From the above given waveforms of 7 to 2 compressor with and without sleep transistor, we can see that spikes appearing during off state of output did not appeared in case of architecture with sleep transistor this shows that leakage current is also reduced.

Table 1 shows that the same architecture of compressor consumes less power with sleep transistor than the compressor without sleep transistor on an average it is about 47.35 % and the Table 2 shows that delay is also reduced with sleep transistor concept. The results are also presented in graphical format to make it more clear (Fig. 10).

Table 1 Power consumption of same architecture with and without sleep transistor

Compressor	Without sleep transistor (μW)	With sleep transistor (μW)	Percentage difference (%)
3 to 2	86.97	47.26	45.65
4 to 2	173.58	90.47	47.88
5 to 2	259.62	136.67	47.36
7 to 2	428.86	220.66	48.54

Table 2 Delay of same architecture with and without sleep transistor

Compressor	Without sleep transistor (ns)	With sleep transistor (ns)
3 to 2	20.44	20.43
4 to 2	24.27	22.42
5 to 2	31.059	29.838
7 to 2	25.672	22.107

Fig. 10 Graph showing the power consumption comparison

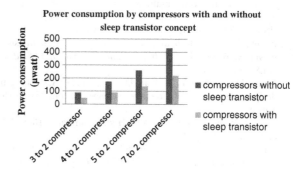

4 Conclusion and Future Scope

This paper presents analysis of different compressors architectures with and without sleep transistor. From this analysis, we came to conclusion that power consumption done by same architecture with sleep transistor is less compared to without sleep transistors on average by 47.35 %. Also the leakage current reduces in case of architecture with sleep transistor. We can see that the waveform of 7 to 2 compressor is shown in paper. While many key results for compressor designing have already been obtained, there are many issues that remain to be addressed. An important issue is how to further decrease the power consumption and delay. Future work may also involve partitioning of large circuits and addition of concept of sleep transistor in them.

References

1. Shi, K., & Howard, D. (2006). Sleep transistor design and implementation—simple concepts yet challenges to be optimum. 1-4244-0180-1/06/$20.OO ©C2006, IEEE.
2. Caruso, G., & Sclafani, D. D. (2010). Analysis of compressor architectures in mos current—mode logic. 978-1-4244-8 157-6/ 1 0/$26.00 ©2010, IEEE.
3. Kim, J. B. (2009). Low-power MCML circuit with sleep-transistor. 978-1-4244-3870-9/09/ $25.00 ©2009, IEEE.

4. Henzler, S., Georgakos, G., Berthold, J., Eireiner, M., & Schmitt-Landsiedel, D. (2006). Activation technique for sleep-transistor circuits for reduced power supply noise. 1-4244-0303-4/06/$20.00 ©2006, IEEE.
5. Brauer, E. J., Hatirnaz, I., Badel, S., & Leblebici, Y. (2006). Via-programmable expanded universal logic gate in MCML for structured ASIC applications: Circuit design. 0-7803-9390-2/06/$20.00 ©C2006, IEEE.
6. Sukhavasi, S. B., Sukhavasi, S. B., Madivada, V. B., khan, H., & Sastry Kalavakolanu, S. R. (2012). Implementation of low power parallel compressor for multiplier using self resetting logic. *International Journal of Computer Applications (0975–888), 47*(3).
7. Vijayasalini, P., Nirmal Kumar, R., Dhivya, S. P., & Tamilselvan, G. M. (2013). Design and analysis of low power multipliers and 4:2 compressor using adiabatic logic. *International Journal of Emerging Technology and Advanced Engineering, 3*(1). Website: www.ijetae.com (ISSN 2250-2459, ISO 9001:2008 Certified Journal).
8. Rajani, H. P., & Kulkarni, S. (2012). Novel sleep transistor techniques for low leakage power peripheral circuits. *International Journal of VLSI design & Communication Systems (VLSICS), 3(4)*.
9. Chiou, D. S., Chen, Y. T., Juan, D. C., & Chang, S. C. Sleep transistor sizing in power gating designs. Department of Computer Science, National Tsing-Hua University, Hsinchu, 30013, Taiwan.
10. Delican, Y., & Morgul, A. (2009). High performance 16-bit MCML multiplier. In *Proceedings of ECCTD '09*, pp. 157–160.
11. Srinivasan, V., Ha, D. S., & Sulistyo, J. B. (2004). Gigahertz-range MCML multiplier architectures. In *Proceedings OJ ISCAS '04*, pp. 785–788.
12. Caruso, G., & Macchiarella, A. (2010). A methodology for the design of MOS current-mode logic circuits. *IEICE Transactions on Electronics, E93-C*(2).
13. Alioto, M., & Palumbo, G. (2006). Power-aware design techniques for nanometer MOS current-mode logic gates: A design framework. *IEEE Circuits and Systems Magazine*, AO-59.

Design and Analysis of Ultracompact Four-Band Wavelength Demultiplexer Based on 2D Photonic Crystal

Chandraprabha Charan, Vijay Laxmi Kalyani
and Shivam Upadhyay

Abstract In this paper, we demonstrate an ultracompact four optical-band wavelength demultiplexer (FBWD) for demultiplexing four optical wavelengths 1.31, 1.48, 1.56, and 1.625 μm. The structure has been designed using silicon rods with refractive index 3.47 which are suspended in air. The device has an area of 13.1×8.6 μm^2 which is built with $r = 0.17a$ and $a = 0.59$ μm. The band-gap of structure is calculated by plane wave expansion (PWE) method and the finite difference time domain (FDTD) method has been used for all other simulation works like coupling measurement, electric field distribution method, and crosstalk measurement. Proposed structure uses different type of defects within the waveguide and filtering nature of photonic crystal to improve transmission efficiency (above 95 %) with low crosstalk.

Keywords Photonic crystals (Phcs) · Plane wave expansion (PWE) method · Photonic band-gap (PBG) · Finite difference time domain (FDTD) method

1 Introduction

Photonic crystals (Phcs) are special materials with a periodicity in their dielectric constant under certain conditions [1]. Phcs can create a photonic band-gap, i.e., a frequency/wavelength window in which propagation through the crystal is prohibited. Periodicity of Phcs breaks by introducing line or point defect in structure and PBG of structure confine the light inside it. Several kinds of optical devices can

C. Charan (✉) · V.L. Kalyani · S. Upadhyay
Government Mahila Engineering College, Ajmer, India
e-mail: charancharanchandraprabha@gmail.com

V.L. Kalyani
e-mail: kalyanivijaylaxmi@gmail.com

S. Upadhyay
e-mail: shivamupadhyay920@gmail.com

© Springer Science+Business Media Singapore 2016
S.C. Satapathy et al. (eds.), *Proceedings of International Conference on ICT for Sustainable Development*, Advances in Intelligent Systems and Computing 408, DOI 10.1007/978-981-10-0129-1_28

be constructed based on Phcs [2]. Phcs devices have received many interests in scientific communities because of their high capacity, high speed, high performance, long life, and compactness which make them suitable for ultra-small integration [3, 4]. In recent years two-dimensional (2D) Phcs captures attention over three-dimensional (3D) ones. Since they are much easier to produce and promising especially for near-infrared and visible spectrum applications [5]. Two-dimensional (2D) Phcs can be used for designing a lot of structures such as lasers, modulators, detectors, splitters, multiplexers/demultiplexers, optical amplifiers, etc. [6].

2D Phcs-based wavelength demultiplexers are useful and essential elements in ultra-dense integrated circuits also used as a key component in passive optical networks [7]. Traditionally, wavelength demultiplexers are realized using arrayed waveguide grating, fiber-Bragg grating or thin-film filter. However, these conventional optical demultiplexers suffer from the disadvantages of their large size of the order of centimeters to millimeters. This is not suitable for photonic integration purpose. The optical demultiplexers based on Phcs structure can overcome the above-mentioned limitations. They can offer compactness and serves as a promising candidate for realization of high density optical integration [8].

A four-band wavelength demultiplexer (FBWD) is device for separating four optical wavelengths which are essentially in optical-communication. FBWD have wide applications in optical-communication networks and play a very important role in fiber-to-the-home (FTTP) systems also used in wavelength division multiplexed system(WDM) [9]. The proposed structure uses waveguiding phenomenon based on PBG property of Phcs for designing of FBWD. The proposed FBWD separates the four optical-communication wavelengths, i.e., 1.31, 1.48, 1.56, and 1.625 μm which are corresponding to O, S, C, and L wavelength band, respectively. The performance characteristics and band-gap is calculated by 2D FDTD and PWE method, respectively.

2 Structure Design

The proposed FBWD is based on wavelength selective filtering nature of Phcs structure. The FBWD structure uses hexagonal lattice structure with silicon rods suspended in air. Various parameters are responsible for determining the characteristics of photonic crystal waveguide [10]. They are mainly the material, lattice structure, band-gap, size of defects, number of output ports, etc. [11]. The design process mainly comprises of following steps.

2.1 Selection of Material

The material Silicon is chosen because of its high refractive index (RI = 3.47). This high RI helps in getting a wide wavelength span (1190–1950 nm). So the structure is suitable for designing demultiplexer for four wavelengths that lie inside this wavelength span.

2.2 Selection of Lattice Structure

The hexagonal lattice structure is chosen because its geometry provides a smaller angle (60°) for bending the light. The smaller angle results in lesser scattering of light and in turn lower losses [12].

2.3 Selection of r/a Ratio

The r/a ratio of the lattice plays an important role in determining its band-gap, here a is the lattice constant of FBWD structure and r is the radius of silicon rods. To make design useful for practical purpose, we need a band-gap region that includes the commonly used wavelengths of optical transmission. It was found by simulation that hexagonal lattice structure with $r/a = 0.17$ provides a wide band-gap where $a = 0.59$ µm is chosen.

2.4 Adding Defects

Defects are the key parameters in the proposed structure for selecting wavelength in FBWD device. The defects can be added by changing the radius of the Silicon rods. The losses for wavelengths 1.31, 1.48, 1.56, and 1.625 µm have been measured for different radii, respectively. The transmission efficiency is found maximum at r'/a ratio of 0.22, 0.25, 0.29, and 0.12 for wavelengths 1.31, 1.48, 1.56, and 1.625 µm, respectively (Fig. 1) where r' is changed radius for each defect. Also the radius of border rods plays an important role in improving transmission efficiency so the radius of border rods are taken as 0.071, 0.06, 0.071, and 0.09 µm for wavelengths 1.31, 1.48, 1.56, and 1.625 µm, respectively.

2.5 Layout of Proposed Structure

After taking all parameters in consideration, the layout of FBWD is finalized as shown in Fig. 2. The radius of the border rods in a black box section is decreased for prohibiting transmission of wavelength 1.625 µm and restricts this wavelength to cross-couple with other wavelengths. The border rod radius in black box section

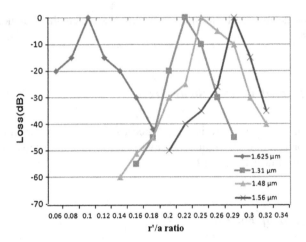

Fig. 1 Transmission loss plots for various r'/a ratio for FBWD

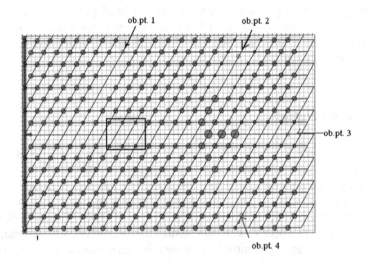

Fig. 2 Layout of proposed FBWD structure

is taken as 0.071 μm. This section act as a band-pass filter [13] for 1.31, 1.56, and 1.48 μm wavelengths signals.

Also seen in Fig. 2 that a Triplexer structure is used for separating remaining three wavelengths. Thus the wavelengths 1.625, 1.48, 1.56, and 1.31 μm are filter out by 1, 2, 3, and 4 output port, respectively. For improving the transmission efficiency, the radius of border rods is decreased along each output port this is indicated in Fig. 2.

3 Simulation and Results

After finalizing the layout structure of proposed design, we simulate it using the Optiwave software. The results are demonstrated below.

3.1 Band-Gap Calculation

The PBG is calculated by PWE band-solver. Figure 3 depicts the PBG plot for proposed structure. It is seen that Phcs with radius of silicon rods $r = 0.17a$ have two band-gap 0.3025–0.4954 (a/λ) and 0.6440–0.6483 (a/λ).

3.2 Electric Field Distribution

Figure 4 represents the electric field distribution for the proposed FBWD structure for four wavelengths.

3.3 Coupling Measurment

Figure 5 represents coupling for the wavelengths 1.625, 1.48, 1.56, and 1.31 μm, respectively.

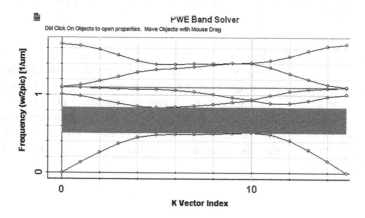

Fig. 3 Photonic band-gap (PBG) diagram for the proposed FBWD structure

(a)   (b)

(c)   (d)

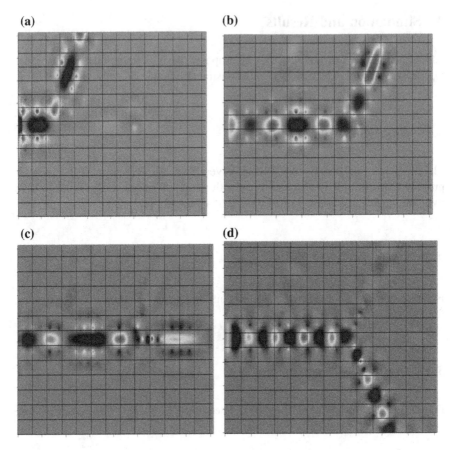

Fig. 4 Steady-state electric field distributions for wavelength signals of **a** 1.625 μm, **b** 1.48 μm, **c** 1.56 μm, **d** 1.31 μm respectively

It is clear from the above figures, that the proposed structure posses high transmission efficiency (above 95 %). Again for designing a wavelength division, demultiplexer crosstalk is a major issue, i.e., an unwanted leakage of a signal from one channel to another channel within the device so minimum value of crosstalk is desired. In this work, the structure is optimized to possess minimum crosstalk between output waveguides. As shown in Fig. 5, wavelength 1.48 and 1.56 μm are associated with low crosstalk, whereas 1.625 and 1.31 μm exhibit negligible crosstalk (Table 1).

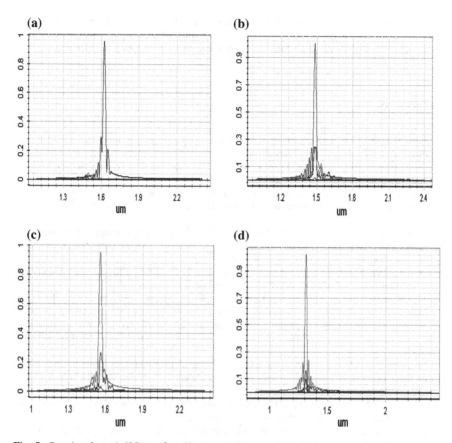

Fig. 5 Couping for **a** 1.625 μm, **b** 1.48 μm, **c** 1.56 μm, and **d** 1.31 μm wavelength

Table 1 Transmission efficiency by optimization

Wavelength (μm)	1.31	1.48	1.56	1.625
Transmission efficiency (%)	98.65	96.30	96.06	96.03

4 Conclusion

In this work, we have proposed a four-band wavelength demultiplexer (FBWD) based on Phcs structure. The demultiplexing action is involved with four wavelengths 1.31, 1.48, 1.56, and 1.625 μm wavelengths 1.48 and 1.56 μm are associated with low crosstalk, whereas 1.31 and 1.625 μm exhibit negligible crosstalk with high transmission efficiency. The proposed FBWD device can be utilized for designing optical switches add-drop filters, and many more application for high-speed optical-communication system.

Acknowledgments The author would like to thanks Mrs. Vijay Laxmi Kalyani Assistant professor in Electronics and communication Department of Govt. Mahila Engineering College Ajmer for her helpful contribution and guidance and also thanks to references for their literature support.

References

1. John, S. (1987). Strong localization of photons on certain disordered dielectric super lattices. *Physical Review Letters, 58*, 2486–2488.
2. Joannopoulos, J. D., Meade, R. D., & Winn, J. N. *Photonic crystals—molding the flow of light*, 2nd ed. pp. 2–5, 192.
3. Fan, S. H., Johnson, S. G., Joannopoulos, J. D., Manolatou, C., & Haus, H. A. (2001). Waveguide branches in photonic crystal. *Journal of the Optical Society of America, 18*, 162–165.
4. Kim, H., Park, I., Park, B. O. S., Lee, E., & Lee, S. (2004). Self-imaging phenomena in multi-mode photonic crylstal line-defect waveguides: Application to wavelength de-multiplexing. *Optics Express, 12*, 5625–5633.
5. Wilson, Rab, Karle, T. J., Moerman, I., & Krauss, T. F. (2003). Efficient photonic crystal Y-junctions. *Journal of Optics A: Pure and Applied Optics, 5*, S76–S80.
6. Masud Parvez Amoh, Md., Hasan Talukder, A. B. M., Omar Faruk, Md., Tisa, T. A., & Mahmood, Z. H. (2012). Design and simulation of an optical wavelength division demultiplexer based on the photonic crystal architecture. In *IEEE/OSA/IAPR International Conference on Infonnatics, Electronics and Vision* 978-1-4673-1154-0112/$31.00 ©20 12 IEEE.
7. Mansouri-Birjandi, M. A., & Rakhshani, M. R. (2013). Wavelength demultiplexer using heterostructure ring resonators in triangular photonic crystals. *TELKOMNIKA, 11*(4), 1721–1724, e-ISSN: 2087-278X.
8. Shih, T.-T., Wu, Y.-D., & Lee J.-J. (2009). Proposal for compact optical triplexer filter using 2-D photonic crystals. *IEEE Photonics Technology Letters, 21*(1).
9. Sinha, R. K., & Rawat, S. (2008). Modling and design of 2D photonic crystal based Ytype dual band wavelength demultiplexer. ©springer science+business medi, LLC.
10. Johnson, S. G., Villeneuve, P. R., Fan, S., & Joannopoulos, J. D. (2000). Linear waveguides on photonic-crystal slabs. *Physical Review B, 62*, 8212–8222.
11. Koshiba, M. (2001). Wavelength division multiplexing and demultiplexing with photonic crystal waveguide couplers. *Journal Lightwave Technology, 19*(5), 1970–1975.
12. Frandsen, L. H., Harpoth, A., Borel, P. I., Kristensen, M., Jensen, J. S., & Sigmund, O. (2004). Broadband photonic crystal waveguide 60 bend obtained utilizing topology optimization. *Optics Express, 12*, 5916–5921.
13. Lee, S. G. (2008). Analysis of photonic crystal filter structures and its application : Wavelength demultiplexer. In *OECC/ACOFT 2008—Joint Conference of the Opto-Electronics and Communications Conference and the Australian Conference on Optical Fibre Technology*, 07/2008.

Designing and Optimization of Nano-ring Resonator-Based Photonic Pressure Sensor

Shivam Upadhyay, Vijay Laxmi Kalyani
and Chandraprabha Charan

Abstract In this paper, we designed a pressure sensor based on a linear waveguide coupled with nano-ring resonator using two-dimensional photonic crystals. The ring resonator minimizes the external parameters effect such as temperature, humidity, etc. The proposed pressure sensor works on the principle of resonance wavelength. Due to the applied pressure, the refractive index of a sensor is changed and thus resonance wavelength of a sensor is also shifted. The simulation results show that the resonance wavelength of a pressure sensor shifts between the ranges 1.3–1.9 μm. The quality factor of sensor is 5179 and sensitivity of a sensor is 7.7 nm/GPa. Pressure sensor has good resolution in the range of nN and minimum detectable pressure is 0.10 nN range. The band-gap of structure is calculated using plane wave expansion (PWE) method and all other computation work such as transmission power, electric field distribution are performed using finite difference time domain (FDTD) method.

Keywords Photonic crystals (Phc) · Linear waveguide · Ring resonator · Photonic Band-Gap (PBG) · Photonic sensor · Finite difference time domain method (FDTD)

1 Introduction

In 1987, photonic crystal concept is given by Sajeev John and Eli Yablonovitch. This research increases the attention to develop the nano-structure devices. Photonic crystal has the potential to provide very compact devices in ultracompact

S. Upadhyay (✉) · V.L. Kalyani · C. Charan
Government Mahila Engineering College, Ajmer, India
e-mail: shivamupadhyay920@gmail.com

V.L. Kalyani
e-mail: kalyanivijaylaxmi@gmail.com

C. Charan
e-mail: charanchandraprabha@gmail.com

© Springer Science+Business Media Singapore 2016
S.C. Satapathy et al. (eds.), *Proceedings of International Conference on ICT for Sustainable Development*, Advances in Intelligent Systems and Computing 408, DOI 10.1007/978-981-10-0129-1_29

range. These ultracompact components are used for the designing of optical integrated circuits. These components are based on photonic crystal structure and photonic band-gap (PBG) concept of dielectric structure. Photonic crystals are the novel class of optical materials. It is a periodic dielectric structure with lattice parameters in the order of wavelength of propagated electromagnetic wave. Such type of photonic crystal allows a limited number of electromagnetic wave to propagate in it. These allowed electromagnetic fields are called modes. The main advantage of photonic crystal is its "light confinement properties" or "control over the flow of light." This feature of photonic crystal (Phc) gives an opportunity to use it in various smart applications such as photodetector, modulator, lasers and sensors [1].

A device that detects the input from the physical environment and responds according to them is called sensor. The applied input may be light, heat, motion, moisture, pressure, or any other physical or biochemical parameters. Photonic crystal-based optical sensors also attract the researchers due to their unique features such as compact size, low cost, high sensitivity, and immunity to electromagnetic interference. It is used to sense or detect temperature, magnetic field, chemical, gas, liquid concentration, surrounding refractive index, curvature, etc. [2, 3].

Photonic crystal structure is mainly classified into two categories, i.e., "dielectric rods in air" and "air holes in slab (photonic slab)." For the layout designing of photonic crystal-based devices both structure is implemented. But researches shows that the devices which are based on the "rods in air" type structure has more out of plane losses due to insufficient vertical confinement. This type of structure gives transverse magnetic (TM) band-gap. On other side, the devices which are designed using "air holes in slab" type structure has very low losses, good sensitivity, and its fabrication is simple. The optical sensors are further classified in two class, i.e., physical sensors and biosensors [4].

Recently Levy et al. have reported a novel displacement sensor based on two planar photonic crystal waveguides (PHWG). Its sensing principle is based on the light intensity of output. Output light of a sensor is strongly dependent on alignment accuracy. Any deformation in structure leads to misalignment and thus intensity of output light is reduced. They also proved this concept by giving some simulation results [5]. In Pursiainen et al. have proposed a three-dimensional photonic crystal (Phc) using self assembly fabrication process [6]. Stomeo et al. proposed a force sensor having photonic crystal (Phc) waveguide coupled with the microcavity. The applied force, shifts the operating wavelength in the transmission spectrum thus forces is detected [7]. Olayee et al. have proposed a pressure sensor based on high resolution and wide dynamic range. In this, a linear waveguide is coupled with the nanocavity. This sensor has quality factor 1470 and it detect pressure between 0.1 and 10 GPa with 8 nm/GPa sensitivity [8]. Abdollahi et al. have demonstrated a high sensitive double holes defects refractive index sensor. The sensor design has two waveguides coupled with the microcavities. It is based on the resonance wavelength principle. With the variation in refractive index of sensor material, the resonance wavelength of sensor is also shift. It has sensitivity $s = 500$ nm/RIU and refractive index resolution $\Delta n = 0.0001$ [9]. Lee et al. have designed a novel nano mechanical sensor using silicon 2D-photonic crystal, it also follows the resonance wavelength shifts. In this when any external mechanical force is applied, the shape of air holes and defect length are varied and wavelength of output spectrum is shifted [10]. A.A. Dehghani et al. (2009) have proposed a novel pressure sensor using high-quality photonic crystal cavity resonator. In this the sensor has a linear waveguide coupled with the nanocavities. When any external pressure is applied on the sensor surface, the operating or resonance wavelength of sensor is shifted. It is operating in the range 1450–1550 nm wavelengths. It has sensitivity 11.7 nm/Gpa and minimum detectable pressure 13 nN [11].

The earlier designed pressure sensor results are affected by external parameters such as temperature, noise, adsorbed humidity, etc. In this proposed structure, the effect of external parameters has been minimized using ring resonator with linear waveguide. The proposed pressure sensor works on the principal of resonance wavelength. The quality factor and sensitivity of designed structure is obtained about 5179 and 7.7 nm/Gpa, respectively.

2 Sensing Principle of Pressure Sensor

All photonic crystal sensors are based on the principle of either intensity reduction of output peak or resonance wavelength shift. In pressure sensor, the applied hydrostatic pressure changes the optical and electronic property of sensor. The optical properties like photoelastic, piezoelectric, and permittivity are changed with the pressure. These modified optical properties of a sensor also modifies refractive index of Si, thus PBG of a structure is increased; because PBG of a structure strongly depends on refractive index, radius to lattice constant (r/a) ratio and lattice constant [12]. If any parameter is varied then PBG of a structure is also varied. The relation between pressure and refractive index is given by

$$n = n_0 - (c_1 + 2c_2)\sigma \tag{1}$$

where c_1 and c_2 are defined as

$$c_1 = n_0^3 (P_{11} - 2V \cdot P_{12})/2E \tag{2}$$

$$c_2 = n_0^3 (P_{12} - V(P_{11} + P_{12})/2E \tag{3}$$

where E is Young's modulus, for silicon its value is 130 GPa, V represent Poisson's ratio, V is 0.255, and P_{ij} denotes the strain-optic constant. In pressure sensor, applied pressure has linear relation with the refractive index of a sensor structure. For 1 GPa pressure, the refractive index of a sensor is incremented by 0.03985.

3 Design and Modeling of Pressure Sensor

The proposed pressure sensor is based on square lattice structure with silicon rods suspended in air. Designing steps of a sensor is as follows:

3.1 Selection of Material

In recent years, heteropitaxial layer-based planar Phc sensing devices are designed. In which organic compounds and polymers are also used as a material. The SOI technological platform is more suitable for fabrication of ultracompact and ultra-high performance-based integrated sensors. In the proposed design, Si is used as a substrate material with refractive index RI = 3.46. Thus the complete structure has very high refractive index contrast with air and provides a large band-gap.

3.2 Selection of Lattice Structure

Photonic crystal lattice structure is two types "hexagonal" and "square." In the proposed design, square lattice structure is used.

3.3 Designing the Unit Cell of the Lattice

The photonic band-gap (PBG) of a structure strongly depends on r/a ratio of lattice, where 'r' is radius of Si rods and 'a' is lattice parameter or constant of structure. The proposed sensor structure has r/a ratio 0.1694 with lattice constant $a = 590$ nm.

3.4 Adding Defects

For the propagation of light inside the structure, some defect is created. The proposed pressure sensor is designed by creating linear waveguide coupled with the ring resonator. Ring resonator is introduced inside the structure for minimizing external effects such as effect of increasing temperature or humidity. It acts like a filter.

For the propagation of plane wave, linear waveguide is introduced by removing Si rods. The input plane wave is excited by Gaussian continuous wave optical source with the resonance wavelength 1605.68 nm.

Fig. 1 2D-PC-based pressure
sensor layout

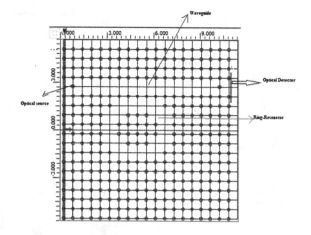

The applied hydrostatic pressure will change the refractive index and optical properties of a sensor, thus operating wavelength of sensor is shift. The output of a sensor is detected by optical detector. The designed layout of ring resonator-based pressure sensor is shown in Fig. 1.

4 Simulation Results

A 2D FDTD simulation tool is used for simulation work of pressure sensor. The transverse electric (TE) polarization mode is selected for the propagation of light inside the sensor structure. In TE polarized wave, the electric field is normal to the axis of rods. Good performance of a sensor is achieved by getting high-quality factor and high transmission efficiency. The concept of proposed design is proved by simulating it at different pressures from 1 to 5 GPa.

The simulation results are given below to prove the concept of proposed device:

4.1 Band-Gap Calculation

The photonic band-gap of photonic crystal acts like a stop band. It does not allow the propagation of light inside the photonic crystal structure. By perturbing the internal architecture of photonic crystal structure or creating defects, this gap is used as a pass band for propagation of light.

In proposed sensor, the plane wave expansion method (PWE band solver) is used for the calculation of photonic band-gap (PBG) of structure. A broad transverse band-gap with transverse electric (TE) mode is found between 1302.2 and 1909.1 nm. Figure 2 shows the photonic band-gap of pressure sensor.

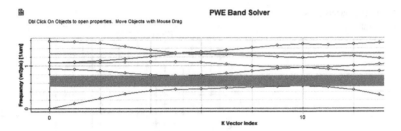

Fig. 2 Band-gap for TE mode

Fig. 3 2D electric field
distribution

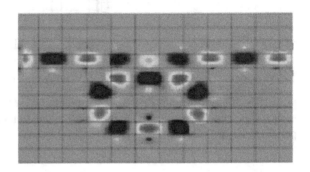

4.2 Field Distribution

The 2D electric field distribution of line defect with ring resonator-based pressure
sensor at the resonance wavelength $\lambda_0 = 1.60568$ μm and 0 Gpa pressure is shown
in Fig. 3.

The 3D electric field distribution of line defect with ring resonator-based pressure sensor at the resonance wavelength $\lambda_0 = 1.60568$ μm and 0 Gpa pressure is
shown in Fig. 4.

Fig. 4 3D electric field
distribution

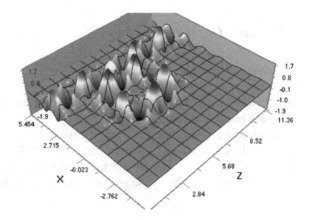

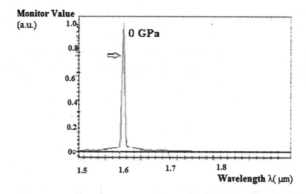

Fig. 5 Normalized transmission spectra of sensor at 0 Gpa pressure

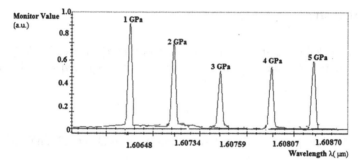

Fig. 6 Normalized transmission spectra of the sensor at different pressures

Figure 5 shows the normalized transmission spectrum of sensor. When no pressure is applied on the sensor, i.e., at 0 GPa pressure, no stress is distributed on the surface of pressure sensor; thus refractive index of sensor remains same and optical properties of sensor do not chang. The resonant wavelength of sensor is 1.60568 µm and transmission power is 90.11 %, respectively.

Figure 6 represents the normalized transmission spectrum of pressure sensor at different pressures such as 1, 2, 3, 4, and 5 GPa. In this condition, the resonance wavelength of sensor is shifted with the applied pressure.

5 Analysis of Pressure Sensor

The pressure sensor results are calculated at different pressures and obtained results are analyzed. The performance parameters of sensor, i.e., quality factor, sensitivity, and minimum detectable pressure are also calculate and listed below:

5.1 Quality Factor

Quality factor is defined as the ratio of resonance wavelength (λ_0) to the full width at half maximum ($\Delta\lambda$) of the resonator's Lorentzian response, i.e., $\lambda_0/\Delta\lambda$ [13]. It is represented by 'Q'.

$$Q = \frac{\omega_0}{\Delta\omega} = \frac{\lambda_0}{\Delta\lambda}$$

where ω_0 is resonance wavelength and $\Delta\omega$ is the full width at half maxima (FWHM) response.

The proposed pressure sensor has very high-quality factor 5179.

5.2 Sensitivity

Sensitivity is defined as the ratio of minimum shift in resonance wavelength to the change in refractive index [13]. It is represented by 'S'. Sensitivity is calculated by

$$S = \frac{\Delta\lambda}{\Delta n}$$

In this design, pressure sensor having sensitivity of about 7.7 nm/GPa.

5.3 Minimum Detectable Pressure

The minimum detectable pressure of sensor is calculated using sensitivity and resolution of sensor [13]. The designed pressure sensor detects minimum pressure up to 0.10 nN.

The complete analysis of pressure sensor at different pressure is given in Table 1. In which at different pressure, the effective refractive index, resonance wavelength, transmission power, and wavelength shift is given.

Table 1 Analysis of pressure sensor

Applied pressure (GPa)	Effective refractive index	Resonance wavelength (μm)	Transmission power (%)	Wavelength shift ($\Delta\lambda$) (nm)
1	2.53985	1.60648	87.4	0.8
2	2.5797	1.60734	72.9	1.66
3	2.61955	1.60759	50.9	1.91
4	2.6594	1.60807	54.6	2.39
5	2.69925	1.60870	55.3	3.02

By analyzing all the results of pressure sensor, it is observed that the applied pressure is distributed on the surface of sensor in the form of stress or strain. This stress changes the refractive index of sensor material, thus optical properties of sensor is varied which modified photonic band-gap of sensor and shift the operating wavelength. In this way, sensor shows sensitivity toward the applied pressure.

6 Conclusion

In this paper, we have proposed a photonic crystal-based pressure sensor for measuring nanopressure. Photonic sensor is a smart and compact sensing device that can provide facility to use it in magnetic field environment. This sensor is based on linear waveguide coupled with the ring resonator. The ring resonator reduces the effects of external parameters. Pressure sensor is designed using square structure with the lattice constant $a = 0.59$ μm, the radius to lattice constant ratio r/a ratio is 0.1694, and the refractive index of slab is 3.46. The designed sensor has dielectric rods suspended in air structure with the dimensions of 11.5×11.4 μm^2. It is observed that applied pressure shifts the resonance wavelength of the sensor. In the absence of pressure (at 0 Gpa), the resonance wavelength and output power is 1.6056 μm, 90.11 %, respectively. Sensor has 7.7 nm/GPa sensitivity with quality factor 5179.

Acknowledgments The author would like to thank the referees for their constructive research work which helped to improve the quality of this paper. I also thank to Vijay Laxmi Kalyani, (Assistant Professor, Department of Electronics and communication, Govt. Mahila Engineering College, Ajmer) for her input on many aspects of this work.

References

1. Joannopoulos, (2008). Photonic crystals molding the flow of light. 2nd ed. Princeton University of press.
2. Yasuhide, T. (2014). Photonic crystal waveguide based on 2-D photonic crystal with absolute photonic band gap. *IEEE Photonics Technology Letters, 18*(22).
3. Liu, Y., & Salemink, H. W. M. (2012). Photonic crystal-based all-optical on-chip sensor. *J. Optics Express, 20*(18), 19912–19921.
4. Olyaee, S., & Dehghani, A. A. (2013). Ultrasensitive pressure sensor based on point defect resonant cavity in photonic crystal. *Journal of Sensor Letters, 11*(10), 1854–1859.
5. Levy, O., Steinberg, B. Z., Nathan, N., & Boag, A. (2005). Ultrasensitive displacement sensing using photonic crystal waveguides. *Journal Applied Physics Letters, 98*, 033102.
6. Pursiainen, O. L. J., Baumberg, J. J., Ryan, K., Bauer, J., Winkler, H., Viel, B., & Ruhl, T. (2005). Compact strain sensitive flexible photonic crystals for sensors. *Journal Applied Physics Letters 87*, 101902–101902.
7. Stemeo, T., Grande, M., Passosea, A., Salhi, A., Vittorio, M. D., & Biallo, D. (2007). Fabrication of force sensor based on two-dimensional photonic crystal technology. *Journal of Microelectronic Engineering, 84*(5–8), 1450–1453.

8. Olyaee, S., & Dehghani, A. A. (2012) High resolution and wide dynamic range pressure sensor based on two-dimensional photonic crystal. *Journal of Photonic Sensors, 2,* 92–96.
9. Shiramin, L.A., Kheradmand, R. & Abbasi, A. (2013) High-sensitive double-hole defect refractive index sensor based on 2D-photonic crystal. *IEEE Sensors Journal, 13*(5).
10. Lee, C., Thillaigovindan, J., & Radhakrishnan R. (2008). Design and modeling of nano-mechanical sensors using sillicon 2D-photonic crystals. *Journal of Lightwave Technology, 26*(7).
11. Olyaee, S., & Dehghani, A. A. (2012). Nano-pressure sensor using high quality photonic crystal cavity resonator. In *IET International Symposium on Communication Systems*, Poznań, Poland, July 18–20, 2012.
12. Shanthi, K. V., & Robinson, S. (2014). Two-dimensional photonic crystal based sensor for pressure sensing. *Journal of Photonic Sensors, 4*(3), 248–253.
13. Lavin, A., & Casquel, R. (2011). Efficient design and optimization of bio-photonic sensing cells (BICELLs) for label free biosensing. In *INNBIOD (REF:IPT-2011) under the Spanish-Ministry of Economy and Competitiveness*.

Classifying Nodes in Social Media Space

Kirti Thakur and Harish Kumar

Abstract Social media provides a platform to interact among people where they share or exchange idea and information. Social network analysis is one of the widest research area used in economics, behavioural, social, political, organizational sciences, etc. Today, maximum information is available online thus a smart system is required to interpret the data. The analysis of information is based on human interaction and the perception of user-generated content. The interpretation fluctuate person-to-person thus automated system is required. In this paper, a methodology is proposed for the classification of node linked with official Panjab University Facebook page.

Keywords Social media · Text classification · Cyber · Information retrieval · Sentiment analysis · Opinion mining · Facebook · Machine learning

1 Introduction

Social media provides a platform to interact among people where they share or exchange idea and information. A wide range of data in different form gets exchanged over internet [1]. Therefore, as the wideness of subjective information over the Internet, it becomes a hotspot in many research fields to understand and interpret the information. Text classification is widely used as a tool to organize and handle text automatically. Text classification is also known as text mining or sentiment analysis. With the help of sentiment analysis, it is easy to explore the activities, feelings, opinion and thought of web users [2].

K. Thakur (✉) · H. Kumar
University Institute of Engineering and Technology, Panjab University,
Chandigarh, India
e-mail: kirtithakur1991@gmail.com

H. Kumar
e-mail: harishk@pu.ac.in

© Springer Science+Business Media Singapore 2016
S.C. Satapathy et al. (eds.), *Proceedings of International Conference on ICT for Sustainable Development*, Advances in Intelligent Systems and Computing 408, DOI 10.1007/978-981-10-0129-1_30

279

Different frameworks are proposed by many authors in order to enhance basic techniques for sentiment analysis. The classification is performed at three levels: sentence, document and aspect. In sentence-level, sentiment is calculated in each sentence as positive or negative opinion. Before analysis, first step is to check whether the sentence is objective or subjective [3]. Document-level sentiment analysis classify whole document in term of positive or negative sentiment. Document-level is based on sentence-level classification, accordingly sentence treated as small document [4]. Aspect-level first identify entities and their aspects then sentiment classification proceeds. Results are derived based on feature set selection and classification technique. Different classification algorithm provides varied result based on application thus it may state that particular algorithm is not the best.

In this paper, main aim is to make an intensive study of different strategies for opinion mining. A model is proposed to classify the social media (Facebook) users as node in sub-categories based on classification of comments. Sentence-level classification will choose to figure out the sentiment. The main focus is on feature sets and classification algorithms. The remaining part of this paper is organized as follows. The formal definition of need of online classification and review of various approaches used by different researchers for sentiment analysis is summarized in Sect. 2. A methodology is proposed to classify nodes based on text mining in Sect. 3. Section 4 summarizes the study and enlists future scopes.

2 Literature Review

Social media platform allows a user to create, share and exchange information. Use of social media has grown massively over the past decade [1]. Social network analysis is one of the widest research area used in economics, behavioural, social, etc [5, 6]. Nowadays, maximum information is available online. Thus a smart system is required to interpret this information. The analysis of information is based on human interaction and perception of user-generated content. To identify personality of a user, human interpretation may vary thus an automated system is required for the analysis of text. [7] predicted the personality of a social media user by training the system with machine learning algorithms. According to Fig. 1, there are several classification methods under two main approaches [8, 9]: lexical and machine learning (ML). Another technique is hybrid approach which is combination of both methods.

(1) **Lexical approach starts with pre-processing** A dictionary is prepared for lexicons with polarity. To calculate the polarity of text, sum of each word polarity score is taken to get overall polarity in term of score. Score is positive if lexicon matched with dictionary word otherwise it is negative. In [10] Lexicons based classification of social emotion such that prediction is carried out toward existing topic.

Lexical approach variants: Baseline approach: dictionary has limited number of tagged words as positive and negative. Part of speech tagging: words are

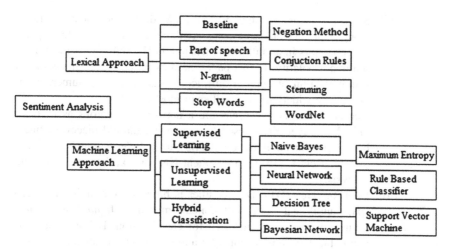

Fig. 1 Sentiment classification techniques

tagged as adjective, adverb, noun, verbs, etc. N-gram: common phrases find out by traversing consecutive words. Longer phrase gives best result. Stop words: The non-semantic words like pronoun, articles and preposition are eliminated during pre-processing. WordNet: find synonyms for all words in favour of better result. Stemming: get rid of suffixes and prefixes. Conjunction Rules: extract exact meaning from sentence using grammar rules. Negation method: not change the evaluation of text.

(2) For **ML** methods, the accuracy of result depends on the feature selection. ML approach further classified as:

(2.1) *Supervised Learning*: Count on the existence of labelled training text. For sentiment analysis subsequent some of the mostly used classifier:

(2.1.1) Naïve Bayes (NB) Classifiers: NB widely used in field of classification. The processing of done at document-level assuming words are independent to each other. [11] An improved NB classifier was proposed to solve the positive and negative classification problem.

(2.1.2) Maximum Entropy (ME) Classifier: The ME maps labelled feature sets to vector using encoding. For each feature set, weight is calculated with encoded vector such that most likely label is determined. [12] ME classifier classify parallel sentence among different language pairs.

(2.1.3) Bayesian Network (BN): BN is base on two assumptions, one in dependent of feature sets and other fully dependent feature. Second assumption leads to the representation of random variables as nodes and edges as conditional dependencies. It provides an absolute model for variables and relationships among them. According to [13], computation complexity for text mining is extremely luxurious; therefore, rarely used.

2.1.4 Support Vector Machine (SVM): SVM determine linear separators along with vary classes to separate them. It constructs a hyperplane for separators in an infinite dimension. [14] Used as classification of sentiment polarity. In this paper, authors proposed a framework for summarization of opinion on micro-blogs in term of numeric. It is one of the widely used classifier in most of fields.

(2.1.5) Neural Networks (NN): NN are a network of neurons as its basic unit. Multilayer NN is used to construct manifold piecewise linear boundaries for nonlinear boundaries, to include regions belonged to particular class. [15–17] implement NN for text data and sentiment analysis.

(2.1.6) Decision Tree (DT): DT classifier is based on the division of data space until leaf node will contain minimum figures of records. Future, these records are used for classification. DT also stated as hierarchical decomposition of training set into small unit of attributes. [18–20] implement DT in context of text classification.

(2.1.7) Rule-Based (RB) Classifier: RB classifies the data space with set of rules. Two sides of rules represent as left-hand side represent condition on feature set and class label by right-hand. Conditions based on term present and absent. To generate rules, two most general criteria are support and confidence. The support is total number of training data set instance appropriate to the rules. The confidence is conditional probability such that right-hand side of rule is satisfied if left-hand is satisfied.

(2.2) Unsupervised Learning: Text classification is to classify documents into some category based on the feature set and base learner. In unsupervised learning, a large number of labelled training documents are used. It is easy to create unlabelled training documents rather labelled. Unsupervised approach adopted by many authors as [21–23] for sentiment polarity detection based on selected feature set. In [24], author proposed a probabilistic modelling framework as unsupervised model based on latent dirichlet allocation (LDA) for sentiment analysis.

(2.3) Hybrid Classification: In [25], Hybrid scheme of classification is used to resolve sparsity issue and to increase the classification accuracy.

Although the above-disused classification techniques provide a wide solution in vary research areas. Previous studies related to sentiment analysis are enlisted in Table 1. An approached based on these classification algorithms and feature sets is proposed to overcome the drawback of individual classifier is known as ensemble learning approach. Ensemble technique forms an integrated output by combining different base classifier outputs, become an efficient classification technique for many area [31, 32]. Ensemble methods include boosting, stacking, SVRCE, random

Table 1 Sentimental analysis studies: a summary

Study	Year	Feature set	Base learner (classification techniques)	Dataset
Alm et al. [30]	2005	POS-tagger	Snow	22 fairy tales
Wilson et al. [20]	2006	N-gram, Syntactic feature	Decision tress	MPQA dataset
Xia et al. [2]	2011	Pos and word relation	NB, ME, SVM (using ensemble method)	Five document-level dataset
Moraes et al. [26]	2013	Bag-of-words	SVM, NB	Four dataset (movie review and 4 product review)
Carolina et al. [27]	2014	Big five model (OCEAN)X	NB, SVM, multilayer perceptron neural network	Twitter dataset (ODM, Sander, semeval2013)
Deng et al. [28]	2014	ITD, ITS	SVM	Three dataset (CMR, MDS, SLMR)
Dam and Veldan [29]	2015	Clustering	K-means	Facebook dataset

subspace, bagging, bagging random subspace, etc. Fixed rules and trained methods are two types of ensemble methods. Individual outputs are combined in fixed manner using fixed rules as voting rule, sum rule, max rule and product rule. Trained methods include meta-classifier (the output for all class viewed as new feature) and weighted combination (for each class or for all classes, each classifier has exactly one weight) [2].

3 Methodology

In this section, a methodology of proposed system is explained. A new dataset and a number of classifiers are proposed to be used for sentiment analysis.

3.1 Data Set

To verify the effectiveness of techniques for sentiment analysis for social media, a new database will be generated. It consists of comments extracted from an official page (Panjab university Facebook page) with all legal consents of higher authorities. Facebook provides a protected environment with security features like login notifications, remote session management, last activity session and one-time password. Thus it is difficult to fetch data from a page without having permission or login access. It provides an interface as Graph API, is the key way to get data in and out of Facebook's social graph.

An access token is a string that recognizes a user, app, or page and provides temporary access to Facebook APIs. Because of privacy checks, the majority of API calls on Facebook need to include an access token. Tokens are of two types: short-lived and long-lived. Short-lived tokens are those generated by web login. These tokens can be upgraded to long-term tokens which last for 60 days. There are various sorts of access tokens as

(a) User access token: This access token provides permission to APIs to post, change or read data on user's behalf by application. This is the most commonly used token. This token can be obtained by user login and subsequent permission to obtain it.
(b) Page access token: This access token is similar to user access token. However, it is required when Facebook page data is to be written, modified or read by application using APIs. To obtain this token, a user access token must be obtained followed by manage pages permission. Using Graph API, page access token can be obtained thereafter.
(c) Application access token: Using server-to-server call, application access token can be obtained. It is used to change or read application settings.
(d) Client token: It can be placed in desktop applications to identify your application. Therefore, this is not a secret identifier. It is used to access a small subset of APIs which are application-level.

In the proposed methodology, page access token will be used. The procedure to obtain user access token is described in Fig. 2.

3.2 Experimental Procedure

The experiment can be performed using data mining toolkit WEKA (Waikato Environment for Knowledge Analysis) version 3.6.12. The WEKA provides a platform to solve data mining problems. It includes a collection of machine learning algorithms. In this study, it is proposed that the performance of 10 different methods including NB, DT, SVM and K-star can be compared. To minimize the variability in the training set, tenfold cross-validation can be performed on dataset for sentimental analysis. The dataset is divided into 10 subsets having equal size and

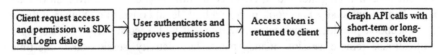

Fig. 2 Steps to obtain access token

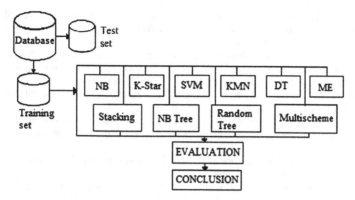

Fig. 3 Experimental procedure, NB as naive Bayes, SVM as support vector machine, KNN k-nearest neighbour, DT as decision tree, ME as maximum entropy

distribution. Then nine subsets can act as the training set and remaining as the test set.

After passing the training and test set through each of classification techniques, results can be evaluated. Accordingly, nodes classified into different groups having same results can be obtained. The proposed model for classification is shown in Fig. 3.

4 Conclusion and Future Scope

Social media have become a vital component of daily life for millions of people, such platform provides a wide range of applications. Sharing of information is one of the major components used by users. The proposed model is for the classification of node with the help of sentiment analysis techniques. The hybrid model may also be used for the tasks of text classification. Feature selection for syntactic relations may also be an important issue. Maximum classifiers are language-oriented. Language independent or conversion of different languages into one stream language can enhance the results.

References

1. Luchman, J. N., Bergstrom, J., & Krulikowski, C. (2014). A motives framework of social media website use: A survey of young Americans. *Journal of Computers in Human Behavior, 38*, 136–141.
2. Xia, R., Zong, C., & Li, S. (2011). Ensemble of feature sets and classification algorithms for sentiment classification, *Journal Elsevier*, 1138–1152.

3. Pang, B., & Lee, L. (2004). A sentimental education: Sentiment analysis using subjectivity summarization based on minimum cuts. In *Proceedings of the Association for Computational Linguistics* (pp. 271–278).
4. Liu, B. (2012). *Sentiment analysis and opinion mining*. Morgan & Claypool Publishers.
5. Hsu, T. S., Liau, C. J., & Wang, D. W. (2014). A logical framework for privacy-preserving social network publication. *Journal of Applied Logic, 12*, 151–174.
6. Barbier, G., & Liu, H. (2011). Data mining in social media. In C. C. Aggarwal (Ed.), *Social network data analytics* (pp. 327–352). US: Springer.
7. Carolina, A., Lima, E. S., & Castro, L. N. (2014) A multi-label, semi-supervised classification approach applied to personality prediction in social media. *Journal Neural Networks, 58*, 122–130.
8. Bhadanea, C., Dalalb, H., & Doshi, H. (2015). Sentiment analysis: Measuring opinions. *Procedia Computer Science, 45*, 808–814.
9. Medhat, W., Hassan, A., & Korashy, H. (2014). Sentiment analysis algorithms and applications: A survey. *Ain Shams Engineering Journal, 5*, 1093–1113.
10. Bao, S., Xu, S., Zhang, L., Yan, R., Su, Z., Han, D., & Yu, Y. (2012). Mining social emotions from affective text. *IEEE Transactions on Knowledge and Data Engineering, 24*, 1658–1670.
11. Hanhoon, K., Joon, Y. S., & Dongil, H. (2012). Senti-lexicon and improved Naïve Bayes algorithms for sentiment analysis of restaurant reviews. *Expert Systems with Applications, 39*, 6000–6010.
12. Kaufmann, J. M. (2012). JMaxAlign: A maximum entropy parallel sentence alignment tool. In *Proceedings of COLING'12: Demonstration Papers*, pp. 277–288.
13. Aggarwal, C. C., & Zhai C. X. (2012). *Mining text data*. Springer.
14. Li Y. M., & Li T. Y. (2013). Deriving market intelligence, from microblogs. *Decision Support Systems*.
15. Ruiz, M., & Srinivasan, P. (1999). *Hierarchical neural networks for text categorization*. In *ACM SIGIR Conference*.
16. Tou, N. H., Wei, G., & Kok, L. (1997). Feature selection, perceptron learning, and a usability case study for text ategorization. *ACM SIGIR conference*.
17. Rodrigo, M., Francisco, V. J., & Wilso, G. N. (2013). Document-level sentiment classification: an empirical comparison between SVM and ANN. *System Applications, 40*, 621–33.
18. Li, Y., & Jain, A. (1998). Classification of text document. *Journal of Computer, 41*, 537–546.
19. Yi, H., Li, W. (2011). Document sentiment classification by exploring description model of topical term. *Computer Speech Language, 25*, 386–403.
20. Wilson, T., Wiebe, J., & Hwa, R. (2006). Recognizing strong and weak opinion clauses. *Computational Intelligence, 22*, 73–99.
21. Yulan, H., & Dey, Z. (2011). Self-training from labeled features for sentiment analysis. *Information Processing and Management, 47*, 606–616.
22. Fu, X., Liu, G., Guo, Y., & Wang, Z. (2013). Multiaspect sentiment analysis for Chinese online social reviews based on topic modeling and how net lexicon. *Knowledge-Based Systems, 37*, 186–195.
23. Montejo-Raez, A., Martınez-Camara, E., Martın-Valdivia, M. T., & Urena-Lopez, L. A. (2012). Random walk weighting over sentiwordnet for sentiment polarity detection on twitter. In *Proceedings of the 3rd Workshop on Computational Approaches to Subjectivity and Sentiment Analysis* (pp. 3–10).
24. Lin, C., & He, Y. (2009). Joint sentiment/topic model for sentiment analysis. In *Proceedings of CIKM* (pp. 375–384).
25. Khan, F. H., Bashir, S., & Qamar, U. (2009). TOM: Twitter opinion mining framework using hybrid classification scheme. *Journal of Decision Support Systems, 57*, 245–257.
26. Moraes, R., Valiati, J. F., Wilson, P., & Gaviao, N. (2013). Document-level sentiment classification: An empirical comparison between SVM and ANN. *Expert Systems with Applications, 40*, 621–633.
27. Carolina, A., Lima, E. S., & Castro, L. N. (2014). A multi-label, semi-supervised classification approach applied to personality prediction in social media. *Neural Networks, 58*, 122–130.

28. Deng, Z. H., Luo, K. H., & Yu, H. L. (2014). A study of supervised term weighting scheme for sentiment analysis. *Expert Systems with Applications, 41*, 3506–3513.
29. Dam, J. W., & Velden, M. (2015). Online profiling and clustering of Facebook users. *Decision Support Systems, 70*, 60–72.
30. Alm, C. O., Roth, D., & Sproat, R.(2005) Emotions from text: machine learning for text-based emotion prediction. In *Proceedings of Human Language Technology Conference and Conference on Empirical Methods in Natural Language Processing (HLT/EMNLP)* (pp. 579–586).
31. Ho, T., Hull, J., & Srihari, S. (1994). Decision combination in multiple classifier systems. *IEEE Transactions on Pattern Analysis and Machine Intelligence, 16*, 66–75.
32. Kittler, J. (1998). Combining classifiers: a theoretical framework. *Pattern Analysis and Applications, 1*, 18–27.

A Novel Hexagonal Shape-Based Band-Stop Frequency-Selective Surface with Multiband Applications

Kiran Aseri and Sanjeev Yadav

Abstract With the aggrandize in the usage of wireless communication, interference has also been elevated. This paper presents a frequency-selective surface (FSS)-based band-stop filter, which leads to minimize the interference hence the impedance bandwidth, gain, and angular stability of antenna has been augmented tremendously. The incorporated design has a square-shaped patch of dimension 21.25 mm × 21.25 mm with two loop elements and a hexagonal slot embedded within it. The proposed multi-band structure is realized in CST microwave studio for three stop bands such as 2.24–2.70 GHz (Wi-Max), 3.41–3.61 GHz (Wi-Max), and 5.05–6.01 GHz (WLAN). Insertion loss at the three resonating frequencies 2.51, 3.51, and 5.47 GHz are −45.80, −40.02, and −52.67 dB, respectively. This paper incorporates different sections including the design of unit cell and finite array, simulation results of the structure, and conclusion.

Keywords Loop elements · Hexagonal-shaped · Multi-band frequency-selective surface (FSS) · Band-stop filter

1 Introduction

Today the world is being prone to move in a direction where everything has become wireless, the main concern is interference and security. Frequency-selective surface is a renowned name to solve this problem [1]. It works as a band-stop filter so that a particular band is stopped and rest of the frequencies will be passed as conventional antenna has not been good reflectors, FSS has been used for augmentation of impedance bandwidth, gain, and angular stability [2, 3]. Many algorithms are used

K. Aseri (✉) · S. Yadav (✉)
Government Women Engineering College Ajmer, Ajmer, India
e-mail: kiranrids18@gmail.com

S. Yadav
e-mail: sanjeev.mnit@gmail.com

© Springer Science+Business Media Singapore 2016
S.C. Satapathy et al. (eds.), *Proceedings of International Conference on ICT for Sustainable Development*, Advances in Intelligent Systems and Computing 408, DOI 10.1007/978-981-10-0129-1_31

to analyse FSS, such as equivalent circuit method, periodic method, spectral domain method, vector-spectral domain method, finite element method, finite difference time domain method, T-matrix method, and many more [4, 5]. FSS has many applications as band-pass radome for missiles [6–8]. It also has been used in radio astronomy [9, 10]. Its multiband characteristic has been used for deep space exploration vehicle [11]. Other types of frequency-selective surface which are mostly used in today's world are: band-pass frequency-selective surface and frequency-selective surface as an absorber. Band-pass FSS has been used to enhance the directivity and to pass a specific band with low loss. FSS as an absorber has been used to absorb a specific band so that it is neither reflected nor transmitted, it also protects from back reflection.

The proposed multi-band structure is realized in CST microwave studio for three stop bands: 2.24–2.70 GHz (Wi-Max), 3.41–3.61 GHz (Wi-Max), and 5.05–6.01 GHz (WLAN). This paper incorporates different sections including the design of unit cell and finite array, simulation results of the structure, and conclusion.

2 Design of the Superstrate

The proposed unit cell superstrate is a single–sided, square-shaped substrate with dimensions 22.05 mm × 22.05 mm and a square-shaped metallic patch incorporated with it of dimensions 21.25 mm × 21.25 mm. Three variable slots are embedded within the patch for the three stopbands. The loop 1 has been employed for Wi-Max band 2.24–2.70 GHz. The loop 2 has been employed for Wi-Max band 3.41–3.61 GHz. Hexagonal-shaped slot has been employed for WLAN band 5.05–6.01 GHz. List of parameters is shown in Table 1. The front view and perspective view with boundary conditions applied of unit cell is illustrated in Figs. 1 and 2, respectively.

Table 1 List of parameters

Name of dimension	Symbol of dimension	Value in mm
Height of the substrate	h	2
Width of the substrate	W	22.05
Width of the patch	W1	21.25
Width of the Loop 1	W2	19.4
Length of the Loop 1	L	21
Width of the Loop 2	W4	14.75
Length of the Loop 2	L1	14.1
Thickness of the Loop 1 and the Loop 2	L2	0.25
Width of the hexagon	W3	13.806
Length of the hexagon	L3	13.1

Fig. 1 Front view of unit cell

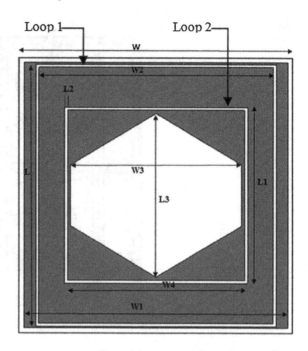

Fig. 2 Perspective view of unit cell with boundary condition applied

Fig. 3 Front view of finite
array

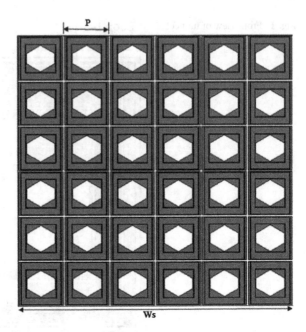

On the basis of the unit cell, a finite array of 6 × 6 element (unit cell) has been
also designed which has width of substrate (Ws) of 132.3 mm. Each element of the
array has been repeated after the periodicity (P) of 22.05 mm as illustrated in Fig. 3.

3 Results

In this section the insertion loss, surface current, and parametric analysis have been
investigated for the mechanism of the proposed FSS.

3.1 Transmitted Electric Field Versus Frequency of Unit Cell

The proposed structure works as a band-stop filter for three bands: 2.24–2.70 GHz
(Wi-Max), 3.41–3.61 GHz (Wi-Max) and 5.05–6.01 GHz (WLAN). Insertion loss
at the three resonating frequencies: 2.51, 3.51, and 5.48 GHz are −45.80, −40.02,
and −52.67 dB, respectively. All the three slots which are embedded affect each
other so by adjusting and controlling the dimensions of the structure desired frequency response has been achieved as shown in Fig. 4.

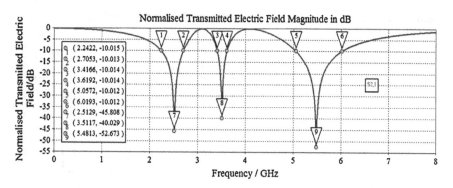

Fig. 4 Simulated insertion loss of unit cell

3.2 Transmitted Electric Field Versus Frequency of Finite Array

A finite array of 6 × 6 elements has been designed on the basis of unit cell. The electromagnetic properties of finite array of FSS have been investigated. Insertion loss at the three resonating frequencies 2.49, 3.50, and 5.46 GHz are −45.59, −50.44, and −53.98 dB, respectively, as shown in Fig. 5.

3.3 Surface Current Distribution at Resonating Frequencies

Surface current distribution has been investigated to study the electromagnetic behavior of the structure. At 2.51 GHz, maximum current is concentrated on the edge of loop 1 as shown in Fig. 6. At 3.51 GHz, maximum current is concentrated

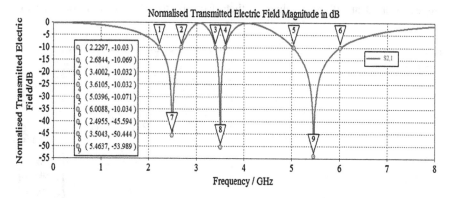

Fig. 5 Simulated insertion loss of finite array

Fig. 6 Surface current at
2.51 GHz

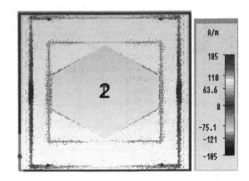

Fig. 7 Surface current at
3.51 GHz

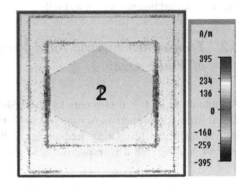

on the edges of loop 2 as shown in Fig. 7. At 5.48 GHz, maximum current is concentrated on the edge of hexagon as shown in Fig. 8.

3.4 Parametric Analysis

Different parameters are varied to get the desired frequency response. The incident angle is varied at different values of theta from 0° to 90°, which shows that the

Fig. 8 Surface current at
5.48 GHz

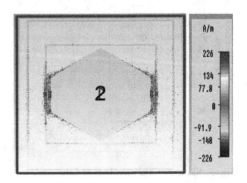

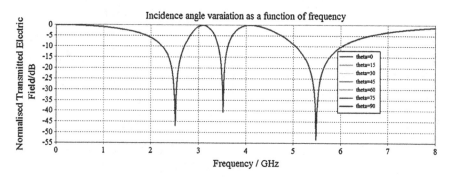

Fig. 9 Simulated incidence angle variation of unit cell

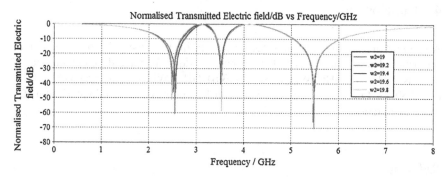

Fig. 10 Simulated W2 variation of unit cell

impedance bandwidth and insertion loss does not vary with different incident angles and provides a stable structure as shown in Fig. 9. Other parameters were also varied which has been illustrated in Figs. 10, 11 and 12.

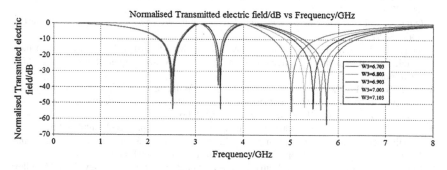

Fig. 11 Simulated W3 variation of unit cell

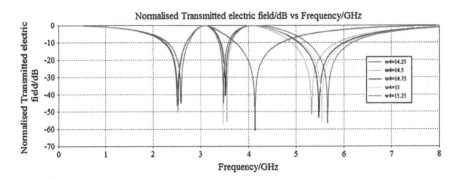

Fig. 12 Simulated W4 variation of unit cell

4 Conclusion

A compact and a novel FSS has been proposed in this paper with multiband characteristics. Three different slots have been embedded for three different stop-bands such as 2.24–2.70 GHz (Wi-Max), 3.41–3.61 GHz (Wi-Max), and 5.05–6.01 GHz (WLAN). In future the proposed work can be optimized with miniaturization for space exploration.

References

1. Munk, B. A. (2000). *Frequency selective surfaces: Theory and design*. New York: Wiley-Interscience.
2. Chen, H.-Y., & Tao, Y. (2011). Bandwidth enhancement of a U-Slot patch antenna using dual-band frequency-selective surface with double rectangular ring elements. *MOTL, 53*(7), 1547–1553.
3. Yadav, S., Peswani, B., & Sharma, M. M. (2014). A novel band pass double-layered frequency selective superstrate for WLAN applications. *IEEE conference (Confluence)*, pp. 447–451.
4. Qing, A., & Lee, C. K. (2010). *Differential evolution in electromagnetics*. Heidelberg, Berlin: Springer-Verlag.
5. Yadav, S., Peswani, B., Choudhury, R., & Sharma, M. M. (2014). Miniaturized band pass double layered frequency selective surface superstrate for Wi-Max applications. *ISWTA, IEEE symposium*.
6. Sung, G. H.-H, Sowerby, K. W., Neve, M. J., & Williamson, A. G. (2006). A frequency selective wall for interface reduction. *Wireless Indoor Environments. Antennas and Propagation Magazine, IEEE, 48*(5), 29–37 (Oct 2006).
7. Werner, D. H., & Lee, D. (2000). A design approach for dual polarised multiband frequency selective surface using fractal elements. In *IEEE International Symposium on Antennas and Propagation Digest* (vol. 3, pp. 1692–1695). Salt Lake City Utah (July 2000).
8. Wu, T. K. (1995). *Frequency selective surface and grid array* (pp. 5–7). New York: A Wiley Interscience publication.

9. Das, S., Sarkar, D., Sarkar, P. P., & Chowdhury, S. K. (2003). Experimental investigation on a polarization independent broadband frequency selective surface. In *Proceedings of National Conference held in KIIT*, Bhubuneswar (March 2003).
10. Wu, T. K. et al. (1992). Multi ring element FSS for multi band applications. *Paper presented at the International IEEE AP-S Symposium*, Chicago (1992).
11. Chatterjee, A., Biswas, S., Sarkar, C. D., & Sarkar, P. P. (2011). A: Polarization independent compact multi-band frequency selective surface. In *International Conference on Current Trends in Technology, NUiCONE*.

AXSM: An Automated Tool for XML Schema Matching

Dhaval Joshi and S.V. Patel

Abstract For data migration, management, restructuring, and reorganizing operations, there is a growing demand of XML schema matching techniques. Most of the XML schemas are heterogeneous as they are independently designed by schema designers for individual customer requirements. Schema matching becomes a difficult task when schemas are large and hence there is a need for automated and accurate XML schema matching solutions to ease the process of mapping and migration. Considerable research has been done for XML schema matching automation. However, fully-automated solution is yet to be achieved. This paper makes an attempt in the direction by presenting an automated tool to match heterogeneous XML schemas with better results as compared to a standard tool COMA++.

Keywords XML schema matching · Heterogeneous schemas · Similarity measures · Joint word segmentation

1 Introduction

XML schema matching is always a challenging task for large-scale applications as they are designed by distinguished schema designers for different requirements of customers. Though schemas are designed for homogeneous customers, it contains heterogeneities at different levels.

D. Joshi (✉)
Veer Narmad South Gujarat University, Surat, Gujarat, India
e-mail: joshi.dhaval@hotmail.com

S.V. Patel
Department of Computer Science, Veer Narmad South Gujarat University,
Surat, Gujarat, India
e-mail: patelsv@gmail.com

© Springer Science+Business Media Singapore 2016
S.C. Satapathy et al. (eds.), *Proceedings of International Conference on ICT for Sustainable Development*, Advances in Intelligent Systems and Computing 408, DOI 10.1007/978-981-10-0129-1_32

When the problem of XML schema matching arises, researchers have made semiautomated and automated solutions to get better results for various business needs like data migration, management, restructuring, reshaping, etc., to perform operations on data. However, they provided good solutions for XML schema matching, we always strive for more automation to make schema matching simple yet accurate.

So, we present here a new automated model for XML schema matching, which provide better results compared to standard tools like COMA++.

This research paper is organized in six sections further. Section 2 provides overview of various schema matching measures for matching XML schemas. Related approaches developed by other researchers are described in Sect. 3. Our automated XML schema matching technique (AXSM) is explained in Sect. 4. Section 5 contains experiments performed on AXSM and COMA tools using different XML schemas and comparison results. Outcomes are discussed in Sect. 6 using charts and actual values of comparison results. Section 7 contains conclusion and future work.

2 Schema Matching

Schema matching is performed using various measures such as name similarity, data type similarity, constraint similarity, annotation similarity, element similarity, and structural similarity.

Name similarity measure is based on two different types of measures, such as syntactic measure and semantic measure. Syntactic measure compares by character sequence. If a group of characters of one name is a part of other name, it defines them as matching names. (e.g., Name and PersonName). Semantic measure is based on natural language processing techniques like synonym match, semantic distance match, etc., where words are compared with their semantic meaning. (e.g., trainee and apprentice).

Data type similarity compares elements by matching their data types though their names are different. XML schema standard has 44 built-in types, which also play important roles for XML schema matching.

Constraint similarity is based on cardinality dependency on elements as minimum occurrence and maximum occurrence of an element of one schema in XML document to compare with other schema.

Annotation similarity is based on content description matching of two different XML schema elements by ranking with the threshold value to generate the matching result. To perform description content matching, syntactic, and semantic comparisons play a major role.

XML element similarity is based on its characteristics like name, data type, range, constraint, etc., to compare XML elements. Generally, name similarity measure and data type similarity measure are widely used by many researchers.

XML document schemas are also compared with their structures. First two nodes are compared by element similarity and then their ancestor, siblings, and decedents are compared. Based on the comparison values, XML schema structures are finalized.

3 Related Work

Do and Rahm [1] developed the COMA schema matching system as a platform to combine multiple matchers in a flexible way. They provided a large spectrum of individual matchers aiming at reusing results from previous match operations, and several mechanisms to combine the results of matcher executions. They used COMA as a framework to comprehensively evaluate the effectiveness of different matchers and their combinations for real-world schemas. The results obtained by COMA show the superiority of combined match approaches and indicated the high value of reuse-oriented strategies.

Rahm, Arnold, and their team [2] narrated the evaluation of COMA tool since its basic design and development in 2002 and it became a commercial matching tool [3]. They highlighted positive features of COMA tool such as multi-matcher architecture, generic approach, effective configuration, GUI, customizability, advanced matching strategies, and repository developed during a decade long evolution journey, which makes schema matching more effective.

Thang and Nam [4] presented a solution for XML schema automated matching problem which produces semantic mappings between corresponding schema elements of given source and target schema which is based on combining linguistic similarity, data type compatibility, and structural similarity of XML schema elements. Their solution contains two important tasks, like modeling XML schema and element similarity measure. In modeling XML schema, directed labeled schema graph with constraint sets is defined over both nodes and edges with the help of ideas presented in [5–7] which also classify schema graph nodes into atomic nodes and complex nodes. Each leaf node in the schema graph has an atomic value (string, integer, date, etc.), list value, or union value and each internal node in the schema graph has a complex content, which refers to some other nodes through directed labeled edges. They computed element similarity by combining linguistic similarity measure, data type compatibility measure, and structural similarity measure. For linguistic similarity, they combined two basic solutions among them one is presented in [5] and second is presented in [8] as Hirst and St-Onge algorithms with the concept of an allowable path to exploit WordNet. For data type compatibility measure, they used a data type compatibility table that gives a similarity coefficient between two given XML schema built-in data types, as defined in [5]. For structural similarity measure, node context matching is performed for two nodes using path similarity measure and context similarity measure. Their algorithm requires long computing time for schema matching as it passes through various similarity measures as well as their XML modeling is too complex for execution.

Khalid et al. [9] introduced a technique for large-scale schema matching using tree mining. They investigated scalability with respect to time performance in the context of approximate mapping where tokenization, abbreviations, and synonyms were used for the linguistic matching of node labels. The matching strategy was hybrid and optimized for schemas in tree format. In their technique, they labeled each node and assign values to it, which is a complex formation of tree and also needs more computation for schema matching.

Rajesh and Srivatsa [10, 11] presented a decision tree-based approach for XML schema matching. They defined similarity characteristics exhibited by the elements at different levels as leaf–leaf, root–root, interior–interior, interior–root and interior–leaf similarities by creating their match filters. Linguistic, root, descendant, and sibling similarity threshold were defined to tune the system for evaluation. But their solution requires a lot of combinations to generate decision tree and by giving various comparison results with Cupid [5], Similarity Flooding [6], COMA [1–3], QMatch [12], and their approach [11] using schema sample sets. Also they state that COMA generates better results compared to other well-known matchers for their sample sets but their approach give 1 % improvement in precision using same sample sets.

Aruna and Veena [13] presented XML schema matching using path similarity measure that contains four types of measures like longest common subsequence, string position, breaching string token values, and alteration in the element position. They defined four positive parameters ranging between 0 and 1 to collaborate all four measures to calculate overall similarity. But the value of parameters need to be decided as per application requirement and comparison needs to be done by all four measures, which require more computation with complexity.

It can be seen that, these tools and techniques for XML schema matching are automatic or semiautomatic. Their limitations have been mentioned above in the discussion of the respective tools. To overcome such limitations we need a better solution with high accuracy. Therefore, we present a new automated model which is capable enough to match multi-named elements with better accuracy for XML schema matching.

We have also compared performance of our technique with COMA, a standard tool for schema matching, on various datasets and found it more accurate for the same level schema heterogeneity.

4 AXSM—Automated XML Schema Matcher

We have modeled an automated XML schema matching using combined approach and measures; and constructed a tool named AXSM (Automated XML Schema Matcher). Our model is diagrammatically described in Fig. 1 which contains five main phases and three parsers which are described in detail with examples.

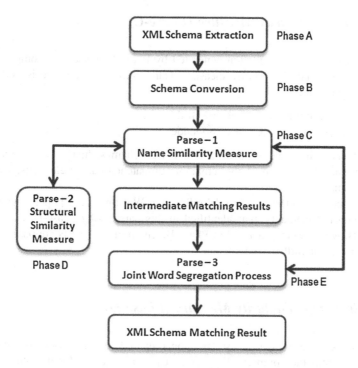

Fig. 1 Automated XML schema matcher (AXSM) model

4.1 XML Schema Extraction (Phase-A)

To compare the schema of two XML documents we need to extract schema contents from it and that contents are passed on to the next phase for systematic text formation as specified in Phase-B.

4.2 Schema Conversion (Phase-B)

The extracted schemas in Phase-A of both XML documents need to be converted to a specified text with predefined separator such that all schema elements are identified by complete path of the document. This path identification structure is used for comparison of two different XML elements by parsing the list with various comparison techniques specified from Phase-C to Phase-E. For, e.g., Event#Purpose.

4.3   Name Similarity Measure (Phase-C)

In this phase, text representation generated by Phase-B is used for comparison of elements using name similarity measure. Name similarity measure is performed using two measures as:

- Syntactic similarity, e.g., Date ⇔ BirthDate
- Semantic similarity, e.g., Sex ⇔ Gender

For every pair of elements from two XML schemas, first syntactic similarity measure is used, and then on the remaining unmatched pair of elements, semantic similarity measure is used.

For semantic similarity we use WordNet and Watson [11, 12] open source API. Each compared elements' pair identified in this parse using Phase-D, is removed from lists, which were generated in Phase-B. The remaining lists are passed to the next Phase-E for further comparison.

4.4   Structure Similarity Measure (Phase-D)

In this phase, the compared list and noncompared lists are used for structural similarity. From the compared list, select a pair and search for the other related elements from noncompared lists using element similarity measure to match those elements. Related elements from the noncompared lists are considered from any of parent, child, sibling, inner attribute, etc. This phase is used by Phase-C to identify structural similarity for finalizing the matching result.

4.5   Joint Word Segregation Process (Phase-E)

Phase-C cannot identify joint words as they are not compared by syntactic or semantic similarity measures. Decomposition of joint words needs to be performed for comparison. For decomposition of element name, we use m-gram algorithm to get combination of words less than joint word length. Each identified word is checked using dictionary and meaningful word is only used for matching using name similarity measures as defined in Phase-C. After receiving matched elements from Phase-C, the final matching result is published to user.

e.g. Phone Number ⇔ Mobile Number and Opening Date ⇔ Start Date

Hence, our model contains three important parsers for XML schema matching, we receive more accurate results compared to other standard XML schema matching tools like COMA++. Experiments and results are shown in Sect. 5.

5 Experiments and Results

We have tested AXSM tool and COMA++ tool with many sample sets of heterogeneous XML schemas. Here we present only three sample sets with results generated by both. We used accounting, academic, and event sample schema sets.

Accounting XML schemas matching results generated by COMA++ and AXSM are shown in Figs. 2 and 3, respectively, which shows that we can get more number of matches in COMA++ results as it compares one to many attributes like Traders -> Address ⇔ (Dealers -> Address and Customer -> Address) for accounting schema. For the same schema, AXSM generates only one-to-one and accurate results as Traders -> Address ⇔ Dealers -> Address only due to Phase-D.

Similarly, we used academic and event XML schemas to compare matching results by COMA++ and AXSM tools and found more accurate results by AXSM.

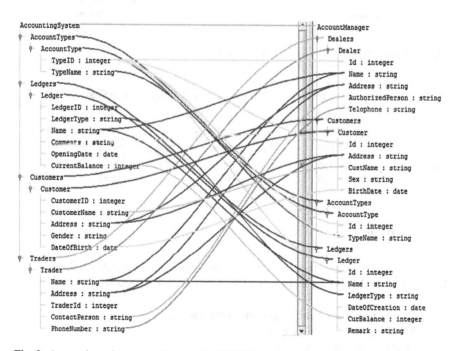

Fig. 2 Accounting schema matching result of COMA++

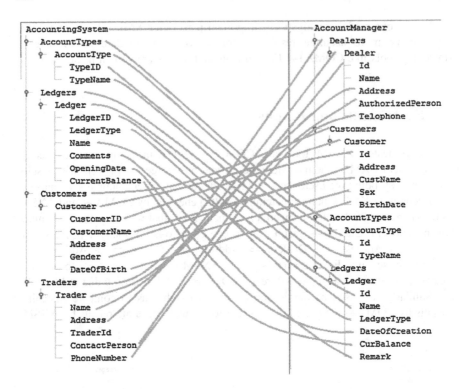

Fig. 3 Accounting schema matching result of AXSM

Table 1 Precision, recall, and F-measure values table for COMA++ and AXSM

Project Type	Precision		Recall		F-measure	
	COMA++	AXSM	COMA++	AXSM	COMA++	AXSM
Accounting	83.33	96.00	76.92	92.30	80.00	94.11
Academic	86.95	91.30	76.92	80.76	81.63	85.71
Event management	70.73	97.29	78.37	97.29	74.35	97.29

6 Discussion

We calculated precision, recall, and F-measure values using formulas 1, 2, and 3 for COMA++ and AXSM are shown in Table 1 and their graphical representations are shown in Fig. 4.

Notations used for calculating all measure values:

M_E Mapping produced by experts

M_M Mapping generated by tool

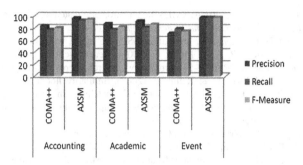

Fig. 4 Precision, recall, and F-measure values for accounting, academic, and event Schemas of COMA++ and AXSM

$$\text{Precision}(P)P = |M_E(\text{intersect})M_M|/M_M \tag{1}$$

$$\text{Recall}(R)R = |M_E(\text{intersect})\,M_M|/M_E \tag{2}$$

$$\text{F-measure}(F)F = (2*P*R)/(P+R) \tag{3}$$

By analysing the results of both, we found AXSM generates more accurate results for the given schema sets compared to COMA++.

7 Conclusion and Future Work

In order to provide a better automated XML schema matching tool, we presented an approach with linguistic matcher that used structural similarity and joint word segmentation processing. Compared to a well-known generic schema matching tool COMA++, our approach generates results with better accuracy. We have been extending this work for multilevel structural heterogeneity.

References

1. Do, H. H., & Rahm, E. (2002). COMA—A system for flexible combination of schema matching approaches. In *Proceedings of the 28th International Conference on Very Large Data Bases Conference (VLDB-02)*, pp. 610–621.
2. Massmann, S., Raunich, S., Aumueller, D., Arnold, P., & Rahm, E. (2011). Evolution of the coma match system. In *Proceedings of the Sixth International Workshop on Ontology Matching*. Bonn, Germany.
3. Rahm, E. (2011). Towards large-scale schema and ontology matching. In *Schema Matching and Mapping*. Springer

4. Thang, H. Q., & Nam, V. S. (2010). XML schema automatic matching solution. *International Journal of Electrical, Computer and System Engineering, 4*(1). World Academy of Science, Engineering and Technology, *4*(3). International Science Index. waset.org/publications/3391.

5. Madhavan, J., Bernstein, P. A., & Rahm, E. (2001). Generic schema matching with Cupid. MSR Tech. Report MSR-TR-2001-58. http://www.research.microsoft.com/pubs.

6. Melnik, S., Garcia-Molina, H., & Rahm, E. (2002). Similarity flooding: A versatile graph matching algorithm and its application to schema matching. In *Proceedings of the 18th International Conference on Data Engineering*. http://dbpubs.stanford.edu/pub/2001-25. (Extended Technical Report, 2001).

7. Boukottaya, A., & Vanoirbeek, C. (2005). Schema matching for transforming structured documents. In *DocEng'05*, pp. 101–110. Bristol, United Kingdom.

8. Budanitsky, A., & Hirst, G.: Semantic distance in WordNet: An experimental, application oriented evaluation of five measures. In *Workshop on WordNet and Other Lexical Resources, Second Meeting of the North American Chapter of the Association for Computational Linguistics*, pp. 29–34. Pittsburgh, PA.

9. Saleem, K., & Bellahsene, Z. (2007). Ela hunt: Performance oriented schema matching. In *DEXA'07 Proceedings of the 18th International Conference on Database and Expert Systems Applications*, pp. 844–853. ISBN:3-540-74467-3, 978-3-540-74467-2.

10. Rajesh, A., & Srivatsa, S. K. (2009). Learning to match XML schemas—A decision tree based approach. *International Journal of recent Trends in Engineering, 2*(1), 58–62. ACADEMY Publisher.

11. Rajesh, A., & Srivatsa, S. K. (2010). XML schema matching—Using structural information. *International Journal of Computer Applications, 8*(2), 34–41.

12. Tansalarak, N., & Claypool, K. T. (2007). Qmatch—Using paths to match XML schemas. *Data & Knowledge Engineering, 60*, 260–282.

13. Tiwari, A., & Trivedi, V. (2012). Estimating similarity of XML schemas using path similarity measure. *International Journal of Computer Applications & Information Technology, 1*(1), 34–37.

Development of Web Map Service for OpenStreetMaps (OSM) Data

Vinod Kumar and Vivek Anand

Abstract Location-based information plays an important role in many defense applications. OpenStreetMap (OSM) data are very vast and freely available. The information contained in the OSM data cannot be made intelligent without visualizing in a good web mapping interface. The commonly available file format to read the OSM data is .shp or .kml. There are some format converters available but they result in loss of attributes. Hence, the need for the development of an interface which can read and render the OSM data without losing the vital attributes arises. This development effort fulfills this need effectively.

Keywords Map · Cartography · Openstreetmap (OSM) · Geoserver · PostGIS · Style layer descriptor (SLD) · Open source · Google map

1 Introduction

As we all know that today in this twenty-first century almost for everything computer science plays an important role. The role of computer science can be direct or indirect, i.e., in order to make everything more accurate, precise, and fast. We are using computer science, e.g., for calculation calculator was invented, to get the accurate weight of any object many digital weighing machines have evolved, in medical field also every equipment are digitalized from ECG machines to whole body diagnostic machines. Many people are developing expert systems in order make machines more intelligent. This research paper focusses on how are we using computer science for digitalization of maps.

V. Kumar (✉) · V. Anand
Department of Computer Science, Central University of Rajasthan,
Ajmer, Rajasthan, India
e-mail: vinod_cs@curaj.ac.in; vinod242306@gmail.com

V. Anand
e-mail: vivekanand22389@gmail.com

© Springer Science+Business Media Singapore 2016
S.C. Satapathy et al. (eds.), *Proceedings of International Conference on ICT for Sustainable Development*, Advances in Intelligent Systems and Computing 408, DOI 10.1007/978-981-10-0129-1_33

Maps are important to visualize the location of a particular area. Maps are being used by peoples around the world from ancient times. In ancient history also we can see that how important is map. Geographic maps of territories have a long tradition and exist from ancient times. In computer science, whole world can be viewed as a large grid of points and each point is denoted by longitude and latitude. If we join some neighbor points then we will get a line data, and if we form a connected line then it forms polygon data. And due to this in computer science a map is formed by combining point, line, and polygon data.

There was a time when defense people used to carry paper maps and area models for their military operation. With this older way of rendering the area makes their job tedious and time consuming as to search any location on a paper map waste not only time but also lots of energy.

Military operations are never simple operations; they should be quick and accurate. For some operations, they have to spend many days in forests, deserts or hilly areas where anyone can easily lose their location. In sensitive border areas, they have to take care of country's border areas. The traditional way of paper maps cannot help them in such areas as after few days the paper maps will lose data on its folded area, and can misguide anyone so it become useless. For this reason all these soldiers need to render some digitalized map which will not only be helpful but also save their time.

Some of example uses in military are tracking the military convoy on maps, which may carry arms or military goods and transfer it from one unit to another unit. Another example is to locate some location of a sensitive area on the map and for all these, open street map data are perfect data because of its dense and accurate information.

2 OpenStreetMap Data

OpenStreetMap is a combined and open project of editable map. This project was started by Steve Coast from the United Kingdom in 2004, with a revolutionary idea that why we should pay for map data. Before the start of this project Google was the only map data source, which was charging for its map data. But Steve's concept was that if we all contribute geospatial data of our neighborhood, then soon we will have a huge and dense geospatial data for our map which will be free.

The geospatial data of the world for any reason, we can get freely from this project. And these data come in *.osm file format. This *.osm file contains three different data.

Node data: These are the point data, which contain latitude and longitude information with its name and other information. A node can represent the location of school, collages, temple, offices, houses, etc.

<node   id="1880367997"   visible="true"   version="1"   changeset="12840803"   timestamp="2012-08-24T05:37:07Z" user="PlaneMad" uid="1306" lat="26.4734485" lon="90.8673631"/>

Way data: These are the line data which contain reference id of different node data. They also contain name and other information of way data. Way data can represent different roads, rivers, areas, etc.

```
<way id="126713934" visible="true" version="4" changeset="16750278" timestamp="2013-06-29T10:52:13Z" user="Ashim"
uid="398735">
  <nd ref="1403968697"/>
  <nd ref="2365493527"/>
  <nd ref="2365493525"/>
  <nd ref="2365493523"/>
  <nd ref="2365493522"/>
  <nd ref="2365493498"/>
  <nd ref="2365493496"/>
  <nd ref="2365493494"/>
  <nd ref="714907850"/>
  <nd ref="2365493492"/>
  <tag k="electrified" v="no"/>
  <tag k="gauge" v="1676"/>
  <tag k="railway" v="rail"/>
  <tag k="source" v="landsat"/>
  </way>
  </way>
```

Relation data: These are collections of line and point data, which contain reference id of node and way data. They also contain other important information with these data. A relation data can contain a collection of airport, collection of roads, etc.

```
<relation    id="958721"    visible="true"    version="143"    changeset="19767512"    timestamp="2014-01-02T14:31:43Z"
user="Oberaffe" uid="56597">
  <member type="relation" ref="2810382" role=""/>
  <member type="relation"ref="2881251" role=""/>
  <tag k="name" v="National Highways"/>
  <tag k="network" v="IN:NH"/>
  <tag k="type" v="network"/>
  </relation>
```

3 Style Layer Descriptor

Geospatial data contain useful data used for making maps. But it does not contain any visual component to make maps visually esthetic but it must be styled, means it must styled to add color, thickness, and other visual attributes. Without styling even after processing geospatial data, we will not get complete information from the map.

In GeoServer by developing proper sld file we can differentiate between different attributes. For example, road and river both are line data only but to differentiate between both road and river using sld we can make our task easy, as sld only helps to view our geospatial data in map properly. Using sld we can easily differentiate road and river.

In GeoServer, geospatial data for rendering can be viewed into three different classes, i.e., points class, lines class, and polygons class. Lines class data are used to represent line data like roads and river and in sld its representation can be modified using line symbolizer. Polygon class data can be data which represent area, boundary data. Polygon symbolizer is used in sld to represent these data. And last class is the point class which used to locate location data like any school, temple, etc. And this can be symbolized in sld using point symbolizer.

And one important symbolizer that used in sld in wide range is text symbolizer which helps to visualize text information on map like name of any road, building, etc. Different features can be styled using zoom level. For filtering features in sld, different logical operator can be used such as And, Or, and Not operators.

4   Problem Statements

We can see that after digitalization of any equipment that equipment will give more accurate and fast results. For this reason to make rendering of map more accurate, fast, and portable with dense information, we need to digitalize the map.

The other important problem is to use map offline. There are lots of digital map available on the net like Google map, which we can use for rendering but the scope of its use is limited as we always need an Internet connection for using map.

The last problem because of which we proposed this paper is that GeoServer can easily read different geospatial data like *.shp files, *.kml files but there is no way to read *.osm data and render these data as a digital map on GeoServer.

5   Solutions

In order to read and render *.osm data on GeoServer, we are using PostGIS database as an intermediate geospatial database. In this approach, we will read a *.osm file and retrieve its important data and information sand store in PostGIS. So by creating PostGIS store in GeoServer, we can easily render its information on OpenLayer of GeoServer.

6   Algorithm to Process OSM Data

Step 1 Create three different tables to store node, way, and relation information
All three tables initially contain common column id, timestamp, user id, users, version, change set with location (point data type), way (line data type), and relation column (geometry collection data type), respectively.

Step 2 Read the OSM data sequence way and set status, node_key, way_key, relation_key value as null

Step 3 if first element is OSM then continue reading else go to Step 5

Step 4 Read the next element

If next element node then

Update status value into node then insert id, timestamp, changeset, user id, user and longitude, and latitude value into node table

Go to Step 4

If next element is way then

Update status value into way and then insert insert id, timestamp, changeset, user id, user value into way table

Retrieve the location value from Node table using reference id and store in Way table in ways column

Go to Step 4

If next element is relation

Then update status value into relation and then insert id, timestamp, changeset, user id, user value into Relation table

Retrieve the location and ways value from node and way table respectively and store in relation column of Relation table

Go to Step 4

If the next element is tag then check status value

If it is node then

Read the attribute K value and check its presence in container node_key

If its presence is false then update the Node table with K value as column name and add this value into node_key and also update its value with value of attribute V

And if its presence is true then simply update its value with value of attribute V

If it is way then

Read the attribute K value and check its presence in container way_key

If its presence is false then update the Way table with attribute K value as column name and add this value into key_way and also update it value with value of attribute V

And if its presence is true then simply update its value with value of attribute V

If it is relation then

Read the attribute K value and check its presence in container relation_key

If its presence is false then update the Relation table with attribute K value as column name and add this value into key_relation and also update its value with the value of attribute V

And if its presence is true then simply update it value with value of attribute V

Go to Step 4

If next element is equal to start element

Then go to Step 5

Step 5 Stop reading OSM data

Note: status is used to keep track of the status of element whether it is node, way, or relation, which will help us in updating the table in PostGIS and node_key, way_key, and relation_key are the container type data

7 Flow Diagram to Read *.osm File

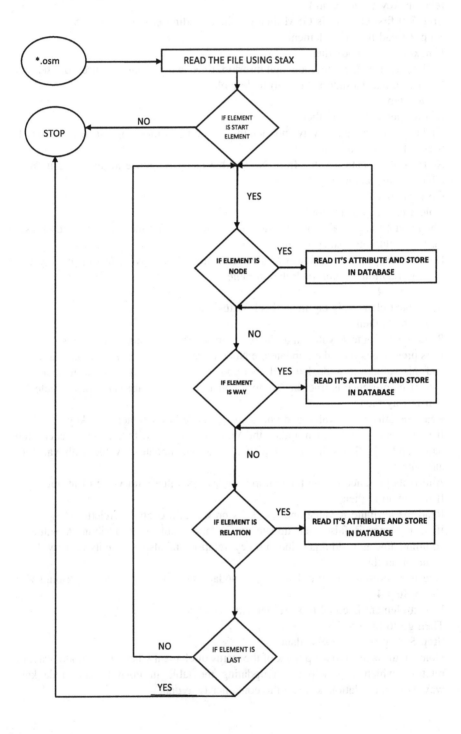

8 Results

After using the developed interface for OSM data, we can easily use any *.osm file of any particular area and after processing it we can view the map of that area on any geospatial server. In our work, we are rendering data on GeoServer (Figs. 1 and 2).

Map without proper style layer We can see that how much it is important to make sld. Because without proper sld we can render the layer but all roads, rivers, buildings look like the same and due to which we are not able to get meaningful information (Figs. 3 and 4).

Fig. 1 Karnataka state's map after processing Karnataka state OSM

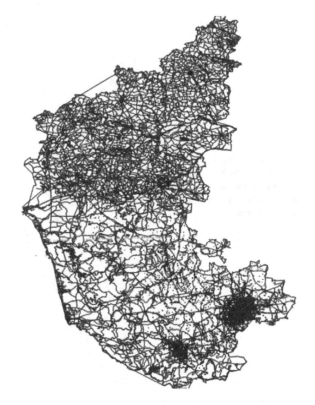

Fig. 2 Karnataka state's
zoom map

Fig. 3 Karnataka state's map
after processing Karnataka
state OSM data using
developed interface and using
developed point and line sld

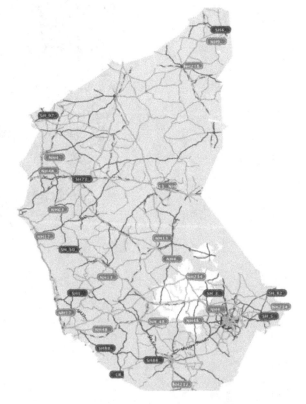

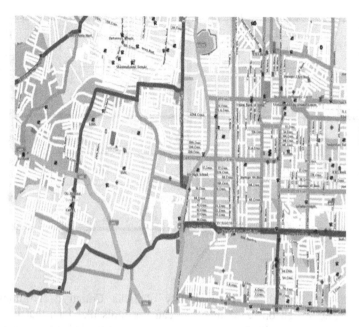

Fig. 4 Zoom map of Karnataka state

9   Conclusion

This paper proposes the development of an interface to read *.osm files and renders it on GeoServer. It also shows the development of a sld file for standard view of map which we get while rendering it on GeoServer. It also shows the importance of development of SLD file while rendering.

10   Future Work

In this world for every work we always have some chance for improvement because any algorithm or task is never 100 % efficient. That is why we can see some hope of development in this use of *.osm files. For fast rendering of the maps, we can use tiling. Tiling is a way of rendering maps into 256 × 256 sizes of jpeg image tiles and together all tiles form a complete map. To implement tiling we can use Mapnik tool, GeoWebCache and many more tool, which can be used to develop tile-based maps (Fig. 5).

By implementing ontology, we can search any location on map using general query for example to search location of Jaipur city on map we can then make query

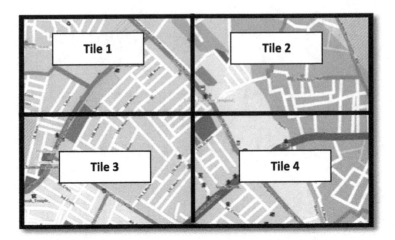

Fig. 5 Example of tiling map

like "capital of Rajasthan" or "Pink city of India Country." To implement ontology-based search on OSM map, we can use tools like Jena that is used to populate the instances with Java. We can use geoname predefined ontology.

References

1. Boundless. http://opengeo.org/technology/openlayers/.
2. Boundless. http://opengeo.org/technology/postgis/.
3. GeoServer. http://docs.geoserver.org/trunk/en/user/styling/sld.
4. GitHub. https://github.com/mbasa/GeoThematics.
5. GISE Advanced Research Lab. www.gise.cse.iitb.ac.in.
6. GISE Advanced Research Lab. http://www.gise.cse.iitb.ac.in/iitbgis/jsp/welcome.jsp.
7. OGC Making location count. http://www.opengeospatial.org/standards.
8. OpenStreetMap. http:// www.openstreetmap.org.

Improved Indoor Positioning Using RSS and Directional Antenna Integrating with RFID and Wireless Technology

Amol D. Potgantwar and Vijay M. Wadhai

Abstract In today's era finding the location and navigation of an object is essential for many reasons. The primary reason is locating the object, and navigation helps systems to make decisions faster. Secondary reason is we may detect fault and posses faster processing and operation. Another side of development and research is for cost-effective techniques involving indoor and outdoor systems. Many researchers are working on indoor positioning and developing lightweighted solutions using radio frequencies. Managing and monitoring radio frequencies in an indoor system have physical barriers; they also have some technological limitations where we can provide solutions. The solutions for indoor localization and tracking are GPS signal, Bluetooth, infrared, RFID, and wireless LAN. All these radio frequencies deal with a crucial issue of gaining RSS (Received Signal Strength). In this work, we are focusing on RFID-based indoor environment where RSS is our main attention, we also target on gaining efficiency and accuracy, stability, robustness. The proposed algorithm of this work is mainly targeting with directional antenna by which we can overcome the physical and other barriers of radio frequency. These kinds of indoor environments such as health care, loyalty management system, automatic parking allotment system, tracking services for older people, or customers inside living communities, mobile robot, logistics system, etc., can be facilitated using directional antenna and RFID tagging for navigation indoor positioning.

Keywords RSS · Directional antenna · DAWN · RFID · RF · Wi-Fi

A.D. Potgantwar (✉) · V.M. Wadhai
Department of Computer Science, Sant Gadgebaba Amravati University,
Amravati, MS, India
e-mail: amolp639@gmail.com

© Springer Science+Business Media Singapore 2016
S.C. Satapathy et al. (eds.), *Proceedings of International Conference on ICT for Sustainable Development*, Advances in Intelligent Systems and Computing 408, DOI 10.1007/978-981-10-0129-1_34

319

1 Introduction

Ranging procedure plays a fundamental role in the localization system, in which received signal strength (RSS) based ranging technique gets the most attraction [1]. To forecast the position of an unspecified node, RSS computation is a simple and dependable method for distance estimation. In indoor context, the accuracy of the RSS-based localization technique is affected by strong variation, specially usually containing considerable amounts of metal and other such reflective materials that influence the propagation of radio frequency signals in nontrivial ways, creates multipath effects, dead spots, noise, and interference. This paper presents improved indoor positioning using RSS and directional antenna integrating with RFID and wireless technology, which is especially more convenient to support context-aware computing.

The registration process is exempted in this system and does not need any new infrastructure, however, the devices provided with WiFi technology can be reused, so they are cost-effective also. Indoor environment can be facilitated using radio frequency, high level of accuracy, precision, cost-effective, and robustness. A vital difficulty in location-aware computing is determining the physical locations. Various approaches in identifying the location can be divided into internal and external location determination. Various wireless technologies available for external positioning systems are Cell Identification (Cell ID), OTD (Observed Time Difference), TOA (Time of Arrival), and GPS (Global Positioning System). But among these technologies GPS is widely used in external positioning system, whereas Bluetooth, WLAN (Wireless Local Area Networks), and RFID (Radio Frequency Identification) are used for internal positioning systems [2, 3].

In the associated work, RADAR is one of the indoor locating systems, which is based on the strength of Wi-Fi signal. This system was developed by Microsoft Research Lab using LF Systems [4]. Template matching algorithm [5] has been used by this system for the comparison of real-time user RSSI and set of already surveyed RSSI database systems. This comparison is done to find out the position of user. In the past few years, indoor positioning system [5] uses Wi-Fi signal-based technology and infrared (IR) technology to identify the current location. The active badge system which is designed at AT&T lab uses diffuse infrared (IR) system, IR technology-based system in which multiple sensors work together with IR code in a regular time interval of 15 s and it uses data based on triangulate theory. As IR technology has some limitations like limited short-range for transmission and line-of-sight (LOS) it is not possible to apply it in large and complex buildings. LANDMARC, an indoor location identification system uses active RFID tag by SpotON RFID sensing device [6], the system setup includes RFID tags, RFID reader, and the communication in the form of signals between these two. Radio signal strength information from RFID tag to RFID reader plays an important role in this system. It uses a classifier algorithm to project the user location after receiving information.

The contributions of this work are: First, the positioning precision and accuracy were improved. Second, the system is more robust, scalable, and adaptable in the sudden change of dynamics, environment noise, and environmental factors. Third, the approximate positioning is continuous as well cost/affordability is maintained. The remaining paper is arranged as follows. Sect. 2 outlines system model and give the detailed procedure of our approach.

And simulation results are provided in Sect. 3. Finally, Sect. 4 concludes the paper.

2 Overview of System Model

2.1 Directional Antenna and Received Signal Strength

In radio equipment, an antenna plays a vital role for trans-receiving the signals. The transmitter antenna sends high-frequency energy into space while the receiver antenna receives this energy and converts it into electricity [7]. Radio location using signal strength is a popular method. RSS-based systems are easy to implement, but the accuracy is dependent on environment and distance. In indoor localization system directional antennas can be used to boost the capacity of wireless networks. In localization systems antenna can be classified into two main categories: Omnidirectional antenna and directional Antenna. Directional antenna can transmit or receive signals to its desired direction rather than like an omnidirectional antenna, which radiates or receives signals uniformly in all directions [8] (Fig. 1).

RSS is a measurement of the energy available in a received radio signal. In the existing Radio-based measurement technique RSS has been used widely in indoor positioning system. RSS determines the nature and characteristics of location fingerprints. The circuit voltage is measured between transmitters and receiving device which varies due to path interfaces and it is measured by received signal strength indicator (RSSI). The distance measure is estimated by the receiving node from the source node by determining the energy of received signal strength. It uses the path

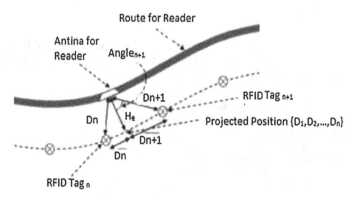

Fig. 1 Calculation of angle for directional antenna

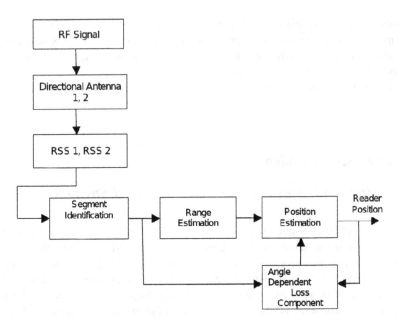

Fig. 2 Block diagram of the positioning system

loss factor to convert RSS into the distance measure and works on the principle that the signal power decreases if distance between nodes increases (Fig. 2).

- Calculate angle: This function converts an angle from radian to degree. To calculate an angle from direct ray, a function atan $2(Y, X)$ is used which returns values in between the interval [−pi, pi], according to the values of Y and X. These values are in radian, to covert it into degree simply multiply an angle with 180 M_P I.

//Convert angle from radian to degrees angle atan $2(y, x)$
Angle angle*180/M_PI

End

- Calculate reception gain: This function calculates peak gain for high performance by increasing the reception of signals in a specific direction. Function to calculate reception gain

```
              If Type equal to zero then
// multiply the gain with the ratio of the solid angles solidAngleRatio
2/(1- cos(M_PI*Width_/360))
Gain Gr*solidAngleRatio Else
gain (gainVals[((int)diff_angle)%360] + gainVals
                  [(((int)diff_angle)+1)%360])/2.0;

Gain      pow(10.0,gain/10),
End If
End
```

- Calculation of transmission gain: Transmission gain is the boosted strength of a signal during transmission. This function evaluates peak gain for transmission.

```
                        If Type equal to zero then
// multiply the gain with the ratio of the solid angles solidAngleRatio
2/(1- cos(M_PI*Width_/360))
Gain    Gt*solidAngleRatio


        (gainVals[((int)diff_angle)\%360] + gainVals
        [(((int)diff_angle)+1)\%360])/2.0;
        pow(10.0,gain/10)


Else
gain
Gain
End
```

- Calculate antenna type: This function checks the valid and invalid antenna types. Valid antenna type is in between Type 0 and Type 8 otherwise it is an invalid antenna type. For valid antenna types, initialize the radiation pattern.

```
        Function Calculate Antenna Type
If Type > 8 || Type < 0 Then
 Invalid Type
If Type != 0 Then
Call initialize_radiation_pattern
End
```

- Initialize radiation pattern: According to valid antenna type plot the gain for that angle.

```
// According to type initialize gain for that angle If type+1 ==
                Type Then
                gainVals[angle]  gain
        End
```

There are following functions are in RSS algorithm.

- Segment Identification: Segment identification [9] is selecting the segment corresponding to the biggest two mobile units when the difference between two RSS values exceeds the limit 1 dB, i.e., consecutive RSS values increase.

```
Pass input Pr
If abs ((RSS_1,RSS_2)) > le6 then
Segment
End if
End
```

- Range Estimation: Ranging-based localization is used to recognize the position of nodes in a network which is based on estimation of the distances between them, called range estimates. RSS is the widely used modality for range estimation in wireless technologies as its information can be obtained with no extra cost.

```
If RSS_1 && RSS_2 != 0 then
    D1 = (MAC *)mac) ->netif() ->Call Calculate distances
    RSS_1, RSS _2 with Txpr Print value od
    D1,D2,RSS_1,RSS_2,Txpr
Call position_esitimation(d1,d2) d_[d_count] <- X
--- call to abs_((d1-d2)) increment d_count
if d_count > 100000 then d_count=0 end if end
    if
end
```

- Position Estimation: Position estimation in wireless sensor network is the problem of estimating the location of an object from a set of noisy environments. Position estimation means to determine location of node and to estimate them. In position-based estimations, relative nodal positions are used for correct and efficient route estimation. Location of a node in a wireless network can be calculated from the waves traveling in between them and number of reference nodes.

```
Pass input : d1,d2
calculate coordinates : (x1,y1) and (x2,y2) Print
values of (x1,y1) and (x2,y2) Delta_n <- abs_((d1-d2))
X_[0] <- delta_n/2 X_[1]
<- delta_n/2
Print values X_[0] and X_[1]
P_[K] <- P_[k_-1]
Pn <- pow(delta_n,2)/12 I <-
P_[0]/pn
Set max = -1
For each i in [0,d_count]
If max< d_[i] then max = d_[i]
End if
Increment I by 1
End for
Calculate: P_[k_] = (I-K[K_] * H[K_%6]) * P_[K_]
End
```

- Angle-dependent loss component function: The energy of the backscattered signal or RSSI values at the reader generated from neighbor tag is mainly dependent on the distance d between reader and tag [9], tag antenna and chip, reader dispatched energy, antenna gains of tag and reader, the reader tag angle dependent loss due to the fact that both the reader and tag antenna are directional.

```
If angle_1 < critical_row If(max(angle_1,angle_2)>critical_row && Y[K_]>
    0.001
            && Y_[K_]-Y_[K_-1] < 0) K_+< 1
End i
If(max_(angle_1,angle_2)<critical_row-) goto out (EOF) End if
If(Y_[k_]<=0.001) goto EOF End if
If(Y_[K_] Y_[K-1]>0 goto EOF
RSS_1 <- RSS_1 (Mac *(mac))->netif()->loss-angle(angle-1) EOF End If
End
```

3 Result and Discussion

Simulation was made to contribute location consciousness of all the nodes inside the network. In this paper, we express the usage of radio frequency identification (RFID) technology in order to accomplish indoor localization of the mobile units inside the system. For simulation, we have used NS2 simulation model. Initially 50 nodes were segregated and their RSS strength without directional antenna and with directional antenna were calculated. In the simulation result after the position initialization process, each node proceeds to the initial position specified by the user. Modification of the starting location of a node is possible in simulation; also the simulated model consists of the centralized controller called as server. Every unit node continuously measure the radio signal strength (RSS) in between itself and each of its neighbors within the limited range, and transmit the information to the server, which then feeds it back to the user according to the previous work [10] on "RFID-based robotic platform" results with directional antenna dramatically changes it also reflect in gaining the signal strengths for indoor environment. Figure 3 shows the simulation of 50 nodes on NS2 platform. Basically, each node will have an angle associated with antenna from server in result to that every node will get RSS and communicates with each other.

The heart of the location-based algorithm is by every 50 ms execution of data from RFID reader and movement sensors. Here we need to monitor directions also. Figure 3 shows the RSS values of a node with direction. Every time the location finding algorithms execute, they read the difference in distance and orientation since the last data access, and clear the last sensed data and store new data (Fig. 4) (Table 1).

Additionally, more accurate position and orientation calibration is performed with RFID tags. Once at least two tags have to be read, the nodes can determine there orientation with the help of the coordinates of each tag (x1, y1) and (x2, y2) and the angle is calculated for fixing the correct node position. Table 2 shows the direction calculated for every node (x1, y1) and (x2, y2).

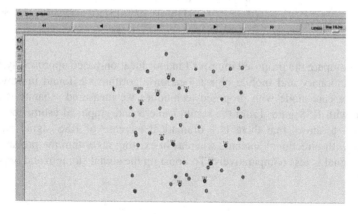

Fig. 3 Simulation on NS2 for the 50 nodes

Fig. 4 RSS without
directional antenna nodes
scenario

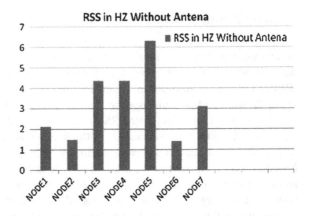

Table 1 RSS without
antenna

Node	RSS in HZ without directional antenna
1	2.13788
2	1.49244
3	4.38551
4	4.38551
5	6.33871
6	1.42361
7	3.10478

Table 2 Direction calculated
for every node (x1, y1) and
(x2, y2)

Sr. no	(x1, y1)	(x2, y2)
1	(107.093, 45.2145)	(107.093, 45.2145)
2	(96.3543, 65.8252)	(16.3466, 18.2039)
3	(6.69791, 117.823)	(16.3466, 18.2039)
4	(40.9957, 101.62)	(1.8003, 98.0942)
5	(284.946, 122.503)	(167.17, 398.087)
6	(309.019, 181.116)	(39.2081, 104.619)
7	(116.103, 221.015)	(39.0198, 328.863)

If we compare the proposed improved indoor location-based approach system to
existing stationary and mobile object location algorithm we found improved the
results. For each node with proposed techniques we measured separate RSS and
compare with RSS gain. From the result Table 3 and graphical estimation on the
result Fig. 5 shows that there is a dramatical increase of RSS signal gain and
accuracy with directional antenna, whereas in existing algorithm the probability of
gaining signal is less comparatively. To boost up the signal strength and sensitivity

Table 3 Direction calculated for each node (x1, y1) and (x2, y2)

Node	RSS in HZ without directional antenna	D1	D2	RSS in HZ with directional antenna	RSS2
1	2.13788	452.404	408.783	9.39458	1.15065
2	1.49244	204.55	455.209	3.03702	9.27917
3	4.38551	201.388	631.9	4.7409	4.81541
4	4.38551	201.388	631.9	4.7409	4.81541
5	6.33871	545.783	661.705	6.4549	4.39138
6	1.42361	377.136	274.829	1.35187	2.54569
7	3.10478	159.656	87.6838	7.54326	2.50087

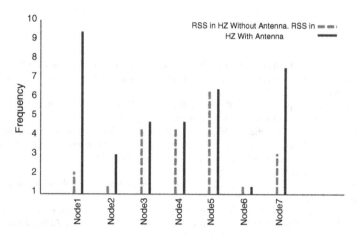

Fig. 5 Comparative analysis of RSS with antenna and without antenna

from reader we have used two different antennae. The reader gets two different RSS, i.e., RSS 1 and RSS 2 form direction one (D1) and direction two (D2). Table 2 shows that RSS gain at different angles associated with tag and calculated with different antennae, i.e., D1 and D2. Here the proposed improved RFID-based localization techniques mitigate the objectives of efficiency and accuracy with precision while locating object.

These experimental setups were performed for an indoor system using NS 2 simulation tool on an open-source platform, considering 50 nodes. According to result of proposed hybrid method we are able to make practical use in real-time applications in indoor environments such as health care, parking, loyalty management, migration services for old people or customers inside the building or shopping malls, robot localization.

4 Conclusion

Indoor RFID based localization system will have basically few challenges namely Inference and RF colision, Tag sensitivity, Tag Spatiality, Tag Orientation and Reader Locality. These challenges are minimized with a improved approach by considering directional antenna system and RSS. In the obstacle cluttered environments, the estimated heading direction of the moving device could be shifted by uncertain environmental effects. Localization algorithm is associated with the RFID tag and calculate location accuracy X and Y. The proposed localization algorithm closely approximate the actual objects positions.

References

1. Joana Halder, S., & Wooju, K. (2012). A fusion approach of RSSI and LQI for indoor localization system using adaptive smoothers. *Journal of Computer Networks and Communications.*
2. Zhogliang, D., Yanpei, Y., Xie, Y., Neng, W., & Lei, Y. (2013). Situation and development tendency of indoor positioning. *China Communication, 10*(3), 42–55.
3. Potgantwar, A. D., & Wadhai, V. M. (2011). Location based system for mo-bile devices with integration of RFID and wireless technology-issues and proposed system. *International Conference on Process Automation Control and Computing, 2011*, 1–5.
4. http://research.microsoft.com/en-us/projects/radar/.
5. Pahlavan, K., Akgul, F., & Ye, Y. (2010). Taking positioning indoor Wi-Fi localiza-tion and GNSS. *Inside GNSS Journal*, 40–47.
6. Xiong, Z., Song, Z., Sclera, A., Ferrea, E., Sottile, F., Brizzi, P., Tomasi, R., & Spirito, M. A. (2013). Hybrid WSN and RFID indoor positioning and tracking system. *EURASIP Journal on Embedded System* by Springer-Link 2013: 2013(6).
7. Zhai, Y., Xing, W., Lei, L., & Liao Y. W. (2013). An improved ad hoc network communication based on cluster. *Advanced Materials Research*, 184–189.
8. www.hongning.org.
9. Saad, S. S., & Zahi, S. N. (2011). A standalone RFID indoor positioning system using passive tags. *IEEE Transactions on Industrial Electronics*, 1961–1970.
10. Munishwar, V. P. (2009). RFID based localization for a miniaturized robotic platform for wireless protocols evaluation. *IEEE International Con-ference on Pervasive Computing and Communications, 03*(2009), 1–3.

Secured Cloud Data Storage—Prototype Trust Model for Public Cloud Storage

D. Boopathy and M. Sundaresan

Abstract Cloud computing is a business word in information technology world. Cloud-based applications and its related businesses are rapidly growing in the day-to-day world. The small and medium enterprises' concerns are rapidly adopting cloud-based business service models because they are concerned only about the operational cost (OP-EX) and the capital expenditure (CAP-EX) is not essential. The cloud computing has undergone many issues like scalability, reliability, compliance cross-border data storage, multi-tenant, data security, downstream, and regulatory issues. The major issue under high controversy is data storage in cloud. Data confidentiality, data integrity, data authentication, regulation, and legal jurisdictions are the major pests that infect user's business. Taking all these in mind, this paper discusses and proposes a model for data security in cloud storage and its safety measures.

Keywords Secured cloud storage · Cloud issues · Data storage · Cloud security · Data security · Cloud data security

1 Introduction

Cloud computing is rapidly replacing the business era with its wide range of service models. The CSPs' are widely spreading their service with their own and shared business models [1, 2]. The own and shared business models are basically designed and extracted from some open-source models. But the core things of cloud computing have never changed. There are different types of business models available

D. Boopathy (✉) · M. Sundaresan
Department of Information Technology, Bharathiar University,
Coimbatore, Tamilnadu, India
e-mail: ndboopathy@gmail.com

M. Sundaresan
e-mail: bu.sundaresan@gmail.com

© Springer Science+Business Media Singapore 2016
S.C. Satapathy et al. (eds.), *Proceedings of International Conference on ICT for Sustainable Development*, Advances in Intelligent Systems and Computing 408, DOI 10.1007/978-981-10-0129-1_35

so the users are confused to choose the better cloud service providers for their purpose [3]. Naturally the user's data will be stored outside the user's premises so the users easily lose their control over the data, which are stored in online cloud storage [4]. If the users lose their control over their data, automatically they are locked with their cloud service provider.

The word trust will rule the cloud computing era [5, 6]. Unfortunately there are no proper standardized international laws and regulations to protect the user's data [7]. The national laws and regulations are framed by some countries and followed by them strictly as a national security. But this is not sufficient to control the data which are stored across their own country border limit [8, 9]. Many countries are protecting their own user's data with basic laws and regulations in the name of belief. This is not sufficient to protect the data which are stored, processed, and transferred in public cloud storage [10].

2 Cloud Computing and Cloud Models

The users are allowed to log into a network-based service, when the vendor provides and operates all the user-required applications from simple to complex levels in the remote machines owned by themselves or by third party companies. The characteristics of cloud computing are on-demand self-service, broad network access, resource pooling, rapid elasticity, and measured service. Cloud computing is divided into two types. They are cloud service models and cloud deployment models (Fig. 1).

The cloud service model contains three types, they are software as a service (SaaS), platform as a service (PaaS), and infrastructure as a service (IaaS) [2, 3]. The cloud deployment model contains four types, they are private cloud, public cloud, community cloud, and hybrid cloud [2, 3].

Fig. 1 Cloud service models

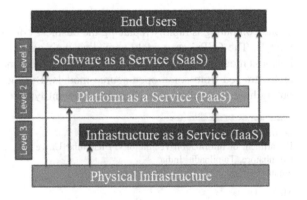

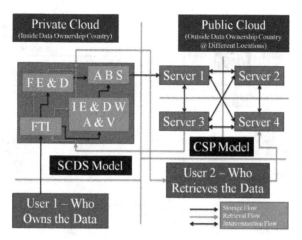

Fig. 2 Secured cloud data storage—prototype trust model

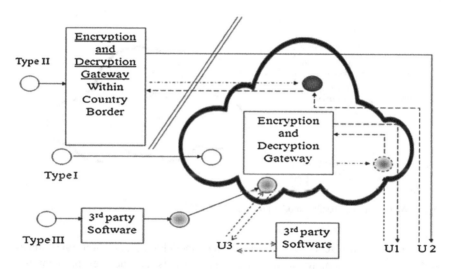

Fig. 3 File encryption and decryption process in SCDS

3 Problem Statement

The cloud computing issues are vast. In this paper, the problems are classified under three different types. The three types are social/user issues, service provider issues, and jurisdiction, regulation, and governance issues.

Social/User issues contain problems such as data protection and security, zero trust mechanism, data confidentiality, data integrity and availability, vendor locked in, and so on. Service provider issues include problem like vulnerabilities of client

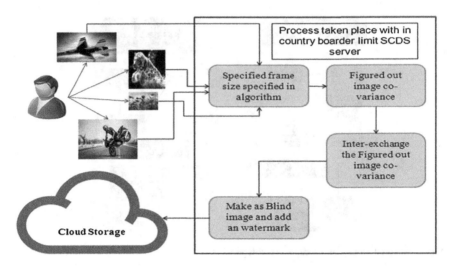

Fig. 4 Image encryption and decryption process using covariance in SCDS

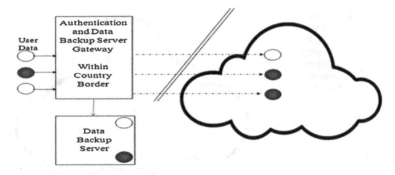

Fig. 5 Automatic data backup model process in SCDS

software, third party vendor issues, multi-tenant model issues, and non-transparency SLA. The jurisdiction, regulation and governance issues includes problem like different jurisdiction limits, no proper standardization and centralized regulations, SLA is eligible to file a suit?—SLA related problems.

4 Objective

The cloud providers are not storing any information within the data owner's country border limit. If the data are not residing within the country limit, then automatically the control on data problem and other data security related problem arises. To overcome the cloud data related issues; creating a new encryption standard and preparing a new regulation for providing cloud service will put a dot for data related issues in cloud storage. Keeping these things in mind, this model was prepared and proposed with four modules to provide the source to destination level security for the data which will store in cloud storage.

5 Secured Cloud Data Storage

When the users' data are ready for transfer to the cloud storage, before that storage of the data needs to cross the checkpoint within their country border limit. The check point process is called Secured Cloud Data Storage process herein after as SCDS. The SCDS contains four modules they are Data Type Identification Module herein after as DTI, Encryption and Decryption Gateway Server Module herein after as E&DGS, Digital Watermark Allocation and Verification Server Module herein after as DWA&VS, and Automatic Data Backup Server Module herein after as ADBS.

The SCDS is a method of pipeline process. It denotes that the modules are designed independently and then merged together to create this SCDS model. So, if this model needs any update or modification, it will take place only at the specified module only. Here pipeline process stands for coupling each module with other modules for the reason of continuous process (Fig. 2).

5.1 File Type Identification (FTI) Model

Once the data enter into the SCDS, first it moves that data into FTI module. The FTI module is a process of verifying the file data type using its file extension. This process is used for identifying the data type for the reason to figure out the data whether it is sensitive one or nonsensitive one. This FTI process will declare all the types of image file as sensitive data and will declare nonimage file as nonsensitive data.

This FTI process will filter the file in sensitive and nonsensitive manners. After the data type identification, the sensitive data will be transferred to the DWA&VS module and the nonsensitive data will be transferred to the E&DGS module.

5.2 Encryption and Decryption Gateway Server (E&DGS) Model

In E&DGS, data will be encrypted using the public key crypto system. The users will encrypt the file using their public key and then the person who needs to access the information, that person needs to use the private key to decrypt the file. The existing public key crypto algorithms are not suggested for use for this E&DGS module. The existing algorithms are revealed algorithms and known by public to all. So, the public key crypto algorithm going to be implemented in E&DGS module, need to be a newly developed algorithm or an enhanced algorithm from the existing algorithm.

The reason for this condition is simple; Snowden reveals that RSA security group weakens in their random number generation algorithms on encryption software, and hardware due to order of their local government and security agencies. And also some of the software and social web network portals fix backdoors with in it to collect and surveillance the user information without their knowledge. If we are going with existing and revealed algorithms, we cannot able to trust the security for the data stored in cloud storage. In that cases cloud security will became myth. So, going with existing algorithm, this E&DGS module cannot give assurance for the cloud trust model (CTM) (Fig. 3).

5.3   Digital Watermark Allocation and Verification Server (DWA&VS) Model

If the data are declared as sensitive data, then the data file is any one of the image format types only. There are 25 different types of image formats available.

Before entering into the Digital Watermark Allocation (DWA) process something needs to be generalized. Because the image may have different frame size and vary from file capacity size. So, each and every image needs to be fixed into specific format frame using the lossless method. Using this method any frame size of image will be fixed into the frame, then the image covariance needs to be calculated. Once the covariance is figured out, it will be inter-exchanged using the public key crypto system, and later that image will be changed into blind image that is after inter-exchanging the covariance, that covariance inter-exchanged image will turn into one specific color. It is helpful at the time of loss of data or leakage of information. The inter-exchanged covariance code will be removed only by the prepared programing method.

The DWA process will be reversed at the time of accessing the information. Once the image is accessed with the prior and authorized user, the watermark will be removed first and then the blind image will processed into inter-exchanged covariance method. Then the inter-exchanged covariance will be processed into normal covariance method to get and give the image to the users. Before the data are sent to the user, this module may reframe that image into original image frame size or it may transfer that image in the fixed frame size too (Fig. 4).

5.4 Automatic Data Backup Model

The DTI identifies the data type and forwards the sensitive data into DWA&VS and nonsensitive data into E&DGS. After the DWA&VS process on sensitive data or E&DGS process on nonsensitive data, the processed data will store one backup copy on Automatic Data Backup Server (ADBS), which is located within the country's border limit. After that processed data will transfer to the cloud service provider's storage and replicated in the cloud service providers multiple servers to avoid some issues like service downtime and service crashes from any one server (Fig. 5).

6 Results and Discussion

The secured cloud data storage (SCDS) model was designed in Network Simulator 2 (NS2) to check whether the model is possible or not.

Figure 6 is the NS2 SCDS designed model and its data flow results between the user storage, storage in multiple servers, data request from user to access the data from cloud are shown in Table 1. These mentioned conditions are simulated in NS2 then outputs and results are showed that this SCDS model is possible to implement in an effective manner.

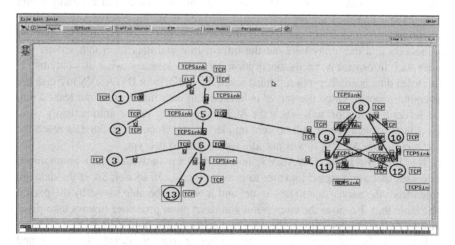

Fig. 6 SCDS—prototype trust model simulated in NS2

Table 1 Process time taken by secured cloud data storage prototype trust model

Time taken to reach each process	Time (ms)	
Data from user to FTI server	100	100
DTI identification time	1.5	1.5
File encryption time	2564	–
Image encryption time	–	2985
ABS time	100	100
Time taken to reach the cloud servers	250	250
Total time taken by SCDS to process a file	3015.5	3436.5

7   Conclusions and Future Scope

Data confidentiality is assured using the Encryption and Decryption Gateway Server (E&DGS) and Digital Watermark Allocation and Verification Server (DWA&VS). The data availability and vendor locked in issues will come to an end using the Automatic Data Backup Server (ADBS). The Data Type Identification (DIT) will avoid the mismatched file formats' uploads, it will avoid the data became vulnerable due to the storage user. The encryption algorithm is in need of enhancement or new level of algorithm development and then the newly developed algorithm will be used only for this SCDS model. If the algorithm is kept as secret, it is not that much easy to reveal and break the algorithm's encryption and decryption systems (like bullrun, hillclimp). The back door fixing, data surveillance on stored data by other country, and illegal use of user data is not possible. The only one issue that may raise is SCDS processed data will be stored on many cloud storages and that data are replicated and mirrored in multiple servers to avoid the downtime issue and server crash issue. But any one of the servers may be attacked by hackers and they may take out the information and data from that server. The encrypted information needs decryption to know exactly what it contains. If the stolen data need decryption it must reach the E&DGS or DWA&VS to finish the reversing process. Once the server is hacked and data information are leaked out, the server information will reach the SCDS processing center automatically. And the data of hacked server will be kept on alert mode. Once the stolen data reach the E&DGS or DWA&VS, it will not allow that data to decrypt.

The DTI, E&DGS, DWA&VS, and ADBS are individual and autonomous processes that are coupled together to provide the SCDS model. So, if the data are stolen from the server due to any issue, and it will not be able to use by the person who stole that. Because the encryption and decryption processes always take place only on SCDS model with in the data owner's country border limit.

Issues like lack of governance on data, different countries jurisdiction issues, and its related issues cannot be solved within rapid manner and also it is not possible in short time. So, preventing the data in effective manner will avoid these types of issues and also avoid the data became vulnerable on cloud storage. Once this level is achieved, this SCDS model will be treated as trust cloud model for data storage in public cloud storage.

References

1. Yu, X., & Wen, Q. (2010). A view about cloud data security from data life cycle. *International Conference on Computational Intelligence and Software Engineering (CiSE)*, 10–12 December 2010 (pp. 1–4).
2. Boopathy, D., & Sundaresan, M. (2014). Data encryption framework model with watermark security for data storage in public cloud model. *International Conference on Computing for Sustainable Global Development*, 5th–7th Mar, 2014 (pp. 903–907).
3. Boopathy, D., & Sundaresan, M. (2013). Location based data encryption using policy and trusted environment model for mobile cloud computing. *Second International Conference on Advances in Cloud Computing*, 19th–20th September 2013 (pp. .82–85).
4. Youssef, A. E., & Alageel, M. (2012). A framework for secure cloud computing. *IJCSI International Journal of Computer Science Issues, 9*(4), 487–500.
5. Sivashakthi, T., & Prabakaran, N. (2013). A survey on storage techniques in cloud computing. *International Journal of Emerging Technology and Advanced Engineering, 3*(12), 125–128.
6. Bhuvaneswari, J., & Vaishnavi, R. (2013). Data security and storage in cloud computing. *International Journal of Emerging Technology and Advanced Engineering, 3*(1), 100–101.
7. Tripathi, A., & Yadav, P. (2012) Enhancing security of cloud computing using elliptic curve cryptography. *International Journal of Computer Applications (0975–8887), 57*(1), 26–30.
8. Huaglory, T. (2012). Security issues in cloud computing. *IEEE International Conference on Systems, Man, and Cybernetics*, October 14–17, 2012, COEX, Seoul, Korea (pp. 1082–1089).
9. Shaikh, R., & Sasikumar, M. (2012). Trust framework for calculating security strength of a cloud service. *International Conference on Communication, Information & Computing Technology (ICCICT)*, Oct. 19–20, 2012, Mumbai, India, pp. 1–6.
10. Ahmad, S., Ahmad, B., Muhammad Saqib, S., & Muhammad Khattak, R. (2012). Trust model: Cloud's provider and cloud's user. *International Journal of Advanced Science and Technology, 44*, 69–80.

A Semicircular Monopole Antenna for Ultra-wideband Applications

Preeti Jain, Bhupendra Singh, Sanjeev Yadav and Ashu Verma

Abstract A novel small effective microstrip-fed monopole antenna, it comprises a semicircular design with ovoid-shaped patch and an effective abbreviated base ground plane, is simulated for UWB applications. The design of the antenna has compact structure with size of 16×18 mm^2. The area of the microstrip antenna with the proposed structure is reduced up to 57 % compared to the conventional monopole antenna. In order to increase the bandwidth of the antenna some slots have been made inside the patch. The proposed antenna design is intended to work within the frequency, which covers from 3.1 to 11 GHz with less than -10 dB return loss. The insertion of a couple of indents on ground plane excites the additional resonances and the bandwidth is increased up to 104 %. Furthermore, in the UWB, good return loss and the high standard radiation patterns are achieved.

Keywords Circular patch · Radiation pattern · Return loss · VSWR · UWB antenna

1 Introduction

These days, wireless system is getting progressively demanding in every field of communication technology. However, the advancements in wireless system need to be enhanced more to fulfill the necessities such as higher rate of data transfer and

P. Jain (✉) · B. Singh · A. Verma
Department of ECE, Amity University, Noida, U.P, India
e-mail: preeti.essare@gmail.com

B. Singh
e-mail: bsingh5@amity.edu

A. Verma
e-mail: verma_ashu92@yahoo.in

S. Yadav
Department of ECE, Govt. Women Engineering College, Ajmer, Rajasthan, India
e-mail: sanjeev.mnit@gmail.com

© Springer Science+Business Media Singapore 2016
S.C. Satapathy et al. (eds.), *Proceedings of International Conference on ICT for Sustainable Development*, Advances in Intelligent Systems and Computing 408, DOI 10.1007/978-981-10-0129-1_36

339

greater resolution. Due to this UWB coating frequency from 3.1 to 10.6 GHz was presented by the Federal communication commission in 2002 and FCC is still working on it. For a long time, different types of antenna design strategies for wideband operation have been concentrated on improving the communication system and radar technology. The outline of wideband antenna is an extremely troublesome assignment particularly for agenda incurable after the entire bargain betwixt shapes, size, expense, and straightforwardness must attained. In ultra-wideband wireless system, one of the main issues is the configuration of a conservative antenna while it is giving large frequency range trademark covering the entire working band. Because of their engaging highlights of large data transmission, basic structural design, omnidirectional radiation pattern, and simplicity of development, a few wideband monopole designs, for example, roundabout, square, circular, pentagon, and hexagon shapes are being proposed for ultra-wideband applications. In any case, the design is not suitable enough to be incorporated with printed circuit sheets, after all these are not having planar structural shape. Accordingly, a microstrip-fed antenna has respectable possibilities for coordination with agenda incurable to its alluring highlights, for example, lower profile, lower cost, and light weight.

A planar monopole is widely used for manufacturing the ultra-wideband antenna because it has various attractive features like lower profile, lower cost, small size, and easy to fabricate. Nowadays, many researchers are working on size reduction and bandwidth enhancement techniques. Bandwidth enhancement techniques in planar antenna are obtained by E-slot patch, H-slot patch, U-shaped slot patch, shorting pin, and slit loaded. Defected ground structure is also responsible for addition resonances to enhance bandwidth. A paper on defected-ground plane in monopole antenna was presented. Many trade-offs are taken to optimize the geometry of an antenna.

Through this paper, we present a novel small size ultra-semicircular structure with rectangular patch in microstrip-fed antenna especially suitable for ultra-wideband applications. To attain the maximum impedance bandwidth, many indents are introduced in the indented metallic patch and some slots are indented in the bottom ground plane. Reenacted and trial simulated results are shown for exhibiting the execution of the proposed antenna design. The proposed antenna design is easy to fabricate on the top surface of a thin FR4 substrate of small size of 16×18 mm^2. The antenna id was designed to work within the frequency range from 3.1 to 11 GHz with less than -10 dB return loss value.

The rest of the paper is structured as follows: in Sect. 2 we discussed a detailed description of design of the proposed antenna. Simulation results are presented in Sect. 3. Finally, Sect. 4 gathers the conclusion of the proposed work.

2 Antenna Design

The suggested antenna is having dimensions of 16 mm × 18 mm, built on FR4 substrate with the relative dielectric constant or permittivity of 4.4 and thickness of substrate of 1.6 mm. The width Wf of the feed line is fixed at 2 mm and it is located on the top surface of the substrate, a semicircular patch of radius 4 mm is combined with ovoid (rectangular) patch with dimension 7 mm × 13.5 mm. The ovoid patch has a separation of 2.5 mm to the ground and is imprinted on the bottom of the substrate. By cut a couple slot with $W_1 \times L_1$ indents of respective dimension at the micro strip patch, it is found that a quite upgraded impedance bandwidth could be accomplished for the designed antenna. The changes made on the grounds and the slotted indents influence the electromagnetic coupling between the semicircular design with the ovoid patch and the ground surface. Moreover, for attaining a greater wideband impedance matching of the proposed antenna, partition L_3 from the ovoid patch and the indented ground plane is introduced. The altered effective ground plane dimension is used as an impedance matching component to maintain the standard impedance matching of an antenna. The measurement of the notch $W_2 \times L_2$ in the truncated ground in the proposed antenna and the feed gap horizontally L_3 are important parameters used for determining the sensitivity of impedance matching (Fig. 1).

The ideal measured dimension for an outlined antenna are as per the following: $W_{sub} = 16$ mm, $L_{sub} = 18$ mm, $W_p = 7$ mm, $L_p = 13.5$ mm, $r = 4$ mm, $W_1 = 1$ mm, $L_1 = 2$ mm, $W_2 = 7$ mm, $L_2 = 1.5$ mm, $L_g = 2.7$ mm, $L_f = 2.5$ mm, $a = 2.5$ mm, $b = 1$ mm, $c = 2$ mm, $d = 4$ mm, $e = 1.5$ mm, $f = 3$ mm, $W_3 = 2$ mm, and $L_3 = 1.5$ mm. It has been discovered that the outlined antenna fulfills all the necessities in ultra-wide frequency ranging from 3.6 to 11.0 GHz. The diameter of the composed design antenna is much compact than other proposed antennas.

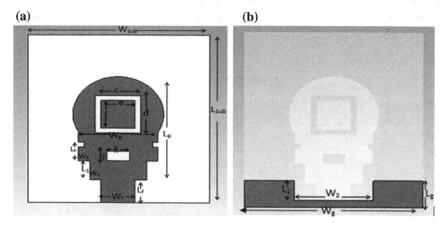

Fig. 1 Front design view (**a**) and bottom design view (**b**) of proposed design with rectangular patch microstrip-fed Antenna

3 Simulation Results

The microstrip-fed semicircular design with monopole antenna with different parameters (L, L_2, and W_2) were developed and concentrated on to exhibit the bandwidth increment. The reenacted imitated repercussion is achieved using the CST software. Figure 2 shows the imitated return loss over the overall frequency range as the slot sizes are changed to $W_2 = 7$ mm, $L_2 = 1.8$ mm in the ground, the impedance bandwidth becomes greater in frequency range. It is also reenacted over the more elevated frequency of antenna and is expressively influenced by the variations in the indented dimension of the ground.

The antenna should have a characteristic impedance of 50 Ω for perfect impedance matching. When the impedance at the generator is not same as impedance at load, impedance mismatching occurs. Due to this, the reflections are produced at the load side and travels from load to generator. This is measured by the return loss.

Figure 2 described the frequency range from 3.6 to 11.0 GHz with antenna bandwidth of 7.8 GHz. This figure is also shows that the return losses of the antenna are −31.95 dB at 4.4 GHz and −22.396 dB at 9.5 GHz with impedance matching at 44.5 Ω.

VSWR describes the efficiency of the antenna in terms for matching. For perfect impedance matching, the VSWR should have a value less than 2.

Figure 3 shows that at each resonant frequency, the VSWR is less than 2.

The radiation pattern is the graphical representation of radiation characteristics. The radiation pattern is the combination of E-plane far-field pattern and H-Plane far-field pattern. The direction of maximum radiation is known as major lobe and 3 dB beam width is calculated at this. Figures 4 and 5 shows the radiation pattern at 4.4 and 9.5 GHz frequencies.

Figure 4 shows the radiated E-plane far-field pattern and radiated H-plane far-field pattern are described at resonant frequency of 4.4 GHz. The radiated E-plane pattern has the main lobe, which has a magnitude of 2.5 dB. The direction of main lobe is 180°. The H-pattern is omnidirectional and has main lobe magnitude of −2.4 dB.

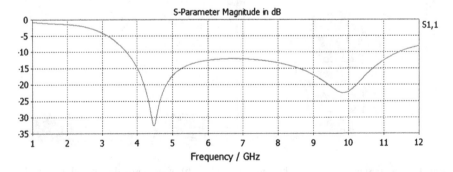

Fig. 2 Return loss of the proposed antenna

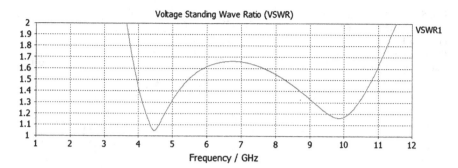

Fig. 3 VSWR of the proposed antenna

Fig. 4 Radiated E-plane
far-field pattern (**a**) and
radiated H-plane far-field
pattern (**b**) of proposed
microstrip-fed antenna at
4.4 GHz

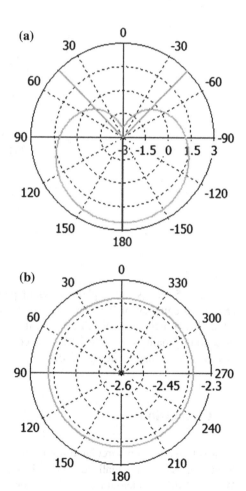

Fig. 5 Radiated E-plane far
radiation pattern (**a**) and
radiation H-plane far pattern
(**b**) of proposed microstrip-fed
antenna at 9.5 GHz

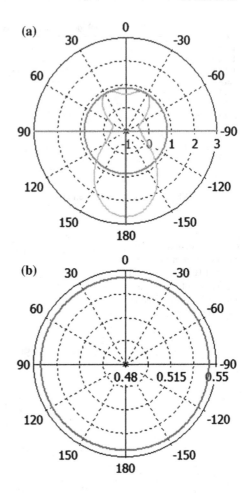

Figure 5 shows the E-plane far-field pattern and H-plane far-field pattern at resonant frequency of 9.5 GHz. The radiated E-plane pattern has the main lobe, which has a magnitude of 2.6 dB with the magnitude of side lobe is −1.8 dB. The direction of main lobe is 180°. Omnidirectional pattern is achieved at H-plane and has main lobe magnitude of 0.5 dB.

4 Conclusion

A novel smaller semicircular design with rectangular microstrip-fed antenna has been proposed and actualized for UWB applications like satellite communication/ WLAN/WiMax/Bluetooth, etc. The insertion of a couple of indents on ground plane excites the additional resonances and the bandwidth is increased up to 104 %.

At each frequency, VSWR < 2 shows a better impedance matching. To get the wide data transmission rate, a couple of indented slots are located at the lowermost edges of the patch and the analysis of ground plane dimension is also varied by parametric analysis. The composed antenna achieved the required −10 dB return loss prerequisite from 3.6 to 11.4 GHz and gives standard radiation patterns. Trial results demonstrated that the proposed design can be a great contender for several UWB applications.

Bibliography

1. FCC. (2002). First report and order on ultra-wideband technology.
2. Young, J., & Peter, L. (1996). A brief history of GPR fundamentals and applications (pp. 5–14).
3. Daniels, D. J. (1996). Surface-penetrating radar (Ser. 6, pp. 72–93).
4. Chen, Z. N., China, M. W. Y., & Amman, M. J. (2003). Optimization and comparison of broadband monopole. *150*(6), 429–435.
5. Agrawall, N. P., Kumar, G., & Ray, K. P. (1998). Wide-band planar monopole antennas. *IEEE Transactions on Antennas and Propagation, 46*(2), 294–295.
6. Antonino-Daviu, E., Cabedo-Fabre's, M., Ferrando-Bataller, M., & Valero-Nogueira, A. (2003). Wideband double-fed planar monopole antennas. *39*(23), 1635–1636.
7. Su, S. W., Wong, K. L., & Tang, C. L. (2004). Ultra-wideband square planar monopole antenna for IEEE 802.16a operation in the 2–11 GHz band. *42*(6), 463–466.

Security Analyses of Different LFSR-Based Ciphers to Propose a Novel Approach Compatible with Parallel Computing Platform, Providing Resistance Against Various LFSR-Based Attacks

Trishla Shah and Darshana Upadhyay

Abstract Development of a framework, for generating sets of random numbers which are highly nondeterministic and the dimensional distribution of which is strong, is need of the hour. Applications of pseudorandom numbers are widespread in areas of keying, re-keying, authentication, smart phone security, etc. Their use is central in the network security domain. Hence, developing a cryptographically secure pseudo-random number generator (CSPRNG) would be beneficial. The proposed generic model is constrained to linear feedback shift registers (LFSR), owing to its good statistical properties, large period, well suited to low power or high speed requirements. The use of pseudo-random numbers are high in hardware areas like wireless devices, smart phones, etc. and in stream ciphers, protocol design, etc. in software areas. Hence, the CSPRNG design is compatible to both–hardware and software applications. For software development of the cipher, a parallel computing environment has been chosen because in today's computing trends, multicore processors are superseding the sequential ones; hence the primary engine for processor performance growth is to increase parallelism rather than increasing the clock rate. The paper thus presents the CSPRNG model based on hardware and software co simulation, using a generic approach. A mathematical model of the PRNG is designed based on above specifications and is mathematically proven to be resistant against various LFSR-based attacks.

Keywords CSPRNG · GSM · Attacks · Keys · Generic · Co-simulation · LFSR

T. Shah (✉)
CSE Department, Institute of Technology, B.H. Gardi College
of Engineering and Technoogy, Rajkot 360005, India
e-mail: tpshah@gardividyapith.ac.in

D. Upadhyay
Department of Computer Science and Engineering, Institute of Technology,
Nirma University, Ahmedabad 382481, India
e-mail: darshana.upadhyay@nirmauni.ac.in

© Springer Science+Business Media Singapore 2016
S.C. Satapathy et al. (eds.), *Proceedings of International Conference
on ICT for Sustainable Development*, Advances in Intelligent Systems
and Computing 408, DOI 10.1007/978-981-10-0129-1_37

Abbreviation

CSPRNG   Cryptographically Secure Pseudorandom Number Generator
GSM   Global System Mobile Communications
LFSR   Linear Feedback Shift Register
NIST   National Institute of Standards and Technology
CUDA   Compute Unified Device Architecture
OpenCL   Open Computing language

1   Introduction

In today's era, use of networks and its applications are growing in a rapid speed. Users often reveal critical informations such as account numbers, bank passwords, personal and financial details, important transaction details, etc., over the Internet. But apart from its use in the Internet, much vulnerability, like password theft, virus attacks, spoofing, message confidentiality threats, message integrity threats etc., have been found, causing potential loss to the user's private information. Hence it is important to build a secure system providing a perfect balance of confidentiality, integrity and availability (CIA) of user's private data. These security parameters are provided by a mechanism of key generation (public and private keys), random password generation, one-time password (OTP) generation, etc. Implementation of these mechanisms is done through generation of unpredictable sets of random numbers having high uncertainty. These random numbers are called pseudo-random numbers. Hence, pseudo-random numbers are at the core in providing security to network applications like smart phones, wireless local area network (WLAN), network protocols, various web-based applications like online shopping, authentication to legitimate sites, etc. Generation of these numbers is done by a generator, known as pseudo-random number generator (PRNG) [1]. So, the PRNG must be able to generate highly random and unpredictable set of numbers to be secure enough for the above applications. But, if there is a flaw or the PRNG is weak, then it may be prone to various network attacks. Designing of a cryptographically strong, compatible, vulnerable, and threat-resistant generator is the need of hour [2]. Hence, through this research, the development of a general framework of a PRNG, for generating cryptographically strong sets of random number is proposed. The proposal aims to build an in-general framework and a unified model for enhanced security specifically for linear feedback shift register (LFSR). For construction of LFSR-based cipher, a thorough study on already existing LFSR-based ciphers is done. This study aims to extract out the behaviour of different ciphers under different application domains. For designing of a good LFSR-based cipher, it is important to understand the mechanism of LFSR as well as working of different LFSR-based ciphers [3]. This study widens the scope of designing and narrowing down the limitations in the current implementation. For the software implementation, a parallel computing platform, i.e. GPU programming

Table 1 Effectiveness of CUDA for designing existing ciphers

Implemented approach using sequential computing	Suggested approach using CUDA	How is CUDA better
(1) In many ciphers, outputs from two components, LFSR and FSM are independent of each other, yet it is done sequential. (2) A technique called hard-coding is used, to increase the speed of computation, but memory used is high. (3) XORing of outputs of LFSR and FSM, is done sequentially	(1) Using CUDA, generation of outputs from LFSR and FSM can be done in parallel. (2) Hard-coding LFSR is done sequentially; this can be done in parallel. (3) XORing of outputs of LFSR and FSM can be done in parallel	Generation of parallel outputs would save time and increase e efficiency
Mathematical equations derived for SNOW 2.0 are as follows: $(x) = x16 + x14 + 1 \times 5 + 1$ F232 [x], 4 = 233 + 2452 + 48 + 239 MUL[c] = (c23, c245, c48, c239) MUL1[c] = (c16, c39, c6, c64). All above equations are solved sequentially using gcc or Microsoft C++ Compiler	These equations can be solved in parallel like splitting entire equation as x16, x14, 233, etc. with one thread solving one term. All these can then be added in parallel	Computational complexity of matrix multiplication for these equations would decrease exponentially to the base 2
The implementation of RC4 is on CPU with the verification process being sequential, leading to overheads	The same approach can be done in parallel, leading to low overheads of cycles	With CUDA the entire algorithm can be optimized

is chosen, for increasing throughput. The current paper is specifically constrained to the area of designing of n-bit LFSR-based cipher. Also, a strong comparison is done among common parallel computing platforms. Designing a mathematical model for PRNG is the main area of work (Table 1).

2 Literature Survey

For designing of LFSR-based stream cipher, a thorough literature study on already existing LFSR-based ciphers is done. The survey boils down to two analyses: *hardware-based ciphers and software-based ciphers*. After a thorough literature survey, the hardware analyses of following ciphers is done:

1. GRAIN-128 [1]
2. GRAIN-128a [2]

3. SNOW 2.0 (both h/w and s/w) [3]
4. SNOW 2.0 modified [4]
5. RFID (AES) [5]

Various software ciphers have also been studied. The software ciphers studied are:

- SNOW 1.0 [6]
- SNOW 2.0 [7]
- RC4 [8]

i. **Linear Feedback Shift Register**

LFSR is a mechanism for generating a sequence of binary bits. The register consists of a series of cells that are set by an initialization vector that is, most often, the secret key. The behaviour of the register is regulated by a clock and at each clocking instant, the contents of the cells of the register are shifted left by one position, and the exclusive-or of a subset of the cell contents is placed in the rightmost cell [4]. A basic LFSR consists of three components: the input sequence (initialization vector), the feedback (tap sequence) and the output [5]. As the name suggests, the feedback gives a linear relationship between the input and the output. Let the input sequence of length n be $(s_0, s_1, \ldots, s_{n-1})$. The feedback is thus a linear function $f(s_0, s_1, \ldots, s_{n-1})$ defined by $f(s) = c_0 + c_1 x + \cdots + c_{n-1} x^{n-1} + x^n$, where c_0, $c_1, \ldots, c_n$ are constant coefficients. The output of the LFSR is determined by the initial values $s_0, s_1, \ldots, s_{n-1}$ and the linear recursion relationship:

$s_{k+n} = (\mathrm{Sigma}(C,X))$, $k \geq 0$ or equivalently, $= 0$, $k \geq 0$ where $c_n = 1$ by definition [9].

ii. **Challenges**

Many stream ciphers have been designed, for the generation of a strong set of pseudorandom numbers, but certain limitations have been examined like: (i) In designing of hardware ciphers, the computational complexity over software performance decreases (ii) Very few ciphers have been designed, working for network security applications in both hardware and software domains (iii) The software implementation is mostly done on sequential basis, i.e. CPU, thus increasing complexities of overhead and time (iv) Ciphers compatible for generating good pseudo-random series on a generic platform for diverse applications have not yet been designed.

3   Proposed Model

To design the model in most appropriate manner, a research methodology adopted is in following manner.

Step 1:   The seed value is generated by a true random number generator (TRNG) for each frame.

Step 2: Second component is a CSPRNG that would produce highly random numbers.

Step 3: The generated pseudo-random numbers are unpredictable, hence are resistant to various LFSR-based attacks like dynamic cube attack, basic correlation attack, refinement attack, guess-and-determine attack, linear approximation attack, algebraic attack, Berlekamp–Massey attack, fast time memory trade-off attack.

Step 4: These pseudo-random numbers are to be used in various security applications based on both hardware and software like key-generation, re-keying.

i. Architecture of CSPRNG

In the above block diagram, the second component is CSPRNG module to produce pseudo-random numbers. The architecture of this CSPRNG is described in following section. It is purely based on LFSR. The architecture of the proposed stream cipher is as follows:

The cipher is designed with multiple LFSR's and multiple feedback polynomials for each LFSR, to keep it secured from the most common attacks on LFSR. An n-bit LFSR-based counter is designed to select multiple LFSR from the pool of n-bit, n-LFSR. An LFSR-based pool is designed, where different sizes of LFSR are coupled into a single unit. At a time, only a single feedback polynomial is used by a group of LFSR. During the next cycle, feedback polynomial is changed. The proposed LFSR-based cipher, is aimed to be generic, i.e. compatible on platforms—hardware and software. Hence, it is designed to be flexible in terms of as below. The working of the cipher is as follows:

(1) Number of ciphers
(2) Number of feedback polynomials
(3) Number of counters
(4) The output function
(5) The finite-state machine.

ii. N-bit LFSR counter

A counter is used to select LFSRs randomly from the LFSR pool. Here, the counter is LFSR-based, because of its randomness property. The feedback polynomial during every cycle would change and also the length of LFSR. The counter will shrink or expand in length as per the application's requirement. If the number of random numbers to be produced is very large, the length of counter would expand and will choose large number of LFSRs from the LFSR pool. Conversely, if the application's requirement is producing lesser number of random numbers, the counter shrinks to a smaller length, choosing lesser number of LFSRs (Figs. 1, 2, 3 and 4.

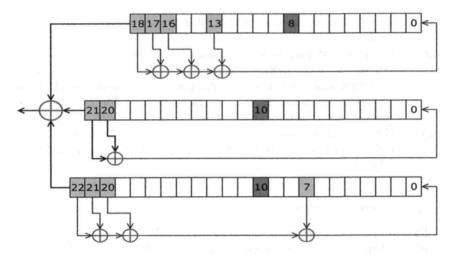

Fig. 1 LFSR of length n

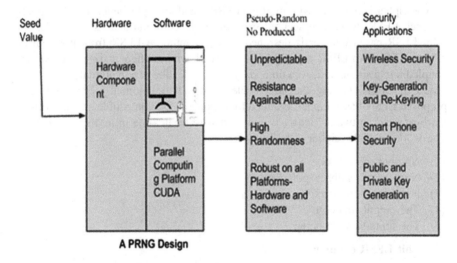

Fig. 2 Block diagram of research methodology

Advantages

- The selection of LFSR is highly random, as the counter itself is LFSR-based. This eliminates the possibility of repetitive selection of LFSR from LFSR pool.
- The argument that the counter is prone to LFSR-based attacks is reduced, as for every iteration the feedback polynomial is changed and the length of LFSR changes too.
- As per the application's requirement, the random number sequence is produced. This makes it flexible and scalable.

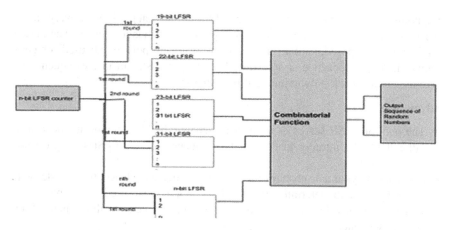

Fig. 3 Architecture of the proposed model

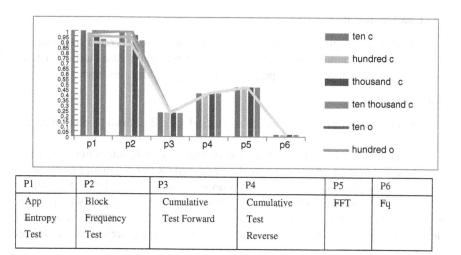

P1	P2	P3	P4	P5	P6
App Entropy Test	Block Frequency Test	Cumulative Test Forward	Cumulative Test Reverse	FFT	Fq

Fig. 4 NIST analysis of CUDA and OpenCL

iii. **LFSR Pool**

It consists of different LFSRs with different lengths. The range of LFSR starts from smallest bit LFSR to largest bit. Every sub-pool of LFSR, i.e. a single sub-pool consisting of all LFSR of same length, is coupled to make a pool of LFSR. During the first cycle, only a single LFSR is chosen from a sub-pool. During, some initial n-bits, some randomly chosen LFSRs are active. In the rest, of n-bits some different LFSRs are chosen. The important part of security is provided in the designing of using feedback polynomial. Every sub-pool is assigned different feedback polynomial, but it remains same for every LFSR within that pool. After every single cycle, the feedback polynomial for that sub-pool would change. Hence, even in one single stream of pseudorandom numbers, the feedback polynomial used for generating the entire sequence is not same. For every sub-stream in the sequence of streams, the feedback polynomial will be different.

Advantages

- The pool of LFSR serves as a mixer to generate highly random numbers. It is difficult for the intruder to decrypt the sequence, as the LFSRs used do not remain constant.
- Feedback polynomials changes for every sub-sequence. Hence, the intruder will never be able to determine the feedback polynomial used for a particular sequence, because by the time a polynomial is determined, it will be changed for another sequence.
- A single LFSR will only be selected from one sub-pool. Hence, the probability of finding another same LFSR is negligible.

iv. Combinatorial Function

A function is designed to perform some operation on the generated output stream to make it more random. This function can be a simple finite-state machine (FSM), selecting some operations to be performed on the generated sequence.

Advantages

It is used to make the output sequence more random and difficult to decrypt by the intruder.

For example, if the input is 128 bits, it is divided randomly into a number of bits say, 38bit-38bit-38bit-14bit, respectively. For initial 38 bits, a random LFSR set is chosen. The feedback polynomial used during the generation of initial 38 bits is same only in the specific pool of LFSR. For the second round, the LFSR's position is changed and feedback polynomials are changed too. This process continues, till the entire 128 bits are generated. Hence, the bits generated are highly random in nature and much difficult to crack.

4   Security Analysis of Proposed Cipher

i. Linear Approximation Attack [10]

In this type of attacks, we try to find that a given sequence of bits is produced by any known cipher or is a truly random stream of bits. Distinguisher is a function which takes a sequence of bits as input and as output, gives either "cipher" or "random", e.g. $z = c \oplus m1 = z$ correct guess, $z = c \oplus m1 = m1 \oplus m2 \oplus z$ wrong guess.

Security

Since the distinguisher identifies an already existing cipher, it will not be able to identify the generated bits as it is completely a new designed cipher [6]. One may argue that a new distinguisher function can be developed for the proposed cipher and using this newly developed distinguisher, linear approximation attack can be done on this cipher. But it is taken into account that the pseudo-random bit sequence generated by this cipher has a very high randomness [7]. As the randomness is very high, it would be very difficult, if not impossible, to distinguish between a bit sequence generated by this cipher and a truly random bit sequence. In other words, a distinguisher for this cipher is very difficult, if not impossible, to develop and hence this particular attack is prevented.

ii. **Correlation Attack** [11]

- It is useful when there is significant correlation between one or more LFSRs and the output Boolean function [12].
- It is also known as "divide and conquer" attack on an LFSR.
- If first order correlation is not there then it may be possible that higher order correlation exist. Correlation between output bit and function of two or more input bits is also possible which is known as higher order correlation [8].
- In the example of a Geffe generator, correlation between $x3$ and F and $x2$ and F can be clearly seen, which 75 % is. So individually $x2$ and $x3$ can be determined by brute-forcing. $X1$ cannot be determined as correlation is 50 %. To avoid this attack, correlation for all should be near to 50 %.

Security

Here the output from the lowest bit LFSR is approximated by its correlation with the generated output set. The feedback polynomial is determined from it. In the proposed cipher, the lowest LFSR (or any LFSR) does not have the same feedback polynomial for generating a sequence of bits. Hence by the time a particular feedback polynomial may be detected, the polynomial is changed for next sub-bits [13]. Moreover, after each round, and also on every use of this cipher, the LFSR pool is chosen randomly and hence, the correlation between the LFSR's bits and of the output bits is a variable and not a constant. Unlike the other well-known ciphers like Geffe generator, the correlation value changes with each round and for each use. Hence, correlation attack is entirely prevented.

iii. **Berlekamp–Massey Attack** [14]

Input and Output of Algorithm

From the given keystream, the algorithm determines the shortest LFSR that can generate it. The algorithm determines the length of this shortest LFSR along with the connection polynomial $C(D)$. The algorithm works only for single LFSR-based stream ciphers.

The Algorithm:
INPUT: A binary sequence Sn = S0, S1, ..., Sn-1 of length n.
OUTPUT: The linear complexity L(Sn) of Sn, 0 <= L(Sn) <= n.
1.   **Initialization:** $C(D) \leftarrow 1, L \leftarrow 0, m \leftarrow -1, B(D) \leftarrow 1, N \leftarrow 0$.
2.   While (N < n) do the following:
2.1.   **Compute the next discrepancy:** $d \leftarrow d = S_N \oplus \Sigma C_i S_{N-i} L_i = 1$
2.2.   if d = 1 then do the following:
          $T(D) \leftarrow C(D), C(D) \leftarrow C(D) + B(D).D_{N-m}$
          If L <= N/2 then L $\leftarrow$ N+1-L, m $\leftarrow$ N, B(D) $\leftarrow$ T(D)
2.3.   N $\leftarrow$ N+1
3.   Return (L)

Security
Berlekamp–Massey attack, works well only for a single LFSR and the proposed
model is using n-bit LFSR at the same time. As per the equation to determine
length: $L' >= N + 1 - L$, the length of current bits, is equated from previously
generated bit sequence. This would prove unsuccessful as in current design; in a
single output bit, n-different feedback polynomials are used.

One may argue that Berlekamp–Massey algorithm may work on ciphers having
more than one LFSR but in this particular case, in addition to having more LFSRs,
the cipher also contains a factor of choosing an LFSR pool randomly. As explained
in the previous attack, the LFSR and its feedback polynomial are variables and not
constants. Hence, this algorithm will never be able to generate a generic LFSR with
a feedback polynomial, which always generates same output as the randomly
chosen pool of LFSRs [15].

5   Experimental Basis

i. Analysis of Parallel Computing Platform

Software implementation is done on parallel computing platform, i.e. GPU programming. CUDA is chosen to reduce the trade-off between time and computation.
In today's computer trends, multicore processors are superseding the sequential
ones; hence the primary engine for processor performance growth is to increase
parallelism rather than increasing the clock rate. Increased parallelism would
increase the efficiency of random number generation. Various parallel programming
platforms are available like CUDA, OpenCL, etc. The survey analysis to find a
better platform is done. The following graph shows the performance metrics of
CUDA over OpenCL in terms of throughput, timings, overhead, etc.

ii. *Survey of Default Generating Libraries*

6 Observations

On the basis of the above literature survey, following conclusions have been made for the proposed cipher.

(i). A hybrid of word-oriented and bit-oriented cipher is to be implemented for designing of LFSR. This would best optimize the initial cycles as well as increase efficiency in software-based ciphers.

(ii). A cipher is to be designed keeping in mind its basic utility, i.e. security over communication with multiple messages using a single common key, and in telecommunication scenario for recovery from frame loss of sync messages. To design the above features, modes can be designed in the cipher.

(iii). To increase efficiency, the component structure needs to work independently, i.e. their output must be independent of each other and only the final output must be XORed. This can be best fitted in CUDA.

7 Discussion

A primary objective of this paper is to design, implement and evaluate the cryptographically secure PRNG on parallel computing platforms. Towards the realization of this objective, the short-term goals of this proposal are to:

(i). Investigate vulnerabilities and security mechanisms in LFSR-based stream ciphers.

(ii). Design wireless interface and techniques for stream ciphers vulnerability modelling, and evaluate security requirements for each component network.

(iii). Design the proposed algorithm for PRNG using a hybrid of various methods (shrinking generator, nonlinear filter, generator and alternating step generator) to break the predictability of LFSRs.

(iv). Comparative analysis of different parallel computing environment, namely OpenCL and CUDA.

(v). Analyse and design proposed algorithm using VHDL (very high speed integrated circuit hardware description language) on hardware platform FPGA-SPARTAN 6 and using CUDA on software parallel platform.

(vi). Identifying the hardware utilization using Spartan-6, FPGA, measurement of execution speed using parallel computing software—CUDA, evaluate randomness of keystream using the NIST statistical test package.

8 Technical Requirements and Feasibility

As the project is focused on both hardware and software implementation, it confines its technical requirements in both these domains. *Hardware Requirements*: VHDL—Very High Speed Integrated Circuit Hardware Description Language Analysis and designing of the proposed algorithm is done using VHDL language. FPGA-SPARTAN 6 The simulation of the proposed algorithm is to be done, using FPGA-Spartan 6 toolkit. *Software Requirements*: CUDA- Compute Unified Device Architecture is a parallel computing platform to parallelize the given algorithm, developed by NVIDIA. The GPU used is Geforce 480.

9 Conclusion and Future Work

Through this paper, a precise review on different network applications, hardware and software ciphers, parallel computing platforms have been done. The study thus enforces the need to build a generic cipher which works efficiently both on hardware and software platforms. From the literature and the experimental basis, designing of the strong cipher is quite clear and easy. An n-bit LFSR cipher, customized for different application and different requirements of computation capacities is proposed.

Acknowledgments This work has been awarded by Computer Society of India—National level project competition, wherein secured first rank for the project titled "LFSR-based Cryptographically Secured Key Stream Generator" on 14 March 2015 at Chennai.

References

1. Hell, M., Johansson, T., & Meier, W. (2007). Grain: A stream cipher for constrained environments. *International Journal of Wireless and Mobile Computing, 2*(1), 86–93.
2. Agren, M., Hell, M., Johansson, T., & Meier, W. (2011). A new version of grain-128 with authentication. In *Symmetric Key Encryption Workshop*.
3. Ekdahl, P., & Johansson, T. (2000). Snow-a new stream cipher. In *Proceedings of First Open NESSIE Workshop, KU-Leuven* (pp. 167–168).
4. Ekdahl, P., & Johansson, T. (2003). A new version of the stream cipher snow. In *Selected areas in cryptography* (pp. 47–61). Springer.
5. Ekdahl, P., & Johansson, T. (2003). A new version of the stream cipher SNOW. In *Selected areas in cryptography*. Berlin Heidelberg: Springer.
6. Fluhrer, S., Mantin, I., & Shamir, A. (2001). Weaknesses in the key scheduling algorithm of RC4. In *Selected areas in cryptography*. Berlin Heidelberg: Springer.
7. Baldi, M., Bianchi, M., Maturo, N., & Chiaraluce, F. A physical layer secured key distribution technique for IEEE 802.11 g wireless networks. *IEEE Wireless Communications Letters*.
8. (2007, 14 December). *Some Results on Distinguishing Attacks on Stream Ciphers, Ph.D. Thesis*. Lund University: Hakan Englund.

9. *Study of the Berlekamp Massey Algorithm and Clock-Controlled Generators*. Ong Eng Kiat, Department of Mathematics, National University of Singapore.
10. Fang, J., Varbanescu, A. L., & Sips, H. (2011). A comprehensive performance comparison of cuda and opencl. In *Parallel Processing (ICPP), 2011 International Conference on* (pp. 216–225). IEEE.
11. *Cuda toolkit documentation-Developer zone.*
12. *A statistical test suite for random and pseudorandom number generators for cryptographic applications.*
13. Upadhyay, D. P., Sharma, P., & Valiveti, S. (2014). Randomness analysis of A5/1 stream cipher for secure mobile communication. *International Journal of Computer Science & Communication, 3*, 95–100.
14. Shah, T., Upadhyay, D. P., & Sharma, P. (2014). *A comparative analysis of different LFSR based ciphers and Parallel computing platforms for development of generic cipher compatible on both hardware and software platforms*. Jaipur.
15. Upadhyay, D. P., Shah, A., & Sharma, P. R. (2014). *IEEE international conference on computational intelligence and communication networks*. Udaipur.

Microstrip Patch Antenna for IEEE 802.11a WLAN (5.25 GHz) Application

Charu Tyagi, Ira Joshi and Shalini Porwal

Abstract This paper presents a simple rectangular microstrip patch antenna for 5.25 GHz IEEE 802.11. The proposed antenna is designed on Roger TMM 3 substrate with dielectric constant 3.25 and thickness of 2.4 mm. The size of antenna is compact with patch dimension 14×16 mm^2. The antenna parameters such as return loss, VSWR, gain, and directivity are simulated and optimized using commercial computer simulation technology microwave studio (CST MWS). The return loss is about −17 dB, gain is 7.9 dB, directivity is 6.3 dBi, and VSWR is 1.27 at the operating frequency 5.25 GHz. The main advantage of this antenna is that the designed structure is very simple compared to other proposed WLAN antennas and the cost for making this antenna is also low.

Keywords Microstrip antenna · Monolithic microwave integrated circuits (MMIC) · Planar inverted-F antenna (PIFA) · Wireless local area network (WLAN)

1 Introduction

With rapid development of wireless communication, the demand for devices that can operate in different bands is increased. However, multifrequency antennas have the advantages of surveying multiple frequencies with one antenna but the cross

C. Tyagi (✉) · I. Joshi
Electronics and Communication Department,
Rajasthan College of Engineering for Women, Jaipur, India
e-mail: charu.tyagi.india@gmail.com

I. Joshi
e-mail: erirajoshi2007@gmail.com

S. Porwal
Electronics and Communication Department,
Government Engineering College, Ajmer, India
e-mail: shalinip49@gmail.com

© Springer Science+Business Media Singapore 2016
S.C. Satapathy et al. (eds.), *Proceedings of International Conference on ICT for Sustainable Development*, Advances in Intelligent Systems and Computing 408, DOI 10.1007/978-981-10-0129-1_38

talk from the neighbor bands makes them a weak choice [1]. WLAN is a flexible data communication system which is implemented as an alternative to wired LAN. WLANs are becoming popular in a number of vertical markets such as retail, health, care, warehousing, manufacturing, and academia which have profited from use of handheld terminal for real-time information transmission to centralized hosts for processing [2]. WLAN is also being widely recognized as a reliable, cost effective solution for wireless high-speed data connectivity.

There are three operation bands in the IEEE 802.11 WLAN standards:

- IEEE 802.11b/g (2.4–2.484 GHz)
- IEEE 802.11a (5.15–5.35 GHz)
- IEEE 802.11a (5.725–5.825 GHz) [3]

IEEE 802.11a employs the higher frequency bands and these bands are mostly used in business network due to its higher cost. Slot antennas are the very good choice for WLAN because of the ability of various frequencies ease of fabrication and compatibility with monolithic microwave integrated circuits (MMIC).

A number of WLAN antennas have been recently proposed and reported in the literature [2]. For designing compact-sized WLAN antennas, different and interesting methodologies are used, few of them are as follows:

- Bending the monopole to different shapes has been used in [4–9].
- Effective size reduction techniques are the use of an inverted-F structure [10–13]. By inverted-F structure achieving a compact size is a design challenge that has been tackled in [14].
- The direct feed PIFA proposed in [15] was combined with parasitic element that was used to generate the 5.8 GHz band.

This paper proposes a compact and simple microstrip patch antenna. Designed antenna is operating in 5.25 GHz IEEE 802.11 band. The main advantage of this antenna is this designing structure which is very simple and compact as compared to other designed antennas [5–13]. As due to its simple design structure it can be easily designed and fabricated so it reduces cost and time of manufacturing this antenna and as the 5.25 GHz WLAN antenna have indoor applications the reduced cost can promote its applications in other areas too where the budget is not much high. The substrate used for designing this antenna is Roger TMM 3, which is a good substrate material for better results.

In this paper there are three sections. The antenna design is explained in Sect. 2. In Sect. 3, simulated results of designed antenna are mentioned, in results we are including the return loss, VSWR, gain, directivity, and current distribution of proposed antenna. And in Sect. 4 paper is concluded.

Fig. 1 Proposed antenna
design for WLAN

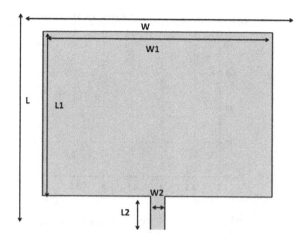

2 Antenna Design

Roger TMM 3 substrate with the dimension of 20×20 mm^2 and the thickness of
2.4 mm is used for designing this antenna and the dielectric constant of this substrate is 3.25. For designing this antenna first we designed a simple rectangular
patch antenna with patch dimensions $14 \times 16 \times 0.035$. The microstrip feeding is
used with dimension L2 = 4 mm and W2 = 2 mm for providing feed to this antenna
as shown in Fig. 1.

By simulating the designed structure it is working for some band for the
5.25 GHz IEEE 802.11a WLAN band. The geometry of the designed antenna is
shown in Fig. 1 and the dimensions of this antenna are listed in Table 1.

3 Result and Discussion

Figure 2a shows the simulated return loss of the proposed antenna. The return loss
gives the band from 5.15 to 5.4 GHz which is less than -10 and -17 dB for the
resonant frequency 5.25 GHz. The frequency 5.25 GHz comes under IEEE 802.11
WLAN standards and it is used in indoor applications.

Table 1 Dimensions of proposed WLAN antenna

Antenna parameter	Value (mm)	Antenna parameter	Value (mm)
L	20	W	20
L1	14	W1	16
L2	4	W2	2

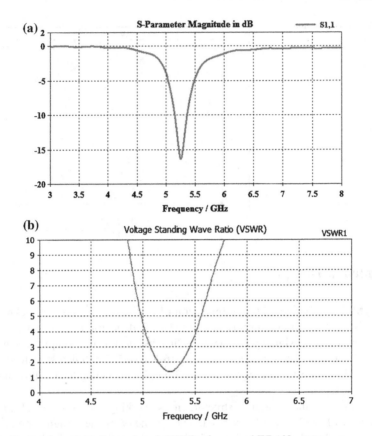

Fig. 2 Simulated results. **a** Return loss. **b** VSWR of proposed WLAN antenna

Figure 2b shows the simulated VSWR results of the proposed antenna. For microstrip antenna acceptable value of VSWR for resonant frequency should be less than or equal to 2. For this antenna VSWR at resonant frequency (5.25 GHz) is 1.27.

Figure 3a, b shows the radiation pattern of this rectangular patch antenna at resonant frequency 5.8 GHz. The antenna is representing the radiation in desired direction. The different parameters that we get from radiation pattern are as follows:

- Angular width (3 dB) = 98.0 Degree
- Gain (dB) = 7.9
- Directivity = 6.3 dBi

Figure 3c is showing the current distribution of this antenna and as we can see that current distribution is maximum on the edges of rectangular patch.

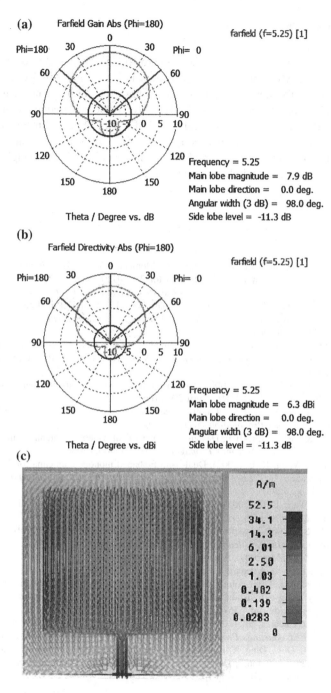

Fig. 3 Simulated results of proposed WLAN antenna. **a** Gain at 5.25 GHz. **b** Directivity at 5.25 GHz. **c** Current distribution at 5.8 GHz

4 Conclusions

In this paper a low-cost compact-sized microstrip patch antenna for 5.25 GHz IEEE802.11a WLAN applications is presented. For designing this Roger TMM 3 material is used. The designed antenna is working for the frequency band 5.15–5.4 GHz with resonant frequency 5.25 GHz. At the resonant frequency, the return loss = −17 dB, VSWR = 1.27, and gain = 7.9 dB. The simulated results are good and its simple planar geometry makes it suitable for microwave integrated circuits. In future this antenna can be converted for multiple bands so that the single antenna can be used for different wireless applications.

References

1. Saghati, A. P., Azarmanesh, M., & Zaker, R., Member, IEEE. (2010). A novel switchable single- and multifrequency triple-slot antenna for 2.4-GHz bluetooth, 3.5-GHz WiMax, and 5.8 GHz WLAN. *IEEE Antennas and Wireless Propagation Letters, 9.*
2. Nayak, P. B., Endluri, R., Verma, S., & Kumar, P. (2013). Compact dual-band antenna for WLAN applications. *2013 IEEE 24th International Symposium on Personal, Indoor and Mobile Radio Communications.*
3. Song, X. D., Fu, J. M., & Wang, W. (2008). Small CPW-fed microstrip monopole antenna for WLAN applications. *Microwave Conference, 2008. APMC 2008.* Asia-Pacific (pp. 1, 4), 16–20 Dec. 2008.
4. Cao, Y., Lu, C., & Zhang, Y. (2008). A compact dual band miniaturized antenna for WLAN operation. In *Proceedings ICMMT,* April 2008 (pp. 416–419).
5. Chang, T. N. & Jiang, J. J. (2009). Meandered T-shaped monopole antenna. *IEEE Transactions Antennas Propagation, 57*(12), 3976–3978.
6. Chu, Q. X., & Ye, L. H. (2010). Design of compact dual-wideband antenna with assembled monopoles. IEEE Trans. *Antennas Propagation, 58*(12), 4063–4066.
7. Yeh, S. H., & Wong, K. L. (2002). Dual-band F-shaped monopole antenna for 2.4/5.2 GHz WLAN application. In *IEEE Antenna Propagation Society International Symposium Digest 2002* (Vol. 4, pp. 7275).
8. Kim, T. H., & Park, D. C. (2005). CPW-fed compact monopole antenna for dual-band WLAN applications. *Electronics Letters, 41,* 291–293.
9. Yildirim, B. S. (2006). Low-profile and planar antenna suitable for WLAN/ Bluetooth and UWB applications. *IEEE Antenna Wireless Propagation Letters, 5,* 438–441.
10. Azad, M. Z., & Ali, M. (2009). A miniature implanted inverted-F antenna for GPS application. *IEEE Transactions on Antennas Propagation, 57*(6), 1854–1858.
11. Gallo, M., Losito, O., Dimiccoli, V., Barletta, D., & Bozzetti, M. (2011). Design of an inverted F antenna by using a transmission line model. In *Proceedings 5th European Conference Antennas Propagation* (pp. 635–638).
12. Liu, D. X., & Gaucher, B. (2005). The inverted-F antenna height effects on bandwidth. *Proceedings IEEE Antennas Propagation Society International Symposium, 2A,* 367–370.
13. Jiang, T. H., Su, D. L., Ding, K. J., Wang, G. Y., & Zhou, Y. (2006). Design of the low-profile inverted-F antenna with multiparasitic elements. In *Proceedings 7th International Symposium Antennas Propagation EM Theory* (p. 14).

14. Razali, A. R., & Bialkowski, M. E. (2009). Coplanar inverted-F antenna with open-end ground slots for multiband operation. *IEEE Antenna Wireless Propagation Letters, 8,* 1029–1032.
15. Wang, H. Y., & Zheng, M. (2011). An internal triple-band WLAN antenna. *IEEE Antennas Wireless Propagation Letters, 10,* 569–572.

Support Vector Machine-Based Model for Host Overload Detection in Clouds

Monica Gahlawat and Priyanka Sharma

Abstract Recently, increased demands in computational power resulted in establishing large-scale data centers. The developments in virtualization technology have resulted in increased resource utilization across data centers, but energy efficient resource utilization becomes a challenge. It has been predicted that by 2015 data center facilities costs would contribute about 75 %, whereas IT would contribute the remaining to the overall operating cost of the data center. The server consolidation concept has been evolved for improving the energy efficiency of the data centers. The paper focuses on support vector machine regression-based novel approach to predict the overload pattern of the servers for better data center reconfiguration.

Keywords Support vector machine regression · Energy efficiency

1 Introduction

Virtualization plays an important role in cloud computing, since it permits appropriate degree of customization, security, isolation, and manageability that are fundamental for delivering IT services on demand. One of its striking features is the ability to utilize computer power more proficiently. Particularly, virtualization provides an opportunity to consolidate multiple virtual machine (VM) instances on fewer hosts depending on the host utilization, enabling many computers to be turned-off, and thereby resulting in substantial energy savings.

In fact, commercial products such as the VMware vSphere Distributed Resource Scheduler (DRS), Microsoft System Center Virtual Machine Manager (VMM), and

M. Gahlawat (✉)
L.J. Institute of Computer Application, Ahmedabad, Gujarat
e-mail: monica.gahlawat@yahoo.com

P. Sharma
Raksha Shakti University, Ahmedabad, Gujarat
e-mail: pspriyanka@yahoo.com

© Springer Science+Business Media Singapore 2016
S.C. Satapathy et al. (eds.), *Proceedings of International Conference on ICT for Sustainable Development*, Advances in Intelligent Systems and Computing 408, DOI 10.1007/978-981-10-0129-1_39

Citirix XenServer offer VM consolidation as their chief functionality [1]. But with the rapid growth in computing demands, the number of data centers grows with the need which leads to more number of servers active at a time. The high active servers' ratio leads to more energy emission and production of Carbon dioxide (CO_2). According to data centers' study, the data centers are not utilized up to their maximum utilization level which leads to more active servers, everyone utilized to less than their total capacity. With this in mind, it is worthwhile to attempt to minimize energy consumption through any means available. Various research agencies and universities have contributed into the research and design of heat dissipation and control in the data center. Virtualization is a technology that contributes to the maximum utilization of the servers by virtual machine (VM) consolidation and VM migration.

The decision of reallocation of virtual machine for VM consolidation depends on the host utilization behavior. The VMs from the under-utilized and over-utilized hosts are relocated to other hosts by packing the VMs on minimum number of hosts. The hosts having no virtual machine are shifted to the passive mode so that the total energy consumption can be reduced. Statistical methods played a great role in predicting the behavior of the host in dynamic manner. The author [2] has proposed various statistical methods for host overload and underload behaviors of the hosts in his thesis. These algorithms take input as the previous or current utilization of the hosts and predict the future based on the previous or current state of the system. He has proposed Local Regression, Median Absolute Deviation, Robust Local Regression, and Markov Chain model for predicting the overloaded hosts [2]. All statistical models cannot be applied to all the environments. The choice of the statistical methods depends on the input data, because every statistical model is based on some assumptions. Markov chain model assumes that the data will be stationary but complex and dynamic environment like cloud experience highly variable nonstationary workload. The author [2] in his thesis modified his model by using multisize sliding window workload estimation method so that it can be suitable for the cloud environment.

The paper proposes a prediction model based on Support Vector Machine (SVM) regression to forecast the overload behavior of the host in the dynamic environment of cloud. The paper is organized as follows. Section 2 explains the basic concepts and modeling approach of the Support Vector Machine. In Sect. 3, the literature review related to Support Vector Machine is presented. Section 4 explains the mathematical formulation of the LS-SVM and DLS-SVM. Section 5 presents the comparison charts and the concluding remarks.

2 Support Vector Machine

Support vector machine is a novel technique based on neural network invented by Vapnik and his co-workers at AT & T Bell Laboratories in 1995. The objective of SVM is to find a generalized decision rule through selecting some particular subset of training data, called support vectors. Training SVMs is equivalent to solving a linearly constrained quadratic programming problem. The main advantage of the support vector machine is that the solution obtained is globally optimal and does not directly depend on the input space. Another key advantage of SVM is that SVMs tend to be resistant to over-fitting, even in cases where the number of attributes is greater than the number of observations. According to Vapnik there are three main problems in machine learning, e.g., Density Estimation, Classification, and Regression. The machine learning algorithms creates a knowledge base based on the training data and then concludes the results based on the knowledge. These algorithms are used in variety of domains these days, e.g., Facebook is using machine learning algorithms for face recognition. Machine learning algorithms are also popular in forecasting future values based on the previous values of a time series. A time series is a sequence of data points usually ordered in time. Machine learning algorithms extract the meaningful data from the time series patterns and predict future values based on the previously observed values. The main focus of this paper is to predict the overload behavior of the hosts in cloud data centers based on the previous load pattern of the hosts in the data center.

The time series prediction using support vector machine is affected by various factors like data is linearly separable or follows nonlinear patterns, the learning is supervised learning or unsupervised learning and on support vector kernels. The points are linearly separable or not are decided by visualizing the points in 2D plane by taking one set of points as being colored red and the other set of points as green. These set of points are considered linearly separable if there exists at least one line in the plane separating all the green points on one side of the line and all the red points on the other side. Usually in practical problems the data points are mapped to the high-dimensional plane and the optimal separating hyper plane is constructed with the help of special functions known as support vector kernels (Fig. 1).

The second point of concern is that the learning algorithm is supervisory learning or unsupervisory. In supervisory learning, the training data is composed of input as well as the output vector (also called supervisory signals) whereas in unsupervisory learning the training data is composed of only input vectors. Supervisory learning produces better results because the output vector is already

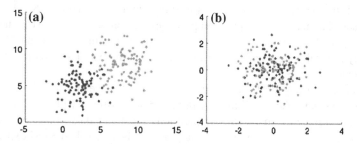

Fig. 1 **a** Linearly separable data. **b** Nonlinearly separable pattern

known and the predicted values by the SVM are compared with the output to learn better for the next step. In unsupervisory learning the output data points are not known and the training depends on the probability to drive better results out of it. SVM comes in the supervisory learning category and the kernel function makes the technique applicable for the linear as well as nonlinear approximation.

3 Related Works

In various practical domains, time series modeling and forecasting have essential importance. An extensive research is going on evaluating the performance of the machine learning algorithms for time series prediction. Various models, e.g., Bayesian neural networks, K-nearest neighbor regression, etc., have been proposed in the literature for improving the accuracy and efficiency of time series modeling and forecasting. The author [1] has compared various time series prediction methods widely used these days. This paper investigated the application of SVM in financial forecasting. Experimental results presented in [3–5] shows that support vector machine is a promising method for time series prediction. The autoregressive integrated moving average model (ARIMA), ANN, and SVM models were fitted to Al-Quds Index of the Palestinian Stock Exchange Market time series data and two-month future points were forecasted. The results of applying SVM methods and the accuracy of forecasting were assessed and compared to ARIMA and ANN methods through the minimum root mean square error of the natural logarithms of the data. Results proved that SVM is a better method of modeling and outperformed ARIMA and ANN.

The author of [6] explains the time series concept and the various methods of predicting the future values based on ARIMA model, Seasonal ARIMA model, ANN model, time lagged ANN, seasonal ANN, SVM for regression, SVM for forecasting, etc., they have also explained the forecast performance measure MFE (Mean Forecast Error), MAE (Mean Absolute Error), MAPE (Mean Absolute Percentage Error), MPE (Mean Percentage Error), MSE (Mean Squared Error), etc. In paper [7], a model based on least squares support vector machine is proposed to forecast the daily peak loads of electricity in a month. In [7] the time series prediction was first used to forecast electricity load. In paper [8], the author has improved the method presented in [7] to derive more accurate results. The author has proposed dynamic least square support vector machine (DLS-SVM) to track the dynamics of nonlinear time-varying systems. The dynamic least square method works dynamically by replacing the first vector by the new input vector to obtain more accurate results.

The author in paper [9] has proposed the modified version of SVM for time series forecasting. The algorithm performs the forecasting in phases. In the first phase, self-organizing maps (SOM) partitions the input data points into several disjoint regions. In the second phase, multiple SVMs, also called SVM experts, are constructed by finding the most appropriate kernel function and the optimal free parameters of SVMs.

The author in [10] presented the algorithm of dynamic least square support vector regression and applied that algorithm to the time series data to predict the future values of the time series accurately. The algorithm considers the input points in a fixed window size for prediction. At each step the algorithm adds one new input point and deletes the last point left with fixed window size. Addition and deletion of input points at each steps makes the algorithm dynamic.

4 Support Vector Machine Regression Formulations

The host utilization is a univariate time series. By analysing the time series patterns, the host overload and underload predictions for a host is possible, because in univariate time series the future values are entirely based on past observations. The goal of the SVM regression is to find a function that presents the most ε deviation from the target values so the maximum allowed error is ε. The future values are predicted by splitting the time series $x_1, x_2 \ldots x_n$ data into training inputs and the training outputs. Given training data sets of N points $\{x_i, y_i\}$, $i = 1, 2 \ldots N$ with input data $x_i \in X \subset R^n$ and output data $y_i \in Y \subset R$. Assume a nonlinear function $f(x)$ as given below

$$f(x) = W^T \phi(X_i) + b \tag{1}$$

w = weight vector, b = bias and $\phi(X_i)$ is a nonlinear mapping to a higher dimensional space. The optimization problem can be defined as

$$\text{Minimize} \quad \frac{1}{2} W^T W$$
$$\text{Subject} \quad \text{to}: \begin{cases} y_i - (W^T \phi(X_i) + b) \leq \varepsilon \\ y_i - (W^T \phi(X_i) + b) \geq -\varepsilon \end{cases} \tag{2}$$

$\varepsilon (\geq 0)$ is a user defined maximum error allowed. The above Eq. (2) can be rewritten as

$$\text{Minimize} \quad \frac{1}{2} W^T W$$
$$\text{Subject} \quad \text{to}: \begin{cases} y_i - W^T \phi(X_i) - b \leq \varepsilon \\ W^T \phi(X_i) + b - y_i \leq \varepsilon \end{cases} \tag{3}$$

To solve the above equation slack variables needs to be introduced to handle the infeasible optimization problem. After introducing the slack variables the above equations take the form

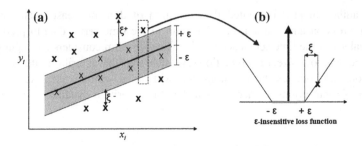

Fig. 2 **a** The accurate points inside ε tube. **b** Slope decided by C

$$\text{Minimize} \quad \frac{1}{2} W^T W + C \sum_{i=1}^{m} (\xi_i^+ + \xi_i^-)$$

$$\text{Subject to :} \begin{cases} y_i - W^T \phi(X_i) - b \le \varepsilon + \xi_i^+ \\ W^T \phi(X_i) + b - y_i \le \varepsilon + \xi_i^- \\ \xi_i^+, \xi_i^- \ge 0 \end{cases} \tag{4}$$

The slack variables ξ_i^+, ξ_i^- defines the size of the upper and the lower deviations as shown in Fig. 2a.

The dual form of the equation [8] after adding the Lagrangian multipliers (α_i^+, α_i^-, ξ_i^+, ξ_i^-) can be rewritten as

$$L_p = \frac{1}{2} W^T W + C \sum_{i=1}^{m} (\xi_i^+ + \xi_i^-) - \sum_{i=1}^{m} (\eta_i^+ \xi_i^+ + \eta_i^- \xi_i^-)$$
$$- \sum_{i=1}^{m} \alpha_i^+ (\varepsilon + \xi_i^+ - y_i + W^T \phi(X_i) + b)$$
$$- \sum_{i=1}^{m} \alpha_i^- (\varepsilon + \xi_i^- + y_i - W^T \phi(X_i) - b)$$
$$\text{s.t.} \quad \alpha_i^+, \alpha_i^-, \xi_i^+, \xi_i^- \ge 0 \tag{5}$$

Applying the conditions of the optimality, one can compute the partial derivatives of L with respect to w, b, ξ_i^+, ξ_i^- equate them to zero and finally eliminating w and ξ obtain the following linear system of equations [10]

$$\begin{pmatrix} o & 1^T \\ 1 & \Omega + C^{-1} I \end{pmatrix}_{(N+1) \times (N+1)} \begin{bmatrix} b \\ a \end{bmatrix}_{(N+1) \times 1} = \begin{bmatrix} 0 \\ y \end{bmatrix}_{(N+1) \times 1} \tag{6}$$

Here $y = [y_1, y_2, \ldots y_N]$, $1 = [1, 1, \ldots 1]$ and Ω with $\Omega(i, j) = k(x_i, y_i)$ ($\forall i = 1, 2, \ldots N$) is the radial base function-based kernel matrix ($K(x, y) = e^{-[x-y]^2 / 2\sigma^2}$). Table 1

Table 1 Symbols used in the mathematical formulation of model

Symbol	Description
W	Weight vector
b	Bias
$\phi(X_i)$	Nonlinear mapping to a higher dimensional space
ε	Maximum error allowed
ξ_i^-, ξ_i^+	Slack variables, defines the upper and lower deviations to ε tube
$(\alpha_i^+, \alpha_i^-, \xi_i^+, \xi_i^-)$	Lagrangian multipliers
$\Omega(i,j) = k(x_i, y_i)(\forall i = 1, 2, ...N)$	Radial base kernel matrix
C	Regularization parameter

shows the symbols used in the formulation. The LS-SVM decision function is thus given by [8].

$$y(x) = \sum_{i=1}^{N} \alpha_i k(x_i, y_i) + b \qquad (7)$$

Host utilization time series data is dynamic, new points are obtained continuously. The new data points should be included in the training data set to track the dynamics of the nonlinear time-varying system and the old points are deleted for constant window size of training data.

5 Experiments

We have used CloudSim for retrieving the utilization of the host based on the workload defined in the PlanetLab folder in CloudSim. It contains the daily virtual machine requirement and the utilization of the host is calculated based on the daily

Fig. 3 Comparison of errors

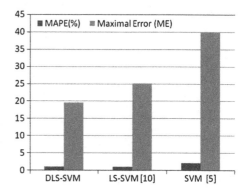

requirement of the virtual machines. After retrieving the utilization of the host's dynamic support vector machine regression model id is applied and the results are compared with [7, 11]. The comparison is based on MAPE (mean absolute percentage error) and Maximal error (ME). The chart in Fig. 3 shows that DLS-SVM produce better forecast for the load pattern of the hosts in the data centers.

References

1. Okasha, M. K. (2014). Using support vector machines in financial time series forecasting. *International Journal of Statistics and Applications, 4*(1), 28–392.
2. Beloglazov, A. (2013). Energy-efficient management of virtual machines in data centers for cloud computing.
3. Kim, K. J. (2003). Financial time series forecasting using support vector machines. *Neurocomputing, 55*(1), 307–319.
4. Gui, B., Wei, X., Shen, Q., Qi, J., & Guo, L. (2014). Financial time series forecasting using support vector machine. In *2014 Tenth International Conference on Computational Intelligence and Security (CIS)* (pp. 39–43). IEEE.
5. Rüping, S. (2001). SVM kernels for time series analysis (No. 2001, 43). Technical Report, SFB 475: Komplexitätsreduktion in Multivariaten Datenstrukturen, Universität Dortmund.
6. Adhikari, R., & Agrawal, R. K. (2013). An introductory study on time series modeling and Forecasting. arXiv preprint arXiv:1302.6613.
7. Chen, B.-J., Chang, M.-W., & Chih-Jen, L. I. N. (2004). Load forecasting using support vector machines: A study on EUNITE competition 2001. *IEEE Transactions on Power Systems, 19* (4), 1821–1830.
8. Niu, D. X., Li, W., Cheng, L. M., & Gu, X. H. (2008). Mid-term load forecasting based on dynamic least squares SVMs. In *2008 International Conference on Machine Learning and Cybernetics* (Vol. 2, pp. 800–804). IEEE.
9. Cao, L. (2003). Support vector machines experts for time series forecasting. *Neurocomputing, 51*, 321–339.
10. Fan, Y., Li, P., & Song, Z. (2006). Dynamic least squares support vector machine. In *The Sixth World Congress on Intelligent Control and Automation, 2006.* WCICA 2006 (Vol. 1, pp. 4886–4889). IEEE.
11. Haishan, W., Xiaoling, C. (2006). Power load forecasting with least square support vector machines and chaos theory. *Proceedings of the 6th World Congress on Intelligent Control and Automation*, Dalian, China, June 21–23, 2006.

A Hybrid Face Recognition Scheme Using Contour and Gabor Wavelet

Bandariakor Rymbai, Debdatta Kandar and Arnab Kumar Maji

Abstract A face recognition system based on the hybrid approach is proposed in this paper. We have combined both the global and local features of a face using Gabor wavelet transform and a face contour. Gabor transform helps in extracting the features of a face, that is, the position of eyes, tip of the nose, and the corner of the mouth, whereas a face contour helps to extract the structure of a face. A combination of these techniques helps to extract the best features.

Keywords Gabor wavelet · Contour · Facial feature · Gaussian · Discrete cosine · Histogram

1 Introduction

A face recognition system is one of the applications used for the authentication process in any organization by identifying or verifying a person from the captured image. The features of the face are selected as fiducial points [1], so that they can be used in the matching procedure where the feature set of an input image is matched with the image in the database. These facial features [2] can be classified as Local Features and Global Features where a Gabor Wavelet Transform helps in filtering the image, so that it will be able to extract the local features of a face such as position of the eyes, nose, and mouth as it reduces the uncertainty values. Gabor Filters are an important process due to their biological relevance such as spatial localization, orientation, and the spatial frequency domain where the features can be extracted.

B. Rymbai (✉) · D. Kandar · A.K. Maji
Department of Information Technology, North Eastern Hill University, Shillong 793 022, India
e-mail: daria26@rediffmail.com

D. Kandar
e-mail: kdebdatta@gmail.com

A.K. Maji
e-mail: arnab.maji@gmail.com

© Springer Science+Business Media Singapore 2016
S.C. Satapathy et al. (eds.), *Proceedings of International Conference on ICT for Sustainable Development*, Advances in Intelligent Systems and Computing 408, DOI 10.1007/978-981-10-0129-1_40

Contour is one of many methods used for images that can extract data of an image shape and the various features can be used in the form of a graph. Contour can produce more precise features.

The contour pixels are normally a small number of pixels representing a shape. The computation cost of feature extracting algorithms can be reduced when we apply a contour method, and also feature extraction procedure arc more effectively when performing a contour as it helps to extract the original shape of an image. Therefore, a contour map is best for the efficacy of feature extraction, which is an essential process to recognize the pattern.

In this section, we discuss two techniques for extracting the important features that can be used in our system, which is the Gabor wavelet transform where we extract the local features of a face and the contour for extracting the global part of a face.

2 Existing Literature

Gabor [3] a Hungarian-born electrical engineer introduced a Gabor function in 1946, which is commonly used for feature extraction, especially in face recognition. The Gabor filters signify a low band-pass filter whose time and frequency response is defined by a Wavelet function multiplied by a Gaussian function. Thus, a 2D Gabor filter [4, 5] consists of complex sinusoidal and modulating with a Gaussian envelope. Therefore, the use of Gabor function helps in localized feature points of a given face, which can be used in the matching procedure.

2.1 Face Contour

Gandhe et al. [6] proposed a system using a Contour Plot as the core of the system. This system treats the whole face as a contour map, where the contour lines of a given face can be generated with the areas of constant gray level brightness.

2.2 Gabor Function

The Gabor filter function [7, 8] can be defined by Gaussian-shaped function which consist of k as the magnitude, (a, b) as the axis, θ as the angle of rotation, $(x0, y0)$ as the peak area, $(u0, v0)$ as the spatial frequency of the sine-wave carrier in Cartesian coordinates, and P as the phase of the sine-wave carrier.

To generate the Gabor filter of an image, here we use the function gaborfilter() comprising of the size of an image I, variance σ, frequency Υ, phase φ, and λ as the wavelet angle

$g(x, y, \sigma, \Upsilon, \lambda, \varphi) = k*\text{Gaussian}(x, y, \sigma)*(\text{Sinusoid}(x, y, \Upsilon, \lambda, \varphi) - \text{DC}(\Upsilon, \sigma, \varphi))$,
where:
$\text{Gaussian}(x, y, \sigma) = (\exp(-\pi*\sigma^2*(x^2 + y^2))$
$\text{Sinusoid}(x, y, \Upsilon, \lambda, \varphi) = \exp(j*(2*\pi*\Upsilon*(x*\cos(\lambda) + y*\sin(\lambda)) + \varphi))$
$\text{DC}(\Upsilon, \sigma, \varphi) = \exp(-\pi*(\Upsilon/\sigma)^2 + j*\varphi)$ and,
$\text{DC}(\Upsilon, \sigma, \varphi)$ compensates the inherent Discrete Cosine formed by the Gaussian-shaped function.

2.3 Contour Generation

The contour generation [9, 10] is serial in eliminating some of the pixels for creating the boundary links to the surface of an input image. As described above, the entire shape of the face is treated as a contour map with the use of constant gray level brightness called contour lines that connect a series of points of equal elevation and are used to define the structure on a map.

Contour lines are arched, straight, or a combination of both lines on a map that describe the connection of actual or fictional shape. The contour plot helps to extract the lines that are of equal values of a given matrix, where the contour plot matrix can be taken as heights with respect to the x and y axes. The matrix must be at least a 2×2 matrix that contains the lower of two different values. The x-values relates to the column indices of the matrix and the y-values relates to the row indices of the matrix. The contour curves are selected automatically and are also called the contour levels.

2.3.1 Level Curve

The function of the two variables $f(x, y) = c$, where c is the constant can be viewed by the help of the level curve.

2.3.2 Level Surfaces

To get the contour surface, a function, $f(x, y, z) = c$ is computed where c is a constant and x, y, z are the variables.

2.3.3 Contour Map

To draw the contour map, the contour curve of $\sin(xy) = c$ is required, where c is a constant and $xy = C$, which is the arcsin of c, and the curve $y = C/x$ are hyperbolas (except $C = 0$, when $y = 0$). The line $x = 0$ is also a contour curve, where the contour map constitutes of hyperbolas and a coordinate axis.

2.3.4 Contour Matching

The contour matching between the input faces and the registered face is used by obtaining the contour matrix c, which contains both the levels and the vertices, which was used by a MATLAB function [c, h] = contour (Z) where h is the contour objects.

3 A New Hybrid Algorithm Using Contour Analysis and Gabor Wavelet

Here we have proposed a combination algorithm to extract both the local and global features that can be used for recognition. Extracting both the features, the selection of fiducial points is easier where it can extract along with the shape of a face

1. The input Image I is converted into grayscale values.
2. The histogram equalization histeq (I) is used to enhance the contrast of an intensity Image I.
3. To filter Image I, a Gabor filtration is used by modulating the wavelet function with the Gaussian envelope.

$$g(x, y, \sigma, \Upsilon, \lambda, \varphi) = k * \text{Gaussian}(x, y, \sigma) * (\text{Sinusoid}(x, y, \Upsilon, \lambda, \varphi) - \text{DC}(\Upsilon, \sigma, \varphi))$$

4. Then the Gabor output is obtained by convolution of the Gabor filter with Image I, using the Matlab function conv2().
5. After Steps 1–4 the contour map is drawn using a MATLAB function imcontour (M), where M, the Magnitude is the absolute Gabor output for extracting only the fiducial points and the shape of an image.
6. The values of the registered image are stored in the database, along with the ID, so that it can be used for the matching process.
7. Then the input image is compared with the images in the database by superimposing the output of an image contour used in Step 5, which consists of levels and vertices.
8. The required threshold value is set for computing the similarity between the input image and the registered image. If the similarity of the input image and the registered image is greater than the threshold value, then the face is matched.

The architecture of the proposed algorithm is discussed in Fig. 1.

3.1 Architecture of the Proposed Algorithm

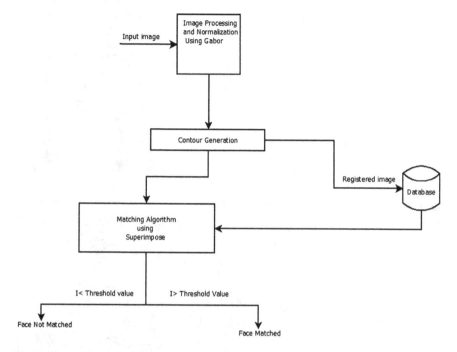

Fig. 1 The architecture of the proposed algorithm

4 Experimental Results

Figures 2, 3, 4 and 5 depicts the experimental results of the said approach. An input image is shown in Fig. 2. First, a contour plot of the input image as shown in Fig. 2 is generated. It can be observed that the contour map selects a large number of features, some of which can be removed further so as to reduce the computational cost as well as time complexity of a matching algorithm. Therefore, in this paper, we present an effective approach for selecting the most suitable features from the image where Gabor filter of the input image is carried out and the output magnitudes' matrix of the Gabor filter obtained is then used for contour plot. Figure 4 shows the Gabor magnitude of the image. Figure 5 depicts contour of the image after applying Gabor filter. Here lies the novelty of the work where we easily observe that the contour generated is much enhanced.

The matching procedure is done by superimposing the contour matrices of the registered and tested image for face verification. Figure 6 shows the output of the superimposed contour matrices of the registered image shown in Fig. 2 and the

Fig. 2 Input image

Fig. 3 Contour map of an input image

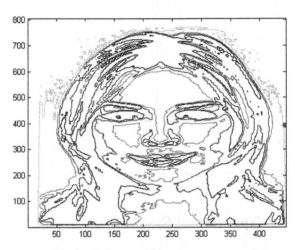

Fig. 4 Gabor filter of an input image

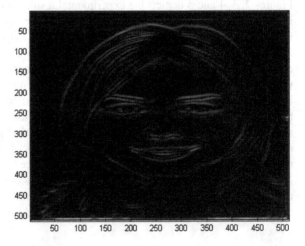

Fig. 5 Contour plot after Gabor filter

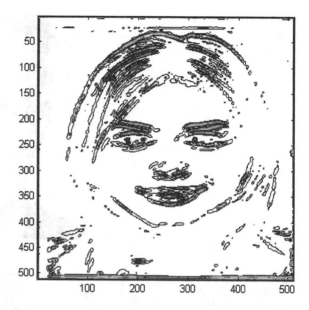

Fig. 6 The output image after superimposed the contour matrices. The image is a recognized image

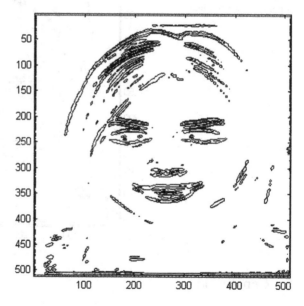

same image when used as a tested image. We see that the matrices superimpose perfectly and thus it can be concluded that the tested image can be recognized.

Another image (Fig. 7) has been used for testing. The result of superimposing its contour matrix with the contour matrix of the registered image of Fig. 2 is depicted in Fig. 8. The two matrices do not result in perfect superimposition, i.e., the tested image cannot be recognized.

Fig. 7 The second tested image

Fig. 8 The output image after superimposing the contour matrices. The image is not a recognized image

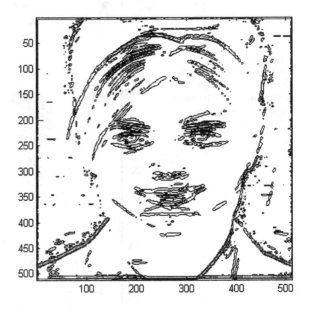

5 Conclusion

In this paper, we have described the hybrid technique where two different techniques, i.e., Gabor filter and contour map are used for extracting the features so that the fiducial points can be easily selected for face verification system. We found that by applying Gabor we can extract the local features of a human face, and along with the use of a contour map for extracting the global or the shape of a face, we can verify the face of a person.

References

1. Wiskott, L., Fellous, J.-M., Uger, N. K., & Malsburg, C. V. (1999). Face recognition by elastic bunch graph matching. In *Intelligent Biometric Techniques in Fingerprint and Face Recognition* (Chapter 11, pp. 355–396).
2. Pantic, M., & Rothkrantz, L. J. M. (2004). Facial action recognition for facial expression analysis from static face images. *IEEE Transactions on Systems, Man, and Cybernetics—Part B: Cybernetics, 34*(3), 14–21.
3. Gabor, D. (1946). Theory of communication. *Journal of the Institution of Electrical Engineers Part III: Radio and Communication, Engineering, 93*(26), 429–457.
4. Ilonen, J., Kämäräinen, J.-K., & Kälviäinen, H. (2005). Efficient computation of gabor features. *Lappeenranta.*
5. Kumar, V., Shreyas, B., & Sarkar, B. (2007). Face recognition using gabor wavelets. In *Conference IEEE Asilomar Conference on Signals, Systems and Computers* (pp. 23–29). IEEE.
6. Gandhe, S. T., Talele, K. T., & Keskar, A. G. (2006). Face recognition using contour matching. *IAENG International Journal of Computer Science, 35*(2), 223–245.
7. Javier, R. *Movellan tutorial on gabor filter.*
8. Knill, O. *Functions of two variables* http://www.math.harvard.edu/archive/21b_summer_04/
9. Hiremath, P. S., Kodge, B. G. (2011). Generating contour lines using different elevation data file formats. *International Journal of Computer Science and Applications (IJCSA), 3*(1), 69–74.
10. Tareque, M. H., et al. (2013, July). Contour based face recognition process. *IJCSET, 3*(7), 244–248.

Data Dissemination Techniques and Publish/Subscribe Architecture in Vehicular Ad Hoc Networks

Vruti P. Surani and Hitesh A. Bheda

Abstract Vehicular Ad hoc Network, also known as VANET, which is a subset of mobile ad hoc network (MANET). It uses mobile connectivity protocols so that communication can easily take place among vehicles and equipments placed beside roads. In this paper, data dissemination technique is used for spreading information over vehicular network. Dissemination of information among vehicles is used for safety and for entertainment as well. This technique increases the quality of driving parameters like time, distance around vehicle, and safety precautions. Publish/subscribe scheme provide services to applications developer to easily design notification service. It enables vehicle/driver to show their interests in certain types of event notifications. (e.g., collision warning note, speed managing note, traffic situation note, etc.). In addition, we have used GPS system to get accurate results. Thus, the key idea behind implementing our paper is to provide good facilities to driver and people.

Keywords VANET · Data dissemination · Publish/subscribe architecture · Broadcast

1 Introduction

A Vehicular ad hoc network is also called as VANET. VANET can be considered as a type of mobile ad hoc network (MANET), it can allow communication between vehicles and nearest steady elements, generally known as road side units (RSU) [1].

V.P. Surani (✉)
Department of CE, School of Engineering, RK University, Rajkot, Gujarat, India
e-mail: vruti.surani@gmail.com

H.A. Bheda
Department of CE/IT, School of Engineering, RK University, Rajkot, Gujarat, India
e-mail: hitesh.bheda@rku.ac.in

© Springer Science+Business Media Singapore 2016
S.C. Satapathy et al. (eds.), *Proceedings of International Conference on ICT for Sustainable Development*, Advances in Intelligent Systems and Computing 408, DOI 10.1007/978-981-10-0129-1_41

Fig. 1 VANET architecture

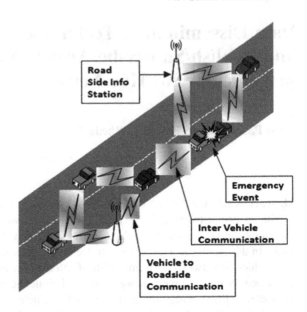

VANET uses cars as mobile nodes to form mobile network. Basically, it supports wireless communication between vehicles as well road side elements to vehicles. VANET can exchange important messages of roads and traffic environments also other perceptible information. Figure 1 shows architecture of VANET. Vehicles and road sided units have limited coverage area. If any vehicles and RSUs are in that range then communication can easily take place. Every vehicle communicates with other vehicles in very dynamic environment of ad hoc network. Road side units provide infotainment services and larger communication services. Nodes of vehicular network are expected to use DSRC protocol which employs the IEEE 802.11p standard for communication. The range of nodes is approximately sated as 100–300 m [1]. Here global positioning system is used to get the accurate location of vehicles.

1.1 Applications and Characteristics

(a) **Applications**
    There are many types of applications available in VANET which can be categorized into following:

- Safety Applications [1]
- Traffic Observation and Control Applications [1]
- Info-Entertainment Applications [2]

(b) **Characteristic**

VANET provides innovative network characteristics [1] that differ it from further type of ad hoc networks [1]. Some important characteristics are specified below [3]:

- Highly dynamic topology
- Frequent network disconnects
- Different communication environments
- High delay constraint
- Interaction with on-board sensors

1.2 Data Dissemination in VANET

Data dissemination is a technique for transmitting and receiving data. Basically, it is used to convey the message from source vehicle to destination vehicles. Use of data dissemination is to increase the quality of driving parameters in terms of time, distance, and safety precautions [1].

The data dissemination approaches [4] in VANET network may be classified on the basis of following category:

(a) V2I/I2V Dissemination (vehicle to infrastructural/infrastructural to vehicle)
(b) V2V Dissemination (vehicle to vehicle)
(c) Opportunistic dissemination
(d) Geographical dissemination
(e) Peer-to-peer dissemination
(f) Cluster based dissemination

1.3 Publish/Subscribe Architecture

Publish/subscribe system provides interaction among publishers and subscribers. In this architecture, subscribers are delivered with the capacity to produce their attention in an event or a set of events aim to get intimation for any type of event generated by publishers [5]. In other words it acts like interaction of producers and consumers. Producers publish or broadcast information to event manager and consumers subscribe to that information which they need to get [6]. The entire process is done by sending notifications.

The architecture of publish/subscribe [7] collaboration service depends on event intimation process. To create an event, a publishers ideally requests a publish() function and subscribers record their interest in an events by typically requesting subscribe() function. Without knowing destination place event manager cannot sent notification and that is why it is persist stored in the event service. The unsubscribe() function is used to terminate a subscription [6].

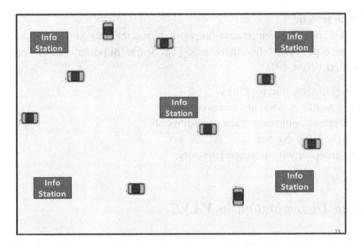

Fig. 2 Publish/subscribe architecture

Figure 2 shows the basic publish/subscribe architecture of vehicular ad hoc network.

2 Data Structure and Algorithm Analysis

In this approach, each vehicle and RSU [8] can obtain the role of publisher, subscriber, or broker. Algorithm for publication, subscriptions, notification, and location determination process is given here. To manage these services each nodes has following type of data structures [7]:

- Publication Table: For storage of active publication
- Subscription Table: For storage of active subscription
- Vehicle ID: For identification of the vehicle
- RSU or Info—station: It shows information about last info-station crossed and time stamp of occurred event

2.1 Algorithm

(a) **Publish Procedure**

   1. Store publication in publication table
   2. if publication is active then check TTL
   3. if any vehicle in 1-hop range
   4. Send publication
   5. else

6. keep on moving for some time
7. go to step 2
8. else discard the publication

(b) **Subscribe Procedure**

1. Store subscription in subscription table
2. if subscription is active then check TTL
3. if any vehicle in 1-hop range
4. Send subscription
5. else
6. keep on moving for some time
7. go to step 2
8. else discard the subscription

Here, TTL represents the time for which publication and subscription is active

3 Simulation Scenario

In our VANET there are number of simulators like OMNet++, Network Simulator, TRANS, NCTUns, GrooveNet, MobiReal, VANET MobiSim, QualNet, etc. In our simulation we used the Network Simulator 2.34 (NS 2.34) with SUMO and MOVE.

3.1 Network Simulator (NS2)

NS2 is an object-oriented Network Simulator. NS2 written in C language. It is an object-oriented version of Tcl script also called as OTcl [9]. Network Simulator is basically useful for local and wide area networks (LAN, WAN) (Fig. 3).

Fig. 3 NS2 simulator

3.2   SUMO (Simulation of Urben MObility)

SUMO is a microscopic traffic simulator which is useful to saw the position of all vehicles on the road. SUMO presents roads and vehicles on a graphical user interface (GUI) [10].

As shown in Fig. 4, we have used SUMO simulator to design topology to display lanes and moving vehicles on the roads.

Here in our simulation we have created communication scenario which is clearly visible in Fig. 5.

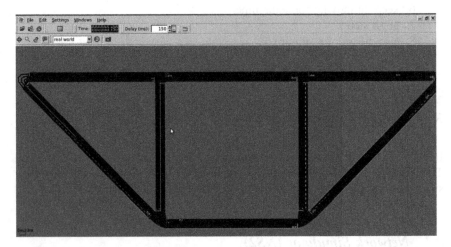

Fig. 4 Road topology using SUMO

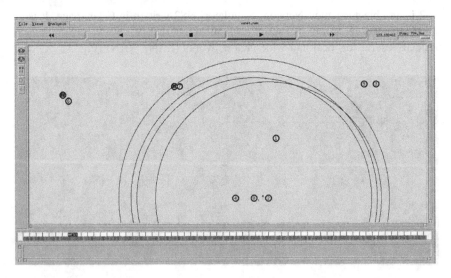

Fig. 5 Communicating nodes

3.3 MOVE

MOVE is a mobility model creator designed for vehicular ad hoc network. It quickly generates practical mobility model for VANET simulation [11].

4 Result Analysis

At the end we have generated results and graphs for our formed simulation. Parameters to be measured for generating graphs are presented below:

4.1 Message Delivery During High Density of Nodes

When the density of nodes is high then the communication take place frequently. Figure 6 shows the simulation result of communication in a high density of nodes on the lanes.

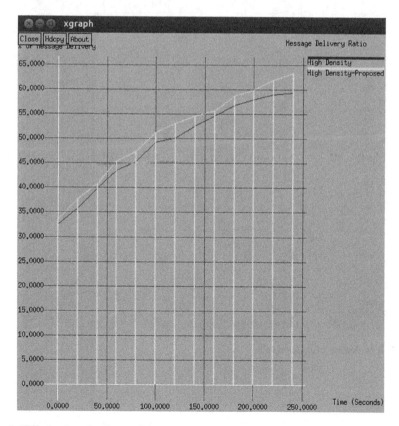

Fig. 6 High density of nodes results

4.2   Message Delivery During Low Density of Nodes

When the density of nodes is low then communication take place rarely. Figure 7 shows the simulation result of communication in low density of nodes on the lanes.

4.3   Delays in Message Delivery

Figure 8 shows the delay occurred during communication of nodes in the existing simulation scenario.

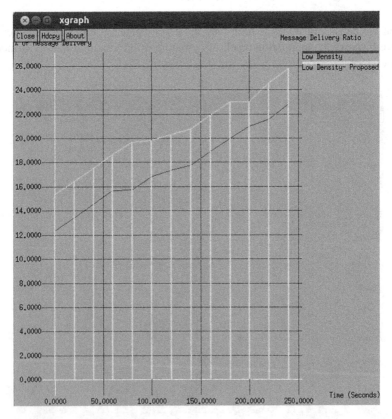

Fig. 7 Low density of nodes result

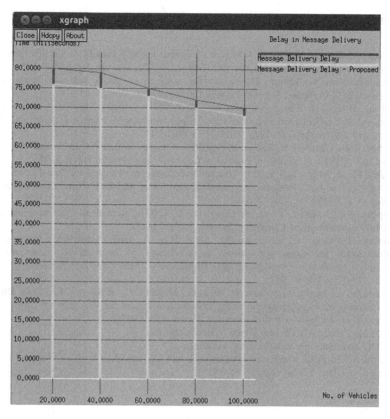

Fig. 8 Delay in message delivery

5 Conclusion

The field of vehicular ad hoc networks has been developed and applied in varieties of applications used day-by-day. Data dissemination is major issue and in demand in vehicular ad hoc networks. Publish/subscribe architecture works well in this functioning of information dissemination. The proposed schema will make the suitable Publish/subscribe architecture. It utilizes the functionality of global positioning system in order to minimize the communication delay. For accomplishing this functionality, it makes use of network simulator with SUMO and MOVE mobility generators for vehicular ad hoc networks. At the end we show the variation of some parameters by generating graphs. Furthermore, we have used GPS system to get results more accurate in less time.

References

1. Kumar, R., & Dave, M. (2012). A Review of Various VANET Data Dissemination Protocols. *International Journal of u- and e-Service, Science and Technology, 5*(3), 27–44.
2. Chandrasekaran, G. *Vanets: The networking platform for future vehicular applications.* Department of Computer Science, Rutgers University.
3. Sabahi, F. (2011). The security of vehicular adhoc networks. In *Third International Conference on Computational Intelligence, Communication Systems and Networks, IEEE Computer Society* (pp. 338–342). 978-0-7695-4482-3/11.
4. Dubey, B. B., Chauhan, N., & Kumar, P. (2010). A Survey on Data Dissemination Techniques used in VANETs. *International Journal of Computer Applications, 10*(7), 5–10.
5. Chitra, M., & Sathya, S. S. (2013). Efficient broadcasting mechanisms for Data dissemination in vehicular ad-hoc Networks. *International Journal of Mobile Network Communications Telematics (IJMNCT), 3*(3), 47–63.
6. Eugster, P. T., Felber, P. A., & Guerraoui, R. (2003). Kermarrec: The many faces of publish/subscribe. *ACM Computing Surveys, 35*(2), 114131.
7. Pandey, T., Garg, D., & Gore, M. M. (2011). A publish/subscribe communication infrastructure for vanet applications. In *Workshops of International Conference on Advanced Information Networking and Applications, IEEE Computer Society* (pp. 442–446). 978-0-7695-4338-3/11.
8. Leontiadis, I., Marfia, G., Mack, D., Pau, G., Mascolo, C., & Gerla, M. (2013). On the effectiveness of an opportunistic traffic management system for vehicular networks. *IEEE Transactions on Intelligent Transportation Systems*, 1524–9050.
9. Network Simulator basic. http://www.isi.edu/nsnam/ns/.
10. SUMO basic. http://en.wikipedia.org/wiki/Sumo.
11. Lan, K. C. MOVE: A practical simulator for mobility model in VANET. National Cheng Kung University, Tainan, Taiwan, R.O.C.

Data Acquisition with FPGA Using Xilinx and LabVIEW

Sanket Mehta, Nirmal Parmar, Jayanee Soni and Arpita Patel

Abstract This paper describes the concept of serial communication using LabVIEW software. Serial communication itself means to establish a communication between a host computer and a programmable device. Serial communication is democratic as most of the computers have one or more serial ports which dilute the usage of extra hardware other than a cable to connect the instrument and the computer. At a time one data bit is send during serial communication. Entire system constitutes of three parts: Xilinx software, programmable device, and LabVIEW software. Xilinx software is mainly used for creating UART communication code in VHDL. Programmable device is used to dump UART code. Logics and programmable blocks have been applied in LabVIEW for communication between host computer and programmable device. A programmable device utilized in the proposed system is FPGA of Spartan family, specifically used Spartan 3.

Keywords UART—Universal asynchronous receiver/transmitter · LabVIEW software · Xilinx software · FPGA—Field programmable gate array · Serial communication · SPARTAN 3 · Papilio loader · Hyperterminal

1 Introduction

Data acquisition is the process to receive data from hardware and display it in appropriate software. The application of data acquisition is controlled by many programs of software which are build up using diverse general-purpose program-

S. Mehta (✉) · N. Parmar · J. Soni
Electronics and Communication Department (ECE),
Charotar University of Science and Technology, Vadodara, India
e-mail: sanketrmehta@yahoo.com

A. Patel
Electronics and Communication Department (ECE), CSPIT,
Vadodara, Gujarat, India
e-mail: arpitapatel.ec@charusat.ac.in

© Springer Science+Business Media Singapore 2016
S.C. Satapathy et al. (eds.), *Proceedings of International Conference on ICT for Sustainable Development*, Advances in Intelligent Systems and Computing 408, DOI 10.1007/978-981-10-0129-1_42

397

ming language, for example, LabVIEW, C, Java, Pascal, etc. A data acquisition system is constituted of three parts: I/O subsystem, a host computer, and the controlling software. In this system, FPGA is used as I/O subsystem, and Xilinx and LabVIEW are used as controlling softwares.

1.1   FPGA

FPGA is the essential part of digital control and processing electronics. Objective of FPGA is to transmit and receive the data bits to/from LabVIEW. Transmitted data will be received by the LabVIEW. UART code must be dumped in FPGA for data acquisition. UART coding would be done in Xilinx. In this system, Spartan 3 is being used as a FPGA.

1.2   LabVIEW

LabVIEW is a short for laboratory virtual instrument engineering workbench. It is a platform for acquiring a system and design for a visual programming language from National instruments.

LabVIEW is normally used in data acquisition, instrument control, and industrial automation at various platforms like Microsoft Windows, Linux, Unix, and Mac OS X. LabVIEW 2014 is the latest version of LabVIEW which is released in August 2014.

1.3   Xilinx

Purpose to use Xilinx software is to prepare UART code. UART code can be made using many softwares like Libero, etc. They are easy to simulate the code and generate programming files in Xilinx. It is necessary to use papilio loader software to load or dump a VHDL code into Spartan 3.

Entire paper is organized as follows:

Block diagram of whole system is presented in Sect. 2. Section 2.1 represents an overview of required component and software. FPGA UART coding is mentioned in Sect. 2.1.4. Acquire and Analyze result is discussed in Sect. 2.1.5. Hardware Testing is represented in Sect. 2.1.6. It is followed by conclusion, future work, acknowledgment, and references, respectively.

2   Block Diagram of System

This section briefly describes a block diagram of entire acquisition system:

As per block diagram it can be easily seen that entire system is reliable on three parts:

1. Xilinx,
2. FPGA–Spartan 3,
3. LabVIEW.

Figure 1 shows how data acquisition system will work. Brief description about entire system has been discussed below. In Xilinx, UART transmit–receive codes have been created and then dumped into FPGA using papilio loader. The FPGA used for entire data acquisition system belongs to the Spartan family name Spartan 3. Papilio loader software must be used to dump any VHDL code. Papilio loader is only used for Spartan family or papilio board. After dumping a code in FPGA, data will be written in LabVIEW using their module like Visa source, Visa write, Vise read, etc. Whatever data written in LabVIEW that data read in LabVIEW itself so that we can check whether written data is correct or not.

2.1   Overview of Required Component and Software

As per mentioned above our entire system comprises three parts:

- Xilinx,
- FPGA–SPARTAN 3,
- LabVIEW.

2.1.1   Xilinx

Xilinx is an American company. It is a supplier of programmable logic device. Xilinx is founded in Silicon valley in 1984. The company headquarter is in California. Xilinx is mostly used to generate a code for particular system. VHDL or

Fig. 1 Block diagram of entire system

Verilog code can be generated using Xilinx. Any system can be made using Xilinx or VHDL Code.

2.1.2   FPGA—SPARTAN 3

FPGAs consist of an array of programmable logic blocks. It allows the blocks to be connected by wire. In our system, Spartan 3 FPGA is being used for data acquisition. There are many important features of Spartan 3 FPGA, which is the reason why many programmers choose SPARTAN 3 FPGA over other FPGA and industries prefer to use SPARTAN 3. SPARTAN 3 is most industry-leading design tool in recent times [1].

Features of SPARTAN 3 FPGA [1]:

- 100 K–1.6 M system gates,
- 66–76 I/O Pins,
- Up to 8 digital clock manager,
- Easy to interface,
- LVDS supported,
- Low cost.

2.1.3   LabVIEW

LabVIEW is a short name of "Laboratory Virtual Instrument Engineering Workbench." LabVIEW is mainly used for two types of programming.

1. *Dataflow Programming*: LabVIEW programming language is also consult as a G, which is a dataflow programming language. In this type of programming, programmer can connect dissimilar function nodes and can make establish the structure of a graphical block diagram [2].
2. *Graphical Programming*: Whatever user have made that can be interfaced into development cycle in LabVIEW. LabVIEW subroutines/programs are also known as virtual instruments. Every virtual instrument has three main components: Block diagram, front panel, and connector panel [2] (Fig. 2).

2.1.4   FPGA Coding

Xilinx is being used for FPGA coding. As per mentioned above, UART Tx–Rx code is created in Xilinx. In this section, simulation result of UART receive code has been shown in Fig. 3 and also some experiment results are shown in Figs. 4 and 5.

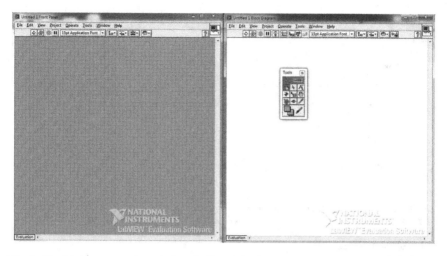

Fig. 2 Graphical programming window (*Left*) and dataflow programming window (*Right*)

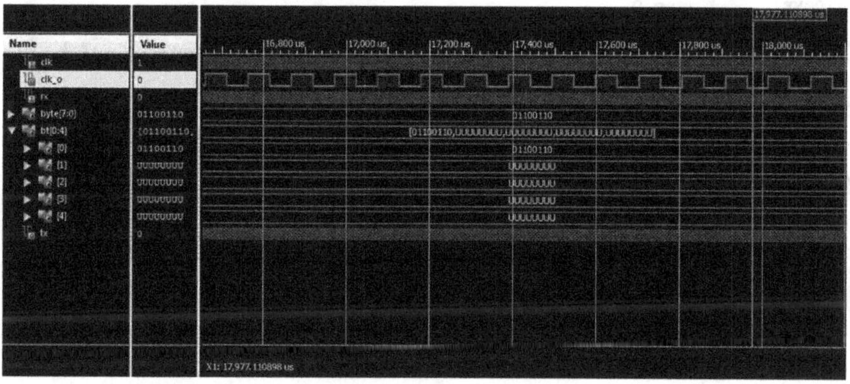

Fig. 3 Simulation results of UART code

2.1.5 Acquire and Analyze Data

Data acquisition using LabVIEW has been described in Fig. 8. Here, visa resource name described the com port of hosting computer in which FPGA is connected. From FPGA, 8 bit data has been transmitted, so in data bits block 8 has been written. There is no need of parity and flow control, so here it will be none. Now, write buffer is used to write data string and read string is used to read whatever data have been written in write buffer. Byte read is counting number of bytes written in write buffer string.

Figures 6 and 7 shows front panel window and data programming window before system started. Figure 8 shows LabVIEW front panel after system started

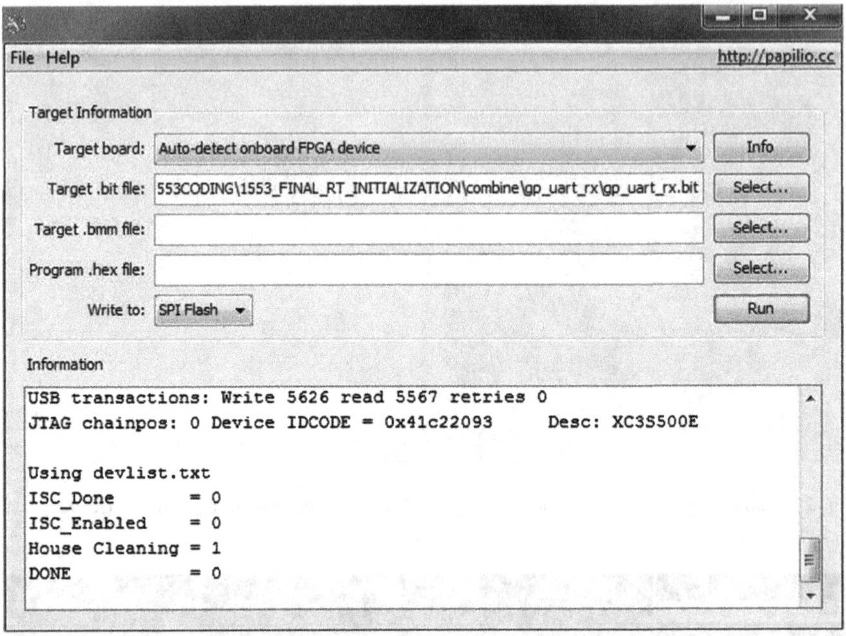

Fig. 4 Dumped UART code using papilio Loader

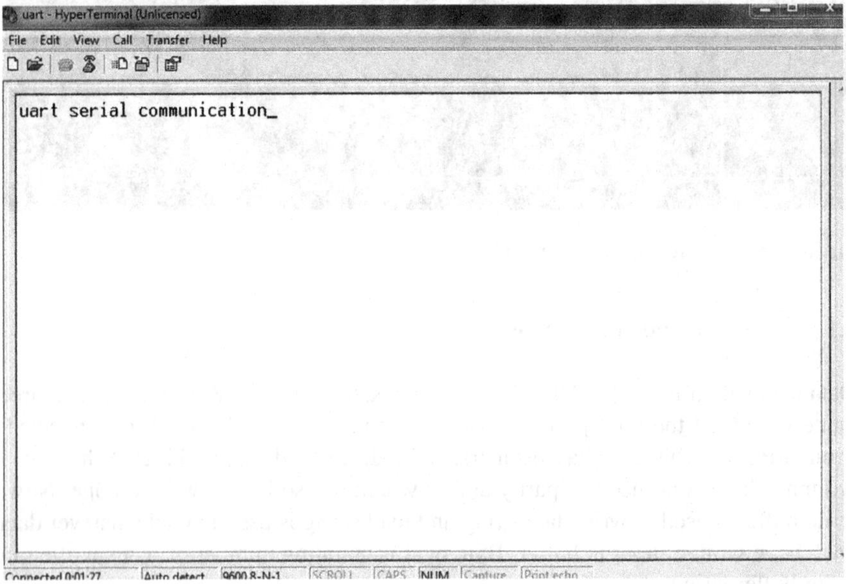

Fig. 5 Hyperterminal output of UART code dumped in FPGA

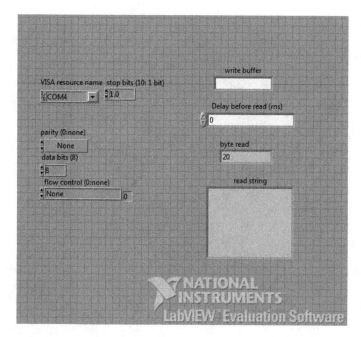

Fig. 6 Front panel before run

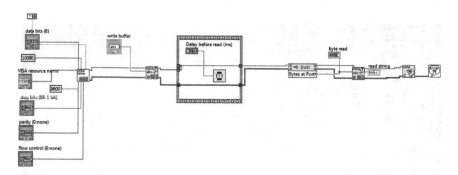

Fig. 7 Data programming window

and it clearly shows that whatever data have written in write buffer it can be read in LabVIEW itself with the use of read string.

2.1.6 Hardware Testing

See Fig. 9.

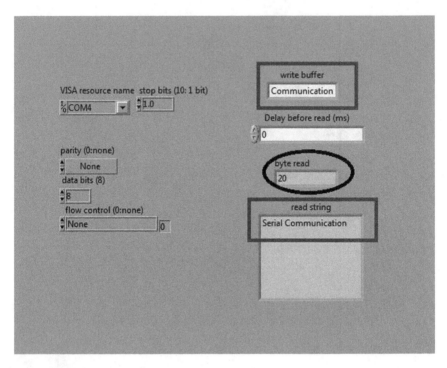

Fig. 8 Data write–read in LabVIEW

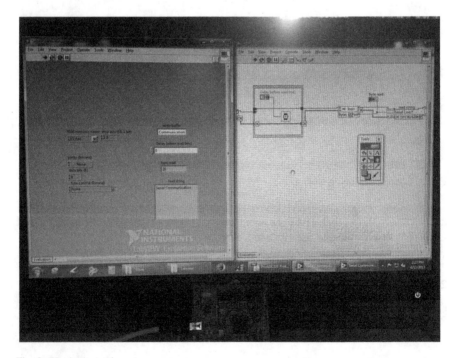

Fig. 9 Hardware testing

3 Conclusion

This proposed system describes the concept of serial communication using LabVIEW software. Entire proposed system consists of three parts: Xilinx software, programmable device, and LabVIEW software. Xilinx software is mainly used for creating UART communication code in VHDL. Programmable device is used to dump UART code. Logics and programmable blocks have been applied in LabVIEW for communication between host computer and programmable device. Experimental results and hardware testing are also explained in this paper.

4 Future Work

If we wish to transfer data from one FPGA to other FPGA, then it could be done using LabVIEW. In that case, whatever data we received in LabVIEW same data can be transferred to the next FPGA via LabVIEW and proposed system can be large and as per our requirement.

Acknowledgments Sometimes words fall short to give treasure, the same pass off with me during this paper. The huge assist and support received from faculty and friends besieged me throughout the paper. I am Thankful to CSTC (Charusat Space Technology Center), in which I have done my work. I am highly obliged to Prof. Arpita Patel for providing me with the necessary information and valuable suggestion and comments on bringing out this paper in the best possible way. I am also thankful to Ms. Khyati Patel, Ms. Anjali Patel, Ms. Jahnavi Kachhia and faculties of CSTC who helped me to successfully completion of this paper.

References

1. Xilinx. (2014). Xilinx Spartan™-3E FPGAs.
2. Halvorsen, H. (2014). *Introduction to Labview*. Telemark University College.
3. National Instruments. (2009). LabVIEW: Getting started with LabVIEW. USA: National Instruments Corporate Headquarters.
4. Anthony Vento, J. Application of labview in higher education laboratories. National Instruments Corp, Austin, Texas.
5. Swain, N. K. et al. Remote data acquisition, control and analysis using labVIEW front panel & real time engine. School of Engg. Technology and Science: South Carolina state University, IEEE.
6. Higa, M. L. et al. (2002). An introduction to labview exercise for an electronic class. IEEE.
7. Kaur, P., & Sharma, R. K. (2014). Labview based design of heart disease detection system. *IEEE Conference (ICRAIE)*, Jaipur, May 09–11, 2014.

SRR and R-CSRR Loaded Reconfigurable Antenna with Multiband Notch Characteristics

Rachana Yadav, Sandeep Yadav and Sanjeev Yadav

Abstract In this paper, a monopole UWB antenna is presented which has reconfigurable multiband notch characteristics. Antenna design is based on split ring resonators, which are introduced on patch side and a complementary rhombic split ring resonator is implemented as a slot on patch. SRR and CSRR are used for notching the different frequency bands. Two pin diodes are mounted on the SRR. Different states of the switches serve different notches in the frequency band, thus leading to band notch reconfigurability. VSWR and gain are analyzed for different modes of operations. CST microwave studio is used to design and simulate the antenna.

Keywords Split ring resonator · CSRR · Satellite communication · PIN diode · Band notches · Reconfigurable antenna

1 Introduction

Multifunctionality is the key feature of a reconfigurable antenna, which plays a advantageous role in today's communication system. Reconfigurability can be of different types (frequency, polarization, radiation pattern). It can be achieved using different techniques like by changing the surface current distribution, physical alteration, or using the different types of switches (pin diodes, varactor diodes, MEMS, etc.).

R. Yadav (✉) · S. Yadav · S. Yadav
Department of Electronics & Communication Engineering,
Govt. Women Engineering College, Ajmer, Rajasthan, India
e-mail: rachanayadav2112@gmail.com

S. Yadav
e-mail: Sandeep.y9@gmail.com

S. Yadav
e-mail: sanjeev.mnit@gmail.com

© Springer Science+Business Media Singapore 2016
S.C. Satapathy et al. (eds.), *Proceedings of International Conference on ICT for Sustainable Development*, Advances in Intelligent Systems and Computing 408, DOI 10.1007/978-981-10-0129-1_43

407

In wireless communication system some specific frequency bands are used only for some specific purposes. To avoid interference with other application an antenna with band-notched characteristics is required. To achieve the band notch function many efforts have been made. SRR is also one of the kinds, which is used to work as a band-stop filter [1–3]. In [4], slot-type split ring resonator is used to notch the WLAN frequency band, whereas in [5], a CSRR is etched inside the circular patch to achieve notched frequency bands. Dual notch bands have been achieved by adjusting the size of the CSRR inside the patch. In [6], a compact coplanar waveguide (CPW)-fed ultra wideband antenna with band notch characteristics is presented. A split ring resonator (SRR) in the circular patch is introduced to achieve a band-notched characteristics in the WLAN (5.15–5.825 GHz) band. SRR and CSRR are also used in reconfigurable antennas [7]. In [8], reconfigurable band notches are achieved using a band-stop filter, which is based on SRR. This antenna has two band notches which are independently controllable and it can be used for cognitive radio system. In [9], two antenna designs are presented. One is based on nested CSRR, while the other one has two identical split ring resonators placed near strip line. In first design switches are used across CSRR slots. Different CSRRs activate by controlling the switch states and then corresponding notch band achieved, whereas in design two, by changing the switching states, UWB and band notch characteristics are achieved. Generally, SRRs are implemented near patch edges or near feedline, while CSRRs are used as slots in ground or patch.

In this paper, an UWB antenna is introduced with reconfigurable band notch characteristics. A complementary rhombic split ring resonator is introduced in patch and two identical SRRs are placed very close to feedline. Two switches are mounted on both SRRs. By changing the state of the switch, different notch bands are achieved.

2 Antenna Configuration and Design

Front view of proposed antenna is shown in Fig. 1. This is a simple patch antenna for UWB application. It has a radiating patch with patch length L and width W and FR-4 substrate of dielectric constant 4.4. Substrate length and height are L_s and W_s, respectively. Table 1 has shown other parameters.

Figure 1b shows the back view of the proposed antenna without switches. Here ground has rounded shape to enhance the bandwidth of the antenna and make it useful for UWB application. When we used rectangular ground of same length, the obtained frequency band was dual instead of UWB, and a tapered microstrip feedline is used for proper matching. Figure 2 shows that large band of frequencies is achieved when width of microstrip line decreases from 1.5 to 1.1 mm.

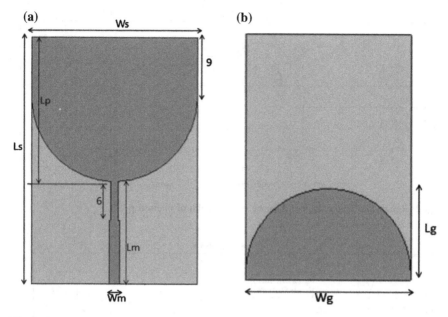

Fig. 1 Antenna structure. **a** Front view. **b** Back view

Table 1 Antenna dimensions of proposed antenna

Parameter	Values (mm)
Width of patch (W)	24
Length of patch (L)	21
Width of substrate (W_s)	24
Length of substrate (L_s)	35
Dielectric constant (ε_r)	4.3
Height of the substrate (h)	1
Height of the patch and ground	0.05
Microstrip feed length (L_m)	14
Microstrip feed width (W_m)	1.5
Length of ground (L_g)	12
Width of ground (W_g)	1.5

Simulated return loss is shown in next figure, a large frequency band is achieved. Here the main motive to gain that large bandwidth was to enhance the chances of getting as more as possible notch bands in that frequency band to use the antenna reconfiguration characteristics more efficiently.

From Fig. 3 it can be seen that 2.5 to 16.6 GHz frequency band is achieved.

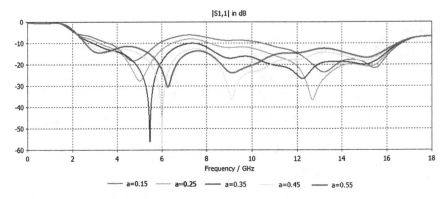

Fig. 2 Return loss with change in tapered section width of microstrip line

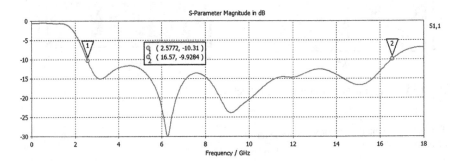

Fig. 3 Return loss of designed antenna

3 Band Notch Antenna Configuration and Design

Now to get band notches, SRR and CSRR are introduced in the basic design. From Fig. 4 it can be seen that single-ring rhombic CSRR is slotted out from patch. And to provide more field coupling, CSRR is placed near to the feeding point.

Here two SRRs are etched near to the microstrip feedline. These two SRRs work as a band-stop filter. Two switches are also loaded in SRRs. Here PIN diode is used as a switch. In ON state, diode is modeled by a 1.5 Ω register and in OFF mode modeled by a .017 pF capacitor.

Figure 5 shows the dimensions of the CSRR. This is a single-ring CSRR and provides a band notch around 10 GHz frequency. Figure 6 shows the dimensions of the SRRs. It can be seen that spacing between SRR and feedline is only 0.25 mm. Each SRR is mounted with switches which are denoted as s1 and s2.

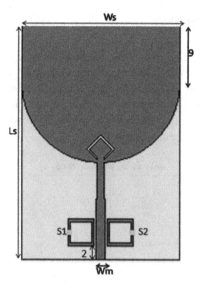

Fig. 4 Proposed band notch antenna

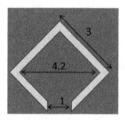

Fig. 5 Dimensions of rhombic CSRR

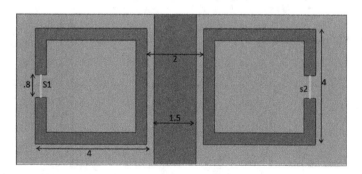

Fig. 6 Dimensions of both SRRs

3.1 Simulated Results

When both switches are ON, only one notched frequency would occurr which is corresponding to the rhombic CSRR structure, while in OFF state of both switches, SRR will start to resonate and provide three-notched frequency bands. Figures 7, 8 and 9 shows the VSWR plot for mode I, II and III and Figs. 10, 11 and 12 shows the gain plot for three modes respectively. Table 2 shows summarized result for proposed antenna for different notch bands.

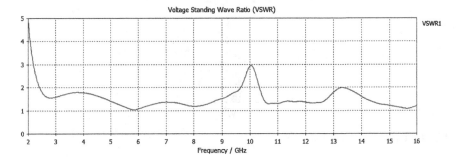

Fig. 7 VSWR plot for mode I

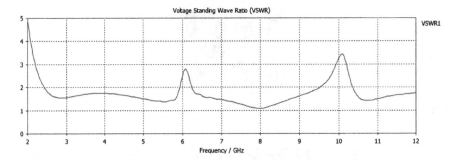

Fig. 8 VSWR plot for mode II

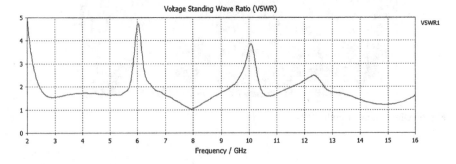

Fig. 9 VSWR plot for mode III

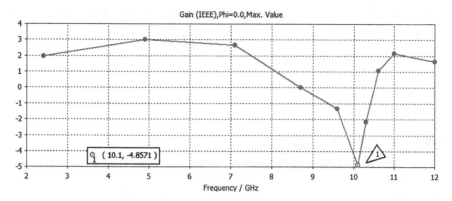

Fig. 10 Gain for mode I

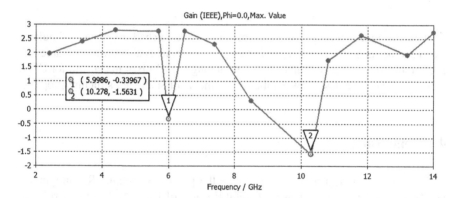

Fig. 11 Gain for mode II

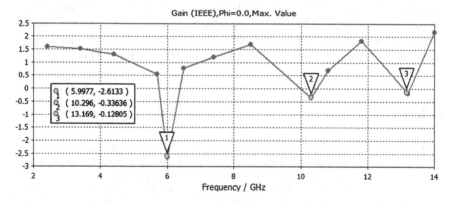

Fig. 12 Gain for mode III

Table 2 Summarized result for proposed antenna for OFF and ON states of switches

Mode	Switch state		No. of notched band
	S1	S2	(bands in GHz)
I	ON	ON	1 (9.6–10.3)
II	ON	OFF	2 (5.9–6.2, 9.4–10.3)
III	OFF	OFF	3 (5.7–6.5, 9.2–10.3, 11.4–12.7)

Table 3 Gain evaluation for different modes

Mode	Mean gain at passband (dB)	Gain at notched frequencies
I	2.34	−4.8 dB at 10.1 GHz
II	2.44	−0.33 dB at 5.99 GHz, −1.5 dB at 10.2 GHz
III	1.23	−2.6 dB at 5.99 GHz, −0.33 dB at 10.2 GHz, −0.12 dB at 13.1 GHz

It can be understood by gain measurement that at notched frequencies simulated gain becomes negative and mean gain value for passband is also shown in Table 3. It is desirable to have min and max gain within range of ±1 db of mean gain for UWB range. This antenna design meets the requirements of stabilized structure properly.

4 Conclusion

CST microwave studio is used to simulate the proposed antenna. Reconfigurable notched bands are achieved using ON/OFF states of the PIN diodes. Here three modes of operation are presented. Notch obtained around 6, 10, and 12.3 GHz center frequency. Band around 6 GHz is used for C-band communication satellites as a transmit frequency, band around 10 GHz is used for radio navigation, while frequency band around 12.3 GHz is used for radio astronomy and fixed satellite communication. Gain for different modes has been analyzed. This antenna has ultra wide band response and can be used in UWB cognitive radio system.

References

1. Bhadra Choudhuri, S. R., Poddar, D. R., Ghatak, R., & Mishra, R. K. (2009). Modulating properties of a microstrip patch antenna using complementary split ring resonator. *The 2009 International Workshop on Antenna Technology (iWAT 2009)* (pp. 1–4).
2. Dong, Y., Toyao, H., & Itoh, T. (2012). Design and characterization of miniaturized patch antennas loaded with complementary split-ring resonators. *IEEE Transactions on Antennas and Propagation, 60*(2), 772–785.

3. Ali, A., Khan, M. A., & Hu, Z. (2007). Microstrip patch antenna with harmonic suppression using complementary split ring resonators. *The 2007 International Workshop on Antenna Technology (iWAT 2007)*.
4. Kim, J., Cho, C. S., & Lee, J. W. (2006). 5.2 GHz notched ultra-wideband antenna using slot-type SRR. *Electronics Letters, 42*(6), 315–316.
5. Liu, J., Gong, S., Xu, Y., Zhang, X., Feng, C., & Qi, N. (2008). Compact printed ultra-wideband monopole antenna with dual band-notched characteristics. *Electronics Letters, 44*(12), 710–711.
6. Sharma, M. M., Kumar, A., Yadav, S., Ranga, Y., & Bhatnagar, D. (2011). A compact ultra-wideband CPW-fed printed antenna with SRR for rejecting WLAN band. *Antenna Week (IAW), 2011 Indian* (pp. 1, 3), 18–22 December 2011.
7. Dakhli, S., Mahdjoubi, K., Floc'h, J.M., Rmili, H., & Zangar, H. (2012). Efficient, metamaterial-inspired loop-monopole antenna with shaped radiation pattern. *Loughborough Antenna and Propagation Conference*, LAPC, Loughborough, England, Proceeding (pp. 1–4), November 2012.
8. Lalj, H., Griguer, H., & Drissi, M. (2014). Reconfigurable multi band notches antenna for cognitive radio applications. *2014 XXXIth URSI, General Assembly and Scientific Symposium (URSI GASS)* (pp. 1, 4), 16–23 August 2014.
9. Al-Husseini, M., Costantine, J., Christodoulou, C. G., Barbin, S. E., El-Hajj, A., & Kabalan, K. Y. (2010). A reconfigurable frequency-notched UWB antenna with split-ring resonators. *Microwave Conference Proceedings (APMC), 2010 Asia-Pacific* (pp. 618, 621), 7–10 December 2010.

Wireless Resonant Power Transmission

Piyush Prasad and Nirmal Thakur

Abstract There has been a tremendous research in wireless technology over the past decade. Cellular phones are taking over the trivial telephones. Bluetooth enabled earphones and other electronic gadgets, and fueled vehicles to electric vehicles. The basic idea is to have power devices without any physical carrier medium. Thus eliminating the use of cords or wire, a simple but powerful concept of electromagnetism has been used to transfer electricity wirelessly. Proposed work includes transmission of power wirelessly at resonant frequency. The range of transmission is highly dependent on resonance. At resonance maximum efficiency is achieved, and all devices resonating at the same frequency can be powered. The operation of wireless power transmission for a resonant frequency of 30 kHz has been carried out successfully. One of the major advantages of such transmission is that one transmitter can power several receiving gadgets. It can also be used to power every single household, industrial, hospital electronic gadgets eliminating the need of physical connection and sockets and wires.

Keywords Resonance · Wireless power · Electromagnetism · Wireless transmission · AWG · Efficiency

1 Introduction

Before the electrical-wire grid was deployed, Nikola Tesla had put in large efforts to transport power over large distances, without any need of a carrier. The resonant transformer designed by Nikola Tesla was termed as the 'Tesla Coil.' This coil was able to generate high voltage, low current and high frequency alternating current.

P. Prasad (✉) · N. Thakur
MPSTME, NMIMS University, Mumbai, India
e-mail: er.piyushprasad@gmail.com

N. Thakur
e-mail: ndthakur123@gmail.com

© Springer Science+Business Media Singapore 2016
S.C. Satapathy et al. (eds.), *Proceedings of International Conference
on ICT for Sustainable Development*, Advances in Intelligent Systems
and Computing 408, DOI 10.1007/978-981-10-0129-1_44

Because of high voltage arcs generated by the Tesla coil, it could not pass through the Government regulations [1].

The main concept of the proposed work is 'one to many', that is, at the receiver end many devices can be powered. It is a non-radiative energy transfer process as it involves stationary fields around the coils and the energy is not wasted.

This work focuses on designing the wireless power transmission prototype based on the basic concept of resonance and electromagnetism.

Resonance is the tendency of a system to oscillate with greater amplitude at a particular frequency than at other frequencies. The two resonant objects carry evanescent waves and when brought close, this evanescent wave coupling allows energy to transfer from one object to the other with high energy transfer efficiency. The current is induced from the transmitting coil onto the receiving coil as both are made to resonate at the same frequency [2].

The principle of electromagnetism is to generate a magnetic field when current is passed through the coil. If another coil is brought in the vicinity of the transmitting coil, current is induced in the receiving coil. Electric to magnetic and back to electric conversion can help transfer electricity wirelessly. Since magnetic fields are used for the transfer, it is 1/100th times lesser harmful as compared to radio waves [3]. The intensity of the field reduces with respect to distance.

The transmitter section designed in the proposed work is able to generate a field of designed resonant frequency, voltage rating with the help of Atmega-16 microcontroller, MOSFET power amplifier and LC tank circuit. Same configuration LC tank circuit is used at the receiver end for resonant matching, whose output is provided to the rectifier and filter combination.

The major advantage of using wireless power transfer is the current need of the world to eliminate wires. Using one transmitter section, many receiver gadgets are powered wirelessly using air as the medium of transfer. This concept can be very well applied in wireless sensor networks [4]. Even though power is transferred wirelessly through air, it does not cause any damage to living beings and there is least interference with the environment [3]. The disadvantage of using wireless transmission is low transmission efficiency and limited range of transmission. The system fails when the separation distance between transmitter and receiver is increased beyond the designed distance [5, 6].

Section 2 discusses related work performed on wireless power transfer. In Sect. 3, design, implementation and working of proposed resonant model is discussed. Results and related discussion are presented in Sect. 4. Finally, Sect. 5 concludes the paper.

2 Literature Survey

Wireless power transmission has been proven to be one of the fastest advancement in technology. Researchers are still working on different methods for the same. Different principles are used varying with respect to the mode of transmission.

One of the concepts used for wireless transfer is mutual induction. Mutual induction between two coils can be used for the transfer of electrical power without any physical contact in between. Low efficiency, the heating of the charging plates and high manufacturing cost are major drawbacks of this method [7].

Another concept used for transferring power wirelessly is the use of electromagnetic waves. The electromagnetic waves tunnel, and not propagate, through the air to be absorbed or wasted. The advantages are as follows: no disruption to any electronic devices or cause physical injury like microwave or radio transmission. Research has proven efficiency up to 5 m of range between transmitter and receiver. Disadvantage of this method is high cost of implementation [8].

The concept of electrodynamic induction is used for transfer of power wirelessly depending on the size of the transmitter. The principle of electrodynamic induction is that when resonant coupling is used and the transmitter and receiver inductors are tuned to a mutual frequency, pulse power transfer takes place over multiple cycles. In this way, significant power may be transmitted over a distance of up to a few times the size of the transmitter. The concept would work perfectly in theory, but converting the radio frequencies into electrical power and electrical power to radio frequencies are two main problems that are withholding this idea to become reality [9].

The use of microwave antenna at the transmitter and receiver end is also used successfully for wireless power transfer. The transmitter consists of a transmitting antenna. A conversion of dc to microwave takes place at the transmitter end. Due to high frequency of microwaves it could be used for large distance applications of wireless power transmission. At the load end, the microwaves are received by the receiving antenna and then the received microwaves are converted back into dc power. The load end is referred to as the rectenna (receiving antenna). Rectenna conversion efficiencies exceeding 95 % have been realized. The major disadvantage of using this method is integration of the antenna with the transmitting section and the power loss caused at low scale production [10].

Electrostatic induction also known as "capacitive coupling" is an electric field gradient or differential capacitance between two elevated electrodes over a conducting ground plane. It is used for wireless energy transmission involving high frequency alternating current potential differences transmitted between two plates or nodes. The disadvantage is the use of plates and limited transmission distance [11].

Electromagnetic waves can also be used to transfer power without wires. Laser beam is used to carry electric energy. This beam is focused on receiving target, such as a solar cell mounted on a small aircraft. Power can be beamed to a single target. This concept is known as "power beaming" [12]. Size reduction has been by far the most important benefit of laser beaming over microwaves. A major issue in space solar power systems employing microwave power transmission is their potential interference [13] with satellite communication systems, which use frequencies in the same multi-gigahertz range that is best suited to microwave power transmission. The use of laser beams for power transmission raises many public policy issues. These include ensuring safety, frequency band allocation and global warming [14].

3 Resonate Model

The major advantage of transmission at resonance is high efficiency. This section focuses on design of resonant model. In addition to the resonant frequency model, receivers have been designed for different frequencies to compare the efficiency with the resonant one.

3.1 Experimental Setup

Input to the working model is AC mains. In case of high power applications, a power factor correction stage may also be included. DC voltage directly from a battery source can also be used to power the digital oscillator. But for the MOSFET amplifier, it is preferable to use supply from AC mains. The digital oscillator which consists of microcontroller needs 5 V for its operation and hence a power supply section consisting of a step-down transformer, rectifier, filter and voltage regulator is designed. A separate power supply section has been designed to generate 18 V needed for the MOSFET amplifier. A step-down transformer is used whose output is provided to the bridge rectifier section. And since MOSFET amplifier IC is used for high power ratings, it is embedded with a heat sink. Output of the amplifier is provided to the LC tank circuit. The LC tank circuit is responsible to create the magnetic field of the desired frequency. Receivers with different inductances are designed keeping the length of the coil constant. The capacitance for all the receivers is chosen to be 0.5 μF.

Figure 1 shows the transmission section block diagram which is designed to resonate at 30 kHz. A frequency up and down switch is used to change the frequency.

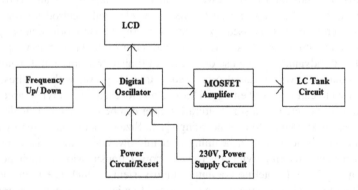

Fig. 1 Block diagram of transmitter

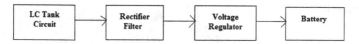

Fig. 2 Block diagram of receiver

Figure 2 shows the receiver section components which contains LC tank circuit resonating at same frequency as transmitter. The receiver also consists of rectifier, filter and voltage regulator which can be fed to a battery or load.

3.2 Working

Atmega 16 is programmed using embedded C language which is used to generate the desired resonant frequency. Resonant frequency is used for power transmission. However, the output of the microcontroller is 5 V and a resistance of 1 kΩ yields a current of 5 mA. For transmission, a much higher voltage and current rating is required and thus amplification is needed, which is provided by the MOSFET amplifier.

MOSFET amplifier increases the voltage and current levels to be given to the LC tank circuit. Wireless power transmission is possible by near-field resonant coupling using high alternating voltages and strong electric fields. Tesla coils [1] utilize a high impedance resonant circuit to produce extreme voltages required for capacitive coupling, and this high voltage creates a problem for small-scale applications like indoor battery charging (corona and arcing caused by strong electric fields can even cause fires).

Magnetic coupling is attractive because it allows fairly good amounts of power to be transmitted without need for high voltages. The basic idea is to have two high-Q resonant circuits, which are now coupled magnetically and are preferred to have as low characteristic impedance as possible.

The transmitter contains only one inductor of 0.051 mH and length of 8 m. The number of turns is 8 and number 15 American wire gauge (AWG) is used. A number 15 AWG has diameter of 0.0571 in. and material used is copper. Figure 3 shows prototype transmitter section.

Fig. 3 Prototype transmitter

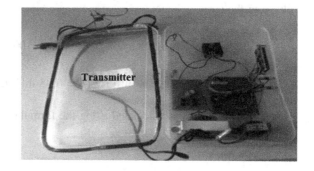

Fig. 4 Prototype receiver

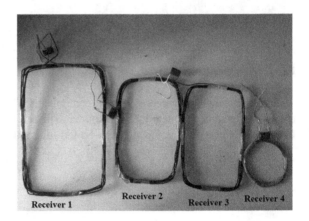

Receiver 1   Receiver 2   Receiver 3   Receiver 4

Table 1 AWG versus diameter

AWG number	Diameter (in inches)
15	0.0571
16	0.0508
17	0.0453
21	0.0285

At the receiver end, the length of the coil is kept constant at 5 m. Receivers of different AWGs are designed with AWG numbers 15, 16, 17 and 21. Increase in gauge numbers denotes decreasing wire diameter. Thus cross-sectional area of each gauge is different and decreasing, which reflects the current carrying capacity. Figure 4 shows the prototype receivers.

Table 1 presents the relation between the gauge number and diameter of wires used in the design of prototype receivers.

4 Results

The transmission efficiency for each receiver is calculated keeping the separation distance between transmitter and receiver same. Efficiency is the ratio of the output voltage to the input voltage:

$$\text{Efficiency}\,(\%) = \frac{V_{\text{out}}}{V_{\text{in}}} \times 100. \tag{1}$$

The transmitter resonates at 30 kHz ($L = 0.05$ mH, $C = 0.5$ μF) and produces voltage of 3 V (V_{in}).

Resonant frequency is calculated using the formula

Table 2 Receiver frequency versus efficiency

Gauge no.	Inductance (mH)	Receiver frequency (kHz)	V_{out} (V)	Efficiency (%)
15	0.051	30	2.14	71.3
16	0.03	41.093	2.03	67.7
17	0.029	41.796	2.01	67
21	0.022	47.987	1.88	62.7

Fig. 5 Efficiency versus frequency

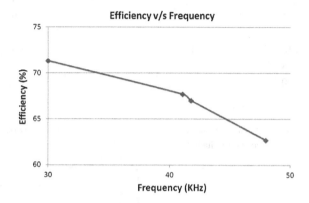

$$f_r = \frac{1}{2\pi\sqrt{LC}} \tag{2}$$

The receivers differ only in terms of inductance value owing to different gauges of wire. The capacitance value of 0.5 µF is used for all receivers and they are kept at a distance of 8 cm from transmitter.

Table 2 shows the variation of efficiency with respect to the receiver frequency for different inductances.

A relationship of efficiency with the resonant frequencies for different receivers is plotted in Fig. 5.

From Fig. 5, it is clear that maximum transmission efficiency is achieved at resonant frequency of 30 kHz for a separation distance of 8 cm. As the receiver frequency shifts from the resonant frequency, the transmission efficiency decreases for the same separation distance.

Table 3 presents receiver voltages recorded at different separation distances for all the designed receivers. This relationship is plotted in Fig. 6. It is clear that for different separation distances also the resonant model (Receiver 1) provides maximum efficiency as compared to other receivers.

From Fig. 6 it is clear that the received voltage decreases with increasing separation distance for all the designed receivers. Receiver 1 (resonant model) still offers more than 60 % efficiency up to a separation distance of 20 cm.

Table 3 Separation distance versus voltage

Separation distance (cm)	V_{out} of receiver 1 (V)	V_{out} of receiver 2 (V)	V_{out} of receiver 3 (V)	V_{out} of receiver 4 (V)
2	2.28	2.09	2.06	1.98
4	2.24	2.07	2.05	1.95
6	2.18	2.05	2.02	1.91
8	2.14	2.03	2.01	1.88
10	2.10	2.00	1.99	1.82
12	2.07	1.97	1.96	1.80
14	2.00	1.95	1.93	1.65
16	1.96	1.93	1.90	1.54
18	1.91	1.90	1.87	1.41
20	1.89	1.88	1.81	1.32

Fig. 6 Separation distance versus received voltage

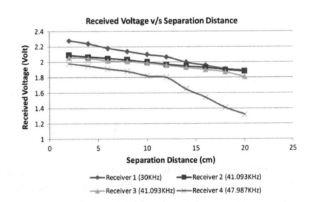

5 Conclusion

The paper discusses about design and implementation of resonant power transmission prototype. Efficient power transmission at resonance has been justified in this paper. For the same separation distance, maximum efficiency is achieved for the resonant model as compared to other receivers. Keeping all the parameters constant in the receiver design, if only AWG is increased, the efficiency decreases. The efficiency also decreases with increasing separation distance for all the receivers, but the resonant model provides maximum efficiency at different separation distances as compared to other receivers. The resonant power transmission has the advantage of one-to-many transmission and a single transmitter has been used to power four receiving gadgets.

References

1. Sibakoti, M. J., & Hambleton, J. (2011). Wireless power transmission using magnetic resonance. *IEEE International Conference on Magnetics* (pp. 2685–2698).
2. Pande, V. V., Doifode, P. D., Kamtekar, D. S., & Shingade, P. P. (2014). Wireless power transmission using resonance inductive coupling. *International Journal of Engineering Research and Applications, 4*(4).
3. Zamanian, A., & Hardiman, C. (2005). Electromagnetic radiation and human health: A review of sources and effects. *High Frequency Electronics* (pp. 16-26).
4. Khssibi, S., Idoudi, H., Van Den Bossche, A., Val, T., & Azzouz Saidane, L. (2013). Presentation and analysis of a new technology for low-power wireless sensor network. *International Journal of Digital Information and Wireless Communications (IJDIWC)* (pp. 75–86).
5. Griffin, B., & Detweiler, C. (2012). Resonant wireless power transfer to ground sensor from a UAV. *IEEE International Conference on Robotics and Automation* (pp. 2660–2665).
6. Kalyan, K., Avaes Mohsin, S., & Suresh, A. (2013). Transmission of power through wireless systems. *International Journal of Engineering and Advanced Technology, 2*(4), 2249–8958.
7. Ho, S. L., Junhua Wang, W. N., & Fu, Mingui Sun. (2011). A comparative study between novel Witricity and traditional inductive magnetic coupling in wireless charging. *IEEE Transactions on Magnetics, 47*(5), 1522–1525.
8. Kalyan, K., Avaes Mohsin, S., & Suresh, A. (2013). Transmission of power through wireless systems. *International Journal of Engineering and Advanced Technology (IJEAT), 2*(4). ISSN:2249 – 8958.
9. Power Transmission. (2014). *International Journal of Technology Innovations and Research, 8*.
10. Awolala Teru, A. Efficient rectenna circuits for microwave wireless power transmission. A thesis submitted in fulfillment of the requirements for the degree of *Master of Science in the Faculty of Science and Agriculture of the University of Forth are.*
11. Wang, J., Ho, S. L., Fu, W. N., & Sun, M. (2011). FEM simulations and experiments for the advanced witricity charger with compound nano-TiO_2 interlayers. *IEEE Transactions On Magnetics, 47*(10), 4449–4452.
12. Dickinson, R. M. & Jerry, G. (1999). Lasers for wireless power transmission.
13. Mehdi, B., & Bolandpour, H. (2014). Analysis of high frequency interferences at wireless communication antennas. *International Journal of Digital information and Wireless Communications, SDIWC* (2014).
14. Girish, C., Rajeev, G., & Narayanan, K. (2010). Policy issues for retailed beamed power transmission. *IEEE Electronic System Design* (pp. 44–49).

Fuzzy Clustering-Based Efficient Classification Model for Large TCP Dump Dataset Using Hadoop Framework

Tarun Budhraja, Bhavya Goyal, Aravind Kilaru and Vivek Sikarwar

Abstract Anomaly exposure is a grave concern in network intrusion classification (IDS). Furthermost anomaly based IDSs employ supervised procedures, whose enactments extremely be contingent on attack-free training statistics. Nonetheless, this generous of training data is problematic to acquire in real-world network environment. Additionally, with fluctuating network setting or services, patterns of typical traffic will be reformed. This leads to high false-positive rate of supervised IDSs. Real-time detection is one of the most imperative issues in intrusion detection. When the network data is becoming enormous with high dimensionality, real-time detection with high detection accuracy and low false alarm rate is challenging for previous methods. The vertical progression in data volume and Hadoop cluster extent make it a momentous encounter to diagnose and localize problems in a production-level cluster setting competently and within a petite epoch of time. Frequently, the disseminated monitoring schemes are not accomplished of perceiving a problem thriving in earlier when a comprehensive Hadoop cluster twitches to depreciate in enactment or becomes unattainable. This research exertion proposed an unsupervised outlier detection based on fuzzy-based model also with typically K-means clustering algorithm and categorize different types of anomaly which can overwhelmed the downsides of supervised incongruity detection. The tryouts were accomplished on KDD'99 intrusion detection dataset to associate the performance and enactment of the proposed procedure with a typical K-means clustering algorithm and perceived reduced computational time and enriched classification accuracy.

T. Budhraja (✉) · B. Goyal · A. Kilaru · V. Sikarwar
School of Computing and Information Technology, Manipal University Jaipur,
Jaipur, India
e-mail: budhraja.tarun123@gmail.com

B. Goyal
e-mail: goyal.bhavya08@gmail.com

A. Kilaru
e-mail: kilaru.arvind@gmail.com

V. Sikarwar
e-mail: sikarwar.vivek@gmail.com

© Springer Science+Business Media Singapore 2016
S.C. Satapathy et al. (eds.), *Proceedings of International Conference on ICT for Sustainable Development*, Advances in Intelligent Systems and Computing 408, DOI 10.1007/978-981-10-0129-1_45

427

Keywords Fuzzy c-means clustering · K-means clustering · IDS · Anomaly detection · Apache hadoop

1 Introduction

Nowadays, internet is a preeminent source to communicate between people through websites, email, and social media, in resultant network information security (NIS) becoming a challenging problem to secure network by hackers because system firewalls are not enough to secure your network from these attacks. Preventing system resources from these unwanted hackers another options is to build a system that give reliable network security by recurrently scan network in certain time and spot unauthorized user because a number of attackers are increasing with number of huge network [1, 2]. Furthermore, IDS triggering mechanism helps to protect our network by spawning alarm once intrusion activity or unauthorized user tries to damage computer resources. Misuse-based triggered mechanisms referred as signature based are those where tarnished attempts can be tracked and anomaly based techniques are based on the assumptions, means abnormal pattern. Under these assumptions, our aim is to build a system which creates a cluster from input dataset and automatically categorized normal and abnormal patterns in dataset and finally use these clusters to classify these datasets. In this paper we are using log data once user attacks on a network by incorrect username, packet transformation from source to destination, etc. Some related work has been reported by fuzzy and K-means based on IDS. In [3], a new variant classification is approached where NSL dataset over original KDD dataset is used with fuzzy clustering neutral network framework as it detects low frequency attacks at a very high rate. Uncertainty between data and redundancy is improved and detection precision is shown above 90 % as compared to KDD '99. In [4], implemented two phase methods over KDD '99 where distance techniques proved meaningful for outlier construction and corresponding to each label, class label is assigned. Clustering for specific patterns of data is being done by fuzzy c-means algorithm, resulted in detection rate increment. In [5], proposed intrusion detection system with J48, fuzzy K-means clustering statistical model and proved out random forest datasets to be best in precision and recall. In [6], proposed K-means clustering technique with naïve Bayes classifier on KDD '99 and proved improvement in precision, accuracy, and false alarm rate. In [7], combined greedy K-means clustering with SVM classifier for network attack detection. In order to select out discriminated feature, information gain (IG), and triangle area-based KNN are applied, resulted in less error rate on training dataset. In [8] analyzed simple K-means clustering on NSL-KDD on WEKA environment and gave completed analysis related to categories of attacks like probe, U2R, R2L, and DOS. Here our research tryouts are to proposed a clustering model that classifies several types of attacks using fuzzy model and reduces computational time to detect these informations. To test the performance of proposed model, we cast off *hadoop* at single-node cluster with small dataset. The rest part of this manuscript consists of some previous fuzzy clustering-based literature survey, KDD dataset description, MapReduce workflow, proposed clustering methodology, experiment results, and conclusion with future road map.

2 KDD Cup Dataset

KDD cup is a popular dataset for evaluation of anomaly detection that consists of nearly 494,010 single-connection vector. In KDD, training dataset consists of 19.69 % normal and 80.31 % attack connection. KDD cup is also an assortment of *TCPdump* data over a network where different 22 types of notorious attack that falls in category of training data out of total 39 which comes in testing phase [9, 10].

Denial-of-service attack Denial-of-service is a type of network attack which attempts to make network resources inaccessible for the user by submerging it with inadequate network traffic and reduce overall transmission rate. In KDD data, denial-of-service attacks have higher percentage with 79.24 %, and we have some specific software which can be installed on user machine to minimize the destruction produced by attacks. Mainly, Dos attacks are "back," "pod," "land," etc.
Probing Probe referred as an object which helps to record the state of network, e.g., hacker can send a blank massage to check the presence of destination source and try to find the weakness of a system which later help to easily attack on system. In probing, attack hackers insert a program on a network at a key pass and get the info related to network commotion. Probe attacks are '*satan*' and 'nmap' [11, 12].
User to root attack U2R attack falls when an end user tries to access the normal user account by making wrong password again and again and each time a log file generated on server by making wrong input. In this attack, hacker tries to gain normal user rights, e.g., *Buffer_overflow,perl*.
Remote to Local attack Unlike trying to get user rights by making wrong password, in R2L attacker sends packet to a machine without having a authorize user account that harm system resources [13, 14].

3 Hadoop MapReduce

Nowadays, growths in social media and mobile device are breeding with a high rate of unstructured data and processing these enormous data is challenging because relational databases are incapable to store these information, and extracting info from these data cannot perform by high-level query languages, e.g. oracle and sol. *Hadoop* is a java-based open-source framework that enables to process data in parallel computer. Hadoop emanated with two main components MapReduce and HDFS where MapReduce consists two key programing structures: one is map and anther one is reduce. Map takes data which exists in distributed file system and breakdown into (key, value) pairs and assign it to reducer as a input for writing further aggregation/merging logic.

A MapReduce program consists of mainly three parts:

- Mapper step-mapper class reads data from input files which exist in HDFS as ⟨key, value⟩ and split master node data into many smaller sub-problems. Worker node processes some smaller problem subset under the control of job tracker node and stores the result in local file where reducer is able to access its store results. Mapper interface takes four generics which define the types of data input key/value and output key/value

4   Clustering Algorithm

Clustering is an unsupervised mining approach in which we place an object into a class according to their similarity that was not defined before, and object that does not belongs to a same category goes to another cluster in a same class. Data clustering helps to identify dense and sparse region; therefore, it makes easy to find an overall distribution patterns as well as correlation among dataset. By this reason clustering can also refer as learning by observations and may also be found as different names, e.g., supervised learning, unsupervised learning, and numeric classification [15]. Clustering found a lot of attention because it is used for exploration of data and discover ordinary grouping of dataset. Data clustering is a common technique which is used in machine learning, data analysis, medical image mining, and business intelligence.

- *Grouping dataset using K-means approach (Background)*

The tenacity of proclaiming clustering algorithm is to classify the dataset and assign it to its similar class. In K-means, clustering idea is to group objects into N-positive integer number cluster. Object groping is performed by taking distance from centroid to data. K-means clustering is also well suited for mutual exclusive clusters [16].

4.1   *Proposed Algorithm for K-Means Clustering*

In proposed work user-definite initial cluster number n where $n > 0$ generate same number of centroid (k), zeros matrix ($) of same input size, and cluster member *new(m)* will produce. Here, zeros matrix helps to update cluster member with minimum centroid on zeros matrix. In step 6 algorithm starts classification data covering each row and column value and it updates centroid value by subtracting actual input from random centroid value. Then we find minimum value with index position of updated centroid. In our classification we are not concerned in value of centroid what we need to take to index position of minimum centroid. In the final step, we will update our previous cluster member *new(m)* with index position that we found by minimum centroid *new(index)*

$$new[index] = new[index] + 1 \tag{1}$$

In our clustering first, we generate random centroid to classification but now we are with zeros matrix of index positions so we can iterate zeros matrix from each row and column and found *actual_centroid* by

$$actual_{centroid}(m) = \frac{updated\ index\ position\ value}{cluster\ member} \tag{2}$$

We run again to check that objects are stirring from one specific cluster to one or more cluster but this time we repeat the classification and the *new_centroid* value which we will find by subtracting input from newly generated centroid and assign next centroid value previous centroid value for some specific iteration, e.g., 1–N

$$actual\_centroid = new\_centroid$$

Proposed Algorithm

Step 1- *Read an input* $q = (N * M)$;
Step 2- Initialize number of cluster $m = n$ *where* $n > 0$
  ➢ *for* $m = 1$ *to* n {
Step 3- *generate random centroid* $= k = random(1 * 1)$; // *math.random*
Step 4- *generate zeros matrix with same as input to update cluster member*
      $\$ = Zeros[N * M]$;
Step 5- *generate cluster member* $new(n) = 0$;
Step 6- calculate row and column size of an input=r, c
  ➢ } // *ending the loop*
// clustering starts from next step//
  ➢ *for* $i = 1$ *to* row {
  ➢ *for* $j = 1$ *to* $column$ {
Step 7- $a[m] = k[m] - q[l * J]$;
Step 9- *find minimum value with an index position of udtated* $a(m)$
Step 10-*update zeros matrix with index position*
      ○ $new[index] = new[index] + 1$;
  ➢ }} *ending the inner loop*
Step 11- initialize $y = 0$ // *iterate zeros matrix to find actual centroid*
  ➢ *for* $v = 1 = 1$ *to* row
  ➢ *for* $b = 1 = 1$ *to* $column$
      ○ $y\_new = \$(v, b, index)$
      ○ $y = y_{new} + y$
  ➢ }} // *ending loops*
Step 11- $actual\_centroid = y/new[m]$;
Step 12- *repeat all the steps to find new_centroid*
      ▪ $a\_new[m] = actual\_centroid[m] - s[i * j]$;
      ▪ *next centroid will form by step* 12
Step 13- *assign next coming centroid value for old for some specified iteration* $actual\_centroid = new\_centroid$

4.2   Proposed Algorithm for Fuzzy C-Means Clustering

Fuzzy clustering is completely reliant on fuzzy concept which refers undecided clustering. Fuzzy clustering overcomes some issues that relate with K-means clustering and enable to capture boundary value objects. Fuzzy logic also permits object to fit in other than one cluster. Fuzzy clustering is also known as relational clustering algorithm [17]. Fuzzy clustering helps to undertaking the problem that falls in decision-making process when we have limited info. Main advantage of choosing fuzzy is with its membership advantage which shows an exact boundary where an object falls (Fig. 1).

Proposed algorithm

Step 1- *Read an input* $q = (N * M)$;
Step 2- Initialize number of cluster $m = n$ *where* $n > 0$
   ➤   *for* $m = 1$ *to* n {
Step 3- *generate random centroid* $= k = random(1 * 1)$; // *math. random*
Step 4- *generate zeros matrix with same as input to update cluster member*
         $\$ = Zeros[N * M]$;
Step 5- *generate cluster member new*$(n) = 0$;
Step 6- calculate row and column size of an input=r, c
   ➤   } // *ending the loop*
// clustering starts from next step//
   ➤   *for* $i = 1$ *to row*{
   ➤   *for* $j = 1$ *to column*{
Step 7- calculate sum of centroid =E[m] =$\sum_{m=1}^{n}(k)$
Step 8- $a[m] = k[m] - q[i * j]$;
Step 9- *membership* $A[m] = a[m] * 100/E[m]$;   *membership for boundary values*
Step 10- *find minimum value with an index position of udtated* $a(m)$
Step 11-*update zeros matrix with index position*
         o $new[index] = new[index] + 1$;
   ➤   } } *ending the inner loop*
Step 12-   initialize $y = 0$ // *iterate zeros matrix to find actual centroid*
   ➤   *for* $v = 1 = 1$ *to row*
   ➤   *for* $b = 1 = 1$ *to column*
         o $y\_new = \$(v, b, index)$
         o $y = y_{new} + y$
   ➤   }} // *ending loops*
Step 13- *actual_centroid* $= y/new[m]$;
Step 14- *repeat all the steps to find new_centroid*
         ▪ $a\_new [m] = actual\_centroid[m] - s[i * j]$;
         ▪   *next centroid will form by step* 12
Step 15-*assign coming centroid value for old for some specified iteration*
         actual_centroid = *new_centroid*

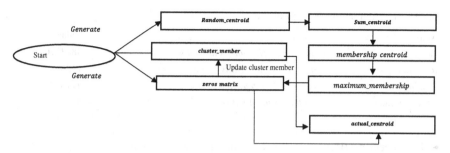

Fig. 1 Proposed methodology

In proposed algorithm now we will update centroid with membership function in terms of percentage value with index position. Membership function of updated centroid is calculated as

$$A[m] = k[m] * 100/E[m] \qquad (3)$$

Here $A[m]$ is the centroid membership in terms of percentage and $k[m]$ is the updated centroid that we found by subtracting from original input. $E[m]$ can be explained as the sum of random generated centroid

$$E[m] = \sum_{m=1}^{n} k[m] \quad \text{where} \quad n > 0 \qquad (4)$$

In proposed work, data clustering is presented deprived of any assistance of distance matrix. Once we initialize cluster number, all the step will run for that specific number and difference will be calculated between centroid and data points. In fuzzy clustering approach main problem is to find membership function for each value but in our proposed work we can see that we evaluated membership in terms of percentage, and instead of using K-means approach now we have a specific range so that this range will take over all the boundary values and enrich classification accuracy of each type of attack over network.

5 Experimental Setup Result

For testing accuracy of algorithm we used small dataset of KDD cup dataset which is the popular raw *TCPdump* dataset. KDD cup has categorized entries by normal values and one attack type (described in Table 1). We ride our code on both algorithms to check classification accuracy of each attack type as well as computational time. Our implementation contains four parts:

Table 1 Types of attacks in KDD dataset

Names	Types	Value
Denial-of-service	*land, back, Neptune, smurf, pod*	3883370
Remote to local	*ftp_write, imap, guess_password, spy*	1126
Probing	*Satan, portsweep, Nmap, ipsweep*	411002
User to remote	*Rootkit, pearl, loadmodeul, butter_overflow*	50

- Preprocessing (Reduce number of entries, labeling dataset),
- Clustering using K-means algorithm based upon MapReduce framework,
- Clustering using fuzzy logic algorithm based upon MapReduce framework,
- Performance analysis,

  - Computational time,
  - Classification accuracy of each type of attack.

(1) **Preprocessing**

KDD'99 cup has approximately five million entries and each with unique label and with redundant records. In this research our objective is to build a fuzzy model that can efficiently classify data using *hadoop* framework and for this purpose, we use the smaller dataset for testing our clustering algorithm containing all types of network attacks, e.g., *normal, guess_password, neptune, back ip_sweep, port_sweep,* and *butter_overflow*. In dataset, first we reduce duplicate entries that origin the unsupervised algorithm to be partial to the frequent entries. While doing this, we found some duplicate records and their invalid values.

(2) **Clustering using K-means algorithm on MapReduce framework.**

After the preprocessing step, we have labeled dataset with unique entries of all possible attacks, and now keeping exact one copy of records in hand we will run proposed algorithm for K-means clustering approach for classifying dataset and find computational time, classification accuracy of each attack, and overall accuracy of classification by first placing data in distrusted file system. Algorithm will be split into mapper and reducer class. Map class performs some common logic for reading the data as key value pair and all data customization will be performed in mapper class and gave key pair output to reduce class for performing some aggregation logic & place output back to HDFS. We tabled top five experiments using K-means methodology (Table 2):

(3) **Clustering using fuzzy logic algorithm based upon MapReduce framework**

Keeping same data in HDFS, we run fuzzy-based algorithm to classify each type of attack on network and tabulated top five experiments (Table 3).

Table 2 Classification accuracy using K-means clustering

Experiment	Classification accuracy normal in %	Classification accuracy guess_pass. and smurf in %	Classification accuracy back attack in %	Classification accuracy ports_weep and Neptune in %	Classification accuracy ipsweep and butter flow attack in %	Overall classification accuracy in %
Experiment 1	90	92	90.9091	85	76.4706	86.8759
Experiment 2	95	90	90.9091	85	76.4706	87.4759
Experiment 3	95	90	90.9091	85	76.4706	87.4759
Experiment 4	90	90	90.9091	85	76.4706	86.8759
Experiment 5	95	92	90.9091	85	76.4706	87.8759
Average	93	90.8	90.9091	85	76.4706	87.3159

Table 3 Classification accuracy using K-means clustering

Experiment	Classification accuracy normal in %	Classification accuracy guess_pass. and smurf in %	Classification accuracy back attack in %	Classification accuracy ports_weep and Neptune in %	Classification accuracy ipsweep and butter flow attack in %	Overall classification accuracy in %
Experiment 1	96	93	98.9091	98	87.4706	93.0759
Experiment 2	98	95	98.9091	98	87.4706	95.4759
Experiment 3	98	93	98.9091	98	87.4706	93.4759
Experiment 4	96	95	98.9091	98	87.4706	95.0759
Experiment 5	98	95	98.9091	98	87.4706	95.4759
Average	97.2	94.2	98.9091	98	87.4706	94.5159

Table 4 Computational time

Computational time	Experiment 1	Experiment 2	Experiment 3	Experiment 4	Experiment 5	Average
K-means clustering	0.0311	0.0209	0.0236	0.0191	0.0316	0.02526
Fuzzy-based clustering	0.0153	0.0173	0.0286	0.0129	0.0112	0.1706

Using fuzzy logic, we evaluated better results with 95 % classification accuracy. It shows that the proposed algorithm efficiently classifies dataset with good accuracy.

Computational Time

For each experiment we intended to catch time length or running time of both algorithms. Results show that our proposed algorithm efficiently classifies the attacks in less. Computational time for top five experiments is given in Table 4 [18, 19]. While running K-means clustering on attack data, it takes average time of 0.02526 because in K-means clustering their data points are given with respect of value from centroid unlike fuzzy clustering which has a specific percentage of data point to centroid and helps to detect boundaries object as well as object assistance in one or more than one cluster.

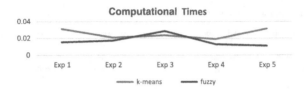

Fig. 2 Computational time

Figure 2 shows variations in computational time. In K-means clustering, highest average time is found to be 0.0311 through top five experiments, and in fuzzy clustering in K-means highest average time we found is 0.0286 for experiment 3.

6   Conclusion

Clustering is a powerful data mining analysis technique which relieves to categorize dense and spare constituencies; therefore, it makes tranquil to find overall distribution pattern and correlations among datasets. In this paper, our key objective is to efficiently classify the data using fuzzy clustering and attempt to figure the relationship among attacks over the network and classify each attack and compare their classification accuracy with respect to computational time. In this paper our second objective is to evaluate performance and enactment of the proposed fuzzy clustering procedure with a typical K-means clustering algorithm and perceived reduced computational time and enriched classification accuracy.

References

1. Song, D., Heywood, M. I., & Zincir-Heywood, A. N. (2005). Training genetic programing on half a million patterns: An example from anomaly detection. *IEEE Transaction on Evolutionary Computation*.
2. Allen, J., Christie, A., Fithen, W., McHugh, J., Pickel, J., & Stoner, E. (2000). State of the practice of intrusion detection technologies. CMU/SEI-99- TR-028, Carnegie Mellon Software Engg. Institute.
3. Zainaddin, D. A. A., & Hanapi, Z. M. (2013). Hybrid of fuzzy clustering neural network over NSL dataset for intrusion detection system. *Journal of Computer Science, 9*(3), 391–403.
4. Songmal, S., Chimphlee, W., Maichalernnukul, K., & Sanguansat, P. (2012). Implementation of fuzzy *c*-means and outlier detection for intrusion with KDD Cup 1999 data set. *International Journal of Engineering Research and Development, 2*(2), 44–48.
5. Bharti, K., Jain, S., & Shukla, S. (2010). Fuzzy K-mean clustering via J48 for intrusion detection system. *International Journal of Computer Science and Information Technologies, 1* (4), 315–318.
6. Banerjee, M., & Soni, R. (2013). Design and implementation of network intrusion detection system by using K-means clustering and Naïve Bayes. *International Journal of Science, Engineering and Technology Research, 2*(3).

7. Takkellapati, V. S. (2012). Network intrusion detection system based on feature selection and triangle area support vector machine. *International Journal of Engineering Trends and Technology, 3*(4).
8. Kumar, V., Chauhan, H., & Panwar, D. (2013). K-means clustering approach to analyze NSL-KDD intrusion detection dataset. *IJSCE, 3*.
9. Siddiqui, M. K., & Naahid, S. (2013). Analysis of KDD CUP 99 dataset using clustering based data mining. *International Journal of Database Theory and Application, 6*(5), 23–34.
10. Tavallaee, M., Bagheri, E., Lu,W., & Ghorbani, A. (2009). A detailed analysis of the KDD'99 CUP data set. *The 2nd IEEE Symposium on Computational Intelligence Conference for Security and Defense Applications (CISDA)*.
11. Nieves, J. F. (2011). Data clustering for anomaly detection in network intrusion detection. Research Alliance in Math & science.
12. Daneshyar, S., & Razmjoo, M. (2012). Large-scale data processing using map-reduce in cloud computing environment. *IJWSC, 3*(4).
13. Shvachko, K., Kuang, H., Radia, S., & Chansler, R. (2010). *The hadoop distributed file system*. MSST, IEEE.
14. Lee, K., Choi, H., & Moon, B. (2011). Parallel data processing with map reduce: A survey. *SIGMOD Record, 40*(4).
15. Panda, M., & Ranjan Patra, M. (2005). Clustering algorithm to enhance the performance of network intrusion detection system. *Journal of theoretical and applied Information Technology*.
16. Chen, Z. (2001). Data mining and uncertain reasoning-an integrated approach. Willy.
17. Srinivasu, P., & Avadhani, P. S. (2011). Implementation of fuzzy c-means and dempster-safer theory for anomaly intrusion detection. *International Journal of Computer Science and Theory*.
18. Sivarthi, S., & Goverdhan, A. (2014). Experiment of hypotheses "fuzzy k-means is better than k-means or clustering. *IJKDP, 4*(5).
19. Singh, T., & Mahajan, M. (2012). Performance comparison of fuzzy C means with respect to other clustering algorithm. *International Journal of Advanced Research in Computer Science and Software Engineering, 4*(5).

System-Related Characteristic-Based Leader Election Protocol for Cognitive Radio Networks

Murmu Mahendra Kumar

Abstract Cognitive radio networks (CRNs) are new communication paradigm that fulfils the today's high demand of radio resources. In CRN, the unused spectrum of privileged (licensed or primary user) node is used by other unprivileged (unlicensed or secondary user) nodes, if need be, in opportunistic manner. The control of the network resources is distributed among the autonomous nodes. In order to improve coordination, we need a leader that can coordinate various activities of secondary user (SU) nodes as well as manage resources for quality of service (QoS) requirements. The present work proposes a leader election protocol for cognitive radio network. The protocol identifies a highest priority value node among the n secondary user node, termed as a leader in CRN. The priority index value is calculated from the degree of nodes, scale strength of signal and average channel per node.

Keywords Cognitive radio network · Leader · Qos · Primary user · Secondary user

1 Introduction

The cognitive radio networks are the smart innovation in the wireless communication field that fulfils the today's spectrum[1] shortage problem. The cognitive radio networks (CRNs) work over unutilized radio spectrum of licensed user. The secondary users (SUs) share spectrum in opportunistic manner in the absence of primary user (PU). In CRNs, the nodes have the capability to adjust its radio parameter

---

[1]The spectrum, channel or links have same meaning in this paper.

M.M. Kumar (✉)
Department of Computer Engineering, National Institute of Technology,
Kurukshetra 136119, Haryana, India
e-mail: mkmurmunitkkr@gmail.com

© Springer Science+Business Media Singapore 2016
S.C. Satapathy et al. (eds.), *Proceedings of International Conference
on ICT for Sustainable Development*, Advances in Intelligent Systems
and Computing 408, DOI 10.1007/978-981-10-0129-1_46

439

with the environments and select most appropriate channel from the available channel set for communication. In case of PU appearance, the switching of channel and redeem computation and communication is not an easy task. Therefore, the QoS information is hard to achieve even though SU nodes have multichannel accessing capability.

In cognitive radio networks, the spectrum allocation policy model is dynamic. The policy allows unlicensed user to share unused portions of the radio spectrum that is already allotted to some other licensed user. This can increase the radio spectrum efficiency and utilization. It has the following steps:

- *Radio spectrum sensing* in order to detect the unused bands.
- *Determination of the best unused frequency band* among the other unused frequency band.
- *Continue to monitor the radio spectrum* to detect the appearance of the licensed user, this leads topology changes.
- *Changing the frequency band* in case the licensed user appearance.

Several cluster head selection algorithms have been proposed in the wireless sensor network (WSN) [1–4], where a cluster head is responsible for the coordination of sensor nodes deployed in the networks. The author of [1] has described a cluster head selection for partially connected sensor networks. The protocol works even when the battery power of sensor node is low. The work proposed in [2] for cluster head selection is based on load balancing by limiting the degree of nodes in WSN. The protocol proposed in [3, 4] is the extension of low-energy adaptive clustering hierarchy (LEACH) protocol and the basic idea about designing protocol is energy efficiency. In CRN, only a few leader election protocols have been proposed in [5–7]. The author of [5] has proposed two leader election protocols for CRN namely deterministic and randomized protocols. The deterministic algorithm uses multiple phases and rounds for leader selection in cognitive radio networks. In the other hand, randomized algorithm uses probability theory to elect a leader for CRN. The works described [6] a leader election protocol for distributed ad hoc cognitive radio network. The election method is based the on the basis of selection factor (SF) parameter. The constraints of SF are residual energy (E), density of secondary user (SU) and the availability of free channel (Z). A node with highest selection factor (SF) value declares itself as a leader in the network. The leader election protocol for CRNs proposed in [7] uses data structures (modes and control messages) for secondary user description as well as for leader selection.

The protocol developed for wireless sensor network is not suitable for leader election in CRNs due to several reasons. The wireless sensor network is infrastructure based, and energy and computing power of sensor nodes are the generic issue. In case of CRN, it is infrastructure less and battery power, computing power, etc. are not an issue for cognitive nodes [8]. Therefore, we can apply leader election protocol of WSN by making it adaptive or by making a fresh and new leader election protocol for cognitive radio networks.

In order to improve coordination among SUs and channel assignments, a leader node of CRNs must have the following properties:

- *Safety*: All the nodes agree on a single node termed as a leader in the cognitive radio network at some time instance and space.
- *Liveness*: Eventually, after termination every connected component has a unique leader within finite time and space.

The proposed protocol is a fresh approach for leader election in cognitive radio network. The leader election parameters include priority value, degree of nodes, energy strength of signal and average channels per node. The priority value of each node is calculated on the basis of assumption made, i.e. degree of nodes, energy strength of signal and average channels per node. The higher priority value node declares itself as a leader in the network.

The work flow of the paper is as follows. Section 2 describes the system model of the proposed work. Section 3 discusses about the algorithm description and the steps of the proposed algorithm. The proof of the correctness is detailed in Sect. 4, and finally we have concluded the paper in Sect. 5.

2 System Model

The system consists of n secondary user nodes and k channels. Each node has a *unique id*. If a network contains n number of nodes, then each node n_i must have unique id_i, where the value of i is 1...n. A node can transmit data to some other node over a frequency band. A usable frequency band is termed as a *channel*. A node can send data to another node over more than one channel. The more the number of channels between two nodes, the more reliable is the connection between them. The signal is *attenuated* over a distance. Much weaker signal is unreliable. The *degree* of a node is the number of distinct nodes with which it is connected through one or more channel. Each node has priority value p_i. The highest priority value node declares itself as a leader (l_i) of the networks.

Identifying a better leader

The cognitive radio networks are the collection of SU nodes. If the number of subordinates is more, better is the chances a node declares itself as a leader. In other words, the more the number of subordinates, the better is the allocation policy for channels. Therefore, we want to maximize the number of subordinates or the degree of the leader.

Also, the more the number of channels between two SU nodes, the better are the chances that two secondary user nodes will remain connected even if one or more channels between them are used by some other nodes (by PU). We want leader node to remain connected with nodes for as much time as possible, although it is not an easy task and we want a leader node (among SUs) to have maximum average number of channels per node.

If the signal strength is weaker, the control packets can be lost during transit. The leader cannot communicate with nodes if such a thing happens. We want a leader node to have the best average signal strength with all its neighbours.

3 Algorithm Description

3.1 Signal Strength and Scaling

In CRNs, the SUs perform sensing operation using [9–11] in radio environment and calculate the signal strength at each node. The signal strength is the ratio of the original signal power divided by the power sensed at the receiver end. The signal strength has values between 0 and 1.

The higher signal strengths are usable but the lower ones are not. We want to scale the strength to a factor such that the lower values get very low values and the middle and higher values get approximately the same values. So we choose the *logarithmic function* to scale the strength. With random experiments, we find the base to be 20.

$$\text{Scale Strength} = \log_{20}(\text{Original Strength} * 20) \tag{1}$$

3.2 Priority Value

In CRN, the higher the degree of node, per node channel allocation and average scale signal strength, the more are the chances of that node to become a leader. Therefore, we introduce a new term, i.e. priority value. Priority value is an index that suggests which node can be a better leader.

$$\text{Priority Value}(p_i) = \text{degree}^2 * (\text{averagechannelspernode}) \\ * (\text{average scale signalstrengthper node}) \tag{2}$$

where degree implies degree of node, which is the number of neighbour nodes.

Average channel per node is the set of average number of channels in a local channel set.

3.3 Algorithm Description

In the process of leader election, each SU node initiates the process by broadcasting a message containing node *id* and *initial signal strength* to its neighbours. For finding neighbour nodes we use existing neighbour discovery protocols as discussed in [7]. Once it receives the message, the receiver node calculates the environmental parameters. On the basis of calculated environmental parameters, each node then calculates the priority value. Now, the receiver node broadcasts new message containing node *id* and *priority value* such as <*id*, Priority value> to the neighbouring nodes from where the message has been received previously. After

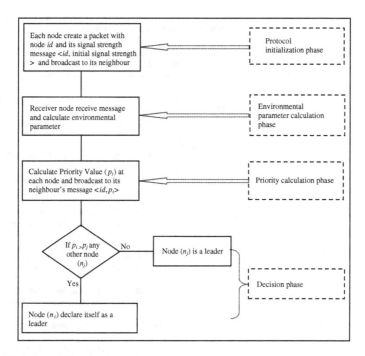

Fig. 1 Work flow of the algorithm (flow chart)

receiving message from the neighbouring nodes, the initiator node now has the set of information about other nodes for leader election in cognitive radio networks. The node with highest priority value index is a leader of the cognitive radio networks. The work flow of the protocol is shown in Fig. 1 and the steps of the algorithm are shown in Fig. 2.

3.4 Proposed Algorithm

The algorithm consists of four phases. We are assuming that no packets are lost in transit and the channels are not FIFO.

3.5 Algorithm Analysis and Its Message Complexity

For each channel two messages are required, one in the information broadcasting phase and the other one in the priority value broadcasting phase. Therefore, the message complexity of the algorithm is 2 * number of channels.

Fig. 2 Leader election
algorithm

> **Information packet broadcast phase:**
> - Each node (n_i) creates a packet.
> - Node (n_i) embeds its unique *id* and initial signal strength in it.
> - This packet is sent to all the channels associated with the node.
>
>   sent_message < *id*, initial signal strength >
>
> **Message processing phase:**
> - Node (n_j) receives packet < *id*, initial signal strength > from other nodes and compute environmental parameters.
>   - a) Node stores all the unique *id's* in a set.
>   degree of node = size of the set
>   - b) channel count = number of packets received
>   - c) strength of the signal = (strength perceived by the receiver) / (initial signal strength in the message)
>   - d) average scale signal strength = $\sum$(scale strength of signal) / (channel count)
>
> **Priority value computation and broadcast phase:**
> - Node (n_i) calculates the priority value according to eq. (2).
> - Node (n_i) broadcasts its priority value (p_i) value to all free channels available.
>
>   broadcast_message <*id*, priority value (p_i)>
>
> **Decision phase:**
> - Node (n_i) receives packet from node (n_j).
>   - a) If its priority is greater than the priority value of other nodes ($p_i > p_j$), the node (n_i) assume itself a leader node (l).
>   - b) Else, the node (n_j) from which it receives the message containing the highest priority value is its leader node (l).

3.6   Simulation Results

The parameters (degree of nodes, average channels per node, and average scale signal strength per node) considered for the algorithm are directly proportional to its priority value, p_i. The highest priority value node is a leader of the cognitive radio network.

In the simulation results Fig. 3a, we have assumed that the degree of node is varying from 1 to 10 and priority value of corresponding node is calculated. The average number of channels and scale strength is static. Similarly, we plot a graph given in Fig. 3b, and at each node we have calculated the priority value (p_i) w.r.t. the average number of channels per node by keeping the degree of node and the scale strength static. Further, we assume that the degree of node and the average number of channels per node are constant and plot the graph of priority value versus scale strength shown in Fig. 3c. The graph showing each parameter value has significant impact on the calculation of priority value index of SU nodes.

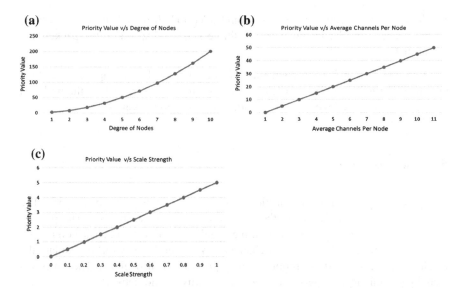

Fig. 3 Simulation results. **a** Graph for priority value (p_i) versus degree of nodes by keeping the average number of channels per node and scale strength is static. **b** Graph for priority value (p_i) versus average number of channels per node by keeping the degree of nodes and scale strength is static. **c** Graph for priority value (p_i) versus scale strength by keeping the degree of nodes and average channels per node is static

4 Correctness Proof

Assumption 1: After the election phase is over, there are two or more nodes that are declared as leader nodes.

Suppose there are two nodes n_1 and n_2 that consider themselves as leader after the election phase is over. Suppose their priority values (p_i) are p_1 and p_2 and their respective id_is are id_1 and id_2.

If that is the case, then

$$p_1 = p_2 > p_i(\text{other than } p_1 \& p_2) \forall i \in (1\ldots n) \text{ and } i \neq 1 \text{ and } 2.$$

Also, $id_1 = id_2$.

However, this contradicts the fact that each node is assigned with the *unique id*. Since two nodes cannot have same *id*, therefore, we cannot have more than one node declared as leader after the election phase is over.

Assumption 2: Suppose after the algorithm is terminated, two nodes n_1 and n_2 consider two different leader nodes l_1 and l_2 as their leader node.

If that is the case, node n_1 received the packet with highest (priority value, *id*) pair from leader node l_1 and n_2 received the same from leader node l_2. If such a situation exists, there can be only two reasons:

- Priority value of leader node (l_1) = Priority value of leader node (l_2) $\wedge$ *id* (l_1) = *id* (l_2), but this cannot happen as proven earlier.

- Node n_2 never received the packet from leader node l_1 or node n_1 never received the packet from leader node l_2, but this can never happen as the packets are never lost in transit.

Occurrence of any of these situations will contradict the assumptions on which the algorithm is based. Therefore, after the algorithm is over, each of the nodes in the cognitive radio networks will consider only one node as its leader.

5 Conclusion

The paper proposed a leader election protocol for cognitive radio network. The election of a leader is based on the priority value (p_i) of each node. The priority value node is considered as a leader of the cognitive radio network. The complexity of the algorithm is 2 * number of channels. We extend our work to simulate the results and the correctness proof of the algorithm is also given. The proposed technique is efficient for leader election in cognitive radio network.

References

1. Mozumdar, M., Gregoretti, F., Vanzago, L., & Lavagno. L. (2008). An algorithm for selecting the cluster leader in a partially connected sensor network. In *3rd IEEE International Conference on Systems and Networks Communications, ICSNS 2008* (pp. 133–138).
2. Mahajan, S., Malhotra, J., & Sharma, S. (2014). An energy balanced QoS based cluster head selection strategy for WSN. *Egyptian Informatics Journal, 15*(3), 189–199.
3. Liao, Q., & Zhu, H. (2013). An energy balanced clustering algorithm based on LEACH protocol. In *International Conference on Systems Engineering and Modeling (ICSEM), 2013*.
4. Ray, A., & De, D. (2012). Energy efficient cluster head selection in wireless sensor network. In *Recent Advances in Information Technology (RAIT), 2012*.
5. Bansal, T., Mittal, N., & Venkatesan, S. (2008). Leader election algorithm for multi-channel wireless networks. *Lecture Notes in Computer Science, 5258*, 310–321.
6. Olabiyi, O., Annamalai, A., & Qian, L. (2012). Leader election algorithm for distributed ad hoc cognitive radio networks. In *IEEE Consumer Communications and Networking Conference (CCNC), 2012*.
7. Arachchige, C. J. L., Venkatesan, S., & Mittal, N. (2008). An asynchronous neighbor discovery algorithm for cognitive radio networks. In *IEEE DySPAN 2008*.
8. Akyildiz, I. F., Won-Yeol, L., & Chowdhury, K. R. (2009). CRAHNs: Cognitive radio ad hoc networks. *Ad Hoc Networks, 7*(5), 810–836.
9. Akyildiz, I. F., Won-Yeol, L., Vuran, M. C., & Mohanty, S. (2006). NeXt generation/dynamic spectrum access/cognitive radio wireless networks: A survey. *Computer Networks, 50*(13), 2127–2159.
10. Zeng, Y., Liang, Y. C., Hoang, A. T., & Zhang, R. (2010). A review on spectrum sensing for cognitive radio: Challenges and solutions. *EURASIP Journal on Advances in Signal Processing, 2*, 5–23.
11. Subhedar, M., & Birajdar, G. (2011). Spectrum sensing techniques in cognitive radio networks: A survey. *International Journal of Next-Generation Networks, IJNGN, 3*(2), 37–51.

Passive Image Manipulation Detection Using Wavelet Transform and Support Vector Machine Classifier

Gajanan K. Birajdar and Vijay H. Mankar

Abstract In this paper, blind global contrast enhancement detection method is proposed using wavelet transform-based features. Wavelet subband energy and statistical features are computed using multilevel 2D wavelet decomposition. Mutual information-based feature selection measure is employed to select the most relevant features while discarding the redundant features. Experimental results are presented using grayscale and G component image database and SVM classifier. Simulation results prove the effectiveness of the proposed algorithm compared to other existing contrast enhancement detection techniques.

Keywords Image forgery detection · Passive authentication · Contrast enhancement detection · Wavelet transform

1 Introduction

Advances in digital imaging technology resulted in sophisticated and cheap, yet powerful tools that enable the generation and manipulation of digital images without leaving any traces of tampering. As digital images are frequently used in military, printing media, medical records, and in court as legal photographic evidence, verifying its authenticity and integrity is important in order to restore trustworthiness of the candidate image. Image forgery detection algorithms are categorized as (1) active methods and (2) passive (blind) methods [4, 10].

G.K. Birajdar (✉)
Department of Electronics and Communication, Priyadarshini Institute
of Engineering and Technology, Nagpur 440019, Maharashtra, India
e-mail: gajanan123@gmail.com

V.H. Mankar
Department of Electronics and Telecommunication,
Government Polytechnic, Nagpur 440001, Maharashtra, India
e-mail: vhmankar@gmail.com

© Springer Science+Business Media Singapore 2016   447
S.C. Satapathy et al. (eds.), *Proceedings of International Conference
on ICT for Sustainable Development*, Advances in Intelligent Systems
and Computing 408, DOI 10.1007/978-981-10-0129-1_47

In active method, known information is embedded using preprocessing such as digital watermark or signature at the time of creating the image. Passive or blind techniques work on the assumption that digital image manipulation may not leave any visual traces of tampering. They may alter the underlying statistics or inconsistency of an image. Image forensic algorithms are of two types: (1) forgery detection and (2) forgery localization. In this article forgery detection is investigated.

Contrast enhancement image processing operation is widely used by the attacker to conceal the cut and paste and spicing forgery in doctored images. In this paper, a blind method is proposed for global contrast enhancement detection based on energy and statistical features extracted using multilevel wavelet decomposition. Contrast enhancement operations can be considered as nonlinear pixel mappings which introduce inconsistencies into the statistical properties of an image. Mutual information-based criterion is employed to choose the most informative features from the original large feature space and removing redundant features. This technique is applicable to large range of gamma values. Results are presented using grayscale and G component image database.

The paper is organized as follows. Section 2 briefly reviews the related work in the area of blind contrast enhancement detection. Feature extraction process is presented in Sect. 3. Feature selection method using mutual information is described in Sect. 4. Proposed algorithm for forgery detection is discussed in Sect. 5. Experimental results and discussions are presented in Sect. 6, and Sect. 7 concludes the paper.

2 Related Work

Various classifier-based approaches are proposed to detect contrast enhancement images blindly. In [1], a technique to capture image features independent of original image content is proposed for blind contrast enhancement detection. LDA is used for classification. Image manipulation detection method based on the neighbor bit planes of the image is presented in [2]. The correlation between bit plane and binary textural properties within the bit plane will not be the same between an original and doctored image.

In [3], a method is proposed to detect doctoring in digital images with 188 dimension feature vector extracted using (1) binary similarity measures (BSMs) (2) image quality measures (IQMs), and (3) higher order wavelet statistics (HOWS). In [7], a method is developed using singular value decomposition (SVD) for image forgery detection. Authors showed that image manipulation changes the linear dependencies of image rows/columns and the derived features can be used to detect image manipulations.

A blind forensic algorithm is developed for detecting the use of local and global contrast enhancement and histogram equalization operations to modify digital images [14]. The algorithm performs detection by identifying unique artifacts introduced into an image histogram as a result of the particular operation examined called as intrinsic fingerprint. In addition to this, a method is proposed to detect the global addition of noise to a JPEG compressed image.

In [6], a method is proposed to detect the global contrast enhancement for both uncompressed and previously JPEG compressed images using features extracted from the zero height gap bins in gray-level histograms. Contrast enhancement detection method is developed using IQM and BSM features extracted from doctored images and SVM classifier [5].

A method is proposed to detect cut and paste image forgery by local contrast enhancement detection in color images using the fact that contrast enhancement operation can disturb the inter-channel similarities of high-frequency components [8]. Local contrast enhancement can be reliably detected even for small block size of 16 × 16. Two modified and improved methods are presented in [9] for contrast enhancement detection in copy and paste forgery. The first method uses a quadratic weighting function instead of a simple cutoff frequency to measure the histogram distortion introduced by contrast enhancements as mentioned in [14]. In the second method, the improvement is achieved by applying a soft threshold strategy to get around the sensitivity of threshold selection compared to the method described in [8].

3 Feature Extraction

In this experiment, two types of features are extracted: first set of energy features from wavelet subband and second set consisting of statistical features. Energy features are derived after the wavelet decomposition from low-pass subbands. First, the 2D wavelet coefficients are transformed into 1D using the row scan order. The normalized energy of the subband C_a containing K_a coefficients is computed as

$$E_{C_a} = \frac{1}{K_a} \sum_{m=1}^{Ka} w_{a,m}^2 \tag{1}$$

where $w_{a,m}$ are the wavelet subband coefficients. Normalized energy feature vector extracts energy distribution among the low-pass subbands. Second feature, the absolute mean of the subband is computed as

$$A_{C_a} = \frac{1}{K_a} \sum_{m=1}^{Ka} |w_{a,m}| \tag{2}$$

Combination of normalized energy and absolute mean forms the energy feature and is used in [11] for texture image retrieval.

Contrast enhancement operation changes the neighborhood pixel correlation and these differences are exploited by extracting various statistical features like mean, standard deviation, and entropy to reveal forgery operation. For achieving better detection, these features should be discernible and independent of each other.

Table 1 Various statistical features

Mean	$\mu = \frac{1}{k}\sum_{i=1}^{k} s_1$
Variance	$\frac{1}{k}\sum_{i=1}^{k} (s_1 - \mu)^2$
Skewness	$\dfrac{\frac{1}{k}\sum_{i=1}^{k} (s_1-\mu)^3}{\left[\frac{1}{k}\sum_{i=1}^{k} (s_1-\mu)^2\right]^{3/2}}$
Kurtosis	$\dfrac{\frac{1}{k}\sum_{i=1}^{k} (s_1-\mu)^4}{\left[\frac{1}{k}\sum_{i=1}^{k} (s_1-\mu)^2\right]^{2}}$
Standard deviation	$\sigma = \sqrt{\frac{1}{k}\sum_{i=1}^{k} (s_1 - \mu)^2}$
Energy	$\sum_{i=1}^{k} s_i^2$
Entropy	$-\sum_{i=1}^{k} s_i \log s_i$
Correlation	$\sum_{i,j} \frac{(i-\mu_i)(j-\mu_j)s(i,j)}{\sigma_i \sigma_j}$

Table 1 depicts various statistical features used in this experiment. Let $s = s_1, s_2, \ldots, s_k$ be the input signals having k number of samples.

From above equations, second set of features is extracted. Final feature vector consists of 8D wavelet energy features and 32D statistical features.

4 Feature Selection Using Mutual Information

In pattern recognition applications, feature subset selection is an important step. Feature selection discards irrelevant and redundant features from the large input data space, selects most informative features, and in turn enhances speed and prediction accuracy of the classification process.

Computing the discrimination power of the individual feature from large feature set is a crucial step in feature selection process. In this article, mutual information-based feature selection measure is employed to choose relevant features. Mutual information is a measure of the amount of information in one variable that can be predicted when another variable is known. It is a measure of the dependence between two random variables and is zero when the variables are statistically independent. The mutual information between the discretized feature values u and the class labels v is evaluated as [12].

$$\mathrm{MI} = \sum_{u \in U} \sum_{v \in V} p(u, v) \log\left(\frac{p(u, v)}{p(u)p(v)}\right) \quad (3)$$

MI measure is a scoring and ranking method of feature selection. Feature subset is generated by selecting the features having the highest values of scores.

5 Proposed Algorithm

Proposed contrast enhancement detection algorithm consists of the following steps:

1. *Preprocessing*: Before feature extraction, the input RGB image is converted into grayscale image using the relation $0.2989 \times R + 0.5870 \times G + 0.1140 \times B$ and G component is extracted from the input RGB image.
2. *Feature extraction*: In this step, wavelet energy and statistical features are extracted. Detailed procedure of feature extraction is explained in Sect. 3. These two types of features are combined to form a final feature vector.
3. *Feature selection*: Most relevant features are selected by discarding the redundant features using mutual information-based feature selection approach.
4. *Classification*: SVM is used as a classifier which classifies input image as authentic or manipulated. Various performance parameters like specificity, sensitivity, and total detection accuracy (T) are computed.

6 Experimental Results

In this section, simulation results are presented using the proposed algorithm discussed in Sect. 5. Database consists of 800 uncompressed color images randomly selected from UCID image database [13]. After RGB to grayscale conversion, contrast enhancement operation is performed with γ values ranging from 0.2, 0.4, … , 3. Similar procedure is followed for G component images.

Wavelet-based energy and statistical features are extracted from these two sets of images database separately, resulting in 40D feature vector. Mutual information feature selection method reduces this to 24D feature vector which is the input to the classifier. Linear kernel SVM classifier is trained for each individual γ that classifies the input images into authentic or contrast enhanced image. 50 % of the images are randomly selected for training and remaining 50 % are used for testing.

Tables 2 and 3 show the specificity, sensitivity, and total detection accuracy (%) as a function of different γ for grayscale and G component images, respectively. Detection accuracy is higher for lower and higher values of γ and it is lowest at 0.8. Changes introduced in the images when $\gamma = 0.8$ are difficult to detect due to small amount of distortions introduced at this level. G component average accuracy (85.45 %) is higher compared to grayscale images (84.74 %).

Figure 1 shows the effect of mutual information measure for feature selection when grayscale image database is used. From the figure it is evident that feature selection reduces the feature dimensions and simultaneously increases the detection accuracy. Contribution of individual feature set on detection accuracy is also investigated. Figure 2 shows detection accuracy using wavelet energy, statistical, and combined features. Accuracy is maximum when both types of features are combined.

Table 2 Specificity, sensitivity, and T detection accuracy (%) as a function of different γ for grayscale images

Γ	0.2	0.4	0.6	0.8	1.2	1.4	1.6	1.8	2	2.2	2.4	2.6	2.8	3
Specificity	90	80	71.5	50	75	86.5	93	95	98	99	99	99	99	99
Sensitivity	99.5	95	92	81.5	69	59.5	74	77	80	82	83	83.5	84.5	85.5
Total	94.75	87.5	81.75	65.75	72	73	83.5	86	89	90.75	91	91.25	91.75	92.25

Table 3 Specificity, sensitivity, and T detection accuracy (%) as a function of different γ for G component images

Γ	0.2	0.4	0.6	0.8	1.2	1.4	1.6	1.8	2	2.2	2.4	2.6	2.8	3
Specificity	89	83	72	48	77	87	94	95.5	98	98	98	98	98	98
Sensitivity	99.5	97	94	81	70	59.5	76	78.5	81	83	84	85	84	86.5
Total	94.25	90	83	64.5	73.5	73.25	85	87	89.5	90.5	91	91.5	91	92.25

Fig. 1 Detection accuracy with and without feature selection

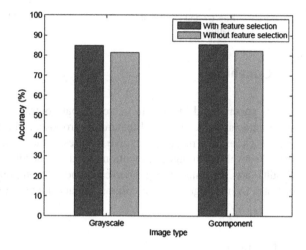

Fig. 2 Detection accuracy using individual and combined feature set

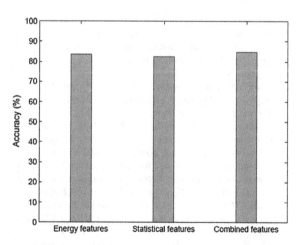

Table 4 shows the comparison of proposed method with existing classifier-based approaches. The method performs better even with low feature dimensionality compared to other existing algorithms.

Table 4 Comparison of various methods

Algorithm	Extracted features	Feature vector dimension	Classifier	Average detection accuracy
[3]	IQM + BSM + HOWS	202	Clairovayent	83
[5]	IQM + BSM	130	SVM	93
[2]	BSM	17	LDA	75.5
[7]	SVD-based features	–	Semi-blind classifier	81
Proposed	Energy and statistical features	24	SVM	85.45

7　Conclusion

In this paper, blind method to detect global contrast enhancement using wavelet subband energy and statistical features is proposed. Mutual information measure is used to choose the most informative features at the same time removing redundant features resulting in higher execution speed and accuracy of classification process. Results are presented using grayscale and G component images. The method operates over a large range of contrast enhancement with good detection accuracy.

References

1. Avcibas, I., Bayram, S., Memon, N., Ramkumar, M., & Sankur, B. (2004). A classifier design for detecting image manipulations. In *International Conference on Image Processing* (pp. 2645–2648) (October 2004).
2. Bayram, S., Avcibas, I., Sankur, B., & Memon, N. (2005). Image manipulation detection with binary similarity measures. In *European Signal Processing Conference* (pp. 1–4).
3. Bayram, S., Avcibas, I., Sankur, B., & Memon, N. (2006). Image manipulation detection. *Journal of Electronic Imaging, 15*(4), 1–17.
4. Birajdar, G. K., & Mankar, V. H. (2013). Digital image forgery detection using passive techniques: A survey. *Digital Investigation, 10*(3), 226–245.
5. Boato, G., Natale, F., & Zontone, P. (2010). How digital forensics may help assessing the perceptual impact of image formation and manipulation. In *International Work-shop on Video Processing and Quality Metrics for Consumer Electronics* (pp. 1–6) (January 2010).
6. Cao, G., Zhao, Y., Ni, R., & Li, X. (2014). Contrast enhancement-based forensics in digital images. *IEEE Transactions on Information Forensics and Security, 9*(3), 515–525.
7. Gul, G., Avcibas, I., & Kurugollu, F. (2010). SVD based image manipulation detection. In *IEEE International Conference on Image Processing* (pp. 1765–1768) (September 2010).
8. Lin, X., Li, C. T., & Hu, Y. (2013). Exposing image forgery through the detection of contrast enhancement. In *International Conference on Image Processing* (pp. 4467–4471) (September 2013).
9. Lin, X., Wei, X., & Li, C. T. (2014). Two improved forensic methods of detecting contrast enhancement in digital images. In *Media Watermarking, Security, and Forensics*, Vol. 9028 (February 2014).
10. Mahdian, B., & Saic, S. (2010). A bibliography on blind methods for identifying image forgery. *Signal Processing: Image Communication, 25*(6), 389–399.

11. Pi, M. H., Tong, C. S., Choy, S. K., & Zhang, H. (2006). A fast and effective model for wavelet subband histograms and its application in texture image retrieval. *IEEE Transactions on Image Processing, 15*(10), 3078–3088.
12. Pohjalainen, J., Rasanen, O., & Kadioglu, S. (2013). Feature selection methods and their combinations in high-dimensional classification of speaker likability, intelligibility and personality traits. *Computer Speech & Language, 29*(1), 145–171.
13. Schaefer, G., & Stich, M. (2004). UCID—An uncompressed colour image database. In *Proceedings of SPIE, Storage and retrieval Methods and Applications for Multimedia,* pp. 472–480.
14. Stamm, M., & Liu, K. J. R. (2010). Forensic detection of image manipulation using statistical intrinsic fingerprints. *IEEE Transactions on Information Forensics and Security, 5*(3), 492–506.

Intrusion Detection System with Snort in Cloud Computing: Advanced IDS

Vikas Mishra, Vinay Kumar Vijay and Satyanaryan Tazi

Abstract Intruders and thieves are important threats to most business and large organizations. These threats and unwanted materials create many disturbances in storing of the data on large scale especially in cloud computing. So maintaining security against these threats is important in any organizations. Security could be of different types like hardware security, software security, malicious behavior of attackers, and many others besides security; many organizations try to introduce many methods which will provide malicious behaviors of attackers so that an alert message will propagate throughout the whole system so that even if some attacker tries to break down the security an alert message is generated. Such a model is the intrusion detection system. An intrusion detection system (IDS) is a hardware device or software application that monitors network and/or system or host activities for malicious activities' policy violations, creates and sends reports to a management station or system administrator which decides whether to take an action on the intrusion or it was only a false alarm. In this paper, we introduce a model or system called snort which is an intrusion detection system based upon rules detection and has the ability to control traffic and matching data with the original database and allows only data to flow which matches with the original database.

Keywords Snort · Detection · Rules · Preprocessor

V. Mishra (✉) · V.K. Vijay · S. Tazi
Government Engineering College, Ajmer, India
e-mail: mishravik126@gmail.com

V.K. Vijay
e-mail: vinay.vijayvargiya@gmail.com

S. Tazi
e-mail: satya.tazi@gmail.com

© Springer Science+Business Media Singapore 2016   457
S.C. Satapathy et al. (eds.), *Proceedings of International Conference
on ICT for Sustainable Development*, Advances in Intelligent Systems
and Computing 408, DOI 10.1007/978-981-10-0129-1_48

1 Introduction

The term cloud is related to Internet where consumption of computer resources is taken place by group of users generally called remote servers. The increasing network bandwidth and highly reliable network connections allow different organizations to store and access data over Internet, and cloud becomes the first choice of various users and organizations that provide services of storing the data over network. Cloud also provides service to move the data within one location to another over a geography distributed network and it is very convenience for user as they does not bother about the hardware connections, and this reduces the complexities, time, and infrastructure capital to build a network system. Hence, cloud computing is an emerging technology in the world which has been envisioned as the important data management architecture which plays an important role in IT enterprise. The cloud computing is an internet-based technology that provides various features and services to the different organizations, storing data and information on demand of customer. As cloud computing mainly focuses on the large pool of resources available over the Internet and also managing and accessing the data and due to all these cloud computing features, advancement of new technology of security of data in cloud computing environment is came into existence. In the recent year, security of data over cloud is the requirement which is raised by every organization which introduces cloud computing as their main part of the system, so in such an environment cloud computing security is must. Intrusion detection system is such a security model which examines the packet over the network and identifies the threats in the network. The IDS models control the network traffic and monitor the individual network connected hosts. It captures all the unwanted packet signals and analyzes the content of individual packets after it identifies malicious packet and generates an alert message to the system. IDS can be classified into two categories HIDS (Host-based intrusion detection system) as name indicates HIDS is working to analyze the data collected by operating system about the actions performed by users and applications and NIDS (network-based intrusion detection system) analyze the data collected by network. In this paper, we introduce snort which is an IDS-based application and is an open-source network intrusion detection system to analyze and detect different attacks [1, 2].

2 Related Work

Intrusion prevention is the process in which first intrusion detection is performed and then attempts are made to stop the detected possible intrusions. The primary objectives of the intrusion detection system are to identify the occurring intrusions, save the information about attacks, attempting to stop them, attempting to stop them, and provide the reports to the security administrator. IDS also examines and analyzes the user and system activities, analyzing system configuration and vulnerabilities, assessing system and file integrity, analysis of abnormal activity patterns, and tracking user policy violations. In this paper, we introduce snort to examine the network traffic and detect different types of attacks. Snort is a lightweight intrusion detection system which works on signature-based methodology of IDS. Signature-based intrusion detection system is used for detecting the known attacks, which is known by the system. Signature could be patterns of strings or characters which are found in payload of packets. This type of detection system introduces the concept of database which is a collection of the previous attacks called as known attacks and when packet comes in the network, the system matches the signature of the packet with the signature of the known attacks stored in database; if matches found, then the system alerts the administrator about the attacks discovered. In this model detection is based on the database of the known attacks, so this method is also called knowledge-based intrusion detection system [3, 4]. The main advantage of this method is that the system administrator does not require any special kind of detection team to detect the attacks as only database of the previous attacks is required but this type of detection method cannot identify the new attacks or intrusions whose patterns do not match with the database, also it is not easy to update the database on regular interval of time [5].

Snort is working on rules which can be written in any language and rules can be easily read and modified. The main work of snort is to check the packet against rules written by user and when this packer is identified it generates an alert and also provides other information about this packet. Snort can be used anytime when you want to have basic security measures in cloud computing that allows you to log and analyze the traffic on the network. Along with a firewall, snort is another fundamental part of the network security system of any organization. Also, internal network detection is important because many of the security issues come from internal network system and snort examines the internal network by comparing the packet data with the preloaded databases [6]. Snort can be run on Linux and windows and available for download at http://www.snort.org and latest available version is 2.9.7.0. Snort also requires third-party library WinPcap driver. The current stable version of WinPcap is 4.1.1 [7, 8].

3 Proposed Architecture

The level of security measures in cloud computing has been evolved in recent years. Intrusion detection system provides various security policies to handle the attacks and assures maximum security in cloud computing. IDS in cloud is implemented in virtual-based environment which requires user data and applications, and cloud service provider's remote server and cloud user have limited control over its data and resources; in this type of IDS model, cloud provider is the central administrative part of the IDS. IDS is generally working in two models: first is signature-based or knowledge-based model, and second is anomaly based or behavior-based model [9] (Fig. 1).

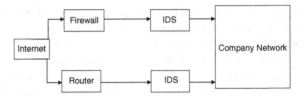

Fig. 1 Location of snort in cloud environment

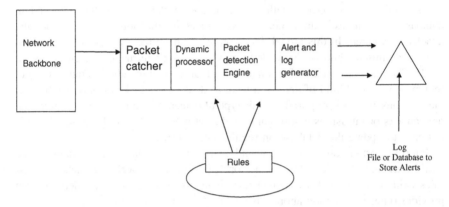

Fig. 2 Architecture of snort

Signature-based model detects the attacks whose patterns match with the known patterns in the database but this model does not identify new attacks whose patterns do not find out or match in saved database, while anomaly based method detects the abnormal behavior of the network.

Snort is the signature-based intrusion detection method which detects malicious packet in the network flow of data from one location to another and it starts with catching packet and follows detection engine model and then finally alerts the message generated as output (Fig. 2).

The architecture of snort contains:

- **Packet Sniffer**—Its main work is to collect packets from network and send to preprocessor.
- **Preprocessor**—its main work is to collect packets from packet sniffer. Preprocessor fragments, arranges, and modifies the packet and sends incoming packets to detection engine for further processing.
- **Detection Engine**—The main work of detection engine is to find the intrusion in the packet with the help of rules defined in snort package. Detection engine does not evaluate the rules in the order that they appear in the snort rule file. In default the order is

1. Alert rule,
2. Pass rule,
3. Log rule.

- **Logging and Alerting system**—It generates the alerts and log file if found any intrusion in the network.
- **Rules**—Rules are generally the language constructs which describe the traffics that are examined by the detection engine and it also has two parts: Rule Header and Rule Options. Rules are generally placed in snort.conf configuration file [10, 11].

4 Implementation and Analysis

Snort is working on network (IP) layer, transport (TCP\UDP) layer protocol, and application layer (Fig. 3).

After successful installation of the snort, open command prompt and type this to open the snort (Figs. 4, 5, 6, 7 and 8).

Snort-V—This command is used to initialize the snort in windows.

Snort-W—This command is used to check different interfaces in network.

After successful implantation of snort, and next to run in IDS mode, we need to modify the snort.conf file. Modify the home network, path to rules, and procedure, and then validate the setting by typing this in command prompt.

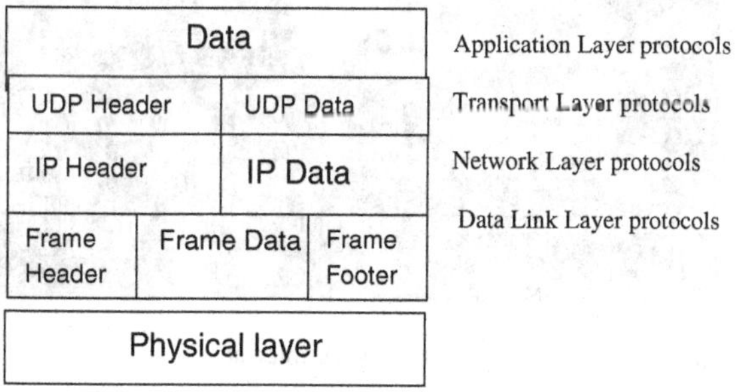

Fig. 3 Implementation of snort in network protocol

Data			Application Layer protocols
UDP Header	UDP Data		Transport Layer protocols
IP Header	IP Data		Network Layer protocols
Frame Header	Frame Data	Frame Footer	Data Link Layer protocols
Physical layer			

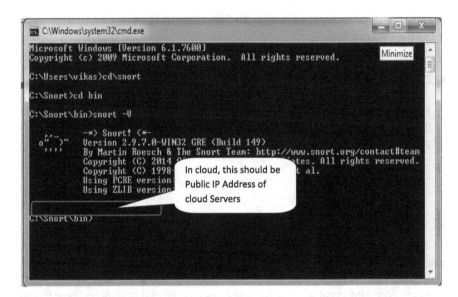

Fig. 4 Shows the initialization of snort

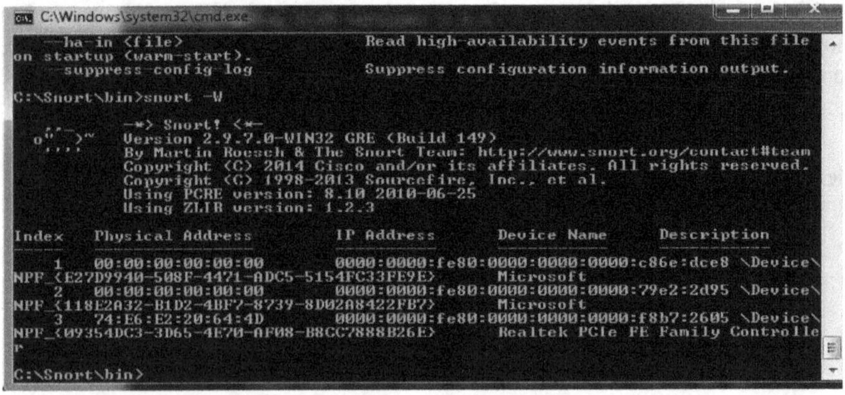

Fig. 5 Shows the different network interfaces

./snort -dev -l./log -h 192.168.1.0/24 -c snort.conf, where snort.conf is the name of your snort configuration file. In this model the rules are applied and configured in the snort.conf file to each packet to decide if an action based upon the rule type in the file should be taken [12, 13].

This changes the path to sub-directory to the specified folder where main snort files are installed.

Finally, validate and run snort in IDS mode as follows.

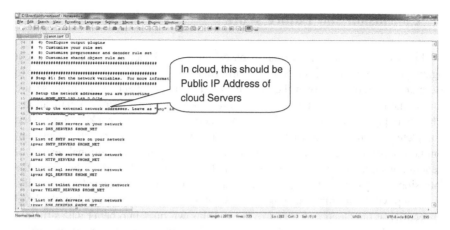

Fig. 6 Shows the steps to change network IP in snort.conf file

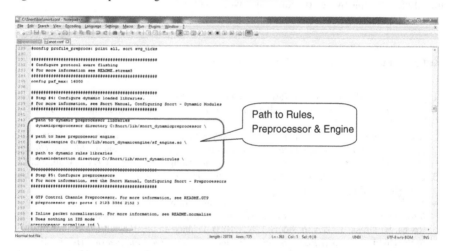

Fig. 7 Shows steps to define the path of rules

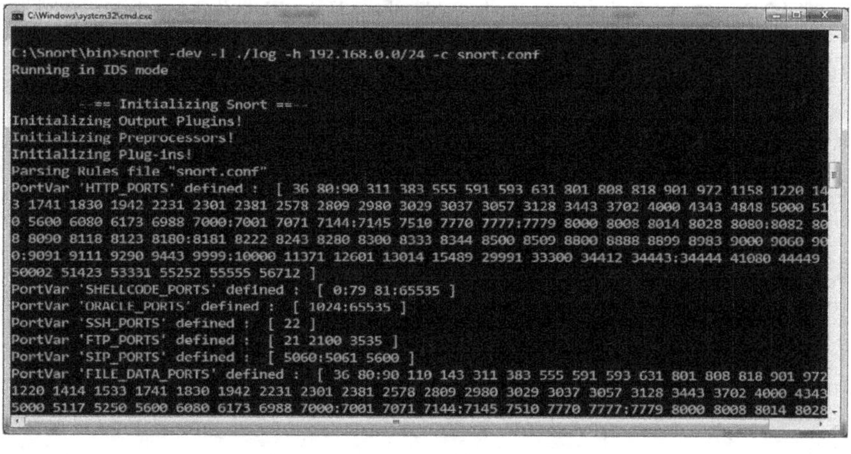

Fig. 8 Shows the output of snort in IDS mode

5 Conclusion and Future Work

In this paper, we introduced a snort-based intrusion detection system to prevent against the malicious packet. Since security in cloud computing is must in order to store and move the data over the internet, snort in IDS is effective and more secure to find the intrusions and proves to be more beneficial to the user or organizations. The snort provides better security to the data packets as the packets are passed completely through the network layer protocols. In future to enhance the security of online activities in snort-based intrusion detection system, an additional feature of generating alarm could be introduced in the system so that after detecting the intrusion the system will ring the alarm and threat could be detected more easily. Next step will be to implementing the new rules in the snort to strengthen the level of security in the cloud and organize the well-known rules into better data structure to achieve better performance. We can also develop a hybrid architecture which introduces the hardware to reduce the workload.

References

1. Dhage, S. N., Meshram, B. B., Rawat, R., Padawe, S., Paingaokar, M., & Misraand, A. (2011). In *International Conference and Workshop on Emerging Trends in Technology (ICWET)*. TCET, Mumbai, India.
2. Rocha, F., & Correia, M. (2011). Lucy in the sky without diamonds: Stealing confidential data in the cloud. In *Proceedings of the First International Workshop on Dependability of Clouds, Data Centers and Virtual Computing Environments*, Hong Kong, ser. DCDV '11, June 2011.
3. Esteves, R., Pais, R., & Rong, C. (2011). *K-means clustering in the cloud—A mahout test* (pp. 514–519), March 2011.
4. Modi, C., Patel, D., Borisaniya, B., Patel, H., Patel, A., & Rajarajan, M. (2012). *A survey of intrusion detection a techniques in cloud.,*. doi:10.1016/j.jnca.2012.05.003.
5. Rocha, F., Gross, T., & van Moorsel, A. (2013). Defense-in-depth against malicious insiders in the cloud. In *2013 IEEE International Conference on Cloud Engineering (IC2E)*. IEEE.
6. *International Journal of Scientific & Technology Research*, *1*(4), May 2012. *International Journal of Information and Computation Technology*, *4*(3), 329–334 (2014). ISSN 0974-2239, © International Research Publications House. http://www.irphouse.com/ijict. html.
7. Takahashi, D. (2010). French hacker who leaked Twitter document to TechCrunch is busted, March 2010. http://venturebeat.com/2010/03/24/french-hacker-who-leaked-twitter-docume nts-to-techcrunch-is-busted/.
8. Danchev, D. (2009). ZDNET: French hacker gains access to twitter's admin panel, April 2009. http://www.zdnet.com/blog/security/french-hacker-gains-access-totwitters-admin-panel/3292.
9. Cole, R. G., Phamdo, N., Rajab, M. A., & Terzis, A. (2005). Requirements on worm mitigation technologies in MANETS. In *Proceedings of the 19th Workshop on Principles of Advanced and Distributed Simulation*, Monterey, CA, June 2005.
10. Sandar, S. V., & Shenai, S. (2012). Economic denial of sustainability (edos) in cloud services using http and xml based ddos attacks. *International Journal of Computer Applications*, *41*(20), 11–16.
11. Labs, Mc Afee. (2013). *McAfee threats report: Second quarter 2013*. McAfee Labs: Technical report.

12. Van Dijk, M., & Juels, A. (2010). On the impossibility of cryptography alone for privacy-preserving cloud computing. In *Proceedings of the 5th USENIX conference on Hot topics in security*, ser. HotSec'10. Berkeley, CA, USA: USENIX Association, 2010 (pp. 1–8). http://dl.acm.org/citation.cfm?id=1924931.1924934.
13. Huang, Y., & Lee, W. (2003). A cooperative intrusion detection system for ad hoc networks. In *Proceedings of the ACM Workshop on Security in Ad Hoc and Sensor Networks*, Fairfax, VA, October 2003.

Rural E-Health Care Model

Rajeev Ranjan Kumar

Abstract Rural e-healthcare model is required to demonstrate the role of information and communication technologies (ICT) for rural development by government of India. The citizens in the villages are deprived from basic health care system. The mortality rate is higher in women and children. The government hospitals are in a very bad shape. Many citizens do not have hospitals around 20 km of the radius. This paper explains design and infrastructure challenges to setup rural health care system. Many states have existing infrastructure and facility called state wide area networks (SWAN). This rural health care model has the aim to use the SWAN facility to reduce the cost and speed of realization. Significant amount of effort is required in designing and developing the ICT solutions through well-managed software engineering process. The project management team will have to use excellent management techniques to ensure sustainability. The goal and idea of this paper is to present available technology, new hardware, software plan, challenges and issues associated with the rural ICT health care model.

Keywords Vital parameter of the body · Village health care system

1 Introduction

Indian rural citizens are suffering from different kinds of health issues. The mortality rates are very high in women and children. There are no government hospitals nearby to the village to get rid of the basic problems. These small diseases are becoming life threatening after certain period of time (Table 1).

ICT has become the backbone to deliver services to citizen conveniently at the door step. The rural health care system using ICT will realize and offer the services of basic first aid, doctor's consultancy and specialized hospitals to the citizens at

R.R. Kumar (✉)
Wipro Limited, Hyderabad, India
e-mail: rajeevranjan.kumar@wipro.com

© Springer Science+Business Media Singapore 2016
S.C. Satapathy et al. (eds.), *Proceedings of International Conference on ICT for Sustainable Development*, Advances in Intelligent Systems and Computing 408, DOI 10.1007/978-981-10-0129-1_49

467

Table 1 Causes of death in children under 5 year 2012

Distribution of causes of deaths in children under-5 2012	
Prematurity	27
Other diseases	16
Acute respiratory infections	14
Birth asphyxia	11
Diarrhoea	11
Neonatal sepsis	8
Congenital anomalies	6
Injuries	4
Measles	2
Malaria	0
HIV/AIDS	0
	0 10 20 30
	Percentage of total

http://www.who.int/gho/countries/ind.pdf?ua=1

their village door steps. The rural health care model will utilize the ICT and digital India Internet of Things (IOT) for offering faster and cost-effective solution. During last 5 years, cost of PC, Internet and electronics equipment has gone down drastically. This is our great opportunity to utilize ICT in low cost to reach remote village location. The proposed model has to use existing telecom infrastructure and SWAN or FTTH (fibre to home), or Wi-Fi for inexpensive connectivity solution.

2 Network Architecture of Rural E-Health Care Model

See Fig. 1.

3 Software Architecture of Rural E-Health Care Model

Software architecture of rural health care system will be based on service-oriented architecture (SOA). Different service provider hospitals will expose their service to the central e-governance server. The client application running on rural health centre will send request to central e-governance server to display list of hospitals available. The client system will choose the available station and send video/audio request to the server. The central e-governance server (service) will connect rural

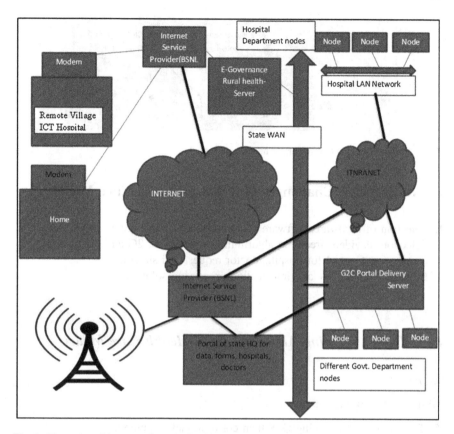

Fig. 1 Network architecture of rural e-health care model

health care centre system to the available hospital network. The video and the audio streams are the major parts of the quality of service. The central e-governance server will keep monitoring quality of service and bandwidth required for video and audio transmissions.

Administrative work will be handled by the G2C server to provide forms, validity of hospitals and doctors. Registration of hospitals and doctors will be taken care by the G2C server. The G2C server will have connectivity to different authorities to provide administer-related government approval and monitoring. The user can access limited service from home by directly connecting to e-governance central server. Mobile and wireless device connectivity will work in similar fashion, and as the wired network 3G and 4G networks are proposed to have excellent video and audio qualities (Fig. 2).

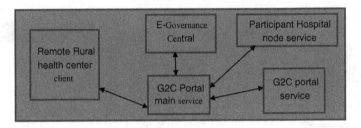

Fig. 2 Software architecture of rural e-health care model

4 Architecture Challenges of Rural E-Health Care Model

1. Selection of hardware, software (preferred open source technology);
2. Selection of video streaming algorithm in broadband, 2G and 3G networks;
3. Congestion Control for hospital/doctor request and streaming;
4. Reliable connectivity and information processing solution.

4.1 Video Streaming Algorithm for E-Health Care Model

See Fig. 3.

Adaptive streaming benefits:

5. Provide better streaming QoS than conventional streaming;
6. Streaming server programme will perform adaptive streaming by changing rate (media and streaming);

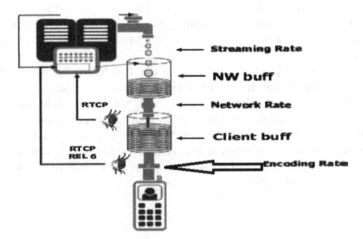

Fig. 3 Video streaming adaptive control flow

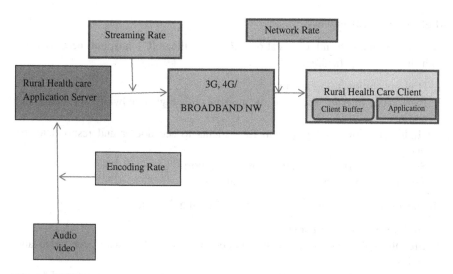

Fig. 4 Audio video streaming model

7. Media rate is same as the rate of media play;
8. There are several approaches to implement adaptive streaming;
9. MPEG-4 helps to provide facility to change number of frame per second by encoding media in several dependent layers and determines OoS by adding and removing layers;
10. H264 SP-Frames: Error-free video streaming can be achieved. Encoding is lossless (Fig. 4).

5 System Architecture of Rural Health Care System

High-level flow of data between sender and receiver:

1. Sender convert video and audio to digital data;
2. Compress the digital data for speedy transfer using compression technique;
3. Send this compressed data using Wi-fi or broadband or ISDN;
4. At receiver decompress video and send it to computer monitor or television;
5. Decompress the audio and send this to speaker.

High-level Requirements:

The remote village health care stations (Remote village ICT hospital) need to setup with the following facilities:

1. High-resolution web camera;
2. Broadband Internet access using modem with high bandwidth for video and audio;
3. High-resolution display system for patients to see doctor and respond to the query;
4. Scanner and printer to send and receive report;
5. Trained person to maintain and operate system.

The hospital department has to have the following facilities:

1. High-resolution web camera;
2. Broadband Internet access using modem with high bandwidth for video and audio;
3. High-resolution display system for patients to see doctor and respond to the query;
4. Scanner and printer to send and receive report;
5. Doctor available on call to support.

The citizens in village will visit rural health care centre for doctor. Trained person managing health care station will adjust the camera and position of the citizen patient. He will connect village system to check the availability of free slot hospital as shown in Table 2.

First, the system will connect to the hospital and check the basic requirement of doctor availability, video and data transmission quality (QOS).

The doctor available in hospital will check vital parameter of the body online with the help of trained professional. Two ways of checking vital parameter of the body can be allowed.

1. Trained professional will check vital parameter;
2. Online automated system to check and display the vital parameters.

Irrespective of the way, the final output of the vital parameter will be displayed on the screen. If this is taken manually, then webcam or scanner will be able to display this information on screen (Table 3).

Table 2 Online hospital selection in rural health care model	Select hospital	Doctor status
	Kamineni Hospital, Hyderabad	Free
	Apollo Hospital, Hyderabad	Busy
	AIIMS, Patna	Busy
	AIIMS, Delhi	Free
	Apollo, DRDO, Hyderabad	Free

Table 3 Online hospital vital parameter display in rural health care model

Risto Disuza Vital parameters	Parameter value
Body temperature	101 F
Blood pressure	101/120

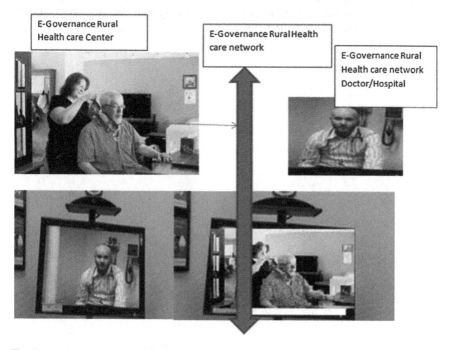

Fig. 5 System architecture of rural e-health care model

Doctor can easily checkup basic problems of fever, cough, and diarrhoea and prescribe medicine. The government has to keep basic medicine at very minimal cost at village (Fig. 5).

6 Conclusion

Indian government has to take advantage of ICT to provide basic health care facility at the door step of every village citizen. Many lives can be saved by providing even first aid at the right time.

The involvement of good hospitals, diverse professionals, embedded medical devices and faster internet service will make rural healthcare systems usable and reliable. New development of IOT (internet of things) will create tremendous capability for connecting heterogeneous devices in cost-effective and flexible way.

In order to achieve this, first requirement is to have participation of well-known software companies to reach this difficult goal.

The key features of rural e-health care system tools include timely availability of doctors to the rural citizen of India. With increase in Internet and IOT, connecting to hospital and diagnostic centre is very much feasible. It empowers village citizen by providing them easy access to best health care system for their lives—LIVE YOUR LIFE TO SAVE THEIR LIFE.

An Implementation Model for Privacy Aware Access Control in Web Services Environment

Rekha Bhatia and Manpreet Singh

Abstract The last decade has witnessed an incremental growth in the number of web service providers as well as web service users who carry out financial transactions online. While this enhanced usage of web to provide financial services has boosted e-business productivity, it has raised significant concerns regarding client's sensitive personal information privacy. In this paper, we have proposed a framework that addresses client's privacy concerns in the context of web services environment. Our approach involves service producers storing their privacy policies in the form of an ontology class and service users storing their privacy preferences in the form of a rule specified in semantic web rule language. Our framework provides automated reasoning techniques for matching the service provider's privacy policies for compliance with the client's privacy preferences. In the event of a policy match, our framework supports automatic generation of the list of service providers who agree to provide service. We demonstrate our approach with the implementation of an example web services scenario.

Keywords Privacy · Access control · Web services · Ontology · SWRL

1 Introduction

The proliferation of web services in every area of our life and their ability to perform an endless number of tasks efficiently, we have allowed them free rein in our lives, with little thought about privacy-related consequences. These days, huge amount of privacy sensitive data about individuals is available in online

R. Bhatia (✉)
Punjabi University Regional Centre, Mohali, India
e-mail: r.bhatia71@gmail.com

M. Singh
Punjabi University, Patiala, India
e-mail: msgujral@yahoo.com

© Springer Science+Business Media Singapore 2016
S.C. Satapathy et al. (eds.), *Proceedings of International Conference on ICT for Sustainable Development*, Advances in Intelligent Systems and Computing 408, DOI 10.1007/978-981-10-0129-1_50

e-government databases, which can be readily accessed by persons having malevolent purposes. The prime cause of this is the open nature of web and web services because of which it is very easy to share and publish information online. This easiness of web availability is giving rise to privacy breaches. The main idea about privacy aware access control is to automate the privacy policy management by the service provider for a better compliance to the needs of the service user so that personally identifiable information (PII) of the service users is accessed not only based on security policies but also on privacy policies [1–4].

In this paper, we have proposed a framework that addresses client's privacy concerns in the context of this emerging web services environment. In order to validate the proposed framework and demonstrate its practical applicability, an implementation of the framework is performed and analysed. In addition, we demonstrate the effectiveness of our framework by varying the number of privacy policies and choices of the number of privacy parameters. Our implemented framework represents the basic elements of privacy policies of service providers and privacy preferences of service users using the ontology language OWL, extended with SWRL for identifying and reasoning about relevant privacy parameters and the corresponding privacy aware access control policies. We have carried out a case study about online good delivery, in order to demonstrate the effectiveness of our framework. To demonstrate the feasibility of our proposed framework, we have conducted a number of experiments on a software-simulated environment. We have quantified the performance overheads of our proposed framework for measuring the inference time, query response time and consistency checking time. The experimental results of the implementation demonstrate the satisfactory performance of our framework and validate the proposed privacy aware access control framework.

The paper is organized as follows: In the next section, related work in ontology-based privacy preserving access control is discussed. In Sect. 3, a description of our proposed ontology-based approach is provided. In Sect. 4, our privacy aware access control model is discussed. In Sect. 5, we have given details of the implementation. In Sect. 6, we have analysed and discussed about the results. Finally, we have concluded the paper with a positive note that by selecting the appropriate web service whose privacy policies are in accordance with the user privacy preferences, the inadvertent and undesired disclosure of user's sensitive personally identifiable information can be avoided.

2 Related Work

The initiation of the Semantic Web and its related technologies, especially the ontologies, has provided access control with new directions. The first language for creating ontologies called Web Ontology Language (OWL) [5] was used to develop policy languages for the Web, such as Rei [6] and to provide interoperability [7–9]. Our work is motivated from the work of [10–12] in which the authors followed a similar approach to match privacy preferences of service users with the privacy policies of service providers using semantic web technologies. The difference lies in which we have incorporated a number of privacy parameters including the incorporation of notion of trust in the policies and preferences. The trust-based approach speeds up the decision making significantly.

3 The Ontology-Based Approach

We have presented an ontology-based privacy aware access control framework to model the relevant access control privacy policies. The prime goal of our framework is to formalize the privacy aware access control concepts using logic-based language. To achieve this goal, we have identified the relevant concepts specified in the next section. In the literature, there are a number of diverse languages developed for specifying machine processable semantics. In the modern age, ontology-based modelling technique has been applied widely and proved as the most suitable logic-based approach for modelling dynamic contexts [10, 11]. The ontology-based modelling approach to achieve privacy awareness [11, 12] is very much beneficial from the reasoning viewpoint, i.e. once certain facts about the universal course of domain have been stated in terms of the ontology, other facts can be inferred using the inference engine through the inference rules.

To model the privacy ontology, we have used the Web Ontology Language (OWL) language to represent the privacy policies of the service providers. The OWL has been the preferred language because of its expressiveness and reasoning ability. In order to support the process of specifying user privacy preferences, we need to define a set of reasoning rules that are associated with the service provider selection by matching user privacy preferences with service provider's privacy policies. In addition, some reasoning rules use mathematical computations, which are not supported by the OWL language. To support such needs, we have used Semantic Web Rule Language (SWRL) rules with Privacy ontology.

4 Privacy Aware Access Control Model for Web Services

The formal notation of the proposed privacy aware access control model in the form of a tuple is

$M = (pii, o, u, ac, p, gl, rp, ec, t)$ where $pii \in PII, o \in O, u \in U, ac \in AC, p \in P, gl \in GL, rp \in RP, ec \in EC, t \in T$.

Here PII is the set of personally identifiable information, O is the set of service providers, u is the set of service requestors, AC is the set of actions allowed on pii, P is the set of purpose for accessing pii, GL is the set of granularity levels for accessing pii, RP is the set of retention period for pii, EC is the set of environmental conditions for accessing pii and T is the set of total trust.

5 Implementation and Evaluation of the Framework

The implementation model is shown in Fig. 1. First of all, we analyse the privacy aware access control framework components and specify constituent elements of privacy ontology. Next, we construct the ontology which contains privacy domain semantics satisfactorily. Then we formulate the rules based on SWRL for specifying user's privacy preferences. Next, we choose an inference engine, and reason our privacy ontology with the SWRL rules. Through the inference process, we can select the service providers whose privacy policies match with the user's privacy preferences. By following this approach, we can preserve user's sensitive PII from disclosure to unwanted service providers.

5.1 Motivating Scenario

We assume a scenario in which a user wants to buy certain goods online. The privacy preferences of the user can be stated as

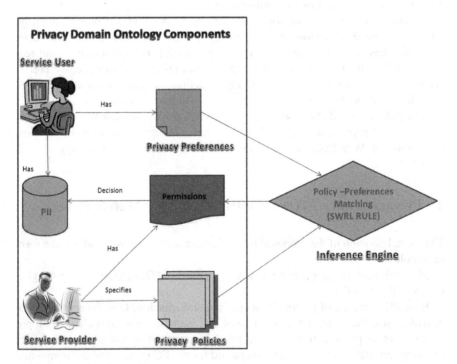

Fig. 1 Privacy domain ontology

Collection of name, address and phone-number is allowed up to high granularity level for delivery purpose. The retention period is – immediate removal of the PII after providing service. The access condition is that goods should be delivered within two days. The trust in the service provider should be greater than 0.8

5.2 Privacy Ontology Description

The Privacy ontology has four main classes: Web_Service_Entity, User_PII, Privacy_Policy and Permission. The Web_Service_Entity class has two sub-classes named Service_Provider and Service_User. The User_PII class contains personally identifiable information of service users like his identification information, contact details, health information, his financial information and other privacy sensitive information. The Privacy_Policy class contains various parameters of privacy policies like granularity, purpose, trust, consent, retention period and access decision. The Permission class specifies various actions which can be performed on user's PII like collection of PII, accessing collected PII, publishing collected PII and sharing collected PII. A number of service providers are registered with the ontology in Service_Provider class along with their corresponding privacy policies in Privacy_Policy class through object property and data property concepts of ontology. An example of a privacy policy of a service provider is

The service provider requires access to the name, address and phone-number of the user in order to deliver goods ordered online. The service provider deletes details of the user PII immediately after providing service. The service provider is able to deliver goods within 5 days. The trust value of the service provider is 0.9.

Next, the privacy preferences of a user's access request are formulated into an SWRL rule and reasoned through an inference engine with the privacy policies of the service providers in order to find out the suitable service provider.

5.3 Experimental Setup for Performance Evaluation

We evaluate the runtime system performance of our developed system, where we adopt our privacy parameters and trust-based decision approach to identify and reason about the relevant privacy preserving policies. With the goal of evaluating the runtime performance of our proposed framework, we have conducted two sets of experiments on a system loaded with Windows 8 operating system running on i3 CPU @ 2.40 GHz with 4 GB of RAM.

The main purpose of this experimental study is to quantify the performance overhead of our proposed approach. Our main measures included number of privacy policies versus query response time and the number of privacy parameters included in privacy policies versus inference time. The first measure indicates how long it took to query the ontology and the second measure indicates how long it took to infer about a user's access permission on a requested service by incorporating privacy parameters (quantity of privacy) into the access control process and making privacy aware access control decisions.

Besides, the other performance measures included in our study are size of the ontology versus inference time, number of privacy policies versus inference time for different reasoning engines and number of policies versus ontology consistency checking time.

6   Results and Analysis

We have examined the performance of the framework and our main finding is that the time for making privacy aware access control decision is acceptable, as it imposes a small, acceptable overhead.

Our first test focuses on measuring the response time of our developed system in the light of increasing number of privacy policies. First, we have selected 10 privacy policies and measured the response time, and then, we varied the number of policies up to 70 and measured the response time. For each of the settings, the average value of the 25 execution runs is used for the analysis as shown in Fig. 2. The test result shows that the average response time of the proposed framework increases when the number of privacy policies increase. The response time varies between 203 and 291 ms for the variation from 10 to 70 privacy policies. We can see from the figure that the average response time seems to be linear. Overall, this type of performance is acceptable.

In the second test performed, we have again evaluated the total response time over different numbers of privacy parameters contained in the privacy policies.

We have varied the number of policies starting from 10 up to 70 with respect to two parameters of privacy (Access condition and Retention period), and then, with

Fig. 2 Number of privacy policies versus average query response time

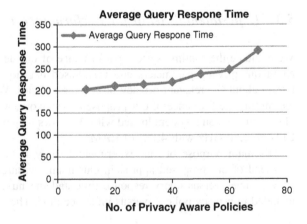

respect to four parameters of privacy (Access condition, Retention period, Granularity and Purpose) and finally with respect to the notion of trust along with all the four parameters of privacy. In order to measure the response time, we have run each experiment 25 times and the average value of the 25 execution runs is used for the analysis. The results of this test are shown in Fig. 3.

We can see that changes to the number of privacy parameters in access control policies increase the response time but this is natural as increase in complexity of the policy will incur overhead on response time and moreover, the increasing trend is linear and thus acceptable. One exceptional result is the introduction of trust as a parameter of privacy disclosure control, which reduces the response time dramatically. Overall, the runtime performance of our privacy aware access control framework is acceptable. The number of privacy policies versus average inference time is also shown in a separate graph in Fig. 4, in which case we have tried to incorporate policies with varied amount of privacy.

In the next test, we have again evaluated the total inference time over various sizes of the knowledge repository. We have varied the number of policies up to 100 to build the ontology of increasing sizes. In order to measure the response time, we have run each experiment 25 times and the average value of the 25 execution runs is used for the analysis as shown in Fig. 5. We can see that, overall, the runtime performance is acceptable for a reasonable-sized ontology.

Next, we have evaluated the performance of our framework with two different reasoning engines. The simplest privacy policies were used for this test. We can see

Fig. 3 Number of privacy policies versus inference time for different privacy parameters

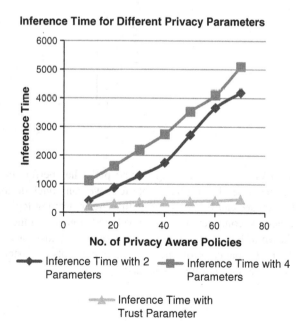

Fig. 4 Number of privacy
policies versus average
inference time

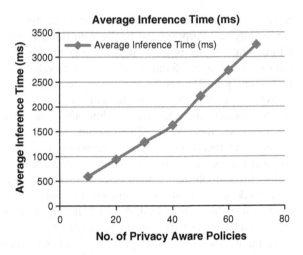

Fig. 5 Size of ontology
versus average inference time

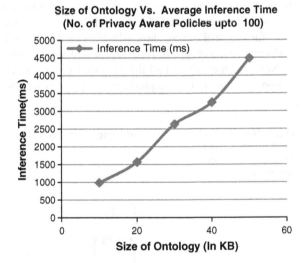

from Fig. 6 that reasoning engine pellet has performed better than the reasoning
engine JESS. From this simple test, we can conclude that the performance of our
framework can be improved with better reasoning tools (Fig. 6).The next test is to
check the consistency of our developed concepts in the ontology. The results are as
shown in Fig. 7. The trend in this graph is almost linear. This test has no effect on
the performance of our framework as concepts consistency is only checked for the
first time, the ontology is loaded.

Fig. 6 Number of privacy policies versus average inference time for different reasoners

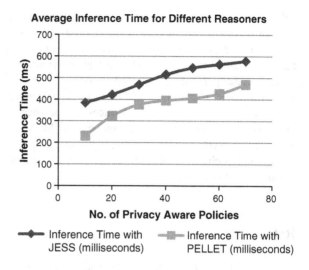

Fig. 7 Number of privacy policies versus consistency checking time

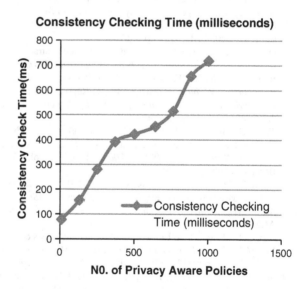

7 Conclusions and Future Scope

As online financial web services and net-based transactions are becoming more and more prevalent among web service users and as users are becoming more aware about web services dealing with their confidential PII, every user wants that his privacy preferences must be adhered to, by the service providers. By selecting the appropriate web service whose privacy policies are in accordance with the user privacy preferences, the inadvertent and undesired disclosure of user's sensitive personally identifiable information can be avoided. In this paper, by reasoning the

SWRL rule-based privacy preferences of the requesting user with the privacy domain ontology, we can select the desired web service provider from a host of service providers and protect the user privacy efficiently as well as effectively. All the experimental results have shown that our framework has satisfactory response as far as the performance is concerned and for the better performance, more powerful machine can be used. Future work will investigate the use of controlled natural language (CNL)-based interfaces for specification of privacy preferences by the users in order to lessen the burden on user. Although the research in this study is grounded in web services paradigm, it can be applied easily to other scenarios having similar characteristics like ubiquitous computing, online social networks and cloud based environments.

References

1. Ardagna, C. A., Cremonini, M., De Capitani Di Vimercati, S., & Samarati, P. (2008). A privacy-aware access control system. *Journal of Computer Security, 16*, 369–397.
2. Casassa Mont, M., Thyne, R., Chan, K., & Bramhall, P. (2005). http://www.hpl.hp.com/techreports/2005/HPL-005-110.pdf.
3. Oberholzer, H., & Olivier, M. S. (2005). Privacy contracts as an extension of privacy policies. In *International Conference on Data Engineering Workshops (ICDEW'05)*, 0:1192, 2005.
4. Byun, J. W., Bertino, E., & Li, N. (2004). *Purpose based access control for privacy protection in relational database systems*. Technical Report 2004-52, Purdue University, 2004.
5. *The World Wide Web Consortium (W3C): OWL Web Ontology Language Overview*, February 2004, W3C Recommendation.
6. Tonti, G., Bradshaw, J., Jeffers, R., Montanari, R., Suri, N., & Uszok, A. (2003). Semantic web languages for policy representation and reasoning: A comparison of kaos, rei, and ponder. In *The SemanticWeb—ISWC 2003*, LNCS (Vol. 2870, pp. 419–437). Springer Berlin/Heidelberg.
7. Mitra, P., Pan, C. C., Liu, P., & Atluri, V. (2006). Privacy-preserving semantic interoperation and access control of heterogeneous databases. In *ASIACCS '06: Proceedings of the 2006 ACM Symposium on Information, Computer and Communications Security* (pp. 66–77). ACM.
8. Pan, C. C., Mitra, P., & Liu, P. (2006). Semantic access control for information interoperation. In *SACMAT '06: Proceedings of the 11th ACM Symposium on Access Control Models and Technologies* (pp. 237–246). ACM, New York, NY, USA.
9. Sun, Y., Pan, P., Leung, H. F., & Shi, B. (2007). Ontology based hybrid access control for automatic interoperation. In B. Xiao, L. Yang, J. Ma, C. Muller-Schloer, & Y. Hua, *Autonomic and Trusted Computing*, LNCS (Vol. 4610, pp. 323–332). Springer Berlin/Heidelberg.
10. Garcia, D., Toledo, M. B. F., Capretz, M., & Allison, D. (2009). Towards a base ontology for privacy protection in service-oriented architecture. In *2009 IEEE International Conference on Service-Oriented Computing and Applications (SOCA)* (pp. 1–8).
11. Ge, Qiang, et al. (2014). The application of SWRL based ontology inference for privacy protection. *Journal of Software, 9*(5), 1217–1222.
12. Kayes, A. S. M., Han, J., & Colman, A. (2014, January). PO-SAAC: A purpose-oriented situation-aware access control framework for software services. In *Advanced Information Systems Engineering* (pp. 58–74). Springer International Publishing.

Comprehensive Review on Eye Close and Eye Open Activities Using EEG in Brain–Computer Interface

Annushree Bablani, Sachin Kumar Agrawal and Prakriti Trivedi

Abstract Brain–computer interface (BCI) aims at providing a communication channel without using any muscular activity. It uses brain signals which are recorded by electroencephalographic activities. These recorded signals are processed using different methods. During recording many disturbances due to muscles movement, noise and eye movements that are called artefacts are generated. In this paper, we have done a comprehensive review of how to detect the artefact generated due to eye movements—eye open and eye closure activities.

Keywords EEG · Brain–computer interface · Artefacts · Feature extraction · Brain rhythmic activities

1 Introduction

In era of technology various ways for communication are available. Communication is possible through various hardware devices available in market. To develop an interacting environment between user and a hardware device, we need an interface. Similarly, human–computer interaction provide a way through which user can interact with computer. Brain–computer interaction is a way in which user and computer interact via brain signals. BCI record intensity of the thoughts in one's mind, analyse them and then specific function is performed. As BCI provides communication channel through thoughts, it proves to be boon for

A. Bablani (✉) · S.K. Agrawal · P. Trivedi
Department of Computer Science and Engineering,
Government Engineering College, Ajmer, Rajasthan
e-mail: annushree.bablani@gmail.com

S.K. Agrawal
e-mail: sachinkumar.aggarwal@gmail.com

P. Trivedi
e-mail: prakrititrivedi@rediffmail.com

© Springer Science+Business Media Singapore 2016
S.C. Satapathy et al. (eds.), *Proceedings of International Conference on ICT for Sustainable Development*, Advances in Intelligent Systems and Computing 408, DOI 10.1007/978-981-10-0129-1_51

485

people having muscular disorder. There are many ways to capture mental condition of user; some of them are invasive and some are non-invasive. Electroencephalography is a non-invasive technique to record one's mental state. During recording many types of brain waves are generated which are known as brain rhythms. Due to eye movements there are disturbances in these brain rhythmic activities. In this paper, we have summarized various ways that are helpful in identifying the eye open and eye close activities.

2   Classifications of BCI

BCI can be classified in any of the following three ways:

2.1   Dependent and Independent BCI

A dependent BCI does not use the brain's normal output pathway to carry the message, but activity in these pathways is needed to generate the brain activity that does carry it. EEG machine uses dependent BCI. For example, when a user move hands, the brain generates the muscular signal and via scalp these signals are recorded. Independent BCI does not depend on muscular activity. The user only thinks about the muscular action that user wants to perform and machine performs that action.

2.2   Synchronous and Asynchronous BCI

"In a synchronous BCI, the analysis and classification of brain potentials is limited to predefined fixed or variable time windows" [1]. Synchronous BCI has predefined methods and follows a queue. Every feature extraction process was performed sequentially. After the first set of feature extraction is processed, then only system allows next set of feature. User thinks and his thoughts are processed sequentially and then machine works. Asynchronous BCI does not occur in a queue. It is a parallel process, that is, one feature extraction method is performed at that time and two or more than two feature extraction methods can be performed or applied.

2.3   Invasive and Non-invasive BCI

Brain–computer interface is the detection of electrical signal from scalp of a certain user or subject. In non-invasive, BCI signals are recorded without surgery and causes no physical harm to subject. Electroencephalography (EEG) is a way of recording electrical activity from the scalp using electrodes. It is a proven method for clinical and research purposes. EEG tools are comparatively cheap, lightweight and easy to use. However, there is need of more accuracy for signal processing and feature extraction technology for eliminating noise or artefacts. Invasive BCI signal recorded with surgery can break the brain skin but it does not damage any neurons. This surgery includes opening the skull through a surgical procedure called as craniotomy and cutting the membranes that cover the brain. Invasive recording techniques combine excellent signal quality, very good spatial resolution and a higher frequency range.

3 Working

To detect the activity of eye open and eye close, we have to first setup a working environment which includes an EEG device and a computer installed with required software. The EEG device is placed on the user brain. This device has electrodes which are used to capture the brain signals. These electrodes are placed according to international 10–20 system. The electrodes are placed as shown in Fig. 1.

They are placed on the brain lobes that are frontal (F), parietal (P), central (C), occipital (O) and temporal (T). They are labelled with numbers from 1 to 8, even number electrode, i.e. 2, 4, 6 and 8 represented electrodes placed in the right hemisphere, while odd number electrodes 1, 3, 5 and 7 represented electrodes in the left hemisphere of the brain. Many types of EEG recording devices are available like EPOC headset, neuro-scan EEG headset, etc. after correctly placing the EEG

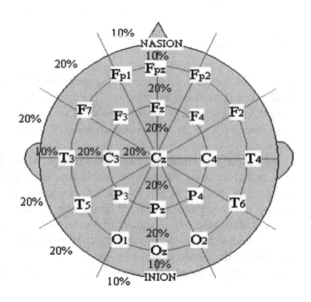

Fig. 1 10-20 electrode placement system [2]

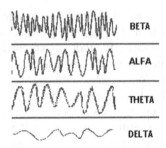

Fig. 2 Brain wave samples with dominant frequencies belonging to beta, alpha, theta and delta band [2]

electrodes according to the 10–20 system, signals generated from brain are recorded, and recording can be done on a single user or multiple. To get precise data we can record data of many users so that error rate can be minimized.

The signals recorded from brain are in the form of alpha waves (8–13 Hz), beta waves (>13 Hz), delta waves (<3.5 Hz) and theta waves (3.5–8 Hz) shown in Fig. 2. These waves have their different characteristics. They show mental state of the person whose brain has been mapped. If beta wave is produced that means person is in excitement state, alpha wave shows normal thoughts, and delta on other hand depicts deep sleep of the object.

Because of environmental noise, muscle movements and eye movements, there is a change in frequency of these waves. Various methods have been proposed in various papers that extract these features from the brain signals. Some of the methods to extract eye open and close movements have been discussed in this review paper.

3.1 Methods

Researchers for their work have either recorded EEG data by hardware or downloaded dataset available on internet. To extract features from the EEG signals a lot of methods have been proposed. Similarly, for extracting feature of eyes open and close detections we are discussing on some of the proposed methods here in our paper:

(a) **Linear Discriminant Function through Alpha Wave**: Detections of eyes open and close using linear discriminant function through alpha wave have been discussed by authors in [3]. In this, authors have used a method for online processing using segments of fixed time interval and showed their overlapping. Figure 3 shows the five structures of the segments during online processing.

This method has used the training step because it is an online processing. Author has divided the database into two datasets: first is training and second is testing. Training set is formed by a second record of each subject, while testing set is

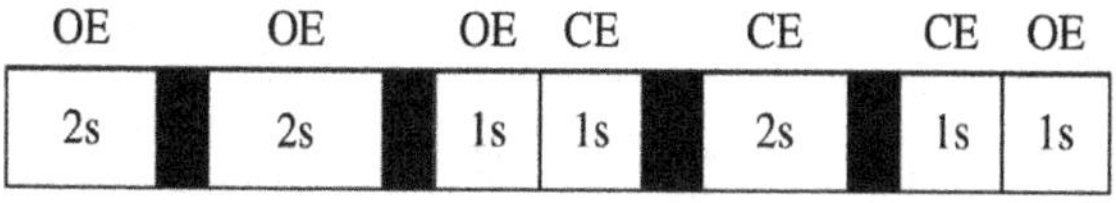

OE		OE		OE	CE		CE		CE	OE
2s		2s		1s	1s		2s		1s	1s

Fig. 3 Five structures of the segments taken during 2 s period overlapping of 1 s period: Open eyes (OE) and close eyes (CE)

formed by four records of each subject. Testing set evaluates performance of the linear discriminant function when classifying eyes events on the EEG records.

(b) KM$_2$O Langevin equation: The KM$_2$O Langevin equation is used for the detection of eyes close activity; only it does not detect activity of eyes open [4]. According to author the EEG signals are weakly stationary during process of insomnia. A signal $X(n)$ can be weakly stationary process if $E[X(n)] = 0$ and $E[X(n) X(m)] = C(n - m))$ ($|n||m| \leq N$) are valid. $C()$ means the covariance matrix of $X(n)$ [4]. It was presumed by author that the EEG's weakly stationary region is broken progressively by the change in the alpha wave which is caused by eye close (EC) because of the amplitude variance of alpha wave deviation. If this irregularity in the weakly stationary region can be found, the alteration in the alpha wave caused by EC might be perceived. Thus, focus was given on the method to find irregularities in the weakly stationary region. The steps to detect EC used by authors are as follows:

i. In the original time series $X(n)$, a region whose length is tows from time k is extracted.
ii. In the region, KM$_2$O Langevin force $v_\pm$ is calculated.
iii. In the $v_\pm$ region, the variance and the Box–Pierce (BP) of $v_\pm$ are calculated as indexes to detect.
iv. Irregularity in the weakly stationary region is detected by determining the threshold of the variance. If the value of the index is greater than the threshold, current time k is regarded as the point where the alpha wave activity changes.

These steps are performed for every sample.

(c) Power Spectrum Density (PSD): In [5] authors have downloaded EEG data in European format and considered eye open and eye close tasks of the subjects. For results they performed experiment on EEGLAB considering eye activities for first 10 s. For computing PSD MATLAB functions, EEGLAB can be used, so authors have used PSD. PSD is the method of signal processing which tells about how the strength of the signal is distributed along frequencies. So it helps in finding out the signals of good strength and weak strength. In [5] authors have calculated that PSD of subjects during eye open and eye close to find dominance of alpha rhythm is more during subjects who are awaken with open eyes or with close eyes. They found out that subjects who are awaken but having eyes close have more dominating alpha waves than the subjects awaken with open eyes. Their result for alpha wave on frontal lobe is shown in Fig. 4.

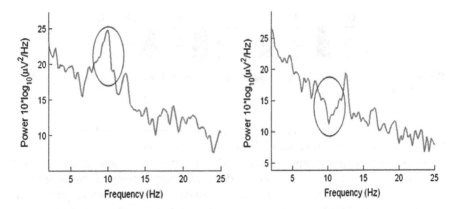

Fig. 4 PSD of Fz with closed eyes in *left panel* and with open eyes in *right panel*

(d) **Statistical method using R**: For recording of data, authors in [6] used mind wave headset and recorded data for one male subject and one female subject. In their study they have used traditional statistical methods using R (it is a statistical programming language and tool). To find the most striking features they have used ANOVA function, and found delta waves as most significant followed by alpha waves. The authors have collected data and then normalized and smoothen the data to get clean data. They found that most of the data points for delta waves were collected near 0 point, so they applied LOESS function to clean the data. After applying LOESS, data was divided into four vertical sections as eye open–eye close; eye open–eye close (4 min each). The analysis of data was done using moving average and double moving average. Figure 5 shows delta wave data points for subject

4 Conclusion

In our paper we have completed a comprehensive review of various methods used for feature extraction with emphasis on eye open and eye closure. In linear discriminant analysis eye movement has been detected using segments of 2 s and this method has not used any threshold value to determine difference between two classes. The performance of the work can be improved if more segments of eye close and eye open are used. The KM_2O Langevin equation has detected only for eye closure to find irregularities in weakly stationary area. PSD has detected dominance of alpha wave by eye movement activity. PSD can be used on occipital lobe as eye movement is easily detected on occipital rather than frontal or parietal which has been used by author. Traditional statistical method using R has been used and delta wave significance has been shown. As delta wave is the wave produced

Fig. 5 a, b Delta wave data point of a subject. **a** Delta waves data points accumulated near 0. **b** Smoothed delta wave points

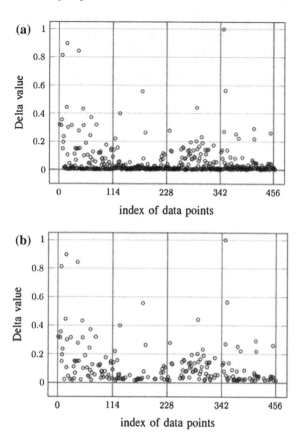

during deep sleep, this method can be applied on alpha and beta waves to get more efficient result. Also, only two subject's data are not sufficient: multi-subject that will result in lower error rate.

References

1. Townsend, G., Graimann, B., & Pfurtscheller, G. (2004). Continuous EEG classification during motor imagery-simulation of an asynchronous BCI. *IEEE Transactions on Neural Systems and Rehabilitation Engineering, 12*(2), 258–265.
2. Teplan, M. *Fundamentals of EEG Measurement.* Institute of Measurement Science, Slovak Academy of Sciences, Dúbravská cesta 9, 841 04 Bratislava, Slovakia.
3. Sakai, M., Wei, D., Kong, W., Dai, G., & Hua, H. (2010). Detection of change in alpha wave following eye closure based on KM$_2$O-Langevin equation. *International Journal of Bioelectromagnetism, 12*(2), 89–93.
4. Delisle-Rodriguez, D., Castillo-Garcia, J. F., Bastos-Filho, T., Frizera-Neto, A., & Lopez-Delis, A. *Using linear discriminant function to detect eyes closing activities through alpha wave.*

5. Valipour, S., Shaligram, A. D., & Kulkarni, G. R. *Detection of an alpha rhythm of EEG signal based on EEGLAB*. Department of Electronic Science, Pune University, Pune, 411 007, Maharashtra, India Department of Physics, Pune University, Pune, 411 007, Maharashtra, India.
6. Oner, M., & Hu, G. Analyzing one-channel EEG signals for detection of close and open eyes activities. *2013 Second IIAI International Conference on Advanced Applied Informatics.*

Design of High Data Rate and Multipath Efficient Underwater Acoustic Communication System Using OFDM–DQPSK

Kosha Mistry and Hardik Modi

Abstract Presently, underwater acoustic communication is an attractive option for underwater communication because of its low attenuation in underwater compared with electromagnetic waves and optical waves. Underwater communication channel is extremely complex in nature. By using multi-carrier communication, in this complex underwater acoustic communication channel, we perceived high data rate. An attractive multiplexing technique denoted as OFDM is one of the most promising multi-carrier communication methods for underwater environment. In this paper, basic knowledge about underwater acoustic communication and OFDM is given. MATLAB is used as functional software for implementing and defining differential quadrature phase shift keying along with orthogonal frequency division multiplexing.

Keywords Underwater acoustic communication · OFDM · QPSK · Multi-carrier communication · MATLAB

1 Introduction

In present, some of the applications of underwater acoustic communication are oceanography, for scanning weather, climate conditions and pollution, and shipping. In underwater environment natural resources are investigated by unmanned underwater vehicles (UUVs) and autonomous underwater vehicles (AUVs). Because of very slow in speed, acoustic wave underwater, and large number of multipath, there is long impulse response and large time delays. Frequency-dependent absorption of acoustic in the water band limits the underwater channel. This bandwidth limitation also limits the maximum achievable data rate

K. Mistry (✉) · H. Modi
Charotar University of Science & Technology, Changa 388421, Gujarat, India
e-mail: koshamistry91@yahoo.com

H. Modi
e-mail: modi8584@yahoo.com

© Springer Science+Business Media Singapore 2016
S.C. Satapathy et al. (eds.), *Proceedings of International Conference on ICT for Sustainable Development*, Advances in Intelligent Systems and Computing 408, DOI 10.1007/978-981-10-0129-1_52

[1]. Underwater acoustic channel has large bit error rate. In underwater acoustic communication, OFDM has been actively prolonged with multi-carrier modulation due to its special potential to get a handle on high data rate communication over long distributive channels [2].

The paper is divided into three parts. In the first and second parts, underwater acoustic communication and OFDM have been discussed, respectively. In the third part, a system for underwater acoustic communication has been defined. From many digital modulation techniques we are going to use orthogonal frequency division multiplexing (OFDM) because of its ability to handle multipath, high utilization of the frequency band, and low inter-symbol interference (ISI). Differential quadrature phase shift keying (DQPSK) will be used as the coding technique for the sake of reducing the system's complexity as well as to improve the data rate. The system will be implemented in MATLAB and simulations will be carried out in order to characterize the OFDM system. Here, the development of such system is that it is capable to work properly in the challenging conditions of an underwater acoustic channel.

2 Underwater Acoustic Communication

Skillfulness of underwater communication is transmitting and receiving below water. The study of sound is acoustic, divided into three parts which are transmitting, communication, and receiving of sound signals [3]. In underwater conditions, such type of communication is adaptable and commonly used proficiency. Acoustic communication has been used effectively for point-to-point communication in deep-water channels. Underwater acoustic is also known as hydroacoustics.

The sound is limited 1500 m/s speed. The speed of sound underwater varies with depth. Sound in underwater undergos very little attenuation in thermally stable deep water. In shallow water acoustic waves are contrived by various phenomenons like Doppler effect, noise, and multipath propagation. Typical frequencies associated with underwater acoustics are between 10 Hz and 1 MHz. Acoustic waves are longitudinal waves and do not have any polarization so that they oscillate along the same direction as the direction of wave.

Underwater acoustic communication is difficult because of multipath propagation [3], propagation delay of channel [4], small available bandwidth [5], large propagation delay [6], high energy consumption [7], path loss, and noise. Over long ranges signal attenuation will be high.

Classifications of underwater acoustic communication system are given as, first is very long range having limit up to 1000 km and bandwidth of less than 1 kHz; likewise second is long range having limit from 10 to 1000 km and bandwidth from 2 to 5 kHz. It also has medium ranging from 1 to 10 km with bandwidth nearly equal to 10 kHz. Short range from 0.1 to 1 km and very short range less than 0.1 km are also ranges of the system with associated bandwidth of 20 to 50 kHz and greater than 100 kHz, respectively [8, 9].

3 Orthogonal Frequency Division Multiplexing (OFDM)

Multiplexing technique is subdivided into two kinds. One of them is frequency division multiplexing where frequencies are divided into corresponding channels. Orthogonal frequency division multiplexing is a category of FDM in which each channel exhibits more than sub-carriers on contiguous frequencies. It is one type of multi-carrier modulation. In an OFDM system sub-carriers are orthogonal. Orthogonality is achieved by placing each of the subcarriers at a multiple of $1/T$, where T refers to the duration of the OFDM symbol [10, 11].

Orthogonality of subcarrier is defined as [12]

$$\int_0^T \cos(2pf_n t) \times \cos(2pf_m t) = d(n - m). \tag{1}$$

Because of orthogonality of subcarrier OFDM does not require guard bands between subcarrier in the frequency domain, which results in savings in useful spectrum and spectral efficiency. In an OFDM system, if the signal passes through multipath channel the orthogonality of the subcarrier is lost, which results in inter-carrier interference (ICI). To overcome these problems, guard intervals (guard band) have to be introduced between each individual channel in the time domain. The guard intervals can be of three different types, namely cyclic prefix, cyclic suffix, and zero pad. In Fig. 1 we demonstrated an OFDM system's frequency domain. Each subcarrier is centered at various frequencies $f_1, f_2, f_3, f_4, \ldots \ldots, f_{N-1}$.

Subcarriers which are overlapped and orthogonal in an OFDM system are shown in Fig. 2. In this synopsis, the channel bandwidth attains 1/Rs. Therefore, when the number of subcarriers becomes infinite then the spectral efficiency of an OFDM system becomes nearly double.

There are three main needs of OFDM system that are delay spread, bandwidth, and bit rate. There is a settlement between these three needs for choosing OFDM parameters. Delay spread means the length of the channel impulse response. The guard time is dictated directly by the delay spread. Two to four times the root mean

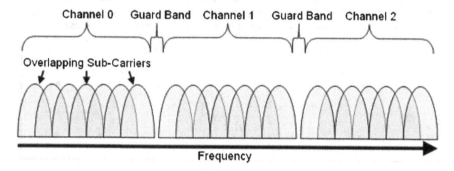

Fig. 1 OFDM system-frequency domain

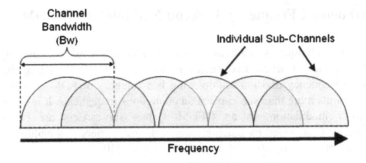

Fig. 2 Channel bandwidth of OFDM

squared of the delay spread should be closely equal to guard time [13]. The number of subcarriers is found. The subcarrier spacing divides bandwidth (BW), which is invert of the symbol duration. Number of subcarrier can also be calculated by the desired bit rate disunited by the bit rate per subcarrier. Symbol rate, modulation type, and coding rate define bit rate per subcarrier [14].

OFDM systems are enforced using a mixture of inverse fast Fourier transform and fast Fourier transform. In the OFDM transmitter, IFFT transform is used to calibrate an input signal overlapped and orthogonal set of subcarriers. At the receiver the transform is used again to process the received subcarriers. By combining the signals from the subcarriers one can estimate the source signal from the transmitter (Fig. 3).

The OFDM transmitter handles the source symbols (e.g. QPSK) in the frequency domain. An IFFT block uses these symbols as input and transforms the signal into the time domain. N symbols are acquired by the IFFT, where the number of subcarriers is denoted by N, and maps each symbol a sinusoidal basis function. The output of the IFFT block is an OFDM symbol which is the summation of the

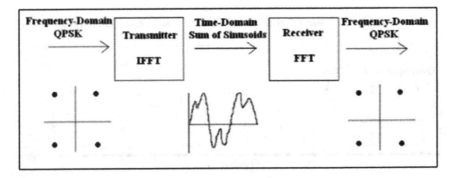

Fig. 3 OFDM system using IFFT and FFT

N sinusoid. NT is the linear measure of the OFDM symbol [15], where the period of the input symbol is T. At the receiver the signal is transformed into frequency domain again by an FFT block and the output is the symbols that were forwarded to the IFFT at the transmitter [16].

In OFDM, inter-symbol and inter-channel interferences are created by multipath propagation. Inter-symbol interference is removed by guard interval.

High data rate transmission over long dispersive channel and large system capacity are achieved by uncommon ability of OFDM system. OFDM has many advantages such as it possesses high data rate of transmission including maximum utilization of frequency band with resisting interfere of narrow band. Due to multipath inter-symbol interference is created. OFDM can deal with ISI without channel equalization. It can also avoid complex channel equalization and it reduces the sensitivity to different types of impulse noise [12]. Sensitivity is critical issue in OFDM especially when frequency shift occurs or phase noise is introduced; in addition to this, another disadvantage of OFDM is that the ratio of peak value to average power is high. It requires the receiver to be very accurately synchronized to the transmitter. OFDM is appropriate for channels which are selected on the basis of frequency and where high data rates are required. Simply, OFDM transforms wideband channel that is frequency selective into small group of narrow-band channels that are non-selective. This property makes OFDM robust by preserving orthogonality against large delay spreads especially in frequency domain [17].

In underwater acoustic communication, usually OFDM is utilized for efficient and reliable transmission of information signals. It is recently in demand for choice of forth coming wireless applications.

4 Underwater Acoustic Communication Using OFDM and QPSK

The use of electromagnetic waves for the underwater communication is very limited. As the sea water is conductive the electromagnetic waves are absorbed by the sea water. So the electromagnetic waves are used for short-distance communication. These acoustic waves possess low propagation velocity that is 1500 m/s compared to that of electromagnetic waves (3×10^8 m/s). The slow propagation velocity cases long channel impulse responses. In underwater communication, acoustic waves create large number of multipath resulting in inter-symbol interference. To avoid the problems of underwater acoustic communication system choice of modulation scheme is OFDM.

Among many modulation schemes available OFDM uses multi-carrier transmission in which frequency spectrum is subdivided into several carriers. Every carrier is modulated with a low rate data stream. Phase shift keying was chosen as the coding technique. This coding technique is simple and produces a constant amplitude signal. By choosing quadrature phase shift keying one can achieve a higher data rate by sending two bits on every subcarrier, but with a slightly higher bit error rate (BER).

In order to keep the complexity of the receiver to a minimum we decided to use differential QPSK. Because of the use of differential QPSK, one avoids complex carrier tracking at the receiver because the data lies in the phase difference and not in the received symbol itself. Differential QPSK can also solve a couple of problems in underwater communications. Doppler spread means a shift of frequency because of affiliated motion of the transmitter as well as of receiver. If the frequency shift is varying slowly, i.e. slow enough for the frequency to be considered constant over two symbol periods, the phase difference should remain unaffected and the data can be retrieved.

4.1 Transmitter

In Fig. 4, a block diagram of the OFDM underwater acoustic communication transmitter is illustrated. The bit stream is coded by a QPSK modulator that means groups of two bits are given a predefined QPSK symbol. QPSK sends two bits on every subcarrier.

In order to prepare the data for differential coding and OFDM modulation the data is transformed from serial to parallel. This transformation is necessary as the input of the OFDM block requires being parallel in contemplation of mapping the symbols the subcarriers. The differential encoder comes after the serial-to-parallel conversion. This is because the frequency shift that differential coding should prevent affects each subcarrier differently, and it will only work if the phase difference is found from two succeeding symbols from the same subcarrier. The differential encoding is done by mapping the QPSK symbols as a complex phase.

In the differential coding block a reference signal is also created. It will act as a starting value for the differential coding and will be the first symbol to be sent. The reference signal will be known at the receiver and will be used in the synchronization at the receiver.

After the differential encoding the OFDM operation is implemented. The OFDM block consists of three operations, zero padding, an inverse Fourier transform, and the insertion of guard interval. Zero padding depends on the sampling frequency f_s. The zero padding is done to achieve over-sampling and to center the spectrum of the data. This is done by simply inserting zeros at the middle of the parallel data.

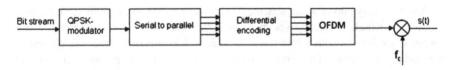

Fig. 4 Block diagram of OFDM communication system transmitter

The inverse Fourier transform, which in reality is an IFFT, takes the zero-padded data as input. IFFT actually maps the data of the subcarriers and creates a time-domain symbol which is the sum of all the subcarriers. This operation also converts the data from parallel to serial again. Then the third OFDM operation preformed is the guard interval that is inserted. A number of zeros are inserted in front of each OFDM symbol as a guard interval. In order to transmit the signal the spectrum of the signal is upshifted from baseband to the predefined carrier frequency, f_c. The shift in frequency is done simply by multiplying the carrier signal by the exponential function, $e^{(i \times 2\pi \times f_c \times t)}$. The OFDM symbols and their guard intervals are then added together creating the total signal $s(t)$, which now are ready to be transmitted on the communication channel.

4.2 Receiver

A block diagram of the OFDM underwater acoustic communication receiver is shown in Fig. 5. The received signal is first downshifted from the carrier frequency f_c, to baseband. In order for the OFDM block to use the correct window for the Fourier transform, the signal has to be synchronized. The synchronized block is essentially a correlation function, which correlates the received signal with the known reference symbol. By locating the maximum of the correlation function one is able to calculate the right window for the Fourier transform and the start of the received signal can be calculated.

The OFDM block at the receiver is also made up of three operations. First the guard interval is discarded, and then the received signal is converted back into the frequency domain by an FFT. The signal is downshifted and transformed into the frequency domain. The down-shifting from the carrier frequency f_c to baseband is done by a multiplication with an exponential function, the same way as the up-shifting at the transmitter, but now with a minus sign in the exponential function, $e^{(-i \times 2\pi \times f_c \times t)}$. The zeros which were inserted at the middle of the data at the transmitter are then removed. Since, now we have the data in the phase difference of the received signal, the next step is to decode the differential coding. The task of the differential decoder is to find and decide this phase difference between two succeeding symbols from the same subcarrier and the output is QPSK symbols. The phase difference is found by multiplying a symbol at a given subcarrier by the

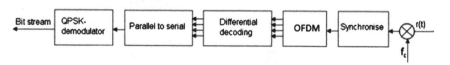

Fig. 5 Block diagram of OFDM communication system receiver

complex conjugate of the previous symbol at the same subcarrier. This is shown in equation below:

$$e^{i\theta_2} \times e^{i\theta_1} = e^{(\theta_2 - \theta_1)} \tag{2}$$

When the value of the phase difference has been decided, it is decoded into QPSK symbols according to Eq. 2. In ideal situations the phase differences found in the differential decoder should be equal to the values shown to the right in Eq. 2. The parallel data consisting of QPSK symbols are first converted into serial data. After conversion it is demodulated. This is done by the QPSK demodulator which demodulates a QPSK symbol into unique digital bits. The result of this is a stream of bits out of the receiver which depending on the communication channel should be equal to the bit stream into the transmitter.

5 Result Analysis and Conclusion

We developed a system that is able to function properly in the challenging conditions of underwater acoustic channel. Using OFDM we achieved efficient and reliable transmission of information signals over UWA channel. Simulation results show that the system explained in this paper works successfully in formidable conditions which are noise, multipath, phase error, or combination of these. This OFDM system is relatively sensitive to synchronization error at the receiver. So it is very promising to use OFDM system for underwater acoustic communication because of its ability to handle multipath, Doppler shifts, inter-symbol interference, and still it is able to maintain high data rate (Figs. 6, 7, 8 and 9).

Fig. 6 The wave file and the bit stream data in

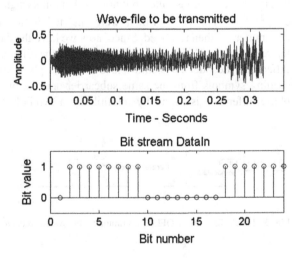

Fig. 7 The signal which is going to be transmitted through the channel

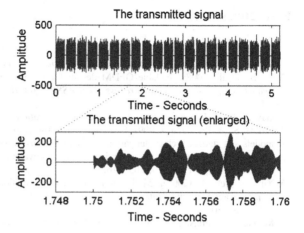

Fig. 8 The first 24 received bits at the receiver and the resulting wave file

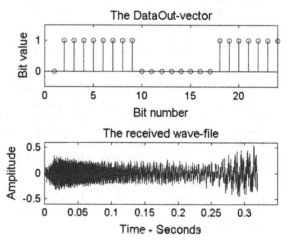

Fig. 9 The received QPSK symbols at the receiver

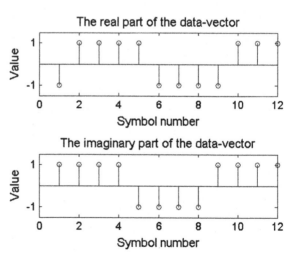

References

1. Proakis, J. G., Sozer, E. M., Rice, J. A., & Stojanovic, M. (2001, November). Shallow water acoustic networks. *IEEE Communications Magazine*, 114–119.
2. Li, B., et al. (2008). Scalable OFDM design for underwater acoustic communications. In *Acoustics, Speech and Signal Processing, ICASSP 2008. IEEE International Conference on*. IEEE.
3. Gayatri, C. H. *Implementation of acoustic communication in under water using BPSK*.
4. Milica, S., & Preisig, J. (2009). Underwater acoustic communication channels: Propagation models and statistical characterization. *Communications Magazine, IEEE, 47*(1), 84–89.
5. Catipovic, J. (1990). Performance limitations in underwater acoustic telemetry. *IEEE Journal of Oceanic Engineering, 15*, 205–216.
6. Yuri, L., et al. (2009). Modulation and error correction in the underwater acoustic communication channel. *International Journal of Computer Science and Network Security, 9*(7), 123–130.
7. Milica, S., & Preisig, J. (2009). Underwater acoustic communication channels: Propagation models and statistical characterization. *Communications Magazine, IEEE, 47*(1), 84–89.
8. Vikrant, A. K., & Jha, R. S. *Comparison of underwater laser communication system with underwater acoustic sensor network*.
9. Xavier, J. (2012). *Modulation analysis for an underwater communication channel*.
10. Debbah, M. (2004). Short introduction to OFDM. In *White Paper, Mobile Communications Group*, Institut Eurecom.
11. Litwin, L., & Pugel, M. (2001). The principles of OFDM. *RF Signal Processing, 2*, 30–48.
12. Xiao, Y. (2003). *Orthogonal frequency division multiplexing modulation and inter-carrier interference cancellation*. Diss. Institute of Automation
13. Kim, B.-C., & Lu, I.-T. (2000). Parameter study of OFDM underwater communications system. In *Oceans 2000 MTS/IEEE Conference and Exhibition* (Vol. 2). IEEE
14. Küsmüs, D. S. (2012). *Modern modulation methods for underwater communication*.
15. Pandharipande, A. (2002). *Principles of OFDM*. IEEE.
16. Thottappilly, A. (2011). *OFDM for underwater acoustic communication*. Diss. Virginia Polytechnic Institute and State University.
17. Li, B., Zhou, S., Stojanovic, M., Freitag, L., & Willett, P. (2008). Multicarrier communication over underwater acoustic channels with nonuniform doppler shifts. *IEEE Journal of Oceanic Engineering, 33*(2), 198–209.

A Hybrid Approach for Big Data Analysis of Cricket Fan Sentiments in Twitter

Durgesh Samariya, Ajay Matariya, Dhwani Raval, L.D. Dhinesh
Babu, Ebin Deni Raj and Bhavesh Vekariya

Abstract Twitter has become one of the most widely used social networks, and its
popularity is increasing day by day as the number of tweets grows exponentially
each day in the order of millions. The twitter data is used widely for personal,
academic, and business purpose. In this paper, we collected real-time tweets from
Indian Cricket Team fans during the eight matches of ICC Cricket World Cup
(CWC) 2015 (here 8 matches means the total number of games India played in
CWC) using the social media twitter. We performed sentiment analysis on the
tweets to test emotions of Indian Cricket Lovers. The analysis is based on the fact
that the emotions of fans change frequently with each event such that when home
country is batting and scoring runs, they will be happy, and for every loss of wicket
they will be sad. When the team is bowling, they will be sad for 'six' and happy for
Wickets. So when Fans are happy, they react with positive tweets and accordingly
when they are sad they react with negative tweets. We analyzed that, when India is
batting, people use fear, anger, nervousness, and tension which are the most frequently used negative words and words like awesome, happy, and love are the most
used positive words. All emotions are entirely dependent on team India's performance. All negative emotions are increased when opponent team hit runs or when

D. Samariya (✉) · A. Matariya · D. Raval · L.D. Dhinesh Babu · E.D. Raj
School of Information Technology and Engineering, VIT University,
Vellore, Tamil Nadu, India
e-mail: samariya.durgesh@gmail.com

A. Matariya
e-mail: ajaymatariya92@gmail.com

D. Raval
e-mail: dhwani384@gmail.com

L.D. Dhinesh Babu
e-mail: lddhineshbabu@vit.ac.in

E.D. Raj
e-mail: ebindraj@gmail.com

B. Vekariya
Department of Computer Engineering, R.K University, Rajkot, Gujarat, India
e-mail: vbhavesh48@gmail.com

© Springer Science+Business Media Singapore 2016
S.C. Satapathy et al. (eds.), *Proceedings of International Conference
on ICT for Sustainable Development*, Advances in Intelligent Systems
and Computing 408, DOI 10.1007/978-981-10-0129-1_53

503

they achieve HIGH SCORE" and is decreased when the Indian team hit runs or when they take wickets'. This paper uses tweets and captures emotions for big data analytics and analyzes emotional quotient.

Keywords Big data analysis · Sentiment analysis · Twitter · Indian cricket fans · Fans emotions · Cricket world cup

1 Introduction

Big data is the hot buzz technology where traditional database technologies cannot be used for processing millions of datasets. The difficulties faced in big data include capturing the data, analysis of data, searching, storage, sharing visualization, and information privacy. This paper makes use of twitter for predictive analysis and to extract values from data. Processing of big data using traditional relational database will incur more time, costs, and resources. Big data can be blended with data lake, machine learning, and artificial intelligence to create analytical algorithms to enhance the big data architecture [1]. Sentiment analysis is the illation of people's views or interests and is also known by the name opinion mining. It is one of the most challenging problems in natural language processing [2].

The information gathered from tweets can be used to classify the tweets as positive tweets and the negative tweets. Used lexicons and machine learning methods are the most common techniques in extracting sentiments from the textual data. The context-based sentimental analysis can be used for regular texts as well as text with a high level of noise in the data [3].

In this paper, twitter data illustrated that fans mood change continuously during the match. The increased popularity of mobile phones has enabled the fans to post their views very quickly on social media. Recent studies illustrate that the usage of social media is so high, and it helps in expressing the emotions in a public platform [4].

2 Related Work

Lexicon sentiment analysis method by Hogenboom, Alexander, et al did sentiment analysis on emotions, and they analyzed over 2080 Dutch twitter tweets and messages, which consisted of emoticons used for sentiment [5]. Veeraselvi, S.J., and C. Saranya proposed sentiment technique using genetic-based machine learning technique [6]. They exposed tweets from twitter and from micro-blogging posts and stated that they can be classified into three categories namely positive, negative, and neutral. The proposed algorithm by Gamallo et al. using Naïve Bayes classification technique is used for detecting the efficiency of English word [7]. They divided words into only two parts, positive and negative, and in order to find any tweet with or without any polarity, then they avoid that tweet and they consider only basic words/texts for polarity. In this algorithm, we pass polarity words like lexicon and multiple words such that we get approximately 62–66 % F-Score [8].

The work by Peng Zhao et al. did sentiment analysis for product and service (for example, two different brands of camera) [9]. The algorithm to analyze system is called opinion analyzer. They get tweets of two same products (Different mobile brands) from twitter and also get customer opinion from tweets and find why customer used or preferred this product and why not use other product.

Another research approach in sentiment analysis in twitter is to extract 1000 tweets from three different websites using polarity prediction and another 1000 for testing purpose [7]. The characteristic properties of tweets such as how tweets can be written, and the types of words used to compose tweets are taken into account for the analysis of the tweets. It also considered like wise re-tweets, hashtags, exclamation marks, and punctuation [10].

3 Proposed Work

Emotion word was used by people to express their mood (Positive and Negative). To find emotion in tweets, we propose a hybrid approach consisting of corpus-based and dictionary-based techniques [11].

We retrieved tweets from twitter's official site via twitter's search API during 2015 ICC Cricket World Cup. For sentiment analysis, twitter API provides access to read and write tweets and provides some percentage of sample data. We design a program to collect tweets in real time using some hashtags like #INDvsPAK, #INDvsAUS, #INDvsSA, #INDvsUAE, #INDvsBAN, #INDvsWI, #INDvsIRE, and #INDvsZIM.

3.1 System Architecture

In the initial phase, the cricket fans' tweets are retrieved from twitter. In the second phase, the tweets are preprocessed using corpus technique to find emotions and using dictionary-based technique, the words are compared and listed as positive words or negative words. From these words, we give a score to each of the words as the positive word "+1" and the negative word "−1". Figure 1 shows the proposed system architecture.

Retrieval of tweets Using twitter search API we retrieve the tweets during World Cup 2015 of Indian Team Fans. We used R statistical tool to retrieve the tweets. We consider that emotions and capitalization are inseparable part of social media. For example one tweet was like RT @narendramodi: Victory and defeat are a part of

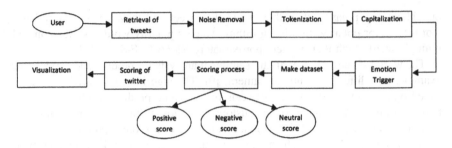

Fig. 1 System architecture

life. Team India played great cricket throughout the tournament. We are proud of them. ☺ #india www.example.com @CWC2015.

Processing We created a data set from tweets that contain Indian Cricket Team fans' positive and negative reactions with emotions.

We processed all tweets as follows:

Remove Re-tweets (RT). Ex: Victory and defeat are a part of life. Team India played great cricket throughout the tournament. We are proud of them. ☺ #india www. example.com @CWC2015.
Remove #hashtags and lexicon Emotions. Ex: Victory and defeat are a part of life. Team India played great cricket throughout the tournament. We are proud of them. Happy www.example.com @CWC2015.
Remove @userName. Ex: Victory and defeat are a part of life. Team India played great cricket throughout the tournament. We are proud of them. www.example.com.
Remove URL. http://www.abc.com Ex: Victory and defeat are a part of life. Team India played great cricket throughout the tournament. We are proud of them.

Scoring After processing, we get only accurate and meaningful tweets. Now tweets are compared to find out whether they have positive and negative words. Positive emotion score is considered as "+1" and negative emotion score as "−1".

Visualization Visualization is a technique that converts Fans' emotion in Graph. For visualization, we have many techniques and many types of graph.

Our analysis was limited; we read tweets posted during the eight matches. Team India played six matches during the pool game, and two additional matches—one match quarterfinal and one match semifinal. For Indian Team total 14,052,167 tweets have been posted [12]. In this Project, we analyzed 1009, 1395, 2100, 3466, 2106, 1567, and 1333 tweets. Table 1 shows tweets emotions and lexicon.

Figure 2a shows the emotion of joy and anticipation during the game. The emotion joy was increased when Indian Player hit SIX or Indian Bowler takes wicket, so every time it is increased and at last Fans assume (Predict) WIN and start celebration. And it also decreases when a wicket falls, or Pakistan hit a SIX. Figure 2b shows the negative emotion of Fans during the match. We considered only three negative emotions namely anger, sadness, and fear. From Fig. 2b we can

Table 1 Sample emotion and lexicon in tweets

Emotions	Lexicon sample	Tweets sample
Joy	Wonderful, happy	Wonderful helicopter shot by @msdhoni
Anticipation	Hope, believe	RT @narendramodi: I believe that we will win. #JayBharat #India
Fear	Worry, nervous	I am afraid, India never win against South Africa in CWC
Anger	Hell, shit, WTF	Shit, Wht the hell virat doing. #INDvsAUS
SAD	Sadness, sad	Very sad, ☺ India lose in semifinal. Only one thing to say good luck for 2019

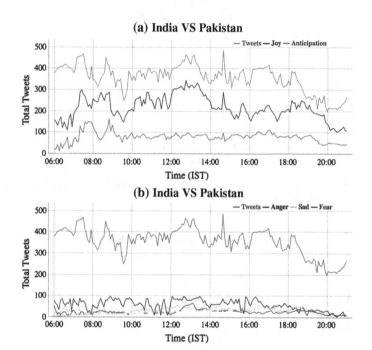

Fig. 2 a and **b** Emotion changes during match between India and Pakistan. Total numbers of tweets were different for every scale

say that anger is the most frequently detected negative emotion when compared to sadness and fear.

Figures 3, 4, 5, 6, 7, 8 and 9 show negative and positive emotion patterns during matches' tweets. Figures 3, 4, 5, 6 and 7 are played under in Pool B, and Fig. 8 is under quarterfinal and Fig. 9 under semifinal match.

Figure 9b depicts that the match had more negative emotions. In this match India lose. However, some positive tweets are there because some Indian Team Fans' motivate the players and some fans' say bad words and making jokes of them, but still we respect them (e.g., @msdhoni we are with u. #India).

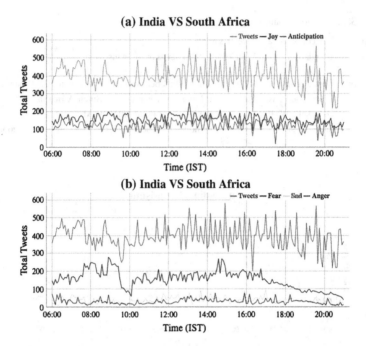

Fig. 3 a and **b** Emotion changes during match between India and South Africa. Total numbers of tweets were different for every scale

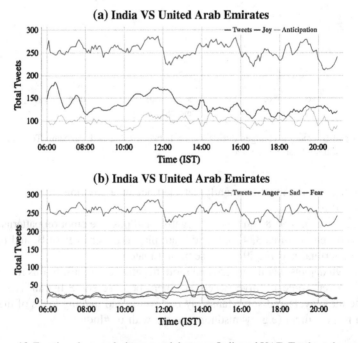

Fig. 4 a and **b** Emotion changes during a match between India and UAE. Total numbers of tweets were different for every scale

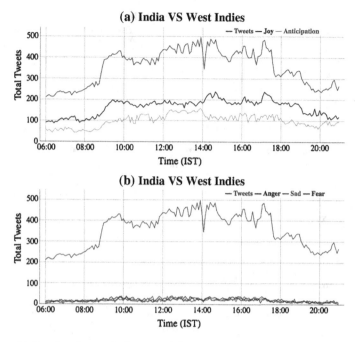

Fig. 5 **a** and **b** Emotion changes during a match between India and West Indies. Total numbers of tweets were different for every scale

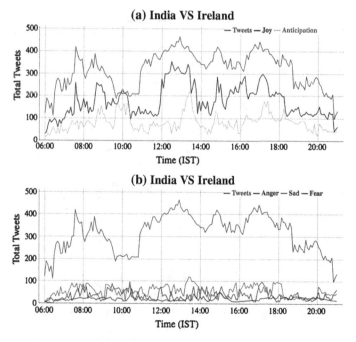

Fig. 6 **a** and **b** Emotion changes during a match between India and Ireland. Total numbers of tweets were different for every level

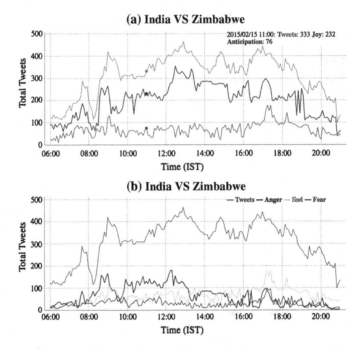

Fig. 7 **a** and **b** Emotion changes during a match between India and Zimbabwe. Total numbers of tweets were different for every scale

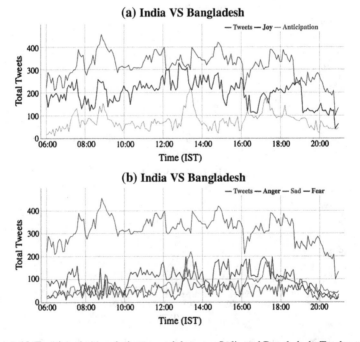

Fig. 8 **a** and **b** Emotion changes during a match between India and Bangladesh. Total numbers of tweets were different for every scale

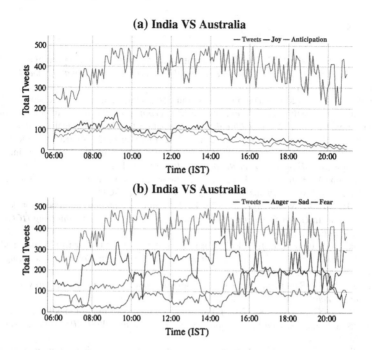

Fig. 9 **a** and **b** Emotion changes during a match between India and Australia. Total numbers of tweets were different for every scale

4 Conclusion

The growth of Internet and social networking sites has made twitter, a widely used micro blogging site for communication. The user shared views, daily activity, communication, and opinions are used for sentiment analysis to derive useful insights. From the derived twitter data, mammoth datasets were created for sentiment analysis and data mining. In this paper, we did a detailed study on sentiment analysis and to discover tweets made by sports fans from India for all Indian team matches in ICC Cricket World Cup 2015. The implemented methods collect tweets automatically from Twitter and clean all tweets using noise classifier. The noise-cleaned tweets are again filtered through words having only meaning full words. The meaningful words were transmitted to the sentiment classifiers that divide tweets into three categories, namely positive category, negative category, and neutral category.

References

1. Mukherjee, S., et al. (2012). TwiSent: A multistage system for analyzing sentiment in twitter. In *Proceedings of the 21st ACM International Conference on Information and Knowledge Management*. ACM.
2. Turney Peter, D. (2002). Thumbs up or thumbs down? Semantic orientation applied to unsupervised classification of reviews. In *Proceedings of the 40th Annual Meeting on Association for Computational Linguistics*. Association for Computational Linguistics.
3. Mohammad, S. M., & P. D. Turney. (2013). *NRC Emotion Lexicon*. NRC Technical Report.
4. Go, A., Bhayani, R., & Huang, L. (2009). Twitter sentiment classification using distant supervision. In *CS224 N Project Report, Stanford* (pp. 1–12).
5. Hogenboom, A., et al. (2013). Exploiting emoticons in sentiment analysis. In *Proceedings of the 28th Annual ACM Symposium on Applied Computing*. ACM.
6. Veeraselvi, S. J., & Saranya, C. (2014). Semantic orientation approach for sentiment classification. In *Green Computing Communication and Electrical Engineering (ICGCCEE), 2014 International Conference on*. IEEE.
7. Barbosa, L., & Feng, J. (2010). Robust sentiment detection on twitter from biased and noisy data. In *Proceedings of the 23rd International Conference on Computational Linguistics: Posters*. Association for Computational Linguistics.
8. Liu, B. (2012). Sentiment analysis and opinion mining. *Synthesis Lectures on Human Language Technologies, 5*(1), 1–167.
9. Zhao, P., Li, X., & Wang, K. (2013). Feature extraction from micro-blogs for comparison of products and services. *Web Information Systems Engineering–WISE 2013* (pp. 82–91). Berlin Heidelberg: Springer.
10. Priyanthan, P., et al. (2012). Opinion mining and sentiment analysis on a twitter data stream. In *ICTer 2012*.
11. Aston, N., et al. (2014). Sentiment analysis on the social networks using stream algorithms. *Journal of Data Analysis and Information Processing*.
12. ICC CWC. http://www.icc-cricket.com/cricket-world-cup.

Performance Analysis of Medical Image Compression Techniques

Nayankumar R. Patel and Ashish Kothari

Abstract A large amount of medical data is generated through advanced medical imaging modalities. The digitization of medical image information is of immense interest to the medical community to reduce transmission time, storage costs, and for implementation of the e-healthcare system like telemedicine (Journal of Medical Imaging and Health Informatics 1:300–306, 2011) [1]. Digital images in their original state require considerable storage capacity and transmission bandwidth. In this paper, an exhaustive comparative analysis of different compression techniques and their applications in the emerging fields of medical science such as telemedicine and teleconsultation has been carried out. The performance of compression algorithm can be measured using objective measures such as MSE, PSNR, SSIM, and correlation.

Keywords Image compression · Compression ratio · Telemedicine · DCT · DWT · SPIHT · JPEG2000

1 Introduction

With the rapid growth of medical services such as e-health, telemedicine, and teleconsultation it is required to develop fast and efficient medical image analysis and compression techniques. It is important to transmit medical images in a com-

N.R. Patel (✉)
Faculty of Engineering & Technology, C. U. Shah University,
Wadhwancity, Gujarat, India
e-mail: nrp264@gmail.com

A. Kothari
E.C. Department, Atmiya Institute of Technology & Science,
Rajkot, Gujarat, India
e-mail: amkothari@aits.edu.in

© Springer Science+Business Media Singapore 2016   513
S.C. Satapathy et al. (eds.), *Proceedings of International Conference
on ICT for Sustainable Development*, Advances in Intelligent Systems
and Computing 408, DOI 10.1007/978-981-10-0129-1_54

pressed and secure form so that efficient medical diagnosis can be performed by the medical practitioner.

For correct diagnosis it is necessary that the compression method preserves all important data. Lossless image compression techniques help to compress the image as well as maintain relevant information, but compromise the compression ratio. Lossy compression techniques are more efficient in terms of compression ratio but have significant loss of image quality and cannot preserve the characteristics needed in medical image processing and diagnosis.

To optimize the above requirements, i.e., high compression ratios and the preservance of relevant information, the ROI-based compression is one of the best options to achieve the optimum compression ratios without any loss of useful information; it is basically done by selecting different important regions of an image along with the background and then compression methodology is applied on these regions separately and not on the whole image. Low compression level is applied on the useful regions while high compression is applied on the unimportant regions and the background. As a result, very high CRs are achieved by this methodology without any appreciable loss of information and clarity of image [2, 3].

This paper, in continuation with this thought, analyzes the suitability of lossy techniques for medical image compression. The research tries to answer the question "Which of the existing lossy or lossless techniques works better for medical images in terms of compression rate and quality?" and answering this question is the main goal of this work. The aim is to analyze techniques and find a compression scheme that can compress medical images quickly and at the same time increase the compression rate while maintaining a good level of visual quality. The various algorithms considered are Discrete Cosine Transformation (DCT), JPEG2000, Vector Quantization, Fractal Compression, Block Truncation Coding (BTC), Discrete Wavelet Transform (DWT), Embedded Zero-tree Wavelet (EZW) coding, and Set Partition in Hierarchical Tree (SPIHT) Algorithm.

2   Basics of Compression

Image compression algorithms may be broadly categorized into two types: lossy and lossless [4, 5].

1   *Lossy Compression*

   In lossy image compression, the redundant pixel data are discarded during the compression process so that the compressed image is only an approximation of the original material. Often, adjusting the compression parameters can vary the degree of lossiness, allowing the image-maker to tradeoff file size against image quality.

2   *Lossless Compression*

   In lossless compression schemes, the compression ratio is low and visually identical to the original image and may be displayed as an exact digital replica of the original. Only the statistical redundancy is exploited to achieve compression. In general, lossless techniques provide far lower compression ratios than lossy techniques with the bonus of preserving all image content.

With recent advances in medical imaging, a large number of data are generated by MRI and CT scans. Although in the past decade the storing capacity of device has increased and cost of storage has decreased, there remains a strong demand for efficient image compression techniques to accommodate the rapid growth of these data and to reduce the associated storage and bandwidth costs [6].

Within a typical digital radiology setup, data are stored in a central server or, in larger schemes, in one of many server nodes. This large repository of archived patient data greatly increases the frequency with which a user can access patient images. However, the client, in many cases, is a low-to-mid range computer with a modest memory and bandwidth. This type of setup imposes constraints that must be addressed:

1. *Choice of compression method—lossy or lossless*: If such data are to be transferred over low-bandwidth networks, efficient compression is essential. Currently, hospitals are reluctant to use lossy compression due to the potential for legal ramifications given incorrect diagnoses. Lossless compression does not carry these ramifications, but does not compress the image data to the magnitude achieved by lossy compression.
2. *Proprietary or open standards*: Should the hospital employ proprietary compression schemes or open standards?
3. *Scalability*: As some clients will be limited in the computer memory, a client receiving scaled data may browse low-resolution versions of the image for selection of a region or volume of interest (ROI, VOI, respectively) for preferred download. Further, reduced resolution viewing decreases the client's dependence on high bandwidth connections while simultaneously decreasing rendering times.
4. *Internet Communication Protocols*: The client–server communication protocol must be generic enough to be easily deployed on a variety of computer operating systems.

3   Typical Review of Image Compression

In telemedicine, the discussion of image compression is divided into three separate categories: Compression before primary diagnosis for rapid transmission, compression after primary diagnosis for long-term archiving, and compression for database browsing where progressive would be useful. In the past two decades several scientific studies have been performed to determine the degree of compression that maintains the physical and diagnostic image quality. Many research studies have been published during that period of time reviewing the existing techniques or presenting new advances in the field of medical image compression in both categories of lossy and lossless [7–9] (Tables 1 and 2).

Table 1 Comparative analysis of image compression techniques

Sr. no	Property	JPEG	JPEG2000	SPIHT	EZW	EBCOT	Hybrid DCT-DWT
1	Targeted compression	No	Yes	Yes	Yes	Yes	Yes
2	Progressive transmission	Yes	Yes	Yes	Yes	No	Yes
3	Low memory	Yes	Yes	No	No	Yes	No
4	ROI modeling	No	Yes	No	No	Yes	No
5	Compression ratio	Poor	Good	High	Better	Better	High
6	PSNR	Poor	Satisfactory	High	Good	Good	Good
7	Perceptual image quality	Poor	Satisfactory	Good	Good	Good	Good
8	Mean square error	High	Satisfactory	Low	Low	Low	Satisfactory
9	Correlation coefficient	Low	Satisfactory	High	Good	Good	High
10	Computational complexity	Low	High	High	High	Moderate	Moderate

Table 2 Literature review of medical image compression technique

Compression technique	Features	Limitations
JPEG	Higher compression ratio	Lossy compression
Wallace [10]	DCT-based technique	Less visual image quality
Vander Kam et al. [11]	Predictive algorithm	Block artifacts
Al-lahan and El Emary [12]	Huffman or arithmetic entropy algorithm	Low throughput
		Incompatible for ROI
JPEG 2000	ROI coding	Lower compression ratio w. r. t. JPEG
Christopoulos et al. [13]	Embedded bit stream	More complex
Tohoces et al. [14]	Lossless compression	More execution time
Zhang and Wu [15]	Multiple resolution representation	Gibbs effect at lower bit rate
Hui and Besar [16]	Error resilience	
JPEG-LS	Near lossless	Incompatible for ROI
Miaou et al. [17]	High speed, less complexity	
Ferni Ukrit et al. [18]	More scalable than JPEG 2000, EZW and SPIHT	
EZW	Good compression ratio	Complex algorithm
Jilani and Sattar [19]	Embedded bit stream	
Shingate and Sontakke [20]	DWT-based technique	
Babu and Alamelu [21]	Predictive algorithm	
	Zero tree coding of wavelet	
Janaki and Tamilarasi [22]	Adaptive arithmetic coding	
	Successive approximation coding	

(continued)

Table 2 (continued)

Compression technique	Features	Limitations
SPIHT	Extension of EZW	
	Lossless compression	
Zhu et al. [23]	Low computational complexity	
Sriraam and Sriraam [24]	Selective ROI coding	
Jindal et al. [25]	Low memory requirement	
CALIC	High compression ratio	Poor performance
Ferni Ukrit et al. [18]		More complex algorithm
Pandian and Sivanandam [26]		
LZW	Row by row pixel classification	More complex process
Yao et al. [27]	Good compression ratio	
DCT based compression	Higher compression	Lossy compression
Saraswathy et al. [28]	Less complex	Low PSNR
Gupta and Garg [29]		High MSE
Breeuwer et al. [30]		Poor image quality
DWT based compression	Good image quality	Low compression ratio
Kharate et al. [31]	High PSNR	
Grgic et al. [32]	Lossless image compression	
Hybrid DCT-DWT	Good compression ratio	More complex
Singh and Sharma [33]	Good image quality	
Telagarapu et al. [34]	Fast execution	

4 Simulations and Results

In this paper, an efficient hybrid DWT-DCT technique for image compression is presented in which the 2-level 2-D DWT is taken, followed by applying the 8-point 2-D DCT. The DCT is applied only to the DWT low-frequency components that result in higher compression ratio (CR) preserving important information. Three cases have been taken into account in which each case depends on the consideration of different subimages of the DWT output. The performance of hybrid DCT DWT technique is compared with the DCT-based compression and DWT-based compression using image quality parameters such as compression ratio, MSE, and PSNR. The Matlab results and graph analysis is shown below. It shows that the hybrid technique performs better than DCT and DWT. The results show performance improvement with least false contouring and a higher compression ratio is achieved compared to other standard stand-alone schemes (Figs. 1, 2, 3 and 4).

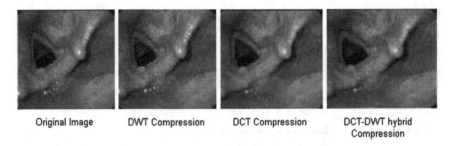

Original Image   DWT Compression   DCT Compression   DCT-DWT hybrid
                                                                     Compression

Fig. 1 Matlab results on endoscopy image

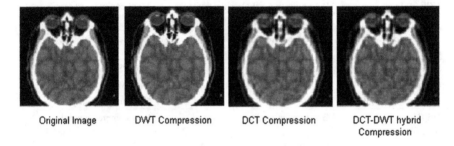

Original Image   DWT Compression   DCT Compression   DCT-DWT hybrid
                                                                     Compression

Fig. 2 Matlab results on CT scan image

Fig. 3 PSNR of test images at 97 % compression for DCT, DWT, and hybrid DCT DWT technique

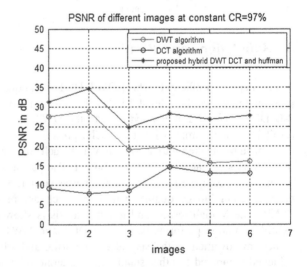

Fig. 4 PSNR of test images at 93 % compression for DCT, DWT, and hybrid DCT DWT technique

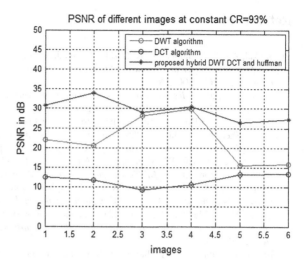

5 Conclusion

From the literature review, it is understood that medical image compression-related works, apart from achieving high compress ratio, also aim at algorithms that are low in complexity and produce visually lossless image quality. Low complexity is important for real-time medical applications and also, poor image quality reduces the understandability of the image, which is important during disease analysis. Moreover, it can be understood from the literature study that among the various techniques proposed such as lossy, lossless, and near lossless developed by JPEG, JPEG-LS, JPEG2 K, SPIHT implemented by DCT and DWT, DWT has more importance due to its manifold characteristics such as multi-resolution property and applicability to a variety of medical images. Many notable algorithms of lossy nature can achieve high compression ratios (50:1, or more), but they do not reconstruct exactly the original version but with some degradation in the image quality. On the other hand, lossless compression techniques permit perfect reconstruction of the original image, but the typical achievable compression ratio is low compared to lossy compression. Thus, it can be concluded from the literature study that new trends of context modeling, that is, ROI-based compression as emerging trends in the medical image compression have to be probed further to produce efficient compression algorithms. The experiment result shows that the hybrid DCT DWT-based compression technique performs better than the existing technique in all respects.

References

1. Sapkal, A. M., & Bairagi, V. K. (2011). Telemedicine in India: Challenges and role of image compression. *American Scientific Publishers: Journal of Medical Imaging and Health Informatics, 1*(4), 300–306.
2. Sayood, K. (2011). *Introduction to data compression* (3rd ed., pp. 423–513). Elsevier.
3. Sonal, K. D. (2000). Study of various image compression techniques. In *Proceedings of COIT* (pp. 799–803), RIMT Institute of Engineering & Technology, Pacific.
4. Li, C., Shen, Y., & Ma, J. (2005). An efficient medical image compression. In *Engineering I004E Medicine And Biology 27th Annual Conference*, 1–4 Sept. 2005. Shangai, China: IEEE.
5. Said, A., & Pearlman, W. A. (to appear). An image multiresolution representation for lossless and lossy compression. *IEEE Transactions on Image Processing.*
6. Rehna, V. J., & Jeya Kumar, M. K. (2011). Hybrid approach to image coding: A review. *International Journal of Advanced Computer Science and Applications, 2*(7).
7. Cyriac, M., & Chellamuthu, C. (2012). A novel visually lossless spatial domain approach for medical image compression. *European Journal of Scientific Research, 71*(3), 347–351. ISSN 1450-216X.
8. Bindu, K., Ganpati, A., & Sharma, A. K. (2012). A comparative study of image compression algorithms. *International Journal of Research in Computer Science, 2*, 37–42. eISSN 2249-8265.
9. Mrak, M., Grgic, S., & Grgic, M. (2003). Picture quality measures in image compression systems. *IEEE* Transactions *in Electrical and Computing*, 233–237.
10. Wallace, G. K. (1991) The JPEG still-picture compression standard. *Communications of ACM*, 30–44.
11. Vander Kam, R. A., Wong, P. W., & Gray, R. M. (1999). JPEG-compliant perceptual coding fora grayscale image printing pipeline. *IEEE Transactions On Image Processing, 8*(1).
12. Al-lahan, M., & El Emary, I. M. M. (2007), Comparative study between various algorithms of data compression techniques. In *Proceedings of the World Congress on Engineering & Computer Science.*
13. Christopoulos, C., Skodras, A., & Ebrahimi, T. (2000). The JPEG2000 still image coding system: An overview. *IEEE Transactions on Consumer Electronics, 46*(4), 1103–1127.
14. Tohoces, P. G., Varela, J. R., Lado, M. J., & Souto, M. (2008). Image compression: Maxshift ROI encoding options in JPEG2000. *Computer Vision and Image Understanding, 109*(2), 139–145.
15. Zhang, C. N., & Wu, X. (1999). A hybrid approach of wavelet packet and directional decomposition for image compression. In *Proceeding of IEEE Canadian Conference on Electrical and Computer Engineering*, May 9–12, (pp. 755–780). USA: IEEE Xplore Press.
16. Hui, T., & Besar, R. (2002). Medical image compression using JPEG2000 and JPEG: A comparative study. *World Scientific: Journal of Mechanics in Medicine and Biology, 2*, 313–328.
17. Miaou, S. G., Ke, F. S., & Chen, S. C. (2009). A lossless compression method for medical image sequences using JPEG-LS and interframe coding. *IEEE Transactions on Information Technology in Biomedicine, 13*(5).
18. Ferni Ukrit, M., Umamageswari, A., & Suresh, G. R. (2011). A survey on lossless compression for medical images. *International Journal of Computer Applications (0975–8887), 31*(8).
19. Jilani, S., & Sattar, S. A. (2010). A fuzzy neural networks based ezw image compression system. *International Journal of Computer Applications IJCA, 2*(9), 1–7.
20. Shingate, V. S., & Sontakke, T. R. (2010). Still image compression using embedded zero tree wavelet encoding. *International Journal of Computer Applications (0975–8887), 1*(7).
21. Babu, D. V., & Alamelu, D. N. (2009). Wavelet based medical image compression using ROI EZW. *International Journal of Recent Trends in Engineering, 1*(3).

22. Janaki, R., & Tamilarasi, A. (2011). Still image compression by combining EZW encoding with Huffman encoder. *International Journal of Computer Applications (0975–8887), 13*(7).
23. Zhu, L., Wang, G. Y., & Wang, C. (2008). Formal photograph compression algorithm based on object segmentation. *International Journal of Automation and Computing, 5*(3), 276–283.
24. Ramakrishnan, B., & Sriraam, N. (2005). Compression of DICOM images based on wavelets and SPIHT for telemedicine applications. In *Proceedings of ICBMP*.
25. Jindal, R., Jinda, S., & Kaur, N. (2010). Analyses of higher order metrics for SPIHT based image compression. *International Journal of Computer Applications (0975–8887), 1*(20), 56–59.
26. Pandian, A.P., & Sivanandam, S. N. (2012). Hybrid algorithm for lossless image compression using simple selective scan order with bit plane slicing. *Journal of Computer Science, 8*(8), 1338–1345. ISSN 1549-3636.
27. Yao, X., Xiao, T., & Mao, S. (2008). Image compression based on classification row by row and LZW encoding. In *Proceedings of the Congress on Image and Signal Processing*, May 27–30, pp. 617–621. China: IEEE Xplore Press.
28. Saraswathy, K., Vaithiyanathan, D., & Seshasayanan, R. (2013). A DCT approximation with low complexity for image compression. In *International Conference on Communications and Signal Processing (ICCSP), 2013*, IEEE.
29. Gupta, M., & Garg, A. K. (2012). Analysis of image compression algorithm using DCT. *International Journal of Engineering Research and Applications (IJERA), 2*(1), 515-521.
30. Breeuwer, B., Heusdens, R., Gunnewiek, K., Zwart, P., & Hass, P. (1995). Data compression of x-ray cardio angiographic image series. *International Journal of Cardiac Imaging, 11*, 179–186.
31. Kharate, G., Patill, V., & Bhale, N. (2007). Selection of mother wavelet for image compression on basis of nature of image. *Journal of Multimedia, 2*.
32. Grgic, S., Grgic, M., & Zovko-Cihlar, B. (2001). Performance analysis of image compression using wavelets. *IEEE Transactions Industrial Electronics, 48*.
33. Singh, H., & Sharma, S. (2012). Hybrid Image compression using DWT, DCT & huffman encoding techniques. *International Journal of Emerging Technology and Advanced Engineering, 2*(10), 2250–2459. ISSN 2250-2459.
34. Telagarapu, P., et al. (2011). Image compression using DCT and wavelet transformations. *International Journal of Signal Processing, Image Processing and Pattern Recognition, 4*(3).

Integrated Framework Using Frequent Pattern for Clustering Numeric and Nominal Data Sets

Aswathy Asok, T.J. Jisha, Sreeja Ashok and M.V. Judy

Abstract Clustering is an exploratory technique in data mining that aligns objects which have a maximum degree of similarity in the same group. The real-world data are usually mixed in nature, i.e., it can contain both numeric and nominal data. Performance degradation is a major challenge in existing mixed data clustering due to multiple iterations and increased complexities. We propose an integrated framework using frequent pattern analysis, frequent pattern-based framework for mixed data clustering (FPMC) algorithm, to cluster mixed data in a competent way by performing a one-time clustering along with attribute reduction. This algorithm comes under divide-and-conquer paradigm, with three phases, namely crack, transformation, and merging. The results are promising when the algorithm is applied on benchmark datasets.

Keywords Frequent pattern analysis · Clustering · Normalization · Sum of squared error · FPMC

1 Introduction

Clustering is the process of identifying the classes of objects with similar characteristics. Data clustering segregates the similarities and variances in the database to form groups of related data as either classes or clusters. Apart from classification,

A. Asok (✉) · T.J. Jisha · S. Ashok · M.V. Judy
Department of Computer Science and IT, Amrita School of Arts and Sciences,
Amrita Vishwa Vidyapeetham, Kochi, India
e-mail: achu2061991@gmail.com

T.J. Jisha
e-mail: jishatj13@gmail.com

S. Ashok
e-mail: sreeja.ashok@gmail.com

M.V. Judy
e-mail: judy.nair@gmail.com

© Springer Science+Business Media Singapore 2016
S.C. Satapathy et al. (eds.), *Proceedings of International Conference on ICT for Sustainable Development*, Advances in Intelligent Systems and Computing 408, DOI 10.1007/978-981-10-0129-1_55

523

clustering is an un-supervised learning method to uncover the causal structures and patterns of a given dataset. It is also known as automatic classification in the sense, and data objects can be treated as an implicit class. The distinct advantage is that it can automatically find the groupings. The clustering methods are mainly divided into the following categories: partitioning, hierarchical, density-based, and grid-based methods. Partitioning method is a popular heuristic method which improves the segregation by moving the objects from one group to another by a local optimum approach. k-means and k-medoids are most commonly used partitioning methods. Hierarchical method breaks down the data objects to various levels of hierarchies. This method has two approaches, agglomerative and divisive. The agglomerative approach builds the hierarchies in a bottom-up fashion, whereas divisive approach does the same in the top down. Density-based method solves the difficulty in finding arbitrary-shaped clusters. The clusters are grown on the basis of density to solve the issue. The high density area is termed as clusters, whereas the sparse areas are used to differentiate the clusters. Grid-based method is one of the high-speed clustering methods, which divides the object space into a number of cells that form a grid structure. The processing time depends upon cells in each dimension of the quantized space [1]. Most clustering algorithms focus on numerical data clustering. However, real data sets are primarily mixed in nature which contains both numerical and categorical data types. Major challenge in mixed data clustering is to find a single clustering solution for both data types with improved performance and reduced complexities.

In this paper, we propose an efficient clustering algorithm for mixed data, FPMC which performs clustering after crack, transformation, and merge phase. In crack phase, the total data set is divided into nominal and numerical packs. In transformation phase, frequent patterns are mined to extract frequency-token which are numerical substitutes to nominal values. Attribute reduction is achieved by converting 'n' number of nominal attributes to a single numerical attribute. Merge phase combines the output of transformation phase and numerical attributes which will undergo normalization before any numerical clustering algorithm is applied.

Section 2 deals with the existing algorithms in mixed data clustering and explains the advantages and disadvantages of each method. Section 3 talks about the proposed work in detail, the process flow along with implementation details and the validation measures used in the work. Section 4 gives the experiment setup and the accuracy of the FPMC algorithm. Section 5 presents the conclusion and the extension of the proposed framework.

2 Related Works

Frequent pattern analysis has been a very interesting and focused area of research in data mining. The frequent pattern analysis finds item sets, subsequences, or substructure that appears frequently in a dataset. The frequency is measured in terms of number of occurrences of a particular sequence in the dataset with two important matrices, support and confidence. Support means the probability of occurrence, whereas confidence is the certainty of occurrence of an event. In frequent pattern analysis, the frequency greater than or equal to the minimum support is checked. For example, following are the patterns obtained of an application:

$$I1, I2, I3 = > 2$$
$$I3, I4, I5 = > 3$$
$$I2, I3, I5 = > 3$$
$$I4, I5, I1 = > 2$$
$$I1, I3, I5 = > 4$$

If the minimum support is set at 3, then the frequent patterns obtained are {{I1, I2, I3}, {I4, I5, I1}, and {I1, I3, I5}}. These are the only patterns that meet the general criteria, i.e., select the only pattern that has support count greater than or equal to minimum support count. The efficiency of association rule generation is different for the various types of frequent item set, namely Eclat, Apriori, and FP growth algorithms.

Eclat algorithm uses vertical database layout where each item of transaction is stored along with its cover in the database. This computes the support by interaction-based approach. The disadvantage is that it is suitable only for small data sets [2]. Apriori algorithm is based on level-wise search. This algorithm begins with the selection of one item and then proceeds by adding the item one at a time and is checked against the support which is required as per the requirement of the application [3]. The FP growth tree is the approach in which the problem of Apriori algorithm is solved by introducing a compact data structure which avoids the need of candidate generation. The execution time of this algorithm is large due to complex data structure and also it is difficult to fit in the main memory [4].

Commonly used algorithms for mixed data are Ralambondrainy [5], k-prototype [6], CLARA [7], and cluster ensemble [8]. Ralambondrainy proposed a new algorithm for mixed data set by converting the categorical attributes into binary and treating the binary as numerical value. By the conversion into a numerical value, k-means algorithm can be directly applied. The drawback of this approach is the space costs and computational complexity with increase in binary attributes in a data set. The other drawback is with 'mean', which do not give the cluster character. The k-prototype algorithm uses squared Euclidean distance as the dissimilarity measure for numerical (Sr) and the number of mismatches between two objects (Sc) for categorical. Sr+∞Sc is the combined dissimilarity measure where ∞ is the weightage given to provide equal importance to both sides, i.e., numeric and nominal attributes. It uses k-modes to update categorical attribute and all other procedures are similar to k-means. As k-prototype is derived from traditional k-means, it is having the same problems that k-means possess. CLARA combines sampling along with clustering program. This method finds objects by k-medoids, so it clusters categorical attributes. This is inefficient for large data set clustering. Cluster ensemble approach splits the data set into nominal and numerical and then applies clustering separately and at last combines the results of clustering. After combining the result, numerical or nominal clustering method is applied. Even though it avoids the problems with other clustering approaches, this approach has got high clustering complexity. The error rate will get multiplied as clustering is conducted three times.

3 Proposed System

Frequent pattern-based framework for mixed data clustering (FPMC) algorithm addresses the major challenges in the existing clustering solutions like performance degradation and space computational complexities by avoiding repeated iterations and optimizing using attribute reduction. Nominal attributes are replaced with the count of frequent patterns that are mined from the data set. For this FPMC uses Apriori due to its easiness in implementation and simplicity in data structure. Apriori is also suitable for large data sets.

The process flow of FPMC algorithm is given in Fig. 1.

First phase of the algorithm is crack stage where algorithm first separates the total data set into nominal and numerical after preprocessing. The crack stage produces two results, one is nominal and other numerical pack. The nominal pack undergoes the second phase, i.e., transformation phase where frequent pattern analysis is done on the dataset to obtain a frequency-token value for each frequent

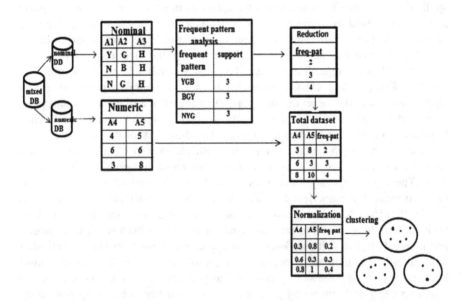

Fig. 1 FPMC flow of execution

item set. This value is obtained by analyzing the nominal pack with the frequent patterns derived from Apriori analysis by evaluating Eq. (1).

Let $P_1, P_2, \ldots P_n$ be the frequent patterns obtained after Apriori analysis:

$$RS_{value} = \left\{ \begin{array}{ll} P_i, & Count + +\,; flag = Valid \\ !P_i, & flag = Invalid \end{array} \right\} \qquad (1)$$

where P_i represents the ith frequent pattern and RS_{value} is the result of row-wise scan. If a match is found, i.e., $RS_{value} = P_i$, increment the count and mark the row as valid; else mark it as invalid.

Third step is the merging phase, where numeric and frequency-token attributes are merged forming a complete numeric dataset. Normalization is done on the dataset for variance stabilization. Normalization is a process in which all attributes are given an equal weight. This is particularly useful for distance measures while used in clustering. There are mainly three types of normalization techniques available namely min-max, z-score, and decimal scaling normalization. Min-max normalization refers to the process of altering the original data into a specified range in a linear fashion. For mapping a v value, of an attribute A from range [min_A, max_A] to a new range [new_min_A, new_max_A], the computation is given by Eq. (2):

$$\frac{v - min_A}{max_A - min_A}(new\_max_A - new\_min_A) + new\_min_A \qquad (2)$$

where 'v' is the new value in the required range.

Z-score normalization is based on mean and median, and it is also called as zero mean normalization. The formula is given in Eq. (3):

$$d^* = \frac{d - mean(P)}{std(P)} \qquad (3)$$

where mean(p) is the sum of the all attribute values of P and Std(P) is the standard deviation of all values of P. Decimal scale normalization is based on the decimal point movement depending on the absolute values of the attributes. The formula is given below in Eq. (4):

$$max(|d|) < 1.[5] \qquad (4)$$

Z-score normalization is used in the FPMC algorithm as it maintains the range and dispersion of the data set, i.e., Standard deviation/variance. After normalization an efficient numerical clustering algorithm is applied. Pseudocode of FPMC is given in Table 1.

Table 1 FPMC Algorithm

Step 1	: Partitioning
	After the replacement of missing values from the dataset it is divided into numerical and nominal packs.
Step 2	: Transformation.
	Step 2.1: Frequent patterns are generated using Apriori.
	Step 2.2: Perform row wise scan on the nominal attributes and execute the operations in *Equation* (1) for all instances. After the row wise scanning of entire dataset, we get the value for frequency-token. For the first frequency pattern i.e. P_1 go to Step 2.3 otherwise Step 2.4.
	Step 2.3: Set frequency-token attribute as count value for all rows marked as valid. And also store a copy of count to init-token.
	Step 2.4: If (init-token < count) Set frequency-token as count value if it is empty, or replace it with count value if it is non-empty.
Step 3	: Merging
	Join the numerical and frequency-token attribute.
Step 4	: Normalization
	Z-score normalization is used in FPMC algorithm as it maintains the range and dispersion of the data set
Step 5	: Clustering
	Perform any numerical clustering algorithm.
Step 6	: Validation of the results
	SSE is used as evaluation criteria for FPMC algorithm. The Sum of squared error for each data point is the distance to the nearest cluster. The clustering produces good results with small value for SSE with minimum number of clusters. Equation (5) gives the formula for SSE calculation where m_i represent the mean of the cluster and x the data point C the cluster.

$$SSE = \sum_{i=1}^{K} \sum_{x \in C_i} dist^2 (m_i \cdot x)$$

(5)

4 Experiments and Results

To evaluate the effectiveness of FPMC algorithm, three different types of real-time data sets were taken into account. The accuracy of the experiment is evaluated using sum of squared error and percentage of incorrectly clustered instances. The experiments consider two algorithms for the analysis. They are simple k-means and cobweb. In simple k-means SSE for a data set without class label and percentage of incorrectly clustered instances for a data set with class label is used. The cobweb algorithm is used to show the percentage of incorrectly clustered instances of data set with class label. We have taken three data sets of different characteristics, for better analysis. They are automobile, labor, and post-operative patient datasets.

Automobile dataset has eighteen numeric and six nominal attributes. The aim of choosing this data set was to cluster mixed data containing equal number of numerical and nominal attributes. Post-operative patient dataset contains one numeric and eight nominal attribute. In this data set, 64 instances belonging to the patients are sent to the general hospital floor; 24 instances represent patients prepared to go home and two instances of patients sent to the intensive care unit. Labor dataset contains eight numeric and eight nominal attributes. Table 2 shows the comparison results of SSE for the data sets with and without class labels. FPMC is compared with simple k-means. We cannot make use of SSE validation in cobweb because it is not based on distance measures.

Table 2 Comparison results of SSE

Number of clusters	FPMC			k-means		
	3	4	5	3	4	5
Labor	13.48	12.78	10.89	119.52	106.30	99.23
Automobile	68.71	55.41	46.79	607.29	560.59	555.31
Post-operative patient	19.81	18.75	5.02	178.69	169.84	150.78

Table 3 Comparison results of percentage of incorrectly clustered instances

Number of clusters	FPMC			k-means			Cobweb
	3	4	5	3	4	5	
Labor	36.73	38.77	42.85	36.84	50.87	54.38	85.55
Post-operative patient	46.47	40.84	43.66	52.22	63.33	71.11	52.63

Table 3 gives the comparison results of the FPMC with simple k-means and cobweb using percentage of incorrectly clustered instances as a validation measure. This validation technique is applicable only for data sets with class label.

5 Conclusions and Future Work

The main objective of clustering is to group similar instances of a data set. The grouping of instances is made on the basis of similarity measures. Even though there are many distance measures available, most of them are applied either on numeric or nominal data. But the real-world data are usually mixed in nature. So we cannot directly apply these distance measures. For this most algorithms for mixed data require partitioning of a dataset into nominal and numeric which increases the complexity and degrades the clustering result. In our proposed work we try to find a solution to this problem by transforming nominal data into numeric. The future plan is to improve the performance using efficient pattern generation and clustering algorithms in multidimensional dataset.

Acknowledgments This work is supported by the DST Funded Project, (SR/CSI/81/2011) under Cognitive Science Research Initiative in the Department of Computer Science, Amrita School of Arts and Sciences, Amrita Vishwa Vidyapeetham University, Kochi.

References

1. Han, J., Kamber, M., & Pei, J. (2012). *Data mining: Concepts and techniques*. Morgan Kaufmann, USA.
2. Hipp, J., Myka, A., Wirth, R., & Güntzer, U. (1998). *A new algorithm for faster mining of generalized association rules*

3. Liu, B., Ma, Y., & Wong, C. K. (2002). Improving an association rule based classifier. *Principles of Data Mining and Knowledge Discovery Lecture Notes in Computer Science, (1910, 2000)*, 504–509.
4. Wang, K., Tang, L., Han, J., & Liu, J. (2002). Top down FP-growth for association rule mining. *Advances in Knowledge Discovery and Data Mining Lecture Notes in Computer Science, 2336*, 334–340
5. Huan, Z. (1998). Extensions to the k-means algorithm for clustering large data sets with categorical values. *Data Mining and Knowledge Discovery, 2*, 283–304.
6. Huang, Z. (1997). Clustering large data sets with mixed numeric and categorical values. In *Asia Conference of Knowledge Discovery and Data*.
7. Ahmad, A., & Dey, L. (2007). *A k-mean clustering algorithm for mixed numeric and categorical Data*.
8. He, Z., Xu, X., & Deng, S. (2005). *Clustering mixed numeric and categorical data: A cluster ensemble approach*.

Image Encryption by Using Block-Based Symmetric Transformation Algorithm (International Data Encryption Algorithm)

Sandeep Upadhyay, Drashti Dave and Gourav Sharma

Abstract Today's world of communication is using image for message exchange also, may be one of the reason is that, text-based data is not that much secure until and unless we perform a good level of confusion-and-diffusion-based encryption. In the case of image, security of image is a serious issue now-a-days because of ever increasing multimedia development, use, and associated brute force attacks. In the proposed paper, we are using IDEA (International Data Encryption Algorithm) as one of the strongest secret-key block ciphers by describing how IDEA can be used for image encryption before modifying the plaintext. The block cipher IDEA operates on 64-bit plaintext blocks, uses 128-bit key to give 64-bit cipher text blocks. We have used the pixel values of image to be encrypted. The proposed method consists of following five stages

1. Getting the pixel values of an image in two-dimensional plane.
2. Expending the 24-bit value of each pixel in 64-bit representation.
3. Applying IDEA algorithm over the 64-bit plaintext.
4. Adding eight bits at RHS to the generated 64-bit ciphertext in 24-bit.
5. Reallocating the color value from that 72-bit resultant in three pixel values.

Keywords Plaintext · Ciphertext · Subkey · Expansion · Bit-Rotation · Pixel · Round · Dummy zero

S. Upadhyay (✉)
Department of Computer Science and Engineering,
Techno India NJR Institute of Technology, Udaipur, India
e-mail: sandeepupadhyay30@gmail.com

D. Dave
Department of Computer Science and Engineering, Central University of Rajasthan,
Ajmer, India
e-mail: drashti2110@gmail.com

G. Sharma
Information and System Department, Regen Powertech Pvt. Ltd.,
Udaipur 313001, Rajasthan, India
e-mail: gouravsharma170@live.com

© Springer Science+Business Media Singapore 2016
S.C. Satapathy et al. (eds.), *Proceedings of International Conference on ICT for Sustainable Development*, Advances in Intelligent Systems and Computing 408, DOI 10.1007/978-981-10-0129-1_56

1   Introduction

The cryptography or encryption is the science of scrambling data to make it indecipherable to all, except the intended person [1, 2]. The main purpose of cryptography is to protect the interests of parties communicating in the presence of adversaries. Symmetric-key algorithms for cryptography uses shared secret key for both encryption of plaintext and decryption of cipher text. Two or more parties that want to communicate with each other uses this secret key to maintain a private information link. The keys used for encryption and decryption are either trivially related, or are often identical.

2   The IDEA Algorithm

The block cipher IDEA operates on 64-bit plaintext blocks, uses 128-bit key to give 64-bit cipher text blocks. The plain text is passed through rounds (eight identical transformation) and an output transformation (the half-round). The encryption and decryption process are similar. The figure [2, 3] below shows the complete process of the algorithm. IDEA is a highly secure algorithm as it uses operations from different groups—modular addition and multiplication, and bitwise eXclusive OR (XOR) [4].

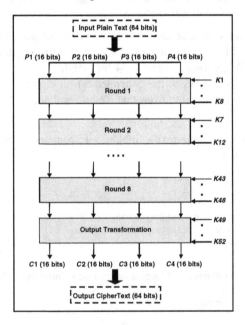

1. We have taken 128-bit key. This 128-bit key is divided into eight keys (K1, K2, K3, K4, K5, K6, K7, and K8). [2] In every round up to eighth round, six keys are used. We have generated a 128-bit key for the image which is implied on every pixel of the image.
2. IDEA is immune to differential cryptanalysis. In fact, there are no linear cryptanalytic attacks on IDEA, and there are no known algebraic weaknesses in IDEA.
3. In the figure [2, 3] below shows the steps to be performed in each round. Instead of Add and Multiply IDEA uses Add* and Multiply*, respectively. Add* and Multiply* are addition modulo 216 and multiplication modulo 216 + 1, respectively. It simply ensures that even if the result of an addition or multiplication of two 16-bit numbers contains more than 16-bits, we bring it back to 16 bits.

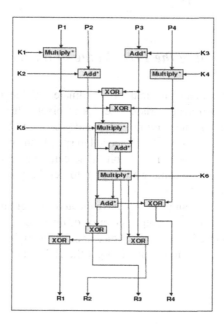

4. The fundamental criteria for the development of IDEA were military strength for all security requirements and easy hardware and software implementation. This is why we have used it on images so the decryption process is not possible without the 128 bit key.
5. By implying this Algorithm on images we can increase the complexity by many times. This eventually results in better security of the confidential messages.

3 Approach for Image Encryption Using Block-Based Symmetric Transformation Algorithm

IDEA works on numeric and textual values (IDEA takes use of Fermat's theorem which applies on numerical as well as textual values), hence, we tried some way to encrypt the image-based data on it by considering all the variations. In this paper, the initial focus is over numeric data then the focus will be on image which is made up on two-dimensional platform using the pixel values.

Use of IDEA is meaningful because in IDEA predicting the decryption key is difficult because its key is 128-bit in length.

We have used the pixel values of image to be encrypted. The proposed method consists of following five stages.

3.1 PHASE-I: Getting the Pixel Value of an Image

Getting the pixel value of an image in three coordinate variable metrics—The image is made up of three coordinate variable metrics of pixel values. As we know that there are three basic colors namely; red, blue and green, i.e., RGB. Each color has 256 shades; which means that 8 bits are needed to specify the precise value of color. This determines that $8 + 8 + 8$, i.e., 24-bit value can represent the color value of each pixel [5]. By doing the same we can plan to get the pixel value of complete image in 2D matrix format.

To get the pixel value of an image we have created a program in java which used to take image as input and gives output in terms of three coordinate pixel values by following the principle of raster scan logic, i.e., left to right and top to bottom. Following is the java code programmed to get the RGB values of an pixel.

```java
package com.sanshti.bo;
import java.awt.Color;
import java.awt.image.BufferedImage;
import java.io.IOException;
import javax.imageio.ImageIO;

public class ImageTest {
public static void main(String ars[]) throws IOException{
BufferedImage hugeImage = ImageIO.read(new
java.io.File("153.jpg"));
    int width = hugeImage.getWidth();
    int height = hugeImage.getHeight();
    int pix,r,g,b;
for (int row = 0; row < height; row++)
{
  for (int col = 0; col < width; col++)
{
   pix = hugeImage.getRGB(col, row);
  Color c=new Color(pix, true);
 /*r = (pix >> 16) & 0xFF;
 g = (pix >> 8) & 0xFF;
 b = pix & 0xFF;*/
 r=c.getRed();
 g=c.getGreen();
 b=c.getBlue();
System.out.println(" (R,G,B) : ("+r+","+g+","+b+")");
}
 }
        // int[ ][ ] result =
convertTo2DUsingGetRGB(hugeImage);
          }
private static int[ ][ ]
convertTo2DUsingGetRGB(BufferedImage image) {
    int width = image.getWidth();
    int height - image.getHeight();
    int[][] result = new int[height][width];

    for (int row = 0; row < height; row++) {
       for (int col = 0; col < width; col++) {
          result[row][col] = image.getRGB(col, row);
      }
   }
   for (int row = 0; row < height; row++) {
      for (int col = 0; col < width; col++) {
         System.out.println(result[row][col]);
      }
   }
   return result;
}}
```

3.2 *PHASE-II: Expanding the 24-Bit Value of Each Pixel in 64 Bit Representation*

To expand the range from 24-bit value to 64-bit value, we need to put a logic of bit expansion. We need to do this because IDEA can work over 64 bit. We can expand to 64-bit plaintext value but we need to do this in 4 levels.

Level-1: Expansion from 24 bit value to 36 bit value.

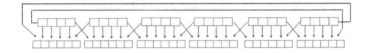

Level-2: Expansion from 36 bit value to 48 bit value.

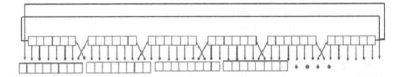

Level-3: Expansion from 48 bit value to 60 bit value.

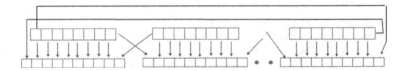

Level-4: Appending 4 zeros at the Low Significant Bit side.

This will give us a 64-bit format which will work as plaintext ready for IDEA encryption algorithm.

3.3 PHASE-III: Applying IDEA Algorithm Over 64-Bit Plaintext

Now we are having the appropriate plaintext in 64 bits over which we can apply the IDEA algorithm covered in 8 rounds using 6 keys where each key is of 16 bit. Suppose we are having following color

Whose pixel value is–(255, 165, 0) which are the three coordinates and hence, they will work as input. The figure mentioned below is the snapshot showing that how to apply the IDEA on the projected plaintext

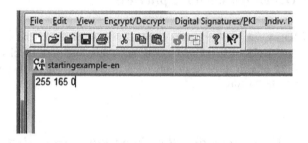

As we apply the IDEA algorithm, it is mandatory to specify the 128 bit key. Suppose we have specified that key is 12 34 56 78 9A BC DE F1 23 45 67 89 AB CD EF

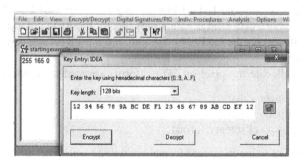

After applying the IDEA algorithm, we will get the encrypted data and encoding of encrypted data, as per below mentioned diagram

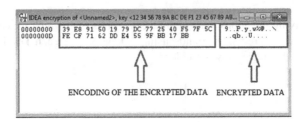

3.4 PHASE-IV: Adding Eight Bits at Least Significant Bit of the Generated 64-Bit Ciphertext

The output of IDEA, i.e., cipher text is in blocks of 64 bits. Append eight dummy zero (0) bits at the RHS of generated cipher text, this will give blocks of 72 bits. Now divide the 72 bits in three segments each of 24 bits. This will give us three pixel values, each in 24 bits.

3.5 PHASE-V: Reallocating the Color Value from that 72 Bit Resultant in Three Pixel Values

As we are now having 72 bits so we segment these 72 bits in 24 × 3 bits. Now truncate the first 24 bits and last 24 bits leaving the middle 24 bits. Now, we are having 24 bits whose first 8-bit segment will specify the RED color part, second 8 bit segment will specify the GREEN color part and third 8 bit segment will specify BLUE color part.

Finally to reallocate the pixel on 2D platform, we need three variables to pass

a. X-axis coordinate value
b. Y-axis coordinate value
c. RGB value of pixel integer format

To get the RGB value in integer format we will be using the standard following formula

$$RGB = (256)2 * RED \text{ part value} + (256)1 * BLUE \text{ part value} + (256)0 \\ * GREEN \text{ part value}.$$

4 Conclusion

Soul work of this paper was to implement the IDEA algorithm for image encryption. We studied the IDEA algorithm and came to know that it can work on numerical data every image is composed of colored pixels and these pixels are composed of three basic colors. We identified that every basic color has 256 shades that means every basic color can be recognized in 8-bit format. This led us to analyze that every pixel can then be represented in 24 bit. By applying the encryption over every pixel, we encrypted the complete image using IDEA. It is different from other image encryption techniques because of its complexity as it follow the principle of confusion and due to increased key length getting used in IDEA which is 128 bit in nature.

References

1. Almasri, O., & Jani H. M. (2013). Introducing an encryption algorithm based on IDEA. *International Journal of Science and Research (IJSR), India, 2*(9).
2. Stallings, W. *Cryptography and network security* (4th ed.). Pearson Education.
3. Kahate, A. *Cryptography and network security* (2nd ed.). Tata McGraw Hill.
4. Sharma, K. D., Verma, H. K., & Kumar, A. (2012). Study and performance analysis of idea with variable rounds. *International Journal of Advanced Research in Computer Science and Software Engineering, 2*(5). ISSN: 2277 128X.
5. Pakshwar, R., Trivedi, V. K., & Richhariya, V. (2013). A survey on different image encryption and decryption techniques. *International Journal of Computer Science and Information Technologies,*4(1). ISSN 0975 9646.

Content-Based Watermarking Using MCA

S. Radharani and M.L. Valarmathi

Abstract This paper discusses content-based watermarking for image authentication using minor component analysis (MCA). The host image is divided into blocks and MCA is applied to each block and the resulting mixing matrix represents the features of the image blocks. From this mixing matrix, different watermarks are generated namely frobenius norm (FN), mean, standard deviation, and combined mean and standard deviation. The middle frequency coefficients of transformations like discrete cosine transform (DCT), discrete wavelet transform (DWT), and combined DCT-DWT are used to embed this watermark. The robustness of the watermarking scheme is tested against various image processing operations like filtering, noise, etc.

Keywords ICT · DCT · DWT · MCA · SVD

1 Introduction

These days, central government as well as state government planned to apply e-governance effectively in various services by using information and communications technology (ICT) to provide better transparency, accuracy, and security of its services to the general public. The information is created, stored, communicated, and managed by using ICTs, which are a diverse set of tools and resources utilized for this purpose. A range of security threats involved in e-government are listed in [1]. In order to overcome these threats, a mechanism called digital watermarking is used. This paper explores the capability of digital watermarking as a secure electronic government framework.

S. Radharani (✉)
Research and Development Center, Bharathiar University, Coimbatore, India
e-mail: radhasaravanan2010@gmail.com

M.L. Valarmathi
Department of Computer Science and Engineering, Government College
of Technology, Coimbatore, India

© Springer Science+Business Media Singapore 2016
S.C. Satapathy et al. (eds.), *Proceedings of International Conference
on ICT for Sustainable Development*, Advances in Intelligent Systems
and Computing 408, DOI 10.1007/978-981-10-0129-1_57

Since, there has been a tremendous and potential growth in the area of information technology; it is easier to manipulate the digital data, efficiently store, and transmitted easily using internet. The digital documents can be simply edited, copied, and send out through internet by existing dominant tools. This rises that, there is a need for protecting ownership details and providing authentication for multimedia document. One of the solutions for this problem is hiding the ownership information in the multimedia document which is called as watermark. It is concealed in a multimedia document in a form, unnoticeable by a human being; it can be identified by a computer.

An application of Public key watermarking algorithm is discussed in [2]. The purpose of security in this paper is maintained by detecting any variation done in the Job-Card. The properties of a digital watermark for e-governance and e-commerce type of data are described in [3].

Watermarking is a method that reduces content piracy in digital media [4–6] and is robust against probable attacks. The procedure of embedding information into digital content, which can be later extracted or detected for the reasons like authentication, copyright protection, broadcast monitoring, etc. is called digital watermarking. Some of the fundamental properties of watermarking are robustness, imperceptibility, fidelity, capacity, security, and time complexity. This work meets all the properties.

There are two types of watermarks namely, visible and invisible. The watermark which is visible can be perceptible to the human observer, whereas invisible watermarks are undetectable by the human observer and do not change the quality of the image. The present work uses invisible watermarking.

Watermarking schemes are applied in various fields like copyright and content protection, authentication, and integrity verification, security (e.g., Passport photos), etc. A watermarked image is subjected to certain manipulations known as attacks. They may be intentional or unintentional. The few possible attacks on watermarks are joint photographic experts group (JPEG) compression, addition of noise, low pass filtering or high pass filtering, histogram equalization, and Gamma correction. The present work is robust against all the above said attacks.

The spatial or transform can be used to embed the watermark. In the spatial domain, the coefficients of the pixels in the host image are directly altered. Transform domains, such as DCT, DWT, singular value decomposition (SVD), Radon transform are widely employed in image watermarking. This paper makes use of different transform domains to embed the watermark.

To obtain better watermarking scheme, the image (feature) content can be used as a watermark which describes the host image. By using the content-based watermark generation techniques, the performance is raised [7]. This technique is followed for color images with RGB model in [8], where FN is taken as watermark and that is derived by applying ICA on host image. This is considered as watermark. This can be embedded in different transforms such as DCT, DWT, and combined DCT-DWT. This method has been improved by using HSV model instead of RGB model in [9] for Combined DWT-DCT domain. It can be seen from the result that the performance is superior than [8]. The watermarking schemes proposed in this paper are content based with different transform domains by applying MCA in HSV model. The intentionally tampered blocks of the watermarked image are located in the present work.

2 Overview of MCA

Minor component analysis (MCA) is an important statistical tool for signal processing and image processing [10]. Using MCA, the directions of smallest variance in a distribution are determined. For dimensionality reduction, principal component analysis (PCA) and independent component analysis (ICA) are normally used. MCA plays a significant role but is not often used. MCA is used for content-based image retrieval system in [11]. Signal energy is uniformly distributed by accomplished PCA method. Principal components describe the global energy of an image; whereas minor components give the details of image. Minor components are extracted using learning algorithm from minor components are extracted using learning algorithm [12] from the principal components, simply by changing the sign of it. The gray scale watermarking using MCA is discussed in [13].

The Eigen vectors associated with the r smallest eigen values of the autocorrelation matrix of the data vector are defined as the minor components, where r is referred to as the number of the minor components. Expressing data vectors in terms of the minor components is called MCA. MCA provides a powerful technique in many information processing fields like least squares, moving target indication, clutter cancelation, curve and surface fitting, digital beam forming, and frequency estimation.

Consider a single linear neuron with the following relation,

$$y(k) = w^T(k)x(k), \ k = 0, 1, 2, \ldots \tag{1}$$

where $y(k)$ is the neuron output, the input sequence $\{x(k) \ / \ x(k) \in R^n \ (k = 0, 1, 2,\ldots)\}$ is a zero mean stochastic process, and $w(k) \in R^n \ (k = 0, 1, 2,\ldots)$ is the weight vector of the neuron. Let $R = E\left[x(k) \ x^T(k)\right]$ denote the autocorrelation matrix of the input sequence $x(k)$, and let λ_i and v_i $(i = 1, 2,\ldots, n)$ denote the eigen values and the corresponding orthonormal eigen vectors of R, respectively. The orthonormal eigen vectors $v_1, v_2, \ldots, v_n$ can be arranged such that the corresponding eigen values are in a non decreasing order: $0 \le \lambda_1 \le \lambda_2 \le \ldots < \lambda_n$.

3 Watermarking Schemes

Most of the authors have used ICA for watermarking. This paper uses MCA to generate the watermark from the host image. The watermarks generated in this paper are frobenius norm (FN), mean (M), standard deviation (SD), and combined mean and standard deviation (CMSD).

3.1 Watermark Generation Using MCA

The procedure for generating the watermark is given below for all the techniques discussed in this paper.

1. The host image 'HI' of size '$n \times n$' is segmented into blocks of size '$m \times m$,' which gives 'K' blocks where $n \geq m$.
2. Every block is performed by MCA and the mixing matrix 'A' is extracted.
3. The content-based watermark 'w' of the block is determined from the mixing matrix. The watermarks generated are FN, M, SD, and CMSD.
4. Repeat steps 2 and 3 for determining the watermark for all the blocks. This set creates the watermark, W.

3.2 Embedding and Extraction of Watermark Using DCT

3.2.1 Embedding Algorithm

1. Host image's DCT is evaluated for every block.
2. The middle frequency coefficient (i, j) is selected in each block.
3. The chosen coefficient is modified by

$$DCT(i,j) = \text{sign}(DCT(i,j)) * (\alpha * w(k))$$ (2)

where 'α' is the embedding strength.
4. Inverse DCT is carried out.
5. The steps 1–4 are repeated for all the blocks. The outcome is the watermarked image WI*.

3.2.2 Watermark Extraction Algorithm and Authentication

1. Computed watermark (W*) is generated by applying the watermark generation algorithm on the received image I'.
2. Every block's DCT is evaluated.
3. The hidden watermark is obtained by,

$$w' = \frac{|DCT(i,j)|}{\alpha}$$ (3)

where 'α' is the embedding strength.

4. This set creates the extracted watermark, W'.

5. The blockwise percentage difference (Δ) between W* and W' is calculated by

$$\Delta = \frac{(|W^* - W'|) * 100 * \alpha}{\max\{W^*\}} \tag{4}$$

3.3 Watermark Embedding and Extraction Using DWT

3.3.1 Embedding Algorithm

1. To decompose the host image, DWT (Haar wavelet) is applied up to four times to get the 4th level subbands, LL_4, HL_4, LH_4, HH_4.

2. Replace the mid-component in the scaled watermark with the same sign.

$$HL_4 = \text{sign}(HL_4) * \alpha * w(k) \tag{5}$$

where 'α' is the embedding strength.

3. Inverse DWT is carried out.

4. The steps 1–3 are repeated for all the blocks. The outcome is the watermarked image WI*.

3.3.2 Watermark Extraction Algorithm and Authentication

1. Computed watermark (W*) is generated by the same way that of DCT-based method.

2. Every block's DWT is evaluated.

3. The hidden watermark is obtained by,

$$w' = \frac{|HL_4|}{\alpha} \tag{6}$$

where 'α' is the embedding strength.

4. The resultant is the extracted watermark, W'

5. The blockwise percentage difference (Δ) between W* and W' is calculated using Eq. (4).

3.4 Watermark Embedding and Extraction Using Combined DCT-DWT

3.4.1 Embedding Algorithm

1. To decompose the host image DWT (Haar wavelet) is applied up to four times to get the 4th level subbands: LL_4, HL_4, LH_4, HH_4.
2. Compute DCT of subband LH_4.
3. Replace the mid-component in the scaled watermark with the same sign as shown in Eq. (3).
4. Inverse DCT and DWT are carried out.
5. The steps 1–3 are repeated for all the blocks. The outcome is the watermarked image WI*.

3.4.2 Watermark Extraction Algorithm and Authentication

1. Computed watermark (W*) is generated by the same way that of DCT-based method.
2. Compute DWT of the host image.
3. Compute DCT of subband HL_4.
4. The hidden watermark from the preferred DWT subband is obtained by applying Eq. (6).
6. The resultant is the extracted watermark, W'.
7. The blockwise percentage difference (Δ) between W* and W' is calculated using Eq. (4).

4 Results and Discussions

The proposed scheme is tested with test images of different sizes. The images are resized into 256×256. In this paper, seven images are taken for testing the implementation of the present methods. Among these, Baboon, Lena, house, and pepper are the standard test images. The value of 'a' is taken for various watermark generation methods are 0.27 for FN, 0.053 for M, 0.085 for SD, and 0.174 for CMSD.

Figure (1) shows the original and watermarked images for DCT, DWT, and Combined DCT-DWT domains with all types of watermarks.

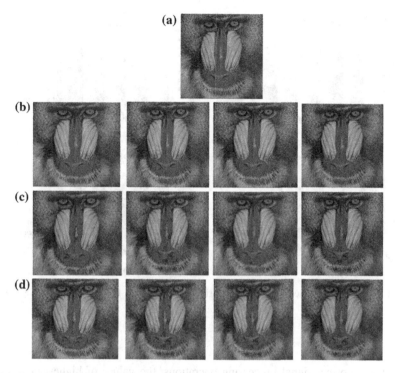

Fig. 1 Original and watermarked image with FN, M, SD, and CMSD—Baboon. **a** Original.
b DCT. **c** DWT. **d** Combined DCT-DWT

4.1 Extraction Efficiency

Figure 2 shows the extraction efficiency of different test images for various
watermarking techniques. The highest percentage difference values are from 3.9 to
6.1 %, for all the test images.

4.2 Quality Metrics

In Fig. 3, the quality metrics of the watermarked images using various watermarks
are compared. It is inferred that, the PSNR value is much better for DCT domain.
The other quality metrics (NCC and IF) are also better for the DCT domain. By
observation, there is no perceptible difference between the host and watermarked
images.

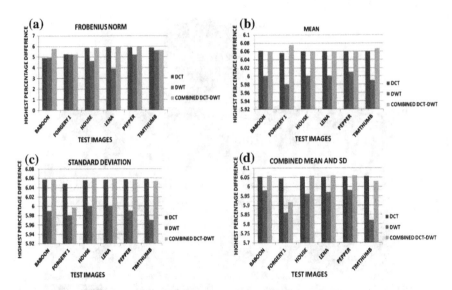

Fig. 2 Results of extraction efficiency for various watermarks. **a** FN. **b** M. **c** SD. **d** CMSD

4.3 Robustness

For all the normal signal processing operations, the values of highest percentage difference (Δ) are from 1.1 to 48.2 % using MCA in the DCT domain, from 0.62 to 18.89 % in the DWT domain, from 0.19 to 6.03 % in the combined DCT-DWT domain. The highest percentage differences in various watermarking techniques are given in Fig. 4 for Lena image. The figure shows the robustness against incidental image processing operations on Lena Image. Except Histogram equalization (48 %) in DCT domain, the remaining attack's percentage differences are less than the threshold 25 % in all transform domains, indicating that there is no tampering.

4.4 Tamper Detection

The watermarked Baboon has been intentionally tampered by adding a random white signal. Tamper detection for various transformation domains on watermarked Baboon image for FN is given in Fig. 5.

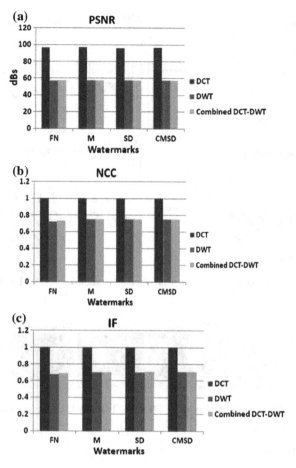

Fig. 3 Qualitiy metrics. a PSNR. b NCC. c IF

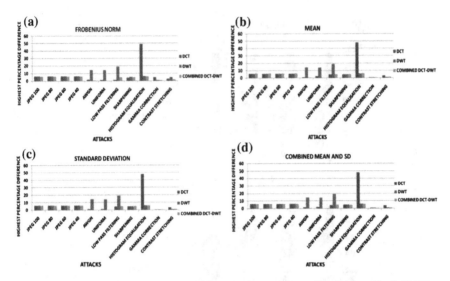

Fig. 4 Results of various watermarks after incidental distortions. **a** FN. **b** M. **c** SD. **d** CMSD

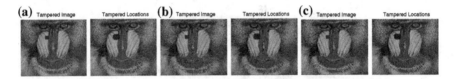

Fig. 5 Baboon with tamper detection using FN. **a** DCT. **b** DWT. **c** Combined DCT-DWT

5 Conclusion

This paper has discussed MCA-based watermarking in transform domin. The watermarks are generated from the host image itself. They are FN, M, SD, and CMSD. The PSNR values in different transforms are compared. The result shows that DCT-based MCA methods are good. All the methods discussed here, are correctly finds tampered regions of the image and correctly authenticates the image.

References

1. Sharma, D. K., Pathak, V. K., & Sahu, G. P.(2007). Digital watermarking for secure e-government framework. *Computer Society India*, 182–191.
2. Sarbavidya, S., & Karforma, S. (2012). applications of public key watermarking for authentication of job-card in MGNREGA. *Bharati Vidyapeeth's Institute of Computer Applications and Management, 4*(1), 435–438.

3. Sherekar, S. S., Thakare, V. M., & Jain, S. (2008). Role of Digital Watermark in e-governance and e-Commerce. *International Journal of Computer Science and Network Security, 8*(1), 257–261.
4. Katzenbeisser, S., & Petitcolas, F. A. (2000). *Information Hiding*. Boston, London: Artech House.
5. Eskicioglu, A. M., & Delp, E. J. (2001). An Overview of Multimedia Content Protection in Consumer Electronics Devices. *Signal Processing: Image Communication, 16*(7), 681–699.
6. Castro, D., Bennett, R., & Andes, S. (2009). Steal these policies: Strategies for reducing digital piracy. *Information Technology and Innovation Foundation*.
7. Parameswaran, L., & Anbumani, K. (2008). Content-based watermarking for image authentication using independent component analysis. *Informatica (Slovenia), 32*(3), 299–306.
8. Radharani, S., & Valarmathi, M. L. (2012). Content based watermarking for color images using transform domain. *International Journal of Engineering Research and Applications, 2*(1), 773–779.
9. Radharani, S., & Valarmathi, M. L. (2012). content based hybrid DWT-DCT watermarking for image authentication in color images. *International Journal of Engineering Inventions, 1*(4), 32–38.
10. Peng, D., & Yi, Z. (2006). A new algorithm for sequential minor component analysis. *International Journal of Computational Intelligence Research, 2*(2), 207–215.
11. Jankovic, M., Zajic, G., Radosavljevic, V., Reljin, N. K. N., Rudinac, M., & Reljin, B. (2006). *Minor Component Analysis (MCA) applied to Image Classification in CBIR systems* (pp. 11–16). IEEE: Neural Network Applications in Electrical Engineering.
12. Chen, T., Amari, S. I., & Murata, N. (2001). sequential extraction of minor components. *Neural Processing Letters, 13*(3), 195–201.
13. Li, Y. M., & Luo, J. (2012). A novel image watermarking scheme based on MCA neural networks. *Applied Mechanics and Materials, 229*, 1874–1877.

Image Enhancement Based on Log-Gabor Filter for Noisy Fingerprint Image

Neeti and Arihant Khicha

Abstract Image enhancement is a technique to improve the quality metrics of a distorted image. In forensic laboratory, if the input is noisy fingerprint image then image enhancement plays the crucial role to authenticate the verification stage. In this work, the author proposed to construct a filter based on Log-Gabor theory for enhancement of high noisy (Gaussian Noise) fingerprint image. To construct the filter, first initialize all filter parameters and to overcome the wrap around effect, first calculate the sine and cosine differences along with angular distance. When the filter is constructed, do the convolution with FFT of input noisy image. The resultant enhanced image attained PSNR and MAE (Mean Absolute Error) 7.9230, 101.728 dB, respectively. The comparison values of quality metrics for Mat lab inbuilt function (wiener and median filter) are 7.3291, 7.3281, and 101.977, 101.986 dB respectively for same test image. In this work, there is a little improvement in PSNR value which play significant role in verification stage.

Keywords Image enhancement · Log-Gabor filter · PSNR · MAE etc.

1 Introduction

In digital image processing, the input is a two-dimensional signals, such as photograph or video frame and output will be post processed image or characteristics parameters of an input image. In forensic laboratory, during the verification of fingerprint images, it follows three stages [1], first is preprocessing stage where the input image prepared for further processing by suitable image enhancement

Neeti (✉)
Manipal University, Jaipur, India
e-mail: neeti.kapoor87@gmail.com

A. Khicha
Rajasthan College of Engineering for Women, Jaipur, India
e-mail: arihantkhicha@gmail.com

© Springer Science+Business Media Singapore 2016
S.C. Satapathy et al. (eds.), *Proceedings of International Conference on ICT for Sustainable Development*, Advances in Intelligent Systems and Computing 408, DOI 10.1007/978-981-10-0129-1_58

techniques, then followed by meta-processing stage and verification stage. So if the input is noisy fingerprint image, then image enhancement plays the crucial role to authenticate the verification process.

2 Image Enhancements

Image enhancement is a technique to improve not only visual perception of a given image but also quality metrics of a distorted or noisy image by using the specific algorithms. The objective of an image enhancement is to reduce the noise of a given input noisy image. Image enhancement deals with the improvement of visual appearance of an input to improve the delectability of objects to be used by either a machine vision system or a human observer. There are mainly two types of image enhancements technique [2]:

1. Image enhancement in spatial domain which only deals with pixel to pixel relation.
2. Image enhancements in frequency domain which first convert the image in frequency domain with the help of Fourier transform and perform operations on it. After finishing the required operations, the operated image reverts to spatial domain by inverse Fourier transform. This presented work fall in frequency domain.

3 Log-Gabor Filter

Gabor Theory was first proposed by Dennis Gabor in 1946 [3]. But in recent times it was seen that Gabor theory provides the numerous applications in the field of digital image processing. Gabor work synthesizes the studies of Nyquist in communications theory and Heisenberg in quantum mechanics, by which he said that Gaussian shape as an optimal envelope for time frequency representation [4]. Log-Gabor filter are logarithmic transformation of Gabor domain [5], which removes the annoying DC components allocated in medium and high pass filters. The 2D equations of Gabor filter are as follows:

$$
\begin{aligned}
\text{symmetric:} \quad & \cos(k_x x + k_y y) \exp - \left\{ \frac{x^2 + y^2}{2\sigma^2} \right\} \\
\text{anti-symmetric:} \quad & \sin(k_x x + k_y y) \exp - \left\{ \frac{x^2 + y^2}{2\sigma^2} \right\}
\end{aligned}
\tag{1}
$$

The Gabor filter space domain and frequency response has shown in Fig. 1a, b.

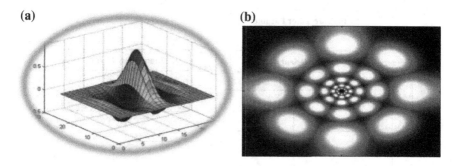

Fig. 1 **a** Gabor filter [4]. **b** Gabor filter frequency response [4]

4 Flowchart of Presented Work

The flowchart of presented image enhancement work is shown in Fig. 2. First initialize the Gabor filter parameters, say filter orientation = 4, sigma = 0.55 which is standard deviation of Gaussian distribution, numbers of wavelets scales = 5 and minimum wavelength, etc. There are some complexity issues during implementation, if we choose the higher orientation and wavelets scales. So, the values of filter orientation and wavelets scales were choose based on better result of input image. After this perform iteration in MATLAB code for maximum filter orientation and wavelet scales.

When all values of specified wavelets scales are finished, then increment the next filter orientation. After that do the fast Fourier transform of the input noisy fingerprint image which convert input fingerprint image to frequency domain and then do convolution with filter function and perform the inverse transform. Calculate, the thresholding based on proper mean and variance values, as computed previously and subtract the noise vector from the IFFT.

The last two steps of the algorithm are continuous until the calculation for all orientation is completed and finally the enhanced version of input noisy image is generated.

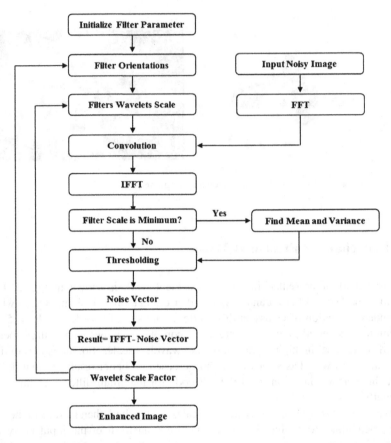

Fig. 2 Flowchart of presented work

5 Results and Analysis

The measurement of quality of an image based on quality metrics of measured image and its values is expressed in dB (Decibels) [6]. The image quality metrics are peak signal-to-noise ratio (PSNR), mean absolute error (MAE), mean square error (MSE), average power (AP), etc. The input noisy image and results of enhanced image based on Gabor filter is shown in Fig. 3a, b.

For the comparison of the previous result, author calculate the quality metrics of the enhanced image based on Median and Wiener filter on the same input test image, the author did not developed these two filter, it is used from MATLAB inbuilt function, i.e., wiener2 and medfilt2. For result of quality metrics of enhanced image for the same input noisy test image in all technique shown in Table 1, where all values are expressed in dB.

Fig. 3 **a** Input noisy finger print image. **b** Enhancement result

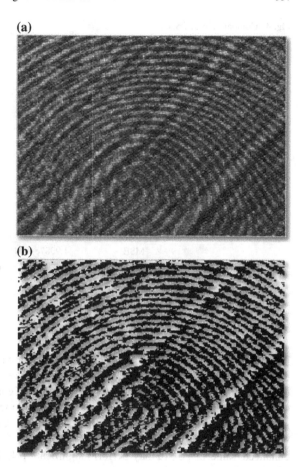

Table 1 Comparison of quality metrics

Quality	Weiner	Median	Presented
PSNR	7.3292	7.3281	7.9230
MAE	101.977	101.986	101.738
SNR	0.0273	0.0262	0.621
MSE	1.2027e+004	1.2030e+004	1.0490e+004
AP	0.2391	0.2300	1.4464e+003

For easy comparison of PSNR which is most important quality metrics of an image it is shown in bar diagram as Fig. 4.

We have tested the second input noisy fingerprint against the Weiner and Median filter and calculated their quality metrics. The quality metrics for second test image are given in Table 2, which again justify that the presented work result is better as comparison to standard median and wiener filter.

Fig. 4 Comparision of PSNR

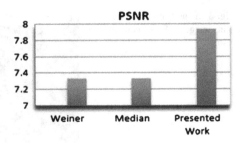

Table 2 Comparison of quality metrics

Quality metrics	Weiner	Median	Presented work
PSNR	5.3235	5.3228	5.8653
RMSE	138.1541	138.1654	129.8004
SNR	0.0374	0.0367	0.5792
MSE	1.9087e+004	1.9090e+004	1.6848e+004
AP	0.4807	0.4740	2.0464e+003

6 Conclusions and Future Scope

It was observed that presented log-Gabor wavelet-based filter give the better image enhancement result in special case of Gaussian noisy finger print image than the other standard filter. As input image have very low signal power so we applied the soft thresholding. The resultant enhanced image attained PSNR and MAE (Mean Absolute Error) 7.9230 dB, 101.728 dB, respectively. Although the input test image was very high noise level, enhanced PSNR value is very less. But this small enhanced value in PSNR may perform significant role for verification process. At all, the presented method showed the better result as comparison with their predecessor techniques; especially in case of high noisy (Gaussian) fingerprint images. As Log-Gabor is non-orthogonal, though during actual implementation, increasing the wavelet scales will increase the computational cost and memory [4]. Future scope of this work would be to achieve orthogonal nature where wavelets are critically sampled, and as a result reducing computation time and memory storage.

References

1. Yang, J., Liu, L., & Jiang, T. *An improved method for extraction of fingerprint features*. Beijing, China: National Laboratory of Pattern Recognition, Institute of Automation, Chinese Academy of Sciences.
2. Gonzalez, R. C., & Woods, R. E. (2009). *Digital Image Processing*. Pearson Education: New Delhi.
3. Gabor, D. (1946). Theory of Communication. *Journal of the Institute of Electrical Engineers, 93*(429), 457.

4. Fischer, S., Redondo, R., & Cristóbal, G. (2009). How to construct log-Gabor Filters. Instituto de Optica (CSIC), Serrano 121, 28006 Madrid, Spain, July 5, 2009.
5. Fischer, S., Sroubek, F., Perrinet, L., Redondo, R., & Cristóbal, G. (2007). Self-invertible log-Gabor wavelets. *International Journal of Computer Vision, 75*(2), 231–246.
6. Jomaa, D. *Segmentation of low quality fingerprint images*. Borlänge: Computer Science, Dalarna University.

Computational Performance Analysis of Ant Colony Optimization Algorithms for Travelling Sales Person Problem

Mahesh Mulani and Vinod L. Desai

Abstract Ant colony-based optimization approach is based on stigmergy behavior of natural insects. Ant Colony Optimization shows promising behavior on dynamic problems like Travelling Sales Person (TSP) problems and other TSP like problems. The paper discusses variants of ACO algorithms as well as measures their performance using TspAntSim simulation tool (Aybars U, Dogan A, J Adv Eng 40 (5), 2009 [1]). TSP is NP-Hard problem which shows fluctuant behavior on instances available in the online library TSPLIB. The paper focuses on behavior of ACO algorithm on such problem instances. The basic idea behind this analytical behavioral study is to help the algorithm design the parameter settings as well as problem instance characteristics. The paper concludes with the remarks on algorithm design criteria of any dynamic problem like TSP and its analytical outcome.

Keywords Ant colony optimization · Dynamic problem · TSPLIB

1 Introduction

Ant Colony Optimization [2, 3] is a population-based approach which is inspired by natural behavior of ant colony's food searching behavior. Ants have an ability to find the closest path from the source place to the destination [2]. Source can be a place where ants reside and destinations are generally food contents. ACO is used for solving combinatorial problems [4] which are NP-Hard in nature [4]. It is used

M. Mulani (✉)
Department of Computer Science, RK University, Rajkot, India
e-mail: mahesh.thackerkskvku@gmail.com

M. Mulani
Department of Computer Science, KSKV Kachchh University, Bhuj, India

V.L. Desai
Government Science College, Chikhli, Navsari, India
e-mail: vinoddesai123@gmail.com

© Springer Science+Business Media Singapore 2016
S.C. Satapathy et al. (eds.), *Proceedings of International Conference on ICT for Sustainable Development*, Advances in Intelligent Systems and Computing 408, DOI 10.1007/978-981-10-0129-1_59

in the applications like Travelling Sales Person problem [4, 5], Quadratic Assignment problems [4], Job Shop Assignment and Scheduling [6, 7], Vehicle Routing [8], Graph Coloring [9], and other scheduling problems. This paper focuses on the performance of available Ant Algorithm on Travelling Sales Person Problem. The problem instances are taken from the TSPLIB online library [10]. The paper is divided into five main phases. Section 2 describes Ant Colony Algorithm overview, Sect. 3 describes Travelling Sales Person Problem, Sect. 4 describes analysis and results of algorithm on selected test instances of TSP, and further sections describes conclusion and future aspects.

1.1 Ant Colony Optimization with Travelling Sales Person Problem

Travelling Sales Person (TSP) is NP-Complete problem in which salesperson has to visit each city exactly once. There are many algorithms available to solve travelling sales person problem to find shortest possible path from source to destination. Earlier TSP [11] is successfully solved by methods like nearest neighbor, greedy algorithm, 2-opt 3-opt algorithm [11], simulated annealing, lin–kernighan, and tabu search. Every algorithm found difficulty in finding optimal or nearer optimal solution. Then there arise a need to find a heuristic-based search method to solve TSP with optimal solution for the given sets of data set. Ant Colony Optimization approach is initiated by Marco Dorigo in 1992 [2, 12]. It is an approach to find the closest path from source to destination which is based on Ants pheromone deposit and evaporation mechanism. ACO approach can be represented using following Fig. 1.

Following parameters are found useful to apply ACO to TSP.

Construction of graph
Constraints specification
Pheromone trails and heuristic value
Solution construction procedure

Fig. 1 Ant's behavior

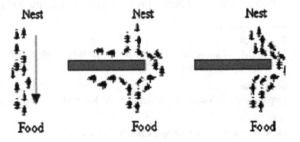

Construction of a graph is a main procedure used for constructing a partial tour. The graph contains edges and vertices where edges are the distance between the cities and edges are source and or destination points. TSP is a problem where all cities must have to be visited and each city must be visited at most once. This is the only constrain specification criteria for TSP. Pheromone trails can be considered as probability of visiting city m immediately after n. The heuristic value is inversely proportional to distance between the city m and n. Each ant visits the unvisited city and it terminates once ants visit all the cities.

There are many variants of ACO algorithm which are discussed in the further sections. They all differ mostly in their way of constructing a solution, tour construction. Pheromone evaporation value also differs in many algorithms.

1.2 Ant System (AS)

Various ant system algorithms were introduced like ant density, ant quality, ant cycle, etc. AS algorithm is divided into two steps. First is solution construction by ants and another is updating of pheromone. The amount of pheromone value deposited by ants in one tour is depends upon the number of ants as well as length of the tour generated using any nearest neighbor heuristics approach.

1.2.1 Solution Construction by Ants

In this algorithm, n_1 artificial ants currently execute a tour for TSP. Ant k uses probabilistic action choice rule to decide which city to visit next at each solution construction step. The rule can be described as follows:

$$\rho^k_{ij} = \frac{[\tau_{ij}]^\alpha [\eta_{ij}]^\beta}{\sum_{l \in N_i^k} [\tau_{ij}]^\alpha [\eta_{ij}]^\beta} \quad \text{if } j \in N^k \tag{1}$$

The parameters used above can be specified as under.

$\eta_{ij} = 1/d_{ij}$ which specifies a heuristic value decided before start of the algorithm

α, β Used determines the relative effect of the pheromone value and the heuristic data

N_i^k probable neighborhood of ant k when being at the current city it is a set of city that kth ant has yet not visited.

Using the stated rule in Eq. (1), the probability of choosing a particular edge increases the value of the pheromone trail associated with that edge. It also affects the information related to heuristics that is η ij.

Every ant maintains a memory, which is a collection of cities that is already visited. It also maintains the order of their visit.

It helps to decide neighborhood which is feasible for the above rule.

1.2.2 Updating the Pheromone

When all the ants have constructed their individual tour, the next step should be the updation of pheromone which ants has deposited during the tour specified. It can be calculated by following rule. It is known as pheromone evaporation rule [?].

$$\mathcal{T}_{ij} \leftarrow (1 - \mathcal{P}) \, \mathcal{T}_{ij} \tag{2}$$

where, $0 < \mathcal{P} \leq 1$ is the rate pheromone evaporation.

After the above phase, the ants need to deposit pheromone on the edges they have passed in their entire tour. The formula used for this purpose is as follows:

$$\mathcal{T}_{ij} \leftarrow \mathcal{T}_{ij} + \sum_{k=1}^{nl} \Delta \mathcal{T}_{ij}^{k}, \quad \forall (i,j) \in L \tag{3}$$

where $\Delta \mathcal{T}_{k}^{ij}$ is the value of pheromone that ant k has deposited on the edges during the visit. $\Delta \mathcal{T}_{k}^{ij}$ can be specified as follows:

$$\Delta \mathcal{T}_{ij}^{k} = \begin{cases} \frac{1}{C^k}, & \text{if edge } (i,j) \text{ belongs to } \mathcal{T}^k \\ 0, & \text{Otherwise} \end{cases}$$

where, k is the total distance covered by the kth ant during the tour $\mathcal{T}^k$.

1.3 Elitist Ant System Algorithm

Elitist ant system [] is considered as the first improvement of Ant System algorithm. The main theme of this algorithm for the best tour found since the startup of the algorithm. It is used to provide additional support to the edges belonging to the best tour. The best-so-far tour is denoted by $\mathcal{T}^{bs}$

1.3.1 Updating the Pheromone

The additional support of tour $\mathcal{T}^{bs}$ can be achieved by adding a value e/C^{bs} to its edges. The parameter e specifies the weight given to the tour and C^{bs} is the total distance covered during the tour. So, pheromone deposit formula needs modification. Which is as follows:

$$\mathcal{T}_{ij} \leftarrow \mathcal{T}_{ij} + \sum_{k=1}^{nl} \Delta T_{ij}^k + e\Delta T_{ij}^{bs} \tag{4}$$

where, ΔT_{ij}^{bs} can be defined as follows:

$$\Delta \mathcal{T}_{ij}^{bs} = \begin{cases} \frac{1}{C^{bs}}, & \text{if edge } (i,j) \text{ belongs to } T^{bs} \\ 0, & \text{Otherwise} \end{cases}$$

1.4 Ant System Based on Rank

The next improved algorithm of AS is addition of rank in the AS algorithm. In this approach the rank is decided based on the amount of pheromone deposited by the ants. Depending upon the rank, the ant the amount of pheromone deposit can be decreased.

1.4.1 Updating the Pheromone

There are two prior steps needs to be implemented for updating of the pheromone. The First step is to sort the ants by their increasing tour length. Second is the value of the pheromone that an ant can deposit is decided based on the rank(r) assigned.

The only condition is, only best-so-far ants are allowed to deposit a pheromone. The pheromone update can be implemented by the following formula:

$$\mathcal{T}_{ij} \leftarrow \mathcal{T}_{ij} + \sum_{r=1}^{w-1}(w - r)\Delta \mathcal{T}_{ij}^r + w\Delta \mathcal{T}_{ij}^{bs} \tag{5}$$

where,

$$\Delta \mathcal{T}_{ij}^r = 1/C^r \text{ and } \Delta \mathcal{T}_{ij}^{bs} = 1/C^{bs}$$

1.5 Max-Min Ant Colony Algorithm (MMACA)

Compared to previous algorithm where only ranks are assigned, this algorithm introduces the range of ranks which are known as Maximum T^{max} and Minimum T^{min}. This is a reason why it is known as Max-Min algorithm. The other important improvement is that the best-so-far found ant can deposit the pheromone only. The other improvement is related to pheromone trails, where the upper limits are used to start a pheromone trails. Finally, it also suggests re-initializing the pheromone whenever the system reaches to immobility phase. Immobility is the phase where ant cannot move any where during the tour.

1.5.1 Updating the Pheromone

For the updating of pheromone following formula is used:

$$\mathrm{T}_{ij} \rightarrow \mathrm{T}_{ij} + \Delta \mathrm{T}_{ij}^{best}, \tag{6}$$

where, $\Delta \mathrm{T}_{ij}^{best} = 1/C^{best}$,

$\Delta \mathrm{T}_{ij}^{best} = 1/C^{bs}$ (in best-so-far case) and $\Delta \mathrm{T}_{ij}^{best} = 1/C^{bt}$ (in iteration best tour)

2 Analysis and Results

In this paper, we have discussed four major algorithm of Ant Colony like Ant System Algorithm (ASA), Elitist Ant System Algorithm (EASA), Ant System based on Rank (ASR), and Max-Min Ant Colony Algorithm (MMACA). These algorithms are tested on a simulation tool known as TspAntSim.

Different Test Instances are selected from a library for TSP instances known as TSPLIB. The following table displays results of the best tour found as well as standard deviation for the optimum path of the travelling sales person problem.

The parameter selected for the algorithm is Alpha = 3, Beta = 3 and population size taken is 10 and number of tour selected is 100 (Table 1).

From the table it can be seen that Max-Min Ant System proves better result among all in most of the test instances. But to find out best tour the standard deviation is more compared to other algorithm.

Table 1 Results of best tour iteration and standard deviation of ants algorithm

Sr. no	Test instances	ASA		EASA		ASR		MMACA	
		Best tour found	Standard deviation	Best tour found	Standard deviation	Best tour found	Standard deviation	Best tour found	Standard deviation
1.	Att532.tsp	109749	869.011	18346	517.105	22179	8.100	17388	188.700
2.	Berlin52.tsp	107792	2949.420	18555	745.905	21625	2.750	18045	2059.113
3.	d198.tsp	17903	1420.012	17585	189.066	21208	360.00	17335	1421.459
4.	eil101.tsp	17861	1590.418	17591	196.548	22360	2813.40	17553	1205.055
5.	eil51.tsp	17853	1359.857	18373	2269.322	20481	–	18073	1446.796
6.	eil76.tsp	18072	455.375	18521	1028.622	20942	140.40	17243	453.215
7.	kroA100.tsp	17575	999.663	18159	720.885	22989	1663.307	17439	1529.595
8.	kroA150.tsp	18112	437.889	18767	26.700	21502	187.500	17268	1485.139
9.	kroA200.tsp	36373	1879.925	18344	517.382	21303	736.959	17695	3153.698
10.	kroB100.tsp	17795	1894.311	18268	86.866	21189	–	18085	–
11.	kroB150.tsp	17895	861.885	17690	42.398	23217	–	17809	1943.170
12.	kroB200.tsp	17899	717.531	18436	626.51	21226	618.300	16943	1863.841

3 Conclusion

The ant colony optimization algorithms [3] Ant System Algorithm (ASA), Elitist Ant System Algorithm (EASA), Ant System based on Rank (ASR) and Max-Min Ant Colony Algorithm (MMACA) behaves differently on different problem instances. Data sets which are taken from TSPLIB tested on TspAntSim tool [1] and their behavior with various parameters are recorded. Their performances are also compared with the other algorithm as well as other data sets. This study and analysis can be useful to the novice researches to understand the swarm intelligence algorithm and their approach towards solving NP-Hard problems.

4 Future Aspects

More AI like algorithm can be tested on these instances as well as more number of instances can generate usefull analysis for further research. The problems can be generated manually and it can be tested on real-time basis. The same approach can be implemented in Linux-based environment to perform customized empirical formulae.

Screen Shots of TspAntSim Tool

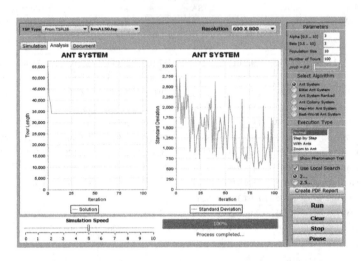

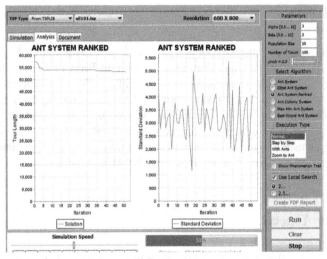

References

1. Aybars, U., & Dogan, A. (2009). An interactive simulation and analysis software for solving TSP using Ant Colony Optimization algorithms. *Journal of Advances Engineering, 40*(5).
2. Dorigo, M., & Stutzle, T. (2004). *Ant colony optimization*. London: The MIT Press.
3. Kumar, B., & Das, G. (2011). Ant colony optimization. A Computational Intelligence Technique. *International Journal of Computer & Communication Technology (IJCCT), 2*(VI).
4. Maniezzo, V., Colorni, A., & Dorigo, M. (1994). The ant system applied to the quadratic assignment problem. Technical Report IRIDIA/94-28, IRIDIA, Universit´e Libre de Bruxelles, Belgium.

5. Shweta, K. M., & Singh, A. (2013). An effect and analysis of parameter on ant colony optimization for solving travelling salesman problem. *IJCSMC, 2*(11), 222–229.
6. Jun, Z., Xiaomin, H., Tan, X., et al. (2006). Implementation of an Ant Colony Optimization technique for job shop scheduling problem. *Transactions of the Institute of Measurement and Control., 28*(1), 93–108.
7. Toley, C., & Bhagat, B. (2014). An application of ant colony optimization for software project scheduling with algorithm in artificial intelligence. *IJAIEM, 3*(2), 149–153.
8. Doctor, S., & Ganesh, K. (2004). Unnamed vehicle navigation using swarm intelligence ICISIP. IEEE (pp. 249–253).
9. Costa, D., & Hertz, A. (1997). Ants can colour graphs. *Journal of the Operational Research Society, 48*, 295–305.
10. University Heidelberg, Institute for Informatics, Germany. http://comopt.ifi.uni-heidelberg.de/software/TSPLIB95/.
11. Applegate, V., & Chv_atal, W. C. (1990). Lower bounds for the travelling salesman problem. In \TSP '90", CRPC Technical Report CRPCTR90547, Center for Research in Parallel Computing, Rice University.
12. Dorigo, M., Middendorf, M., & St¨utzle, T. (Eds.), *Abstract proceedings of ANTS2000 – From Ant Colonies to Artificial Ants: A Series of International Workshops on Ant Algorithms* (pp. 7–9) ANTS. Universit´e Libre de Bruxelles, September.

Evaluating Performance of Reactive and Hybrid Routing Protocol in Mobile Ad Hoc Network

Vaishali V. Mandhare and Ravindra C. Thool

Abstract Recently, a number of researches have been done on Mobile Ad hoc Network (MANET). The network form temporarily without any infrastructure or centralized administration. To give quality of service (QoS) support to MANET, different routing protocols are proposed. Route routing and selection of protocol is the first stage of different applications of MANET. Before selection of protocol, there is a need to compare the protocol. This paper evaluates the performance evaluation of reactive (Dynamic Source Routing) and hybrid (zone routing protocol) routing protocol. Simulation is done using different quality of service metrics like Packet Delivery Ratio, End-to-End Delay, Throughput, Packet Loss, Average Energy Consumption, etc. Parameters use discrete event simulator NS 2 for measuring the performance of protocol. By observing the performance of protocol, this paper concludes that hybrid routing protocols performance is better in terms of delay, packet drop rate, energy, etc. Remaining parameters are better for DSR protocol as packet delivery ratio, throughput, etc.

Keywords MANET · Reactive protocol · Hybrid protocol · DSR · ZRP

1 Introduction

Nowadays, tremendous growth in communication is taking place using different devices like laptop, palmtop, mobile handset, etc. Due to heavy resource requirement, traditional wired network is not suitable for different applications. Wireless network is free, error prone due to dynamic nature of the network; also it is easily

V.V. Mandhare (✉) · R.C. Thool
Department of Information Technology, S.G.G.S.I.E & T, Vishnupuri, Nanded
e-mail: v_mandhare@yahoo.com

R.C. Thool
e-mail: rcthool@yahoo.com

© Springer Science+Business Media Singapore 2016
S.C. Satapathy et al. (eds.), *Proceedings of International Conference on ICT for Sustainable Development*, Advances in Intelligent Systems and Computing 408, DOI 10.1007/978-981-10-0129-1_60

deployable and less cost effective network over traditional wired network. Device is called node and every node wants to communicate with each other for sending data. Sometimes it happens that node is out of range; at that time, intermediate node acts as router and forwards the packet to the destination and seems communication is like multihop wireless network with dynamic technology. This type of network is called Mobile ad hoc network (MANET) [1]. Mobile ad hoc network is Infrastructureless, self-configuring, and self-healing network. Nodes communicate with each other without any administration. The primary challenge in building a MANET is equipping each device to continuously maintain the information required to properly route traffic. Such networks may operate by themselves or may be connected to the larger Internet. They may contain one or multiple and different transceivers between nodes. This results in a highly dynamic, autonomous topology [2]. MANET form temporary network without any centralized administration. Routing plays important role in the communication for sending the data packets to the destination and thus required routing protocol. Recently, huge amount of researchers focus on improving performance of protocol in MANET environment [3]. It also tries to improve performance of routing protocol by improving quality of service. Internet Engineering Task Force (IETF) form working group (WG) for MANET forms the group that mainly focus on improving performance of MANET.

Lots of research has been done to evaluate performance of protocol like AODV, DSR, DSDV, TORA, OLSR, and ZRP. Among PAODV, AODV, CBRP, DSR, and DSDV [4], among DSDV, DSR, AODV, and TORA [5], among SPF, EXBF, DSDV, TORA, DSR, and AODV [6], among DSR and AODV [7], among STAR, AODV, and DSR [8] and among AM Route, ODMRP, AMRIS and CAMP [9]. This paper mainly compares reactive (DSR) and hybrid (ZRP) routing protocol using various parameters.

The rest of the paper is organized as follows. In Sect. 2 descriptions about routing protocols are present. Section 3 describes about simulation environment for routing protocol. Section 4 gives result and discussion of routing protocol. Section 5 present conclusion and future scope.

2 Routing Protocol

2.1 Dynamic Source Routing Protocol (DSR)

Dynamic source routing protocol (DSR) [10] is on-demand routing protocol construct especially to restrict the bandwidth consumed by control packets generated in ad hoc wireless network. This protocol eliminate periodic table update message which is required in table driven approach.

The main difference between this and other on-demand protocol is that DSR is beaconless protocol. To ensure its presence it does not send periodic hello packets. DSR consist of three steps as route request, route replay, and route maintenance phases [11]. It first flooded Route Request (RREQ) packets throughout the network, after receiving RREQ packets nodes who receive this packet give Route Replay (RRPLY) to that node.

When source node who not finding destination flooded RREQ packets. If the node that receives RREQ was not intended destination then it again flood RREQ packets to it neighbor, provided the packet time to live (TTL) counter has not exceeded. Each RREQ packet consists of sequence number. When node receive packet firstly check sequence number before sending it to destination. This reduces the loop formation and repetition of one packet duplication. Destination node after receiving RREQ packets gives RRPLY to the intended source node by using shortest path.

Route cache is also used during route reconstruction phase. If intermediate node receiving RREQ and it has route entries in its cache then it directly replay to source node by giving information about route. Several optimizations have been done using route cache to improve the performance of protocol. If link or route is broken, then that information is deleted from route cache by updating the cache hence it ultimately affect the performance of routing protocol. During network partition, affected node initiates RREQ packets. An exponential back off algorithm is used to avoid frequent RREQ flooding. Disadvantage of this protocol is that route maintenance mechanism dose not locally repair broken link. Also stale route cache entries are creating problem in route construction phase. Connection setup delay is also higher then table driven routing protocol. Routing overhead is directly proportional to path length.

2.2 Zone Routing Protocol (ZRP)

Zone routing protocol (ZRP) [12] is hybrid routing protocol which adapts best feature of both proactive and reactive routing protocols. The main concept using this protocol is that it used proactive scheme inside the zone and reactive scheme outside the zone. First, zone radius is formed and after that communication is taken place. An intra-zone routing (IARP) protocol is used inside the zone and interzone routing (IERP) protocol is used outside the zone. Each node is capable of maintaining routing information within its routing zone by extracting periodic update messages or packets. Hence, the larger the routing zone, higher the routing control messages [13].

The path finding processes may result in multiple route replay packets reaching the source. Amongst all this source nodes use the shortest path as optimal path for sending the packets to the destination. If link is break then configuration is done. Broken link is bypass by searching alternate path for reaching the packet to the destination. After, path update message is send throughout the network to inform about broken link. By combing the both the approaches it reduces control overhead. In absence of query control message ZRP tends to produce higher control overhead then other scheme. Query control packets must ensure that redundant information is not propagated throughout the network. Also, the zone radius must impact on performance of routing protocol.

3   Simulation Environment

Simulation Environment and parameters are very useful for any type of communication. Here we use discrete event simulation NS2 [14]. Carnegie-Mellon university monarch group develop and support for simulating multihop wireless network using different layers model in NS2 [15]. Different parameters are used for simulation environment as radio model used is Two Ray Ground, TCP traffic is used. Two types of protocols are used as reactive (DSR) and hybrid (ZRP) routing protocol. 512 packet sizes are used, Maximum speed for generated scenario is 2 m/s. Area used for scenario is 800 * 800. Total number of node used for scenario is 30. MAC 802.11 standard is used. Different pause time used are 10, 20, 30, 40, 50. Total simulation time used for environment is 300 ns.

4   Result and Discussion

4.1   Packet Delivery Ratio

Figure 1 interpreted that packet delivery ratio of DSR protocol is high as compare to ZRP routing protocol. Reason behind it is that ZRP is hybrid routing protocol and it uses both approaches as proactive and reactive. Hybrid protocol first form zone radius after that it communicate inside the zone. When communication inside the zone is done then it communicates outside the zone. Hence time required for packet to reach to the destination is high as compare to DSR routing protocol. Pause time increases ZRP protocols performance lowers down and in case of DSR protocol for 20 pause times it increases and after that it remains steady for next pause time.

4.2   End-to-End Delay

From Fig. 2 demonstrated that DSR protocol required higher delay than ZRP routing protocol. Time required for packet to reach to destination in DSR protocol is high because when route or path is required it first goes into cache to check that route is available or not. If route is available then it is ok otherwise it again send route request packet. Hence time required for packet to reach to destination is high

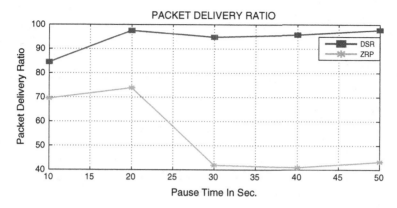

Fig. 1 Packet delivery ratio versus pause time

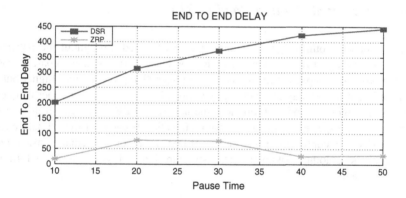

Fig. 2 End-to-end delay versus pause time

which degrades the performance of protocol. In other side ZRP is hybrid protocol which requires less delay as pause time increases. But for DSR protocol as delay is directly proportional to pause time.

4.3 Throughput

Figure 3 stated that throughput of ZRP protocol lowers down as pause time increases but for DSR protocol as pause time increases throughput increases. Throughput of DSR protocol is better as compare to ZRP protocol for different pause time as it increases.

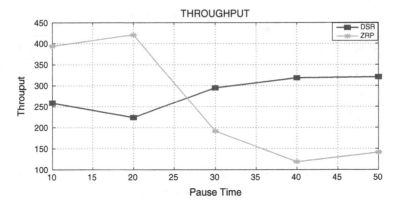

Fig. 3 Throughput versus pause time

4.4 Normalized Routing Load (NRL)

Routing load is nothing but the control packets flooded throughout the network. From Fig. 4 say that routing load of DSR protocol is very less as compare to ZRP routing protocol. As pause time increases load over ZRP protocol goes higher but in DSR protocol it is steady and very less. Reason behind it is that number of query control messages is flooded throughout the network in ZRP protocol. It uses large number of messages to exchange for communication because of hybrid nature and absence of query control message. In case of DSR protocol, on-demand basis routes are established; doing lookup in cache structure hence less overhead in forming route is present.

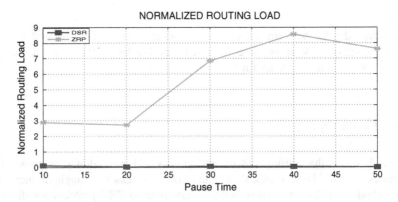

Fig. 4 Normalized routing load versus pause time

4.5 Number of Packets Generated

Packet generated rate of ZRP protocol is high for lower pause time but when pause time increases it goes down shown in Fig. 5. In DSR, protocol rate is change randomly as pause time change or increases (Fig. 6).

4.6 Number of Packets Received

Packet received rate of ZRP protocol goes down as pause time increases but for DSR protocol it change randomly. Because of dynamic nature of topology path or route may break and packet get lost. Received rate of DSR protocol is better as compare to ZRP in increasing pause time situation.

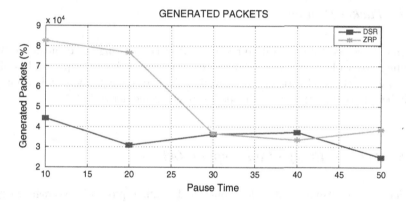

Fig. 5 Number of packet generated versus pause time

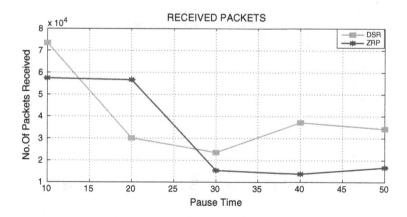

Fig. 6 Number of packets received versus pause time

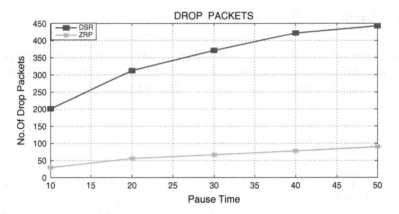

Fig. 7 Drop versus pause time

4.7 Packet Dropped

Figure 7 depicts that packet drop rate of DSR protocol is high as compare to ZRP routing protocol. As pause time increases drop rate of both the protocol is also increases. ZRP uses hybrid approach while DSR is reactive protocol and also large number of control messages is flooded throughout the network.

4.8 Average Energy Consumption (AEC)

Figure 8 demonstrates AEC for considered routing protocol. Average energy required by DSR protocol is high as compare to ZRP routing protocol. As pause

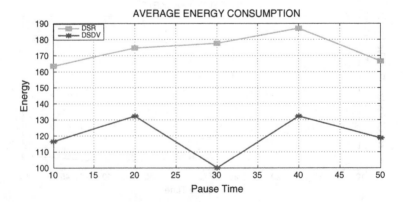

Fig. 8 Average energy consumption versus pause time

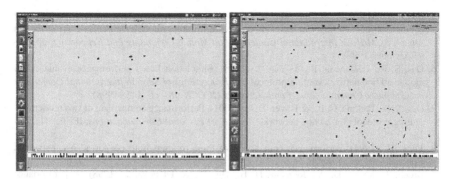

Fig. 9 Snapshot of simulation for DSR and ZRP protocol

time increases both the protocol behave differently using same simulation environment. Battery power is the main issue in MANET routing protocol.

Figure 9 illustrate snapshot for DSR and ZRP protocol simulation for 20 and 50 nodes, respectively, using zero pause time.

5 Conclusion and Future Scope

In this paper, simulation of two types of protocols is presented as reactive (DSR) and hybrid (ZRP) routing protocol using NS2. From the results and discussion, ZRP protocol is better in terms of end-to-end delay, packet drop rate, energy consumption, generated packet rate, etc. But it is worse in case of packet delivery ratio, throughput, and normalized routing load. The DSR protocol performance degrades due to cache problem. To solve the cache staleness issue in reactive routing protocol is the future task.

References

1. Lumpur, K., & Rahman, M. A.. (2010). A simulation based performance comparison of routing protocol on mobile ad-hoc network (Proactive, Reactive and Hybrid). In *International Conference on Computer and Communication Engineering (ICCCE 2010)*, Vol. 1, pp. 1, 5, May (2010).
2. Baraković, S., & Baraković, J. (2010). Comparative performance evaluation of mobile ad hoc routing protocols. In *MIPRO, 2010 Proceedings of the 33rd International Convention*, 24–28, pp. 518, 523, May (2010).
3. Royer, E. M., Barbara, S., & Toh, C.-K. (1999). A review of current routing protocols for adhoc mobile wireless networks. In *IEEE Personal Communication*, pp. 46–55, April (1999).
4. Boukerche, A. (2004). Performance evaluation of routing protocols for ad hoc wireless networks, Journal of *Mobile Networks and Applications*, 9(4), 333–342, (Kluwer Academic Publishers).

5. Broch, J., Maltz, D. A., Johnson, D. B., Hu, Y.-C., & Jetcheva, J. (1998). A performance comparison of multihop wireless ad hoc network routing protocols. In *Proceedings of the 4th Annual ACM/IEEE International Conference on Mobile Computing and Networking*, pp. 85–97, October(1998).

6. Das, S. R., Castaneda, R., & Yan, J. (1998). Simulation based performance evaluation of mobile ad hoc network routing protocols. In *Proceedings of Seventh International Conference on Computer Communications and Networks (ICCCN'98)*, pp. 1–5 (1998).

7. Das, S. R., Perkins, C. E., & Royer, E. M. (2001). Performance comparison of two on-demand routing protocols for ad hoc networks. *Journal of Personal Communications*, IEEE, 8(1), 16–28.

8. Jiang, H. (1994). Performance comparison of three routing protocols for ad hoc networks. *Communications of the ACM*, 37, 1–5.

9. Broch, J., Maltz, D. A., Johnson, D. B., Hu, Y., & Jetcheva, J. (1998). A performance comparison of multi-hop wireless ad hoc network routing protocols. In *Proceedings of the Fourth Annual ACM/IEEE International Conference on Mobile Computing and Networking, MobiCom'98*, pp. 25–30, October (1998).

10. Johnson, D. B., & Maltz, D. A. (1996). Dynamic source routing in ad hoc wireless netwok. Mobile computing, Kiuwer Academic Publishers, Vol. 353, pp. 153–181, May (1996).

11. Almutairi, A., & Hendawy, T. (2011). Performance comparison of dynamic source routing in ad-hoc networks. In *IEEE-GCC Conference*, Dubai, 19–22 February (2011).

12. Hass, Z. J., & Pearlman, M. R. (2003). The zone routing protocol for ad-hoc network, <draft-ietf-manet-zone-zrp-04.txt>, January (2003).

13. Siva Ram Murthy, C., & Manoj, B. S. (2004). *Wireless ad-hoc network- architecture and protocol*, Pearson Publication.

14. Network simulator tutorial. http://csis.bitspilani.ac.in/faculty/murali/resources/tutorials/ns2.htm.

15. Greis, M., Network simulator tutorial. http://www.isi.edu/nsnam/ns/tutorial/.

Sentiment-Based Data Mining Approach for Classification and Analysis

Viral Vashi and L.D. Dhinesh Babu

Abstract Traditionally, individuals gather feedback from their friends or relatives before purchasing an item but today the trend is to identify the opinions of a variety of individuals around the globe using micro blogging data and twitter is such one famous micro blogger where user expresses their view in form of tweets. In this project, we have provided a data mining approach based on how nowadays user provides their view in tweets. This approach basically is around sentiment expressed in tweets to classify them into set of feature and also to analyze tweet to compare service or product-based organization.

Keywords Data mining · Classification · Sentiment analysis · Twitter analysis

1 Introduction

1.1 Data Mining

Data Mining or knowledge extraction is a method well known for pattern extraction and to make data more understandable. Data mining is a root or basic to many sub category in it like business mining, medical mining, music mining, text mining, etc., and every year something new is add as an improvement over previous methods or as an innovation. With these new techniques Data mining has become important element in many fields and people are using it more with all their historic data to get that extra knowledge.

V. Vashi (✉) · L.D. Dhinesh Babu
SITE, VIT University, Vellore, India
e-mail: viral2vashi@gmail.com

© Springer Science+Business Media Singapore 2016
S.C. Satapathy et al. (eds.), *Proceedings of International Conference
on ICT for Sustainable Development*, Advances in Intelligent Systems
and Computing 408, DOI 10.1007/978-981-10-0129-1_61

581

1.2   Text Mining

Text mining is one of data mining variation where as input we take unstructured text and try to find some information from it which is more useful then text. There are many research fields in text mining as well like Academic applications, Online media applications, Opinion mining, Sentiment analysis, etc., our focus here will be on Sentiment analysis (Figs. 1 and 2).

1.3   Sentiment Analysis

Sentiment Analysis is a text mining method and it is a study of someone's sentiments about something like good–bad or positive–negative, etc. People always put sentiment analysis and opinion mining as same but there is difference between and

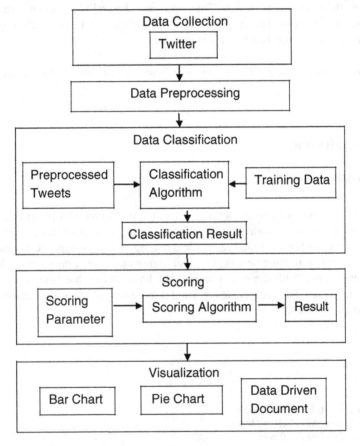

Fig. 1 Sentiment analysis process for tweets as data

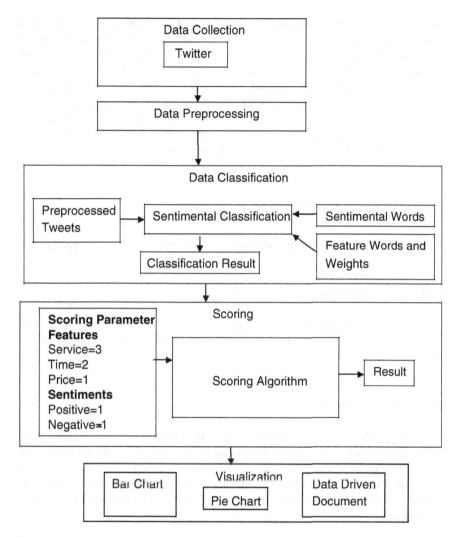

Fig. 2 Proposed sentiment-based classification and analysis flow

many researchers have stated that both have different meaning and opinion mining's focus is on opinion of people where sentiment analysis focuses on their sentiment [1].

1.4 What Is the Need of Sentiment Analysis?

With the boost of internet usage we have different source of information which can be very useful. One such information source is micro blogging websites where user talks about various topics and put their views about it. They also express their felling about things like product they use or service they get and try to put their experience in it. This information is very useful to organization that is proving those service or product to know sentiments of customer as they get direct feedback from customer. This information can also be achieved in real time as it is been observed that people express their view as fast as possible via micro blogging site once such site is twitter.

Thus, sentiment analysis is important and useful methods to most organizations as it helps improve knowledge of customer experience and thus leading to fast and necessary taking of steps as a result of this knowledge.

1.5 Sentiment Analysis Process

From previous researches, I put sentiment analysis process in seven steps [1–6].

1.5.1 Data Collection

At this stage, unstructured text data as input is taken. Source of data can be text file or database containing text or directly from some micro blogging website like twitter.

1.5.2 Data Preprocessing

Data once collected is preprocessed as per requirement of user. Preprocessing usually means making data more meaningful by applying different techniques such as removing punctuations from text [2]. There are no specific steps that needed to be followed in this stage and as per once approach they can decide how they want to do preprocessing. Some times to get better result some preprocessing methods are skipped.

1.5.3 Sentiment Identification/Extraction

Sentiment affects the outcome and type of sentiment depends on type of data. At this stage we try to extract different words from text that are more likely to influence the result [1].

1.5.4 Feature Extraction

Based on kind of data we have different set of features' attached to it are extracted from data. This are used for better understanding of peoples sentiment toward specific feature as when people talk on micro blogger they express their view about one feature more than toward whole product [1, 5].

1.5.5 Sentiment Classification

At this stage we actually classify different text to sentiments they express. There are different classification methods that are used such as naive bayes [1, 7].

1.5.6 Scoring

This is not a mandatory step but it makes data more understandable. Here we give score to text based on sentiment expressed in it.

1.5.7 Visualization

At last, knowledge extracted is visualized for better understanding via different graphs or numeric representation.

2 Related Work

Twitter, from the past few years, has become source of information to both end of market, i.e., customer and product/service providing organization. Customer uses twitter to express their views about product/service and organization to provide information about product/service.

Apart from this information, tweets of twitter can also be used as source for data mining to extract some useful information for customer and product/service providing organization [2].

Many researchers have provided different ways to collect preprocess and mine twitter data in different ways. There is no specific order of step that one should follow for sentiment analysis and there are many areas that needed to be explored [1–6, 8–11].

Unlike other data tweets are not at all of some specific type and they differ on user and his way to express felling. Also researcher have observed that year after year we are having seen drastic change in language people uses in their tweets also with Varity of #tags and emoticons [12, 13, 14].

Thus, in this section from previous researches I have provided different methods researches have used at different stage of sentiment analysis on twitter tweets and finally have provided one come process that is carried out for sentiment analysis on tweets.

2.1 Data Collection

Tweets are directly collected from twitter with respect to on what thing and how someone is doing sentiment analysis. Tweets can be related to some specific user account, e.g., @olacabs to some specific topic, e.g., #CWC2015.

2.2 Data Preprocessing

This is important step in sentiment analysis process as here we clean data and put it in table for efficient access. Basically, at first we try to reduce data to some specific time frame or location of when and from where users have tweeted it; this step is only performed if necessary.

Following are the different tweets preprocessing techniques.

2.2.1 Eliminating Stop Words

Stop words are words that are filtered before using data and have no use in further process. List of stop word depends on what we are doing but most common stop words are the, who, that, this, which, etc. [8].

2.2.2 Stemming

It is process used in natural language process where we replace derived word with its original form, e.g., "argue", "argued", "argues", "arguing," and "argus" reduces to the stem "argu". As specified in [] and also a fact that people tweeting comes from different background and they have their own ways to describe a similar situation and while doing analysis it is necessary that we try to reduce it to one word for better understanding and this is one way of doing it [8].

2.2.3 Spelling Correction

People tweet in informal language and in hurray they usually makes spelling mistake. This needs to be replaced with original spelling and there are number a of dictionaries available which makes incorrect spelling to correct one [8].

2.2.4 Replacing Truncated Words

Tweets usually have more truncation of words as twitter only allows 140 characters in single tweet. This truncated word need to be replaced into original form.

2.2.5 Emoticons Replace

Emoticons or smiley is a truncated way to describe felling, e.g., ☺ for "Happy." They are widely used nowadays and are important factor to sentiment analysis and thus emotions are needed to be replaced to word they express [4, 6, 8].

2.2.6 Filtering

Twitter allows user to highlight a topic with # and some other user mention with @. Basically #tag is important thing but # need to be removed before we use data as word following # will not match actual word like #Happy will not match to Happy. For @ in sentiment analysis specifically is of no use as we are only concerned with kind sentiment tweet is expressing and towards whom is already specified while collecting data so any other user referral in tweet is useless [8].

2.2.7 Identification of Punctuations

People uses punctuation in different ways: to create an emoticon is one way which is useful for our analysis but when used due to over excitement or shock is of no use, e.g., WHAT??????????? Or Congratulations!!!!!!!!!!!!! This "?" and "!" should be removed before using data [4, 8].

2.2.8 Lower Casing

People always try to be creative and use lower-upper case for it in tweets as character is only they have to do so like "HAppY" this bring inconsistency in data. Thus, this step is necessary during preprocessing as it provides consistency around data [4].

2.2.9 Compression of Words

As said tweets contain very informal language and people try different thing to express their felling one such way is to stress word usually this is done to express stronger felling like "happyyyyyyyyy" have more felling than just happy [4].

2.2.10 URL Removal

Tweets can also contain url which people uses to direct others to related topic or expanded topic this url if crawled can provide exact sentiment for that specific tweet but it is costly and researchers usually eliminates url from tweets and uses remaining data for analysis [4].

2.3 Sentiment Classification

At this stage of sentiment analysis we try to predict the polarity of tweet towards sentiment, i.e., telling what sentiment those a tweet suggests. This is done using Classification methods. Researcher has used different classification technique based on their requirement. That is available.

Some widely used classification techniques are as follows:

- Naïve Bayes Classifier
- Max Entropy Classifier
- Boosted Trees Classifier
- Random Forest Classifier

Detailed information on classification algorithm and how they are used can be found in paper [1, 7, 9, 15].

2.4 Scoring

Scoring is numeric representation of polarity of tweets. Scoring can be done in many ways and depends on scoring parameter, e.g., number of positive and negative words used in tweets. Result of this step is also used for visualization for better understanding.

2.5 Visualization

Once classification is done data can be used for visualization directly or after applying some scoring algorithm. Usually different graphs are used to display data such as bar char or pie chart. But recent trends have shown innovative visualization technique which can also work dynamically one such example is data driven document.

3 Proposed Work

In our sentiment analysis process, we have fixed feature as TIME, PRICE, and SERVICE this feature are common in any product or service-based organization also we are classifying tweets in two sentiments positive or negative.

Following are process step that need to be applied to all organization that we are comparing

3.1 Data Collection

Data are collected from twitter directly by using account name of organization user name on micro blogger.

3.2 Data Preprocessing

As we have discussed different data preprocessing methods. In our method, only thing that need to be followed is do emoticon replace before punctuation removal else all smiley's will be removed and thus this need to be done.

3.3 Classification

Our approach uses sentiment expressed in tweets for classification. As we know that people may express their view about more them one feature of product it is necessary to have classification algorithm where we can classify single tweet into more than one class but we also have to be sure that tweet is really about that class thus for that we use sentiment for classification.

Here, we try to find which words related to some feature in tweet is expressed closest to sentimental word in tweet and this we use to classify tweet into classes.

Following are steps for Sentiment-based classification

Step 1: Assign each feature with unique <Fid> and some weight should be less than 140.
Step 2: Find Sentimental words, i.e., Positive or Negative in tweet by comparing each word with existing set of positive and negative words and replace it with <A>.
Step 3: For each feature find words matching in tweet with existing set of words associated with that feature and replace it with feature id <Fid>.
Step 4: For each <Fid> in tweet find nearest <A>.
Step 5: For each unique <Fid> in tweet select one with minimum distance to <A> and store in <MinFid>.
Step 6: For each <MinFid> if value is less then weight associated with <Fid> then Classify tweet to feature associated with <Fid>.

3.3.1 Algorithm

```
SentimentalClassification(tweet[],sentimentalWords[],featureWords[][],featureWeight[])
{
        For ( each t in tweet[])
        {
                For each (word w in t)
                {
                        if (match(w,sentimentalWords[]))
                        {
                                w="A";
                        }
                        Else
                        {
                        For (each f in feature[])
                        {
                                if (match(w,featureWords[f][]))
                                {
                                        w=feature[f];
                                }
                        }

                }

                For each (word w in t)
                {
                        For (each f in feature[])
                        {
                                if (w==feature[f])
                                {
                                        distance[f][]=mindistanceA();
                                }
                        }
                }
                For (each f in feature[])
                {
                        mindis[f]=min(distance[f][]);
                }
                For (each f in feature[])
                {
                        If(mindis[f]< featureWeight[f])
                        {
                                Classify[t][f]=1;
                        }
                        else
                        {
                                Classify[t][f]=0;
                        }
                }

        }
}
```

4   Scoring

Scoring in our method depend upon the scoring parameter as we have three feature and two sentiment we specific number attach to each of them. For sentiments positive is +1 and negative is −1 this is fixed and for feature it depends on one who is doing sentiment analysis. We can put feature in order of 3, 2, and 1 where 3 is most important feature and 1 is list.

Scoring Method used is
$$T() = (Px * Tm) - (Nx * Tm)$$
$$P() = (Px * Pm) - (Nx * Pm)$$
$$S() = (Px * Sm) - (Nx * Sm)$$
$$F() = (T() + P() + S())/3$$

where

T() is function to calculate score for Time feature of User account or organization.
P() is function to calculate score for Price feature of User account or organization.
S() is function to calculate score for Service feature of User account or organization.
F() is to calculate final score of an User account or organization.
Px is number of positive tweet found through classification process.
Nx is number of negative tweet found through classification process
Tm is predefined Time multiplier which was fixed before scoring starts.
Pm is predefined Price multiplier which was fixed before scoring starts.
Sm is predefined Service multiplier which was fixed before scoring starts.
This data can then be used for visualization for better understanding.

4.1   Visualization

From scoring result we can now visualize data in terms of sentiments or score and compare different organization with what score they have in different visualization technique we have. Like bar chart or data driven document.

4.2 Algorithm

```
Sentiment_Analysis(Tm=Time Multiplier, Sm=Serivce Multiplier, Pm=Price Multiplier)
{
X[]=Collected Tweets;
For each x in X[Text]
{
        x=EmoticonReplace(x);
        x=StopWordRemove(x);
        x=PuntuationRemove(x);
        x=URLRemove(x);
        x=Filter(x);
        x=SpellCorrect(x);
        x=LowerCasing(x);
}
For each x in X[]
{
        X[Sentiment]=navi_bayes(x);
}

For each x in X[]
{
        T[]=SentimentalClassification(X[],sentimentalWords[],TIME,Weight);

}
For each x in X[]
{
        S[]=SentimentalClassification(X[],sentimentalWords[],SERVICE,Weight);

}
For each x in X[]
{
        P[]=SentimentalClassification(X[],sentimentalWords[],PRICE,Weight);

}
Pt=Count("Positive",T[]);
Nt=Count("Negative",T[]);
Ps=Count("Positive",S[]);
Ns=Count("Negative",S[]);
Pp=Count("Positive",P[]);
Np=Count("Negative",P[]);

Tv=(Pt*Tm)-(Nt*Tm);//Time Score
Sv=(Ps*Tm)-(Ns*Tm);//Service Score
Pv=(Pp*Tm)-(Np*Tm);//Price Score

Fv=(Tv+Sv+Pv)/3;//Final Score

Visualize();

}
```

5 Result and Discussion

To show how our process works and what about it gives we have used data mining tool R project which works with S or R programming language.

As input we have to give three training data set of historical tweets for TIME, PRICE and SERVICE. This data set is already classified into positive or negative. Also we will fetch 100 current tweets from each of user account which we are comparing using twitter package in R.

This data will then be preprocessed and classified and finally the score will be given. This score then will be represented in visual form.

Following is example of how actual tweet look like before and after preprocessing.

Actual Tweet: "@A: you are a sham! Only promise what you can deliver! To other fellow travelers, #avoidolacabs for hassle free travels! #lateflight ☹" Preprocessed Tweet: "sham Only promise can deliver To fellow travelers avoidolacabs hassle free travels lateflight SAD"

Note: Actual user account name has been changed to A

Now this tweet is classified as negative or positive with respect to each of 3 features. Based on classification scoring is done.

Following are the graphical result of comparison of three cab services normalized to Zero onwards. For this comparison TIME relates to late or early with respect to time of booking, PRICE relates to charges and SERVICE relates to how good or bad service is

Parameter for scoring has value
TIME = 3
SERVICE = 2
PRICE = 1

Figures 3,4,5 6 and 7 shows that C is the best cab service providing company with respect to all TIME, PRICE and SERVICE

Fig. 3 Aggregated score

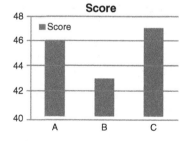

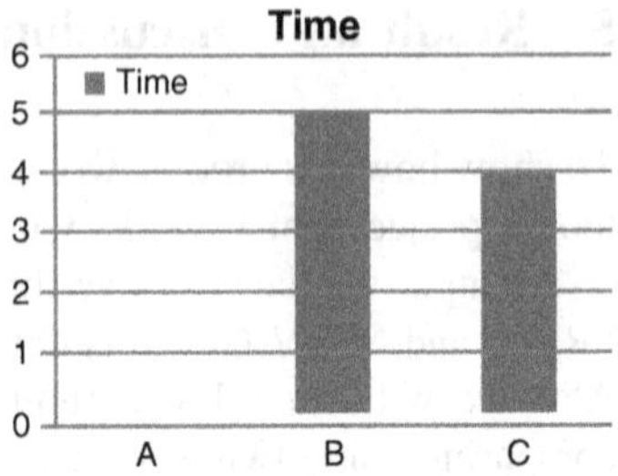

Fig. 4 Time score comparison

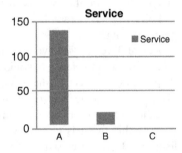

Fig. 5 Service score comparison

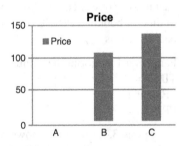

Fig. 6 Price score comparison

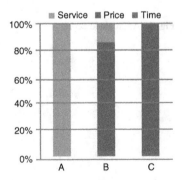

Fig. 7 Time, service, and price comparison in one

6 Conclusion

From our result, we have shown that with single semantic analysis process we can carry out comparison of similar service or product-based and company-based on the parameters we have set.

References

1. Medhat, W., Hassan, A., & Korashy, H. Sentiment analysis algorithms and applications: A survey. Ain Shams.
2. Engineering Journal (2014). 5, 1093–1113.
3. A. Agarwal, Xie, B., Vovsha, I., Rambow, O., & Passonneau, R. Sentiment analysis of twitter data.
4. Vo, B.-K. H. & Collier, N. (2013). Twitter emotion analysis in earthquake situations. *IJCLA 4*(1).
5. Singh, P. K., & Husain, M. S. Methodological study of opinion mining and sentiment analysis techniques.
6. *International Journal on Soft Computing (IJSC)* 5(1), February 2014. doi:10.5121/ijsc.2014. 5102.
7. Hogenboom, A., Bal, D., & Frasincar, F. Exploiting emoticons in sentiment analysis.
8. Gokulakrishnan, B., Priyanthan, P., Ragavan, T., Prasath, N., & Perera, A. S. (2012). Opinion mining and sentiment analysis on a twitter data stream. In *The International Conference on Advances in ICT for Emerging Regions—ICTer* pp. 182–188.
9. Veeraselvi, S. J., & Saranya, C. Semantic orientation approach for sentiment classification.
10. Montoyo, A., Martínez-Barco, P., & Alexandra (2012). Subjectivity and sentiment analysis: An overview of the current state of the area and envisaged developments Andrés. *Decision Support Systems 53* 675–679.
11. Computational approaches to subjectivity and sentiment analysis: Present and envisaged methods and applications. *Computer Speech and Language 28*, 1–6 (2014).
12. Gupte, A., Joshi, S., Gadgul, P., Kadam, A., & Gupte, A. (2014). Comparative study of classification algorithms used in sentiment analysis. *(IJCSIT) International Journal of Computer Science and Information Technologies* 5(5), 6261–6264.
13. Gamallo, P., & Garcia, M. Citius: A naive-bayes strategy for sentiment analysis on english tweets.
14. Pak, A., & Paroubek, P. (2010). Twitter as a corpus for sentiment analysis and opinion mining. In *Proceedings of the Seventh Conference on International Language Resources and Evaluation (LREC'10)*.
15. Zhao, P., Li, X., & Wang, K. Feature extraction from micro-blogs for comparison of products and services.

Offline Handwritten Sanskrit Simple and Compound Character Recognition Using Neural Network

Jyoti Mehta and Naresh Garg

Abstract The intricacy of Sanskrit simple and compound characters makes recognition a laborious task for the probe. For forming a new compound character, we join two or more characters in different ways. In early days of computer wisdom, Optical Character Recognition is one of the working contents of research. In the field of Optical Character Recognition, pattern recognition is an especial feature of Sanskrit script. The prevalence of compound characters in Sanskrit language is more assimilate to other languages scripts. This paper reports a Levenbug–Marquardit algorithm also known as damped least squares method for recognition of handwritten Sanskrit simple and compound characters with different feature sets. Data was collected from people of different age groups categorized as child, young and old. At the first step, we perform pre-processing, second the main notion of the propound technique is to produce the character image using classifier, which will be directly used to recognize the Sanskrit simple and compound characters. Average recognition rate of 78.4, 80.5, and 77.4 % is achieved from our simulated work on noise free images.

Keywords Pre-processing · Segmentation · Feature extraction · Neural network · Character recognition

1 Introduction

An ancient time, the handwritten Sanskrit text written in palm leaves. It is much difficult to conserve them in the same form. This paper constituents various challenges and approaches concerning offline handwritten Sanskrit character recognition.

J. Mehta (✉) · N. Garg
Department of Computer Science and Engineering, GZS-PTU Campus, Bathinda, India
e-mail: jyoti200991@gmail.com

N. Garg
e-mail: naresh2834@rediffmail.com

© Springer Science+Business Media Singapore 2016
S.C. Satapathy et al. (eds.), *Proceedings of International Conference on ICT for Sustainable Development*, Advances in Intelligent Systems and Computing 408, DOI 10.1007/978-981-10-0129-1_62

Fig. 1 Block diagram

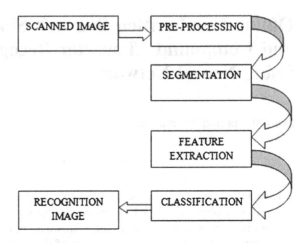

Character recognition can be done either from handwritten or printed documents. Mainly, handwritten character recognition can be done online or offline. Offline is more complicated then online character recognition. In partial, object recognition Sanskrit handwritten is more complicated than other document works, as Sanskrit later has more dihedral angle and modifiers. During recognition process, the researchers face the challenges are attributable to the holes, junctions, several strokes, curves, differing writing styles, sliding and Diagonal images. The steps twisted in handwritten Character recognition (HCR) compose preprocessing, segmentation, binarization, feature extraction, and classification. Figure 1 represents the block diagram of the typical structure of the proposed system.

1.1 Sanskrit Language

Sanskrit is an ancient language. It is the liturgical language of Hinduism, Jainism, and Buddhism. Sanskrit is rarely spoken but written empirical still exists. Sanskrit has 36 consonants, 16 vowels and total 51 characters. The name Sanskrit means

"progressive", "holy", and "sanctified". The corpus of Sanskrit literature comprehends a rich tradition of drama and poetry as well as technical, philosophical and scientific texts. The ancient known text in Sanskrit is "Rig-Veda". In India Sanskrit is one of the 17 official hearth languages (Fig. 2).

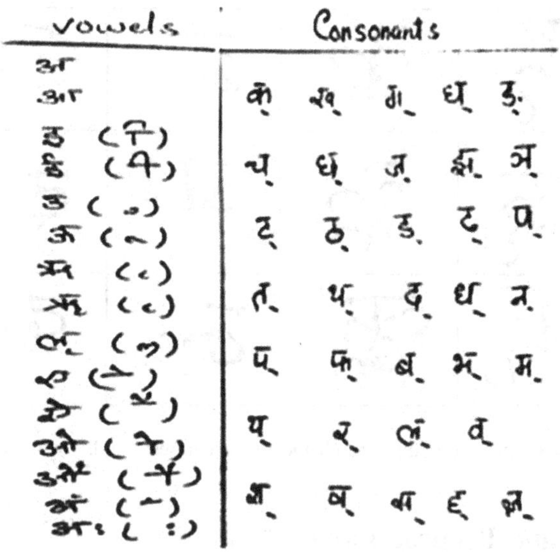

Fig. 2 Vowels and consonants

1.2 Compound Characters

The compound characters in Sanskrit script are derived from Devanagari that lay out following features: The compound characters in Sanskrit are formed by connecting two or more specific consonants. Figure 3 demonstrates some examples.

The consonants in the compound characters are joined in different ways. By removing the vertical head line of a character and then joining another on its prefix side this is the one way of forming the characters, but it is a common approach which can be used by everyone. Another way is to join characters one above the other or lateral by lateral. In certain compound characters, some of the consonants completely changes its conformation and then gets joined to the other.

Fig. 3 Compound characters

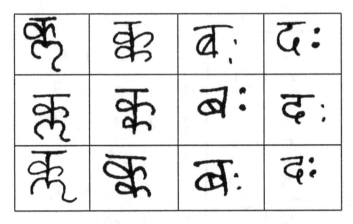

Fig. 4 Dataset of simple and compound characters, categories: child, young, and old

2 Dataset and Preprocessing

Beforehand, 2000 Sanskrit character samples from 1500 different people were collected because there is no standard data set available for handwritten Sanskrit characters. People from different age groups were categorized as child, old and young and were asked to write Sanskrit compound characters on a plain A4 sheet. Since, different writers have different writing styles which lead to variations in data collection and will be useful while training and testing aspect for compound characters. The collected data is scanned by using HP-scan jet at 300 dpi, which provides good quality image with less noise. The digitized images are saved as binary images (0 and 1 pixels) in jpg format. Some of the operations performed during preprocessing are noise removal, size normalization, slant correction, thinning and skeltonization. Thinning and skeltonization operations are very special for feature extraction. The image size is required to be normalized as the size of the character varies from writer to writer and even with same writer from time to time. The input character image is normalized to size of mxn (Fig. 4).

3 Feature Extraction

Converting the input data into set of features is called *feature extraction*. Various Feature Extraction Methods exists such as: Template matching (TM), Deformable templates (DM), Graph description (GD), discrete features, Zoning and Fourier descriptor (FD). Properties of feature extraction are: Aspect Ratio, percentage of pixels above horizontal half point, percentage of pixels to the right of vertical half point, Number of strokes, average distance from image centre, reflected y axis, reflected x axis. Our proposed system gives appropriate results for compound

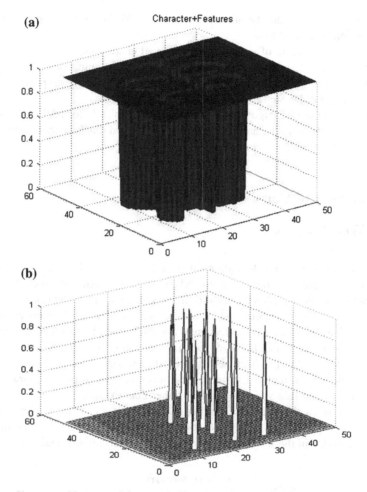

Fig. 5 **a** Character with extracted features. **b** Character with extracted features

character recognition. Figure 5a, b represents the 3D plot of a compound character with end points, holes and junctions. In this figure, red points represents these features.

3.1 Proposed Algorithm

INPUT: Scanned Images. OUTPUT: Recognized Character. Method Begins.

Step 1 Collect the 2000 samples of handwritten images from different age group people.

Step 2 These samples are converted into digitized image.

Step 3 On digitized image Preprocessing operations are performed for removing noise.

Step 4   Morphological operations: Thinning and Skeltonization is performed on preprocessed image.

Step 5   Segmentation of simple and compound characters.

Step 6   Feature Extraction using end points, holes, and junctions.

Step 7   Execute LM Algorithm on 1000 samples for training purposes.

Step 8   Repeat the Step 3–5 for all the database images.

Step 9   Execute LM Algorithm on other 1000 samples is for testing the data.

Step 10 If testing results are matched with our trained database then character is recognized.

4   Summary of Classification

4.1   Neural Network

Neural networks resemble to biological neural networks. The term "neural network" usually refers to models employed in statistics, cognitive psychology and artificial intelligence [1]. Neural network models which emulate the central nervous system are part of [1] computational neuroscience. Neural network architectures can be classified as, feed forward and feedback word networks. In the OCR systems, the multilayer perceptron of the feed forward networks and the Kohonen's Self Organizing Map of the feedback networks most commonly used in neural network. It is more like a real nervous system. It permits solutions to perform Graceful. It follows the ground rule which is implicit rather than explicit.

In pattern recognition, an artificial neural network is a calculus model which is widely used in current days. Neural Network structure in Fig. 6 will determine the number of nodes in the three different layers, i.e., input, hidden, and output.

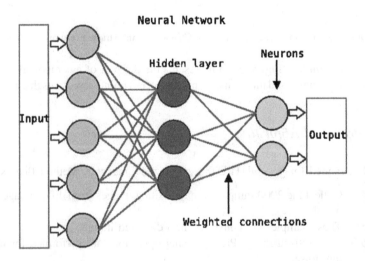

Fig. 6 Artificial network [1]

5 Result

In order to estimate the performance of the propound method, here, Table 1 shows the recognition rate of the different categories (Fig. 7).

Figures 8 and 9 represents the error histogram and best performance during testing the dataset.

Figure 9 represents 4 types of line Red, Blue, Green and Dotted line that show Test, Train, Validation, and Best Performance, respectively.

Table 1 Accuracy data

Category	Recognition rate (%)	
	Accuracy achieved on trained data	Accuracy achieved on tested data
Child	79	78.4
Young	81	80.5
Old	78	77.4

Fig. 7 Graphical representation of accuracy data

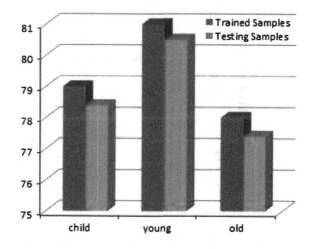

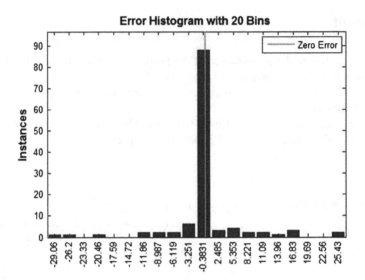

Fig. 8 Graphical representation of error histogram

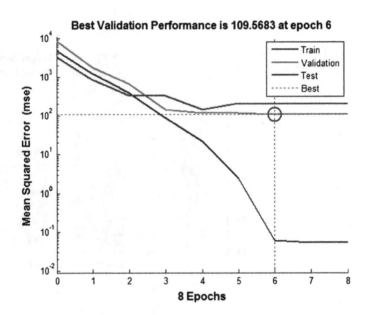

Fig. 9 Graphical representation of best performance

6 Conclusion and Future Scope

In this paper, we have proposed a feature extraction method and neural network classifier for the recognition of the handwritten Sanskrit simple and compound characters. We have acquired a maximum recognition rate of 80.5 % for compound characters in young category using the Neural Network classifier. Besides, we plan to improve the accuracy by increasing the number of samples and to use effective concept of classifier to increase recognition rate.

References

1. http://en.wikipedia.org/wiki/Artificial_neural_network.
2. Dineshkumar, R., & Suganthi, J. (2013). A research survey on sanskrit offline handwritten character recognition. *International Journal of Scientific and Research Publications, 3*(1), ISSN 2250–3153.
3. Magare, S. S., & Deshmukh, R. R. (2014). Offline handwritten sanskrit character recognition using hough transform and euclidean distance. *International Journal of Innovation and Scientific Research, 10*, 295–302. ISSN 2351-8014.
4. Shelke, S., & Apte, S. (2011). A multistage handwritten marathi compound character recognition scheme using neural networks and wavelet features. *International Journal of Signal Processing, Image Processing and Pattern Recognition, 4*(1).
5. Magare, S. S., & Deshmukh, R. R. (2014). Offline handwritten sanskrit character recognition. *International Journal for Research in Applied Science and Engineering Technology, 2*(VI), ISSN: 2321–9653.
6. Starzyk, J., & Ansari, N. (1992). *Feed forward neural network for handwritten character recognition*, IEEE.
7. Dwivedi, N., & Srivastava, K. (2013). Sanskrit word recognition using Prewitt's operator and support vector classification. In *International Conference on Emerging Trends in Computing Communication and Nanotechnology, IEEE*, (2013).

Need of ICT for Sustainable Development of Power Sector

Prashant Kumar, Lini Mathew, S.L. Shimi and Pushpendra Singh

Abstract The existing distribution network configuration is not capable to meet the challenges in present situation and also in near future. In order to meet the future challenges, i.e., increased electricity demand, improve power quality, and reduced cost of electricity, the existing grid should be equipped with advanced metering, modern communication techniques, distributed generation, and energy storage devices. This paper is about how information and communication technology (ICT) relates to power sector for sustainable development. With the help of ICT, the existing electric grid may allow to have electricity generation from distributed energy resources at distribution network level and at consumer ends. There is a need of integrated vision to meet the future electricity demand economically, sustainably and towards empowerment of the consumers with their active participation. This paper has been aligned with the goal of sustainable development and conservation of nature and natural resources for our future generations.

Keywords ICT · Smart grid · Distribution networks · Energy storage · Distributed generation

P. Kumar (✉) · L. Mathew · S.L. Shimi
NITTTR, Chandigarh, India
e-mail: Prashantchahar15@gmail.com

L. Mathew
e-mail: lenimathew@yahoo.com

S.L. Shimi
e-mail: shimi.reji@gmail.com

P. Singh
JK Lakshmipat University, Jaipur, India
e-mail: pushengg@gmail.com

© Springer Science+Business Media Singapore 2016
S.C. Satapathy et al. (eds.), *Proceedings of International Conference on ICT for Sustainable Development*, Advances in Intelligent Systems and Computing 408, DOI 10.1007/978-981-10-0129-1_63

1 Introduction

For the developing nations around the world with increased population and improving lifestyles, electricity with reliability and quality of supply is essential. At present scenario, electricity demand is increasing at faster rate. In order to meet the increased energy demand, following measures may be taken: (i) increase the power generation capacity (ii) reduction in transmission and distribution network losses (iii) motivate the consumers for using energy-efficient electric appliances (iv) increase the penetration of distributed generators in electric utility (v) adoption of demand-side management with energy storage options [1–3].

In the present electrical sector, electricity is being generated at remotely located fossil fuel based power plants, hydro plants, nuclear plants, etc. Fossil fuel reserves are being depleted at faster rate and are also polluting environment by green house gas emissions. The electricity produced from distributed energy resources, i.e., solar PV, solar thermal, wind generators, mini/micro hydro, fuel cell, etc., is distributed and intermittent [4–6]. For accommodating distributed energy resources, electric utility needs of proper information and communication technologies (ICT) along with the electric network. In order to make the electric utility more sustainable, there is a need to integrate ICT with electric power system networks. The transmission network is smartly designed for bidirectional flow of power and operates in an intelligent manner, but distribution networks are designed only for unidirectional flow of power and without utility communication facility with the consumers. In the present system, consumers are not able to know the real-time information, i.e., cost of electricity, power consumption, etc. There is one major drawback of green energy sources that the power generation is intermittent in nature. In order to overcome the challenges imposed by green energy sources, utility may resolve the problems with proper communication between utility and entity, i.e., DGs, distribution network operators, consumers, etc [7–9]. India has many power generation sources, viz., hydro, nuclear, gas, coal wind, solar, etc. Yet there is a wide demand–supply gap. Affecting common people and extracting a huge economic cost by holding back trade and industry are caused by poor power generation capacity [10, 11]. Given the link between per capita income and power consumption, there is an urgent need to push power infrastructure. The energy demand and supply gap in the next few years will change considerably with the change in economy of the countries in the world. There is a need to develop a model to provide affordable, clean, and reliable power to the consumers, in which consumers can play active role to meet the energy demand and also decrease the green house gas emissions [12–15].

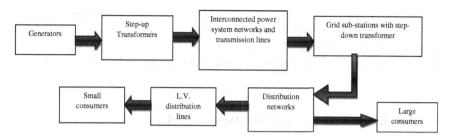

Fig. 1 Power system structure

2 Present Power System Model

An electric power system consists of generating plants, transmission and distribution networks, transformers, etc. Transmission networks connect generating stations and distribution networks, and also connect one area of power system to other areas of power system. Figure 1 shows the structure of a power system, a distribution system transfer power from electric grid to consumers in a particular area. Electricity is being generated at a voltage of 11–33 kV stepped up to higher voltages for transmission and on other side voltage is stepped down as per the consumer requirements.

The distribution system fed from the distribution stations, supply power to the residential, commercial, and industrial consumers [16–18].

3 Smart Grid

Smart grid is a concept to upgrade the grid performance with the help of modern communication technologies, real-time computation techniques, and intelligent electronic devices. Smart grid has many implications for different consumers at different times. With limited reserves of fossil fuel and challenges with nuclear power plants, electric utility realizes to explore the options to generate electrical energy. There is limitation with distributed energy resources, i.e., availability is scattered and energy generation is intermittent [19, 20]. The existing power system is designed to have scheduled generation from remotely located centralized power plants and operated in vertically integrated manner. But due to the presence of distributed energy resources the power system operation will be affected, the supply parameters may not be in permissible limit. At that time, proper communication and advanced sensors and computational techniques are essential for proper power system operation. Smart grid will control the power system parameters in permissible limits with modern techniques [21–23].

As of now the concept of smart grid provides merely the basic structure that envisages new ideas and modern systems of efficient power generation and distribution. The existing power system configuration and distribution developed about 100 years before will not be compatible with the accommodation of distributed generators. This has become necessary as the electric appliances at consumer end are increasingly becoming sensitive to supply parameters [24, 25].

4 Power System Communication Model

The proposed model has been developed with the consideration of communication and information systems foundations for an intelligent, self-healing power system operation. With modern communication technologies, performance of power system can be improved in terms of reliability and optimization of the assets utilization. It helps to deploy monitoring, communications, computing and information technology to address unique business and governing drivers, how to integrate new and existing systems and how to manage and secure the power system operation [26–28]. It will also be helpful to improve the performance of power system in the following ways:

- Consumers receive "pricing signal."
- Stop charging of electric vehicles and pump power onto the grid during peak-load period.
- Increasing set point of air-conditioning by two degrees or turning down the air conditioner during peak-load periods.
- The light at large retail stores are gradually reduced by 20 %.
- Refrigerator and freezer compressors are cycled off.
- Backup generation and storage devices at commercial and industrial facilities come online (Fig. 2).

5 Flowchart of Power System Operation

Repeated interruptions of electricity supply result in reduced efficiency of the industrial sector in terms of production. The energy generated from renewable energy sources can only be fully utilized if operated with the utility grid and given the priority to stay connected [23].

The power generation from distributed energy sources at the consumer end can be fed to the grid, so there will be a reduction in energy generation from the fossil fuel based plants. The operation of the future power system can be proposed by the flowchart shown in Fig. 3 [30, 31].

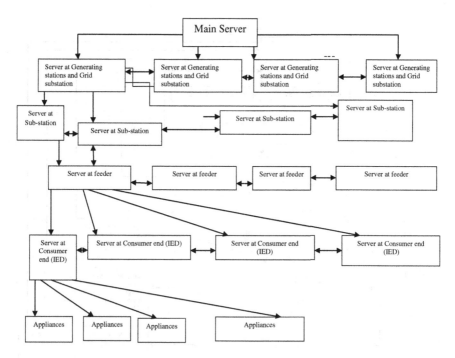

Fig. 2 Power system communication model

6 Future Power System

The applications of advanced sensors, modern communication techniques, and real-time computation techniques and controllers will transform the present grid into a future grid with the following abilities.

- Active involvement of the consumers.
- Accommodating all types of generations and storage devices.
- Improving the power quality.
- Management and utilization of the electrical energy efficiently.

The future power system (i.e., with ICT) will be reliable, secure, economical, efficient, safer, and more environmental-friendly than present power system structure (Fig. 4).

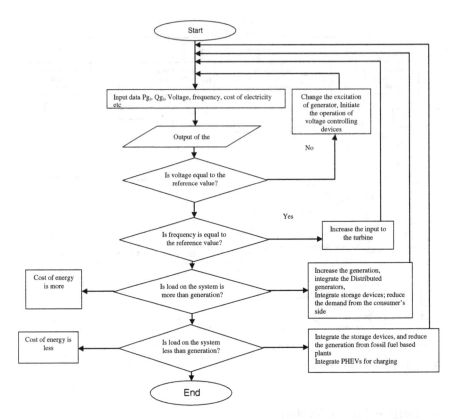

Fig. 3 Power system operation

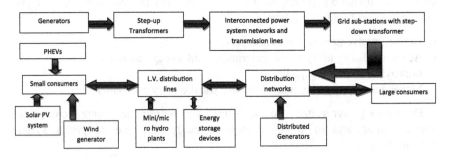

Fig. 4 Future power system structure

7 Conclusion

This paper proposes the future grid, which allows to have distributed generators in distribution networks. The existing power system operates in vertically integrated manner with the power generation from fossil fuel based remotely located centralized power plants. The present grid configuration is modified to operate in horizontally integrated manner. The future grid will be equipped with modern communications technologies, real-time controllers, and real-time computation techniques. With these modern techniques, the existing electric grid will be modified towards sustainable development of the electric sector.

References

1. Singh, P., Kothari, D. P., & Singh, M. (2012). Integration analysis of green energy and conventional electricity generation with interconnected distribution networks. *International Review on Modelling and Simulations 5*(5).
2. Gul Bagriyanik, F., Elif Aygen, Z., & Bagriyanik, M. (2011). Minimization of power transmission losses in series compensated systems using genetic algorithm. *International Review of Electrical Engineering, 6*(2), 810–817.
3. Saint, B. (2009). Rural distribution system planning using smart grid technologies. In *IEEE Rural Electric Power Conference, Repc '09*, pp. B3–8, April 2009.
4. Fan, J., & Borlase, S. (2009). The evolution of distribution. *IEEE Power and Energy Magazine*, pp. 63–68, March/April 2009.
5. Vojdani, A., (2008). Smart integration. *IEEE Power and Energy Magazine*, pp. 71–79, November/December, 2008.
6. Brown, R. E. (2008). Impact of smart grid on distribution system design. *IEEE Power and Energy Society General Meeting-Conversion and Delivery of Electrical Energy in The 21st Century*, pp. 1–4, July 2008.
7. Pipattanasomporn, Feroze, Rahman, (2009). Multi-agent systems in a distributed smart grid: design and implementation. In *IEEE/PES Power Systems Conference and Exposition, PES '09*, pp. 1–8, March 2009.
8. Garrity, T. F. (2008). Getting smart. In *IEEE Power and Energy Magazine*, pp. 38–45, March/April, 2008.
9. Papathanassiou, S. A., (2007). A technical evaluation framework for the connection of DG to the distribution network. *Electric Power Systems Research 77*(1), 24–34.
10. Mougharbel, I., Abdallah, R., El-Fakih, T., & Kanaan, H. (2010). A generic control strategy for interconnected power electronic building blocks based on a universal architecture. *International Review of Electrical Engineering, 5*(2), 793–802.
11. Dag, G. O., & Bagriyanik, M. (2010). Controlling unscheduled flows using fuzzy set theory and genetic algorithms. *International Review of Electrical Engineering, 5*(1), 185–193.
12. Shengyou, X., Chen, M., & Ran, L. (2012). Probabilistic approach to reliability assessment of electric power system containing distributed generation. *International Review of Electrical Engineering, 7*(1), 3478–3485.
13. Wong, L. Y., Sulaiman, M. H., Abdul Rahim, Siti Rafidah, & Aliman, O. (2011). Optimal distributed generation placement using hybrid genetic-particle swarm optimization. *International Review of Electrical Engineering, 6*(3), 1390–1397.
14. Singh, I., & Michaelowa, A. (2004). Indian urban building sector: CDM potential through energy efficiency in electricity consumption. HWWA Discussion paper.

15. Hadjsaid, N., Canard, J. F., & Dumas, F. (1999). Dispersed generation impact on distribution networks. *Computer Application in Power IEEE, 12*(2), 22–28.
16. Kothari, D. P., & Nagrath, I. J. (2011). *Modern power system analysis*, 4th ed. New Delhi: Tata Mc-Graw hill, 2011.
17. Kothari, D. P., & Nagrath I. J. (2007). *Power system engineering*, 2nd ed. New York: Tata Mc-Graw Hill.
18. Wan, Y.-H., & Parsons, B. K. (1993). Factors relevant to utility integration of intermittent renewable technologies. *National Renewable Energy Laboratory*, August 1993.
19. Savier, J. S., & Das, D. (2007). Impact of network reconfiguration on loss allocation of radial distribution systems. *IEEE Transactions on Power Delivery, 22*(4), 2473–2480.
20. Kothari, D. P., Ranjan, R., & Singhal, K. C. (2011). *Renewable energy sources and technology*, 2nd ed. New Delhi: Prentice-Hall.
21. A Report on *Empowering Variable Renewables option for flexible Electricity Systems* by International Energy Agency, July 2008.
22. Lopes, J. A. P., Hatziargyriou, N., Mutale, J., Djapic, P., & Jenkins, N. (2007). Integrating distributed generation into electric power systems: a review of drivers, challenges and opportunities. *Electric Power Systems Research 77*(9), 1189–1203.
23. Massoud Amin, S., & Wollenberg, B. F. (2005). Toward a smart grid. *IEEE Power and Energy Magazine*, pp. 34–79, September/October, 2005.
24. Hamedani Golshan, M. E., & Arefifar, S. A. (2006). Distributed generation, reactive resources and network-configuration planning for power and energy-loss reduction. *IEE Proceedings—Generation, Transmission and Distribution, 153*(2), 127–136.
25. Celli, G., Pilo, F., Pisano, G., Allegranza, V., & Cocoria, R. (2005). Distribution network interconnection for facilitating the diffusion of distributed generation, CIRED.
26. Ozansoy, C. R., Zayegh, A., & Kalam, A. (2007). The real-time publisher/subscriber communication model for distributed substation systems. *IEEE Transactions on Power Delivery, 22*(3), 1411–1423.
27. Kothari, D. P., & Dhillon, J. S. (2011). *Power system optimization,* 2nd ed. New Delhi: PHI.
28. EPRI's Intelligrid Initiative and Intelligrid Architecture. [http://intelligrid.epri.com].
29. Kalam, A., & Kothari, D. P. (2010). *Power system protection and communications*, 1st edn. New Age International (P) Ltd.
30. Kundur, P., Power system stability and control. New York: Tata Mc-Graw-Hill.
31. Wood, A. J., & Wollenberg, B. F. (1996). *Power generation, operation and control*, 2nd ed. New York: Wiley.

Development of Optical Impairment Minimization Technique for Radio Over Fiber Link

Ajay Kumar Vyas and Navneet Agrawal

Abstract In radio over fiber (RoF) links, data speed and the link length is restricted by the number of optical impairment like chromatic dispersion, nonlinearities, phase noise, polarization effect, amplified noises, etc. In the present paper, a technique based on digital signal processing (DSP) has been implemented to mitigate the problem to certain extent, so that the link can find its application for Long-Term Evolution (LTE) and advanced LTE. The accuracy and efficiency of the algorithm has been tested by some standard parameters like error vector magnitude (EVM), symbol error rate (SER), and constellation diagrams. The graphical analysis drawn from simulation software OptiSystem 13.0 reveals the impact of DSP implementation in the optical communication network.

Keywords Constellation diagram · DSP · EVM · RoF · SER

1 Introduction

This worldwide communication infrastructure based on optical networks is indispensable for the economic and cultural enlargement of modern societies [1]. The 3rd Generation Partnership Project (3GPP) LTE (~ 2.6 GHz) is the potential key to meet the exponentially increasing demand of the mobile end user [2]. Deployment of enhanced Node B (eNB) to increase the network coverage has increased the network complexity and cost. Thus, deployment of relay node (RN) with RoF link provides the interface within eNB and RN. Due to high path loss and multipath fading, wireless channel could not be the ideal channel between eNB and RN [3].

A.K. Vyas (✉) · N. Agrawal
Department of Electronics & Communication Engineering, College of Technology
and Engineering, Maharana Pratap University of Agriculture & Technology, Udaipur, India
e-mail: ajay_ap7@yahoo.com

N. Agrawal
e-mail: navneet@mpuat.ac.in

© Springer Science+Business Media Singapore 2016
S.C. Satapathy et al. (eds.), *Proceedings of International Conference
on ICT for Sustainable Development*, Advances in Intelligent Systems
and Computing 408, DOI 10.1007/978-981-10-0129-1_64

The necessity attributable to the growing demand for high-speed broadband connectivity, the radio over fiber (RoF) technology provides endless broadband access to multimedia services with a guaranteed quality of service (QoS) [4–6]. For the betterment of the network the drive towards higher order modulation leads to greater optical impairments, reducing the maximum distance over which increased capacity can be provided. Accordingly, linear impairments such as chromatic dispersion (CD) and polarization mode dispersion (PMD) have to be compensated in the optical domain with dispersion compensating fiber (DCF) and filters. This comparatively low spectral efficiency limits the capacity per wavelength channel to the modulation speed and the electrical bandwidth of transmitter and receiver [7, 8].

Digital signal processing (DSP) is a minimization technique for impairment where phase and polarization are managed, which puts together coherent detection well-built and more practical. In this minimization technique, the received optical signal is fully recovered, allowing compensation of linear and nonlinear optical impairments. Optical impairments in a channel, such as chromatic dispersion, nonlinearities, phase noise, which are predictable as the self-phase modulation of a component that adds to an existing signal as the signal passes through it, is the major general parameter which is used to describe the unsteadiness of the carrier frequency in signal transmission all interacting and restrictive the recital of the link. Extenuating effects of these impairments are conventionally based on techniques in the optical domain, i.e., before the detection [9]. The digital signal-processing unit acts as an adaptive equalization stage at the receiver, predistortion at the transmitter, and electric-field domain signal processing.

2 Optical Impairment

The advance modulation scheme in optical network suffer from the linear, nonlinear, and channel impairments including fiber attenuation, insertion losses, CD, PMD, fiber nonlinearities, polarization effects, and other electro-optics conversion at different segment of link. The prevailing noise source in RoF link is introduced by the erbium-doped fiber amplifier (EDFA), which is used to compensate for fiber loss. Population inversion within the EDFA leads to stimulated emission and amplification of the incoming signal by means of spontaneous emission, generating amplified spontaneous emission (ASE) noise. Single-mode fibers support the transmission of two polarization-modes that are orthogonal to each other. Thus change in the states of polarization (SOP) is also responsible of distortion because temperature fluctuations and random birefringence due to mechanical stress cause the SOP and, therefore, the group velocity to vary with time and across the full length of the fiber.

The nonlinear impairments are self-phase modulation (SPM) and stimulated Brillouin scattering (SBS), which severely deteriorates QoS of LTE–RoF system.

There are many quality parameters to measure the amount of optical impairment within the link. The error vector magnitude (EVM) and symbol error rate (SER) are some of the significant measuring parameters. The EVM is calculated by

$$EVM = \left[\frac{\sum_{l=1,l\neq k}^{M} \frac{d_{kl}^2}{M}}{\sum_{l=1,l\neq k}^{M} \frac{d_k^2}{M}} \right]^{1/2} \tag{1}$$

where dk is the reference distance of region k.

Guard symbols are taken into account when calculating the SER

$$SER = \frac{Error}{Symbol\ Sequence\ Length - 2*Guard\ Symbol} \tag{2}$$

3 DSP and Minimization Technique

The symbiotic combination of DSP, coherent detection, and spectrally efficient modulation formats employing multilevel modulation format is a promising combination for next generation high-speed transmission system due to the high power and spectral efficiencies. With the powerful DSP, coherent optical receivers allow the significant equalization of CD, PMD, phase noise (PN) and nonlinear effects in the electrical domain. However, there are assortments of minimization techniques like using dispersion compensation fiber (DCF), fiber Bragg grating (FBG), etc., yet digital signal processing is the imperative technique for compensating the optical impairments because of number of advantages like signal amplification, delay, and manipulation devoid of debasing the signal quality.

The universal DSP unit performs digital domain impairment compensation to abet in recovering the incoming transmission signal after coherent detection. It can be used with coherent systems designs that utilize QPSK, 16-QAM, and 64-QAM OFDM modulation is used for LTE or advance LTE with single polarization or dual polarization. The basic building block of digital signal processor used in optical communication is shown in Fig. 1. The preprocessing stage includes noise addition, DC blocking, and normalization. Signal is added with noise from a noise source (noise bin) limited to transmission channel bandwidth is converted to the optical sampled signal. Then the DC blocking is applied to offset any imperfectly biased voltages in the modulators. In normalization the received signal is normalized to the M-QAM grid. For the signal recovery the main algorithms stages includes by Bessel filter, resampling quadrature imbalance (QI), CD compensation, nonlinear (NL) compensation, timing recovering, adaptive equalizer (AE), down sampling frequency offset estimation (FOE), and carrier phase estimation (CPE) [10]. The normalized signal passes through the Bessel filter to remove noise out of the optimum bandwidth of the Bessel filter which is 0.75 * symbol rate or 0.75 * bit

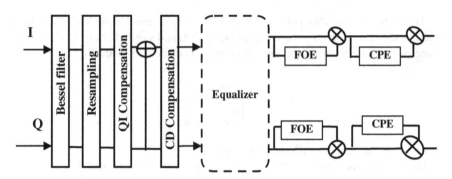

Fig. 1 Different stages of DSP block with preprocessing stage and signal recovery stages

rate/8. These filtered signals are resampled at a rate of 2 samples/symbol. After the resampling of signals, it is passed through QI compensation of the universal DSP unit which mitigates amplitude and phase imbalances within the in-phase (I) and quadrature (Q) signals.

Imbalances could be the resultant of several points along the transmission path and include inappropriate bias voltage settings for the modulators, photodiode responsivity mismatches, misalignment of the polarization controller, and imperfections in the optical 90-degree hybrid. Interpolation is used to adapt the sampled signal waveform to the new sampling rate. The interpolation may be linear, cubic, or step. Cubic interpolation is the suggested interpolation method.

The Gram–Schmidt orthogonalization procedure (GSOP) is used to correct for non-orthogonalization. [11]. CD is a static, polarization-independent phenomenon. Digital filtering can be used to compensate for CD resulting from propagation over fiber the dispersion compensating filter can be implemented in either frequency domain or time domain. Nonlinear impairments, such SPM must be compensated by separate types of algorithms.

Nonlinear compensation is performed using a digital backpropagation (BP) method [12]. The received signals are linearly mapped to the optical field, so that both the optical amplitude and phase become available to the receiver's digital signal processor. The received signal can be digitally propagated through an inverse fiber model to compensate for CD and fiber nonlinearity. BP requires the inverse nonlinear Schrödinger equation (NLSE) to be solved for the parameters of the optical link. Negated spatial domain means the optical link is modeled on a first-in-last-out principle. The first fiber span is the last modeled span and the beginning of each fiber span is the end of each modeled span. The fiber is treated as a series of linear sections and dispersion-less nonlinear sections. Larger number of steps led to more accurate result but increase the computation time [13].

Time recovery is used for synchronization of symbols. For the sampling frequency, the samples are taken at the correct rate. The timing recovery algorithm adaptively determines the correct time to sample the symbol.

Adaptive equalizer is used to compensate for residual CD, PMD, and to reduce inter-symbol interference. In FOE, the mixing with the local oscillator introduces a frequency and phase offset, leading to a rotating constellation diagram.

In CPE, the remaining phase mismatch between the local oscillator and the signal is consequently removed by using blind phase search (BPS) algorithm. The block averaging is used to mitigate the effect of ASE noise, that is, assuming a block of N symbols have the same phase noise [14, 15].

4 Simulation Setup for LTE–RoF Link

Our setup schematic is shown in Fig. 2. Pseudo-random bit sequence generator (PRBSG) data format is used with 10 Gbps data rate. OFDM is a multi-carrier transmission technique, which divides the available spectrum into many orthogonal subcarriers, each one being modulated by a lower rate data stream. OFDM modulator with maximum number of possible subcarriers 1024 with 100 number of prefix point. The average OFDM power is 16dBm and the number of subcarrier per port is 512 with 257 locations. The OFDM modulator has a digital to analog converter with cubic interpolation method. Two low-pass cosine filter (In-phase and Quadrature phase) has cutoff frequency of 0.62 * sample rate (i.e., 3.1 GHz.) The quadrature modulator of $\sim$2.6 GHz carrier frequency modulated the input data. In RoF link, a narrow bandwidth $\sim$0.1 MHz continuous wave (CW) wavelength $\sim$1550 nm) from laser diode is modulated via a LibNo3 Mach-Zehnder Modulator (MZM). To modulate the LTE-Tx with 2.6 GHz RF carrier frequency to the optical carrier with the laser diode signal. The laser CW laser power is varied from 0 dBm to $\sim$5 dBm. The optical signal is transmitted through the SMF with the signal attenuation of 0.2 dBm/km, dispersion of 16 ps/nm/km, and the dispersion slop is 0.9075 ps/nm2/km. To amplify the optical power signal, an EDFA amplifier is used at the end of the SMF link. The transmitted signals are recovered at the receiving

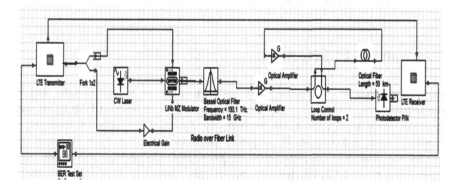

Fig. 2 Setup for LTE–RoF link without DSP unit

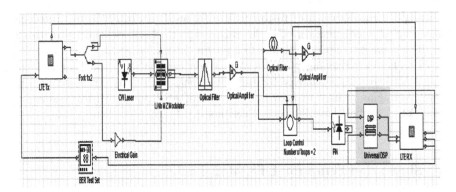

Fig. 3 Setup for LTE–RoF link with DSP unit

end. The PIN photodiode is used as a detector with 10nA dark current responsivity 1A/W, thermal noise 100e-24 W/Hz, and center frequency 193.1 THz. At the LTE receiver, the electrical band pass Gaussian filter is used to minimize the electrical signal noise; group delay becomes constant for all frequencies and prevents the unwanted frequencies. The RF signal is demodulated by the quadrature demodulator, which implements an analog demodulator, using a carrier generator for in-phase and quadrature components; it consists of two low-pass filters. The cutoff frequency of the low-pass filter is configured to 6 GHz; the OFDM demodulator is implemented by the complex point 1024 subcarrier. Those demodulated signals are decoded through the QAM sequence decoder. In the QAM sequence decoder, the bit sequence is split into two parallel sub-sequences; each can be transmitted in two quadrature carriers when building a QAM modulator. In receiving part one band pass filter of 2.6 GHz center frequency is used. Further in case of 16-QAM and 64-QAM modulators, the bit per symbol and power order of symbols is fourth order and eighth order for 16-QAM and 64-QAM QPSK, respectively. All received signals are compared with input data at BER test set. In Fig. 3 the exiting setup promotes adapt with universal DSP unit. The simulation work done for dissimilar length of optical fiber (~ 50 to ~ 500 km).

5   Results and Discussion

Experimental outcome of proposed work is shown in Figs. 4, 5 and 6 with the help of constellation diagrams of transmitted signal, received signal without DSP block and received signal with DSP block, respectively, for different types of modulation scheme. The pseudo-random bit sequence generator input data is ~ 10 Gbps bit stream. The number of bits per symbol for QPSK, 16-QAM, and 64-QAM modulated input signal at ~ 2.6 GHz frequency is transmitted. The red color indicates the signal

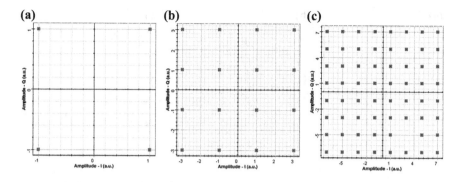

Fig. 4 Constellation diagram of transmitted signal **a** QPSK **b** 16-QAM **c** 64-QAM

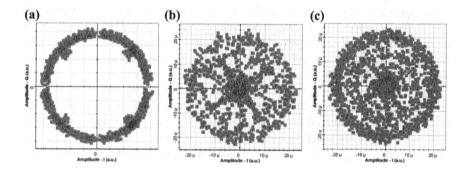

Fig. 5 Received constellation diagram of **a** QPSK **b** 16-QAM **c** 64-QAM without universal DSP

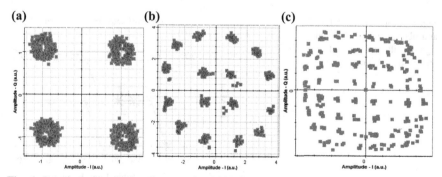

Fig. 6 Received constellation diagram of **a** QPSK **b** 16-QAM **c** 64-QAM with universal DSP block

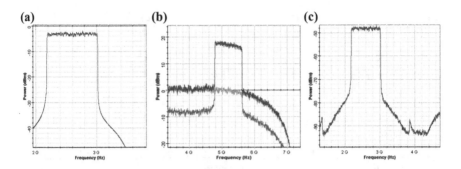

Fig. 7 RF spectrum of **a** input signal **b** output without DSP block **c** output with DSP Block

and the blue characterize the noise signals. It is observed that at the transmission end only the red points are showing and at the receiving end both are presented.

RF spectrum of the LTE signal for the transmitter and receiver are given in Fig. 7. The both spectrum in case of without DSP and with DSP compensation are given in Fig. 7b, c respectively. The transmitter spectrum is noise free signal containing only the blue color whereas the receiving spectrum consisting the red and green signals which also represent the noise and signals.

The EVM depends on signal-to-noise ratio (SNR) of the system. The DSP unit reduces the noise power and improves the signal strength it is evidently mentioned in the Fig. 8. It is observed that when the detected signal passes through the DSP unit, the impairments which were added within the channel were minimized. The value of SER changing with fiber length shows that faultlessness of transmission end.

The EVM and SER are reducing in case of DSP using graphs between EVM and fiber length, and SER and fiber length shown in Figs. 8 and 9. Both the graphs are plotted with higher degree of modulation (i.e.,64-QAM).

Fig. 8 Graph between EVM and fiber length of received signals with and without universal DSP unit

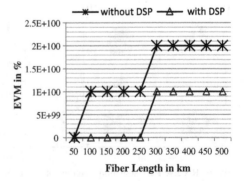

Fig. 9 Graph between SER and fiber length for 64-QAM signals with and without universal DSP unit

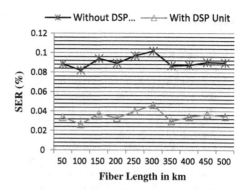

6 Conclusion

In this paper, DSP unit has been implemented to mitigate optical impairment of LTE–RoF link. After constellation diagram is boon to analyze any communication network in term of impact of linear and nonlinear noise. In present investigation has been made for LTE–RoF link with different modulation scheme of QPSK, 16-QAM, and 64-QAM signal as an input to the network. The constellation diagram of receiver signal for 16-QAM and 64-QAM is shown in Figs. 5 and 6. The red dots in the diagram represent the value actually received and blue dots represent noise. The performance of the Universal DSP unit for linear and nonlinear distortion (of RoF link) compensation is analyzed by constellation diagrams, EVM, and SER. It was observed from constellation diagram that adding universal DSP unit improves the link performance significantly by clearly individual indicating the constellation points at the preferred bit positions. EVM is reduced from ~ 48 to ~ 39 % of fiber length of 500 km for 64-QAM signals, while the SER trim down from ~ 0.55 to ~ 0.34 % 64-QAM by the use of same length of fiber. The experiment can be further extended to improve received signals. Authors recommend polynomial predistorter for nonlinearity compensation of the RoF link. This technique considerably reduces the higher order nonlinear impairment with high depth of memory to linearize the RoF system over a high frequency range.

Acknowledgments The authors would like to acknowledge AICTE, New Delhi (India) for financial assistance to procurement of OptiSystem version 13 under MODROB (sanctioned F. No. 9.50/RIFD/MODROB/Policy-1/2013-14 date 17.07.2013)

References

1. Strategic Research Agenda (White Paper). (2012). Innovation in future Network in Europe. eMobility Networld, pp 1–168.
2. Kanesan, T. (2013). Impact of optical modulators in LTE RoF system with nonlinear compensator for enhanced power budget. Technical Digest, Optical Society of America. pp. 1560–1571.
3. Technical Report. (2010). RoF System for Dual WCDMA & LTE System. *NTT DOCOMO Journal, 12*(4), 24–29.
4. Cooper, W. D. (1990). 'Fibre/radio' for the provision of cordless/mobile telephony services in the access network. *Electronics Letters 26*, 2054–2056.
5. Cox, C. H., et al. (1997). What do we need to get great link performance. *International Topical Meeting on Microwave Photonics, 1*(1), 215–218.
6. Vyas, A., & Agrawal, N. (2012). Radio over fiber: Future technology of communication. *International Journal of Emerging Trends & Technology in Computer Science, 1*(2), 233–237.
7. Hu, J., Gu, Y., Han, X., Fang, L., Kang, Z., & Zhao, M. (2012). Additive phase noise effect on M-QAM direct modulation radio over fiber link. *Pacific Science Review, 14*(1), 48–52.
8. Wang, L. X., Zhu, N. H., Liu, J. G., Li, W., Zhu, H. L., & Wang, W. (2012). Experimental optimization of phase noise performance of optoelectronic oscillator based on directly modulated laser. *Chinese Scientific Bulletin, 57*(31), 4087–4090.
9. Fattah, A. Y., & Mohammed, Z. F. (2013). Electronic signal processing for cancelation of optical systems impairments. *IJCCCE, 13*(2), 55–71.
10. OptiSystem Component Library Manual. (2013) pp. 802–601.
11. Oerder, M., & Meyr, H. (1988). Digital filter and square timing recovery. *IEEE Transactions on Communications, 36*(5), 605–612.
12. Ciblat, P., & Vandendorpe, L. (2003). Blind carrier frequency offset estimation for non-circularconstellation based transmission. *IEEE Transactions on Signal Processing, 51* (5), 1378–1389.
13. Sethares, W. A., Rey, G. A., & Johnson, C. R. (1989). Approaches to blind equalization of signals withmultiple modulus. In *Proceedings of the International Conference on Acoustics, Speech, and Signal Processing (ICASSP)* (Vol. 2, pp. 972–975).
14. Godard, D. N. (1980). Self-recovering equalization and carrier tracking in two-dimensional data communication systems. *IEEE Transactions on Communications, COM-28*(11), 1867–1875.
15. Morelli, M., & Mengali, U. (1988). Feedforward frequency estimation for PSK: A tutorial review. *IEEE Transactions on Communications, 9*(2), 103–116.

Mobile Waste Management for Smart Cities: Monitoring Sanitation Through Living Labs

Chandana Unnithan, P.S. Ramkumar, Ajit Babu, Sneha Joseph, Sandeep Patil and Amit Joshi

Abstract Waste management is a priority issue for improving sanitation in urban areas of India. In this paper, we report on current research and ongoing activities in building, piloting, and implementing a mobile application for waste management in two cities. The Living Labs framework is being used as the conceptual foundation for this project, which is stemming from the United Nations Action Team 6 Follow-Up Initiative (International Archives of the Photogrammetry, Remote Sensing and Spatial Information Sciences XL-8, 2014) [5], which in long term leverages ICTs for improvement of public health and environment.

Keywords Smart city · ICT · Living labs · Waste management · Mobile · Social media · Sanitation

1 Introduction

Information communication technologies (ICTs) are being leveraged globally in many ways since their historic growth and ubiquity has continued to change societies. The focus of this paper is on two areas: First, the use of mobile communication technologies; particularly smart phones and mobile Internet. Second,

C. Unnithan (✉)
Victoria University, Melbourne, VIC, Australia
e-mail: Chandana.unnithan@gmail.com

P.S. Ramkumar · S. Patil
Applied Cognition Systems, Bengaluru, Karnataka, India
e-mail: psramkumar@apcogsys.net

A. Babu · S. Joseph
Centre for Advancement of Global Health, Cochin, Kerala, India
e-mail: ajitnbabu@gmail.com

A. Joshi
Computer Society of India, Udaipur Chapter, Udaipur, India
e-mail: amitjoshiudr@gmail.com

© Springer Science+Business Media Singapore 2016
S.C. Satapathy et al. (eds.), *Proceedings of International Conference on ICT for Sustainable Development*, Advances in Intelligent Systems and Computing 408, DOI 10.1007/978-981-10-0129-1_65

the use of social media applications via smart phones. In July 2014, there were a reported 886 million mobile subscriptions and the number of unique subscribers to be 349 million [1, 2]. Smart phones only accounted for 13 % of the total population [1, 2]. Nonetheless, the usability statistics is also significant. 90–95 % of the smart phone users searched for information, products, and services using their devices. Among this, 16 % of the smart phone users were streaming videos, while 11 % used social media applications [1, 2]. It is estimated that 70 % of the Internet page views originate from mobile devices in India [1, 2]. It is evident that mobile connectivity is driving the digital growth in India with majority of the new Internet subscribers accessing the Internet exclusively through mobile phones [1, 2].

Set in this premise, we report on research that is part of a long-term project on improving sanitation, with the ultimate aim of eradicating neglected tropical diseases (NTD) in India [3, 4]. NTDs are essentially caused by lack of sanitation, may it be infrastructure or social behavior. A reported 1.4 billion people are affected by NTDs worldwide with a significant proportion living in India [3, 4]. The research reported in this paper is part of many initiatives that are supported by the UN Action Team 6 (UNAT6) that aims to leverage ICTs and social media for improving public health [5]. In particular, we showcase an application that is being trialed for waste management in a city, in southern India to enable a smart city and implemented through the living labs framework.

The rest of this paper is organized as follows. The next section provides a background and how it is linked to the major project from UNAT6. We also synthesize some of the current initiatives within India for waste management focused on urban areas that may be complementary and also similar to this venture. The following section offers the methodological framework of Living Labs through which the project is being implemented. Subsequently, we discuss the current conceptual development of the mobile application for waste management. In the last section, we offer the current outlook from this project.

2 Context and Collaborators

2.1 Context

In India, the year 2014 has witnessed a significant drive for sanitation [6]. By 2019, the Sanitation Project that has been launched this year has to be completed aimed at cleaning up India. While there are many initiatives under way today in many states within India, the researchers in this project aimed at leveraging ICTs and in particular mobile applications for improving sanitation in urban areas. We focused on two cities in two different states of India, namely Kochi (Kerala) and Bengaluru (Karnataka). The research project itself stemmed from a UN initiative in using ICTs for public health where sanitation is the key to eradicating NTDs in India [3].

We have also studied some initiatives that are being undertaken in various cities of India, while progressing on this project. In the city of Indore, all health officers and sanitation inspectors of Indore Municipal Corporation are given smart phones for digital surveillance and monitoring of garbage disposal system, while the civic body is in the process of launching a mobile-based application to clean the city [7]. In the city of Mumbai, the corporation has reinitiated the project to install RFID tags on dustbins to monitor and manage waste [8]. The Pune municipal corporation has installed global positioning systems (GPS) devices on a trial basis for garbage collection vans [9]. In the city of Chennai, the corporation has launched a mobile application where citizen can post photos of garbage bins and public toilets across the city. The free application based on Android phones, allows citizens to search and rate toilets and garbage bins in their vicinity (based on the mobile number). Any complaints by citizens are handled by corporation officers instantly [10].

2.2 Collaborators

The United Nations Office for Outer Space Affairs (UNOOSA) fosters the use of spatial technologies in public health [5]. Under the auspices of UNOOSA, the United Nations Action Team 6 (UNAT6) supported by WHO (Geneva), has initiated several projects with an international group of scientists and academics collaborating with governments in many countries. While the UNAT6 completed its term in 2012, the initiatives from this team are being followed up by the UN Action Team 6 Follow-Up Initiative (UNAT6FUI) consisting of a group of scientists and academics in Europe, USA, Australia, and India. It is anchored by the Centre for Advancement of Global Health (CAGH) in the State of Kerala, India. Currently, the group has two major projects:

(1) Development of a community-based early warning and adaptive response system for mosquito-borne diseases and
(2) Leveraging the use of ICTs (including mobile technologies) and social media for improving sanitation, aimed at eradicating neglected tropical diseases (NTDs).

The core principles of this group include an open source and community model, where all tools and resources can be used freely; a collaboration of equals who contribute equally to the project; and optimization of resources to the maximal. Any project is deemed successful by this group, when the broader community accepts it for its intended purpose [4, 5].

3 Conceptual Framework: Adaptation of Living Labs Approach

This research is based on and draws from the influential work of Schumacher [11] that adapted from the main principles of living labs framework as per the European Union guidelines. The researchers wish to acknowledge and credit the author for this significant foundation that we have applied to our progressive research work. *A*

Living Lab is an approach based on open innovation environment [11]. It frequently activates in the context of a city or state, integrating contemporary research and innovation processes, and is situated within a citizen–public–private partnership (C3P) [11]. According to Schumacher [11], the concept is based on an organized user-driven *co-creation, exploration, experimentation and evaluation of innovative ideas, scenarios, concepts and related technological artifacts in real-life use cases.* And the use cases often involve user communities, *not only as observers but also as source of creation.*

The main advantage of this approach is in that it allows all involved participants to simultaneously consider two aspects. These are (1) general performance of a product or service; and (2) its potential for adoption. In this research, as suggested by Schumacher [11], the framework is being used by users/citizens, for designing, exploring, experiencing, and refining a mobile application for waste management aimed at building a smarter city. A real-life scenario is currently being piloted for evaluating potential impacts, before a full implementation. The methodological framework is adapted and applied into our project as follows.

Co-creation *bring together technology push and application pull into a diversity of views, constraints and knowledge sharing that sustains the ideation of new scenarios, concepts and related artifacts* [11]. We invited ideas/concepts/technologies for building and sustaining a waste management application.

Exploration *engage all stakeholders, especially user communities, at the earlier stage of the co-creation process for discovering emerging scenarios, uses, and behaviors through live scenarios in real or virtual environments* [11]. We are engaging representative stakeholders (citizens) who create and are responsible for waste, to engage in building waste management systems. Based on the data collected, we will be refining the mobile application and use the data itself for interventions that change social behaviors.

Experimentation *implement the proper level of technological artifacts to experience live scenarios with a large number of users while collecting data which will be analyzed in their context during the evaluation activities* [11]. We are engaging all stakeholders involved in the creation and disposal of waste in experimenting with the new application.

Evaluation *assess new ideas and innovative concepts as well as related technological artifacts in real-life situations through various dimensions such as socio-ergonomic, socio-cognitive, and socio-economic aspects; make observations on the potentiality of a viral adoption of new concepts and related technological artifacts through a confrontation with users' value models* [11]. The application is being piloted, so that all stakeholders are studying the difficulties in potential adoption and refinement is being done.

Thus, in this situation, the Living Lab will be a collaboration of public–private–civic partnerships, where all stakeholders will co-create, in real-life and virtual environment. As a jumping-off point, CAGH has piloted this application in the cities of Cochin and Bengaluru, India. We assumed that the responsibility of waste management has to be shared between community, government, private sector, and educational sector to begin with, as they are all stakeholders in generating waste and contributing to the issue.

From the citizen's point of view, the key issue in waste management is lack of awareness as to waste management facilities, or lack of civic sense. Conversely, from the government's viewpoint, we hypothesized that poor implementation of plans, management of funds, law enforcement are the major issues. From a third perspective, we find that the private sector may be spending funding on unrealistic strategies that may not yield result through their corporate social responsibility initiatives. Lastly, education institutions need to instill the civic sense in students as well as parents, with active participation in ventures.

Before conceptualizing our theme, we considered [12] the physical elements to be addressed for waste management systems for sustainability in the long term namely: public health, i.e., particularly a good waste collection service for maintaining healthy environments; protecting the environment through appropriate waste disposal and treatment; and resource management, i.e., preventing wastage and encouraging recycling. On the soft side [12] of the issue are good governance, inclusiveness (communities and providers), financial sustainability (via private sector funding as well as government), and pro-active policy development. The authors also illustrate varied applications on waste management being used in other parts of the world.

We conceptualized the theme of "*Reduce, Refuse, Reuse, Recycle*" based on potential contribution to solutions, by stakeholders. Specifically, all stakeholders could *reduce* waste generation, *refuse* plastic containers, *reuse* bags and containers, *and recycle* where possible. To action this theme, communities had to participate, become aware of the waste management issues around their vicinity and report them. Governments need to become aware and provide infrastructure/policies and enforce practices. Private sector could support the ventures with commercially viable solutions, while education sector could research, build, and propagate innovations to the issues. With this agenda, along with the collaboration of a nongovernment organization, a private sector partnership and engaging the community as well as educational institutions, we developed a mobile application that can be made available via smart phones, presented in the following section.

4 APGOS-EVM for Waste Management

Our research team developed a new waste management application for cities and is currently trialing this in two cities. Typically, the user needs to create an account which will be verified via their mobile phones. Once the account is successfully created, the application can be used for tracking/reporting waste management issues.

Figures 1, 2, 3 illustrate the mobile application from the citizen's point of view, where they are able to report a problem using the application. There is also the government's point of view (dashboard), where the issues can be tracked and

Fig. 1 The mobile application developed for waste management has the main activity consisting of three fields: *Jurisdiction, Issue, and Solution*. Within the jurisdiction field are two dropdown fields that have 'state,' 'city' as contents. When a state is selected, the corresponding cities are listed. There are two textboxes namely 'area' and 'ward,' in which users need to input the area of the ward where there is a waste management issue

Fig. 2 Issue field consisting of a list of 'Issues/Problems'; Waste Pileup, Trash Collection, Polluting Place. Each issue item selection enables another dropdown in related to the issue selected

solution be implemented. The dashboard provides a short-term view of issues that need to be fixed and a long-term perspective on issues to be resolved through planning and implementation (infrastructure). The dashboard is also a window for

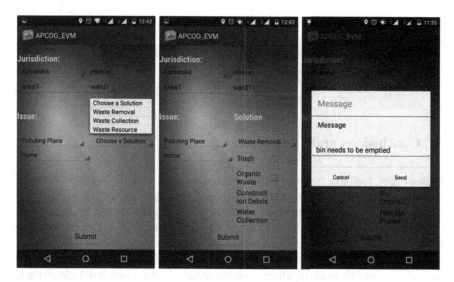

Fig. 3 Solution fields consisting of a dropdown and a multi-select list. The multi-select list items are in relation to the selected solution in the dropdown. A message can be given by the sender as to 'what needs to be done'

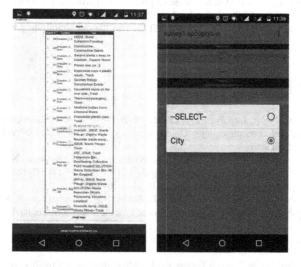

Fig. 4 Web-based tracker system using Google maps where the issue reported can be tracked by the government officials such that the solution required (for example, bin need to be emptied) can be passed on to the garbage trucks. If there are no bins located in a particular area, where there is waste piling up, the government may then need to take a different action by installing a bin or arranging for additional garbage vans that go into the area

private sector and educational institutions to understand the issues and educate the public accordingly (Fig. 4).

Together with civic population, governments, communities, educational institutions, and private sector, in this Living Labs framework, this application can be used for building a smarter city with improved waste management systems.

5 Conclusions and Outlook

In this paper, we have reported on the development of a mobile application that is aimed at improving waste management–one of the major issues for densely populated Indian cities. Appropriate waste disposal facilities may be lacking or underused in cities leading to overflowing dustbins and improper sanitation. The mobile application that is reported and being trialed in two cities currently, leverages current ICTs for offering a sustainable solution for both civic population and the government, to work together to improve the situation and build cleaner and smarter cities for India.

Acknowledgments The authors wish to acknowledge Applied Cognition Systems, Bangalore, India, for developing the application with the members and participants of the Sanitation project via Centre for Advancement of Global Health (CAGH), Kochi, India, members of the UNAT6FUI team and communities involved in this research project. We particularly wish to acknowledge the work of Schumacher on Living Labs that has been adapted to suit our progressive research. In addition, we wish to thank the organizers of ICT4SD for the opportunity to present this work.

References

1. We are Social Report, http://wearesocial.net/blog/2014/07/social-digital-mobile-india-2014/.
2. International Telecommunications Union Statistics, http://www.itu.int/en/ITU-D/Statistics/Pages/stat/default.aspx.
3. Unnithan. C., & Babu, A. (2014). Twitter, Facebook and Living Labs Versus Neglected Tropical Diseases (NTDs) for India. In *Proceedings of the ICTCS'14*, 978-1-4503-3216-3/14/11. ACM, Udaipur, Rajasthan, India, http://dx.doi.org/10.1145/2677855.2677952.
4. Babu, A., Soman, B., Niehaus, E., Shah, J., Sarda, N. L., Ramkumar, P. S., et al. (2014). Community-based EWARS for Mosquito Borne Diseases: An Open Source/Open Community Approach. *International Archives of the Photogrammetry, Remote Sensing and Spatial Information Sciences, XL-8*. ISPRS Technical Commission VIII Symposium, Hyderabad, India.
5. United Nations Action Team 6 Follow Up Initiative, http://at6fui.weebly.com.
6. India Sanitation Portal, http://www.indiasanitationportal.org/19056.
7. Digital Clean up with smartphone weapon, http://timesofindia.indiatimes.com/city/indore/Digital-clean-up-with-smartphone-weapon/articleshow/46967270.cms.
8. BMC to spend Rs 22 Crores on RFID Chips for civic dustbins, http://indianexpress.com/article/cities/mumbai/bmc-to-spend-rs-22-crore-on-rfid-chips-for-civic-dustbins/.

9. PMC installed global positioning system (GPS) devices on garbage collection vans, http://www.eai.in/360/news/pages/14603#sthash.nIpYvfdD.dpuf.
10. Chennai Corporation brings mobile app to clean Chennai, http://timesofindia.indiatimes.com/city/chennai/Corporation-brings-mobile-app-to-clean-Chennai/articleshow/46994423.cms.
11. Schumacher, J., (2013). Alcotra innovation project: Living labs definition, harmonization cube indicators & good practices, Short Guide.
12. Mavropoulos, A., Anthouli, A., & Tsakona, M. (2013). Mobile applications and waste management, recycling, personal behaviour and logistics. D-Waste.Com White Paper. ISSN: 2241-2484.

Optimized Implementation of Location-Aware and Network-Based Services for Power-Efficient Android Applications

Mary Remya Alroy, M. Vishnupriya and Aswathy Asok Nair

Abstract Today's smart phones are comprised of a variety of sensors and hardware, based on which a number of specialized applications have been developed. Unoptimized use of these hardware makes the power consumption of smart phones, a matter of serious concern. Most of the applications installed in smart phones keep track of user location and exchange user data with the servers. Even though a majority of the applications do not need real-time data exchange, they store the data until the cellular data network becomes available and then transfer the data to the servers. This consumes a lot of power from the handset, as cellular data transfer consumes more power than the traditional Wi-Fi networks. For location tracking, most of the applications use GPS, although these applications may not need precise location information. This results in loss of smart phone battery power. In our work, we have modified a health activity tracker app that consumes a lot of power. We tried to optimize power consumption by adding these features: We have replaced simple GPS location tracking by using a combination of network provider information and accelerometer for tracking the activity of the user; Network communication is done, without draining the power of the phone, by transmitting data when it is connected to a Wi-Fi network, but reducing the frequency of transmission when it is connected to a cellular network. The system also identifies the availability of free Wi-Fi by tracking the user's location while connected to the cellular network. Thus the power consumption by the smart phone is reduced.

Keywords Sensors · Wireless connectivity · GPS · Location-based services · A-GPS · Android manifest.xml

M.R. Alroy (✉) · M. Vishnupriya · A.A. Nair
Department of Computer Science & IT, Amrita School of Arts and Sciences, Kochi, India
e-mail: remyaalroy@gmail.com

M. Vishnupriya
e-mail: vishnupriya.chinju@gmail.com

A.A. Nair
e-mail: aswathy.is.in@gmail.com

M.R. Alroy · M. Vishnupriya · A.A. Nair
Amrita Vishwa Vidyapeetham (Amrita University), Coimbatore, India

© Springer Science+Business Media Singapore 2016
S.C. Satapathy et al. (eds.), *Proceedings of International Conference on ICT for Sustainable Development*, Advances in Intelligent Systems and Computing 408, DOI 10.1007/978-981-10-0129-1_66

1 Introduction

Smart phones have become an integral part of the lives of millions of people around the world. It offers various features such as instant messaging, health and fitness, social networking, etc. which made it an essential part of today's life. As they are used regularly everywhere, their energy consumption has become a matter of serious concern. Thus, energy management of smart phones has become the need of the hour. Installation and usage of many third party apps is the main reason for the power drain of smart phones [1].

Smart phone offers third party apps with vibrant variety of features; here we focus on health and fitness applications. Health and fitness apps are of different types like activity trackers, workout guides, nutrition and sleep and exercise gamification apps [2]. Among these the most popular is the activity tracker apps like Runner up, Pedometer, Google fit, etc. [3] Energy management requires a good understanding of where and how the energy is being used [4]. Power draining factors in these apps are:

- Environmental and Motion Sensors [4]: Smart phones contain various sensors like accelerometer sensor, barometer sensor, light sensor, magnetic sensor and proximity sensor. Sensors act like transducers [5] by measuring specific environmental parameters like light, motion, heat, pressure, moisture, etc. and converting them to some human readable output value or by transmitting these signals over a network for further processing [6].
- Wireless Connectivity [7]: These apps periodically update and store the details of user activities in their servers. It can connect to network through Wi-Fi, cellular networks, etc. This causes large power drain as well as data drain [8].
- Location Sensing: Most of the health apps, especially activity trackers, uses GPS for tracking the location when the workout is being done by the users. They depend on GPS for location updates, which consume large power from smart phone.

Our work is structured as follows: Sect. 2 describes the background study performed on location providers and a comparison of these providers. It also includes the types of wireless connectivity possible from smart phone. Section 3 deals with all related works that supported and contributed to our paper. Section 4 has the study of all the relevant health apps available (android) and also an open-sourced app which we considered for study. Section 5 includes the all the details of our proposed system, test cases used and a comparison of our work with other systems, and at last in Sect. 6 we describe conclusion and future work of our study.

2 Background Study

Once an application is installed in the smart phone, its request for location sensing, Internet access, etc. is done using the permissions file, which gets installed along with the application [9]. Various network communication and location services exist alike.

2.1 Network Communication

Any type of communication that a system establishes for data transfer comes under network communication. Applications include the permission for full network access in the permission file [Android manifest.xml in android] using sockets [9].

- *android.permission.INTERNET* [10]: Allows to use the network. Most of the applications use this permission to perform network communication. Since these apps do not need active network communication for its working, it causes heavy data drain and power loss without the knowledge of the user.
- *android.permission.ACCESS_NETWORK_STATE* [10]: Allows to access information about networks.
- *android.permission.ACCESS_WIFI_STATE* [10]: Allows to access information about Wi-Fi networks.
- *android.permission.BATTERY_STATS* [10]: Allows to access the battery statistics.

2.2 Location-Based Services

Most of the applications request the permission for location access while installation; location sensing can be mainly done in three ways [9]:

- GPS
- Assisted GPS
- Cell-ID lookup/Wi-Fi MACID lookup

Location sensing permission is provided using two kinds of permissions in android:

- *android.permission.ACCESS_COARSE_LOCATION*: Accesses approximate location derived from network location sources such as cell towers and Wi-Fi [10].
- *android.permission.ACCESS_FINE_LOCATION*: Accesses precise location derived from GPS, cell towers and Wi-Fi [10].

In our work, we propose a health-tracking system using accelerometer and network provider assisted location tracking. We also purpose a new change in network communication for the application, where data exchange takes place while smart phone is connected to Wi-Fi [8]' which consumes low power. The data to be transferred is backed up in the phone database and not done while connected to cellular network or when battery power is low. When there has been no Wi-Fi connections detected over the period of 24 h, the data is transmitted over the cellular network in order to maintain synchronization with the server. Wi-Fi scanning for available networks drains a lot of system power, but in our system, availability of Wi-Fi networks is checked using the location of the user, which is obtained by cellular communication [11].

3 Related Works

Fehmi Ben Abdesslem, Andrew Phillips and Tristan presented a new view on energy-efficient mobile sensing known as Senseless system, by maximizing the battery with the help of less expensive sensors. This study was concentrated mainly on location tracking and the conclusion was that the power consumption of senseless systems can be reduced by 58 % by replacing GPS-based system with a combination of accelerometer and GPS. Thus this model extends the battery life from 9 to 22 h [12]. According to Xia et al. [11], they proposed a solution to identify the nearest Wi-Fi using the user location obtained from the network provider. This system helps in reducing the number of unnecessary Wi-Fi scanning and avoids long periods of idle state by switching to the nearest Wi-Fi network if it is identified [13]. Their experimental results demonstrate that this scheme can optimize the power consumption effectively. Furthermore, Goran, Iva and Mario [14] present a measurement study on energy consumption of Wi-Fi, 3G and Bluetooth wireless technologies. They proved that the energy consumption of applications that use frequent data transfer can be reduced by collaborative approach, where combination of Wi-Fi and 3G or Wi-Fi and Bluetooth or 3G and Bluetooth can be used. Once the application model is studied this combination approach any of this combination approach can be used. It was proved that power was saved maximum when it used the combination of Wi-Fi/3G for data connectivity.

Location services implemented by Manav Singal and Anupam Shukla [15] using GPS and Web services, put forth a new concept of location tracking with high power efficiency, known as A-GPS system, which helps in very fast location access, but the cost to enable it in handset is high. A-GPS stands for assisted GPS where the source of location information is from the network providers who communicate with the GPS satellites [14]. A-GPS takes into account the GPRS for current location access, which reduces the power required by the GPS for tracking. A-GPS uses proximity cell towers for location fixing. It reduces the time for location fix since location is accessed via cell towers and not by satellites. Comparison paper published by Claas Wilke and team [16] gives a comparison study of energy consumption of similar apps in different domains like health, mail and web browsing, etc. Their study reveals that users are unaware of the fact that most of the apps are designed without considering the hardware components and their power drain, which made energy consumption a matter of serious concern. Studies showed that advertisements coming up in free apps gather much power than the actual app. They also compared energy consumption of various mobile browsers and their energy consumption variation while loading the same pages.

Lehr and Lee [8] published a comparison of Wi-Fi and 3G. Both technologies offer data connectivity, but these studies shown that Wi-Fi provides 100 kbps higher data rates than the 3G. It also suggested methods for integrating Wi-Fi and 3G. This model enhances wireless technology to provide a scalable infrastructure. Results show that the prescribed type of data downloading in this model minimizes the energy consumption. Patrick Gage Kelley, Sunny Consolvo and Lorrie Faith Cranor explain [17] the installation process of applications in android phones and how it is related with permissions and security concerns. User has to take privacy and security decisions for the applications they instal. This study conducted interviews in two cities to know whether users are aware of these kinds of permissions. Prof. Nilima Walde and team published a detailed study [18] of android location services using A-GPS [15] in phones or through web services like GPRS, etc. The studies also reveals that location-aware services provide many other value-based uses like checking the availability of Wi-Fi hotspots, traffic conditions, etc. [14].

4 Existing Systems

There are many health apps available in the market, which use many expensive sensors for both location tracking and network communication. Some of the important apps are:

4.1 Google Fit

Google Fit [18] is a health-tracking app, developed by Google for the android platform. It offers a set of APIs to enable integration and further development. It uses low cost sensors in mobile devices for fitness tracking and activity monitoring. Users can share their fitness data and also delete information as needed.

Advantage It uses less expensive sensors like accelerometer, step detector, step counter, etc. for activity monitoring.

Disadvantage It requires full Internet access for data transfer, without considering the power statistics of the handset. Also it doesn't track the location of the user.

4.2 Endomondo

Endomondo [19] is a famous health-tracking app, which uses GPS signals for location tracking, distance travelled, calories burnt, etc. It has many other features like audio–video playback. It also helps to keep a profile of all workouts done, and to compare the workout done by friends. It provides an efficient training plan.

Advantage It gives the exact location of the user, distance covered, duration and calories burnt. It has automatic pause when stationary. It also enables social networking.

Disadvantage It takes permissions for complete GPS tracking and full network access, which drains battery power.

4.3 Pedometer

The Pedometer App [20] is an open-source application, which has the basic functionalities of activity-tracking app. It checks the distance travelled by the user with the help of less expensive accelerometer sensor. Further, it displays the step count, distance travelled, calories burned and speed. Generally, some phones do not count the step taken by the user when the screen is off, but in Pedometer App, with the help of 'operational level' it also works while the screen is off.

Advantages It is a simple app which uses less expensive sensors like accelerometer. It can be used always since it consumes very less power.

Disadvantages It does not do network communication or location access (Fig. 1).

Our power optimization techniques were implemented by extending this particular app because it offered the minimum set of features for activity tracking, which could be extended by adding location tracking and network communication features.

5 Proposed System

In our proposed system we are actually extending the Pedometer App with new features of location tracking, activity measurement and network communication.

Fig. 1 Screenshot of pedometer App

5.1 Location Tracking and Activity Measurement

As the health app periodically needs the location information, we purpose network-based location tracking. Existing pedometer app use only accelerometer for calculating the distance, we enhanced the system by adding location displaying field which gives the current location of the user with the help of network provider. This can further be enhanced by integrating with Google maps, etc. When the system starts working, it first accesses the location of the user and displays it and accelerometer sensor starts working where the distance covered, calories burned, etc. are calculated with the steps counted by the accelerometer. Once the distance reaches certain kilometre, periodically location is sensed using a network provider. Algorithm and code snippet follows:

Algorithm

Step 1: Start
Step 2: Declare the location access variable
Step 3: Calculate the current user location using network provider
Step 4: Accelerometer sensor starts working
Step 5: Steps counted by the accelerometer is converted to distance
Step 6: If (Distance > 2 km)
       Repeat Step 3
Step 7: Stop (Fig. 2)

5.2 Network Communication

Once an application is installed, it gains the permission for network communication; here we provided this privilege with certain conditions, which lowers the

Fig. 2 Screenshot of extended pedometer app

power drain of the smart phone. Permissions were added for accessing the network state and Wi-Fi connections. As the open-sourced app taken for study did not have server connection, we created the backend PHP server, to establish client server communication. Server communication was allowed, only when Wi-Fi connection was available. But if no Wi-Fi connection was obtained within 24 h, then server communication was allowed irrespective of the network type. Once the pedometer starts working for the day, current system time was saved and then a service which checks for network connections was run. If network connected is cellular, then data communication was not allowed and it parallel checked for the available Wi-Fi networks, using the location of the user. The location of the user its latitude and longitude is calculated then the available Wi-Fi's AP is calculated using the ipaddr and its longitude and latitude are calculated and if the Wi-Fi is in available range it is shifted. Once the Wi-Fi is connected data communication. System date, distance travelled, calories burned and location were sent to the server. Detailed algorithm and some parts of the code snippet are as follows:

Algorithm

Step 1: Start
Step 2: Create and initialize a variable apptime to set as current system time
Step 3: Check whether net connection is available
Step 4: If (net_connected &apptime.get (calender.DAY_OF_MONTH) = current_systime. get (calendar.DAY_OF_MONTH))
Step 4.1: If (connectednetworktype = Wi-Fi)
        Data communication take place
        apptime = current_systime
Step 4.2: Else

        Step 4.2.1: Checks for available Wi-Fi, using the location of the user obtained from network provider.
        Step 4.2.3: If (Wi-Fi available)
                Repeat Step 4.1

Step 5: Elseif(netconnected&apptime.get(calender. DAY_OF_MONTH)=current_systime.get(calendar.DAY_OF_MONTH)))
        Data communication take place.
        apptime = current_systime
Step 6: Else
        No server communication.

Fig. 3 Screenshot of both server and pedometer app when network connected to Wi-Fi

Fig. 4 Screen shot of pedometer app and server when connected to the cellular network

5.3   Test Cases

We developed certain test cases, to check the proper updating of server take place when the network is connected by Wi-Fi. It was found that server was updated frequently when it was connected to Wi-Fi, and was blocked when it was cellular network, But server was updated at once it did not had network communication for the past 24 h. We created textbox, along with the location view to check which provider is connected to the system (Figs. 3, 4 and 5).

6   Comparison of Results

Both the developed pedometer app and Endomondo were tested in power tutor for estimating the average power usage by the application and it was found to be 50 % less power drain by the developed pedometer app when both were used to estimate the same distance (Figs. 6 and 7).

	id	date	distance	calories	location
☐ ✎ Edit ✎ Inline Edit ┇ Copy ● Delete	1	18-04-2015	2.5	50	Emakulam
☐ ✎ Edit ✎ Inline Edit ┇ Copy ● Delete	2	19-04-2015	10	100	Kochi

Fig. 5 Screenshot of the server when pedometer was connected after a day

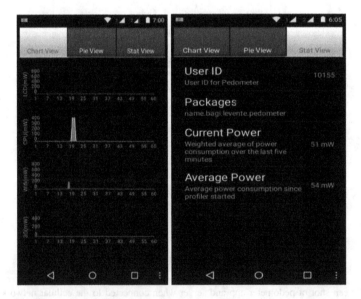

Fig. 6 Screenshot of power usage by the pedometer app in power tutor

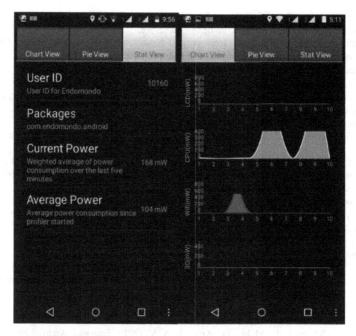

Fig. 7 Screenshot of power usage endomondo app in power tutor

7 Conclusion and Future Work

Power consumption of smart phones was reduced by optimizing location and network services of smart phones. We have presented a study of some leading applications in health and fitness (in android) along with both their pros and cons. We have also mentioned the evolution of new location service A-GPS using both network and GPS, but it grabs very low power from the system.

In future, we purpose an inbuilt system in smart phones that controls the unoptimized usage of permissions like network communication, location access, etc. by the applications. Proper control of these permissions will increase the life of a smart phone. Optimized usage of sensors in the smart phones will also reduce the power drain.

References

1. Power Draining Factors of Android Phones. http://www.brighthand.com/news/a-third-party-app-may-be-draining-your-android-smartphones-battery/.
2. Fitness Apps. http://www.digitaltrends.com/mobile/best-fitness-apps-for-android/.
3. Top Fitness Apps, https://play.google.com/store/apps/category/HEALTH_AND_FITNESS/collection/topselling_free.

4. Sensors Overview. http://www.tutorialspoint.com/android/android_sensors.htm.
5. Sensors. http://whatis.techtarget.com/definition/sensor.
6. Carroll, A., Nicta, G. H. (2010). University of New South Wales and open kernel labs: An analysis of power consumption in a smartphone. In *USENIXATC'10 Proceedings*.
7. Connectivities. http://developer.android.com/guide/topics/connectivity/index.html.
8. Kelley, P. G., Consolvo, S., Cranor, L. F., Jung, J., Sadeh, N., Wetherall, D. (2015, 3 Feb). *A Prof. conundrum of permissions: Installing applications on an android smartphone, compiled list of computer science publications.*
9. Location Services. http://developerlife.com/tutorials/?p=1375.
10. Android Permissions. http://developer.android.com/reference/android/Manifest.permission. htm.
11. Xia, F., Zhang, W., Ding, F., & Hao, R. (2013). A-GPS assisted Wi-Fi access point discovery on mobile devices for energy saving, community-based event dissemination with optimal load balancing. *IEEE Transactions on Computers*.
12. Abdesslem, F. B., Phillips, A., & Henderson, T. (2009, 22 June). Less is more: Energy-efficient mobile sensing with senseless. In *Proceedings of 1st ACM International Workshop on Hot Topics of Planet-scale Mobility Measurements*.
13. Wi-Fi Scans. http://www.cs.purdue.edu.
14. Singhal, M., Shukla, A. (2012, January). Implementation of location based services in android using GPS and web services. *IJCSI International Journal of Computer Science Issues, 9*(1), 2.
15. Difference Between GPS and A-GPS. http://icl.googleusercontent.com/?lite_url, http://www. diffen.com/difference/A-IN&s=1.
16. Lee, W. L., & McKnightb, W. (2003). *Wireless internet access: 3G vs. WiFi telecommunicattion policy* (pp. 351–370).
17. Walde, N., Khatri, P. S., Mehta, D., Avinash, A. V. (2012, January). Android location based services. *IJCSI International Journal of Computer Science Issues, 9*(1), 2.
18. LaMance, J., DeSalas, J. (2002, March). *Assisted GPS, a low infrastructure Approach*.
19. Endomondo Overview. http://fitnesselectronicsblog.com/endomondo-smartphone-app-in-depth-review/.
20. Google Fit Overview. https://developers.google.com/fit/overview.

Power Consumption and Congestion Control of Rendezvous Node for Wireless Biosensor Network

Sagar B. Tambe, Ravindra C. Thool and Vijaya R. Thool

Abstract In biomedical engineering, simulators are imperative tool kits for calculating and analyzing various design choices for wireless sensor network (WSN). The wireless sensors are used in many applications and everyday objects. They are small in size, limited battery, and resource-constrained devices. In a wireless sensor network, limited battery life and unstable power consumption can affect the analysis of the wireless network, flexibility, connectivity as well as the lifetime of the whole sensor network. Various types of issues related to WSN are power constraint, congestion control, fairness, and reliability. In medical applications, large numbers of biosignals are required for analyzing the data that is why we need proper utilization of energy. This paper presents a protocol that will increase the life of the entire wireless network by adjutant's power consumption, speed, and reduce the congestion.

Keywords Wireless sensor network · Power consumption · Congestion control

1 Introduction

Recently, the wireless sensor network has led to progressive research in less cost, time management, and high battery life and power consumption of wireless sensor devices. A consolidation of ad hoc network and cluster protocol is used in IEEE

S.B. Tambe (✉) · R.C. Thool
Department of Information Technology, Shri Guru Gobind Singhji
Institute of Engineering and Technology, Vishnupuri, Nanded, India
e-mail: tambesagar@sggs.ac.in

R.C. Thool
e-mail: rcthool@sggs.ac.in

V.R. Thool
Department of Instrumentation and Control Engineering, Shri Guru Gobind
Singhji Institute of Engineering and Technology, Vishnupuri, Nanded, India
e-mail: vrthool@sggs.ac.in

© Springer Science+Business Media Singapore 2016
S.C. Satapathy et al. (eds.), *Proceedings of International Conference
on ICT for Sustainable Development*, Advances in Intelligent Systems
and Computing 408, DOI 10.1007/978-981-10-0129-1_67

647

Fig. 1 Communication of
wireless sensor network

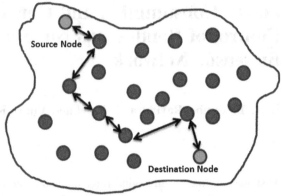

802.15.4 standard for wireless patient monitoring network technology. Different
types of wireless sensor devices to communicate with each other are shown in
Fig. 1. These biosensor devices are sending and receiving the signals form the
patient monitoring system. This wireless biosensor is made from biological material, semiconductors, plastics, and polymer [1]. This biosensor node distributes the
energy consumption and better performance with respective to original routing
protocol. The main aim of this paper is to simulate power consumption and congestion control of wireless sensor network. In a wireless sensor network, the
congestion control is a big issue when congestion happens, coverage fidelity and
network throughput are penalized. In sensor technology, congestion can be categorized as persistent congestion and transient congestion. Persistent congestion is
caused by data sending rate and transient congestion is caused by variations of link
[2].

2   Related Work

When a node moves from a source to the destination location, it can recalculate the
optimal transmission parameters for each node that will maximize its lifetime based
on balance power and target location. The packet size and duty cycle based on
wireless sensor node location can be used for requirements of delay [3]. The author
Dali Wei proposes energy-efficient clustering (EC) and distributed clustering
algorithm that resolution size of cluster depending on the hop distance to the data
sink, while to reduce energy consumption levels, and equalization of node lifetimes
[4]. In wireless sensor technology, the hierarchical sampling techniques are beneficial when the wireless network units are enriched with various sensors with different power consumption and resolution. Triggered sensing technology is used to
activate high-resolution, more power-consuming, low-power, low-resolution sensing units, and for more accurate observation to identify the multiscale sensing for

environmental monitoring applications [5]. Another challenge in wireless sensor networks is to provide efficient communication and reliablity for data transmission [6]. Congestion may occur at sensors that receive more data, which causes packet loss ratio, energy waste, and reduction of throughput. Xiaoyan Yin proposes a rate-based fairness-aware congestion control (FACC) protocol [7], which controls congestion fair bandwidth appropriation for various movements. WSN reduces the sinks node for end-to-end delay, increases success rate, decreases energy dissipation for sinks node [8].

2.1 Duty Cycle

If the sensor nodes are not continuous in nature, then there is no use of putting the node on. To off the node when it is of no use is also an option to reduce the wastage of energy. This power saving technique is called duty cycling. But in this technique, two modes are activated such as sleep mode and wake up time mode. The authors Hesham Abusaimeh and Shuang-Hua Yang proposed a scheme to balance the energy consumption of the network [9].

2.2 Clustering for Sensor Network

All sensor nodes are organized into clusters; each and every node is assigned a cluster called as cluster head. Each separate node in the cluster head transmits to a sink node directly. The main function of cluster head is it can move anywhere in the network. With the help of this cluster head, power consumption is balanced. In [10], authors proposed an energy-efficient clustering scheme for multiple base stations. Traffic management and data dissemination is depending on the increasing number of sink nodes in the network [11, 12].

3 Mathematical Model for Sensor Nodes

Let us consider a set of N sensor nodes, where $N = \{n_1, n_2, ..., n_i\}$.

Let the flow originating from node n_i be f_i and let r_i be the rate at which flow f_i is admitted into the network.

Adaptively assign flow f_i, rate r_i based on queuing theory (M/M/1 Model).

$$BS^i_{\text{unoccupancy}} = BS^i_{\text{max}} - BS^i_{\text{unoccupancy}}$$

where, $BS^i_{\text{unoccupancy}}$ = Current queue length of node n_i

and BS^i_{max} = Maximal buffer size.

Here we need to calculate some values like energy consumption speed, remaining energy, total lifetime and remaining lifetime.

Let the initial energy of the node is *InitEng, RemEng* is the remaining, i.e., current energy level and time period that the node takes to consume energy, then the energy consumption speed of each node is calculated with the following formula.

$$ConsSpeed = \frac{(InitEng) - (RemEng)}{Time} \tag{1}$$

where *ConsSpeed* is the energy consumption speed. The lifetime of the node is the time period for which the node is running, i.e., the node is receiving and transmitting the packets. Each node calculates its energy consumption speed and sets the *R* bit accordingly.

$$\text{if } ConsSpeed < T \text{ then } R = 0 \text{ else } R = 1;$$

where *T* is the threshold value. To calculate the lifetime of the node, the formula is

$$Lifetime = \frac{InitEng}{ConsSpeed}$$

Remaining energy of the node can be calculated by the following formula.

$$RemEng = InitEng - ((PktT \cdot TEng) + (PktR \cdot REng))$$

Where
PktT   = number of packets transmitted by the node
PktR   = number of packets received by the node
TEng = the amount of energy required to transmit packet
REng = the amount of energy required to receive the packet

$$RemainTime = \frac{RemEng}{ConsSpeed}$$

The above formula is to calculate the remaining lifetime left for the node. Total consumed energy can be calculated as

$$TotalConsumedEnergy = InitEng - RemEng$$

and average energy per hop is

$$AverageEnergyPerHop = \frac{TotalConsumedEnergy}{N} \tag{2}$$

where *N* = Number of nodes per hop.

Each time when the node has to send the packet, it can be locally generated data or the packets received from other nodes, the sensor first checks for its energy consumption speed. If the speed exceeds certain threshold then R bit in the packet is set to 1. It indicates to the sink node that the energy consumption speed of the node is more because of more traffic. So when the sink receives any packet, it checks the R bit. If it finds the R bit is 1 then it sends a message to mobile node with the location of the congested node.

4　Results and Discussion

As we know the sensor network is having many-to-one nature, all the sensors propagate their sensed data as well as the data that is received from other nodes toward the sink or base station. From some basic results of network simulation, it is proved that as the packet drop increases, the energy consumption also increases because of the retransmission. But after certain point, as the packet drop increases, the energy consumption decreases. As the reporting rate is low, the average energy consumption is also less. But as reporting rate increases, the average energy consumption gets increased, and after a certain point of the increased reporting rate, the average energy consumption speed gets reduced with the increased reporting rate. There are more chances that near sink nodes may suffer from congestion which may result in packet drop, retransmission ultimately consumes more energy. And if the network is stable then those nodes will die earlier than others. This may cause the loss of network connectivity and also limits the network lifetime. So the use of mobile nodes in the region where the energy consumption speed is high is a good solution. This protocol definitely increases the network lifetime up to certain extent; because the mobile nodes reduce the congestion by moving into the congested region, and forwarding the data. Biosensors face many problems that do not arise in networks: Power consumption, limited hardware, performance, decreased reliability, density of node, and number of sensors nodes than those found in conventional networks are few of the problems that have to be considered when developing a simulation tool for use in biosensor networks. The following Table 1 shows the rendezvous node is nothing but one extra node, which just moves to the intended location and help other nodes in data forwarding.

Before applying the proposed algorithm on sensor nodes, there is a need to see the effect of varying different parameters on the energy consumption. Different parameters are: node density, i.e., number of nodes in the network, reporting rate (RR)—number of packets transmitted per second, packet size, etc.

Table 1 Results of fairness index with varying number of rendezvous nodes

No of rendezvous node	0	1	2	3	4	5
Average energy consumed	0.860	0.772	0.812	0.767	0.828	0.860
Fairness index	0.998	0.998	0.998	0.998	0.998	0.998

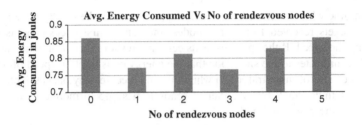

Fig. 2 Avg energy consumed versus number of rendezvous nodes

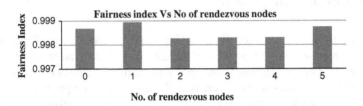

Fig. 3 Fairness index versus number of rendezvous nodes

While using the rendezvous node, it is necessary to find the proper location for the node, proper number of rendezvous nodes, and the speed of the rendezvous nodes. As it is clear that the near sink nodes consume more energy than the nodes which are far from the sink, the near sink nodes need extra resources. So the rendezvous node is placed near the sink node.

Figure 2 shows the graph of the average energy consumed by the network versus number of rendezvous nodes. Figure 3 shows the fairness index with varying rendezvous nodes. The fairness index for one rendezvous node is maximum than other cases. So for this paper work, only one rendezvous is considered.

Table 2 shows the simulation parameters like topology, routing protocol, number of nodes, i.e., 26, initial energy is 2 J, simulation time, i.e., 10 s, the reporting rate is 40 packets per second, i.e., interval is 0.025. For the implementation of the proposed algorithm, the simulation parameters considered are given in the following Table 2. In this new scenario, the rendezvous node is used. And with the use of above simulation parameters, the packet delivery ratio (PDR), packet loss ratio (PLR), and throughput has been calculated for the four cases, i.e., in the basic scenario with the congestion control and varying reporting rate, using the rendezvous node and the last one is the combined use of rendezvous node and congestion control algorithm. These results are shown in Table 3. With congestion control algorithm, the packet delivery ratio is maximum than others. Also the use of rendezvous node provides more packet delivery than the normal scenario. As in the normal scenario, the packet delivery ratio is less, it is clear that the packet loss ratio increases with the use of congestion control algorithm, the packet loss ratio is less among all. The throughput is increased in all the cases, i.e., with congestion control algorithm and the use of rendezvous node than the normal scenario. The throughput

Table 2　Simulation Parameters

Network area	1000 × 1000
Topology	Random
MAC protocol	802.11
Routing Protocol	AODV
Number of nodes	26
Queue length	100
Initial energy	2 J
Simulation time-	10 s
Packet size-	50
cbr start time	1.0 s
cbr stop time	10.0 s
Interval	0.025

Table 3　PDR, PLR, and throughput for the four different test scenarios

Node, ratio, throughput	Number of control	Congestion Ctrl	Rendezvous node	Rendezvous node +Con
Total nodes	26	26	27	27
Packet D. ratio	27.2817	39.504	35.2665	40.0595
Packet L. ratio	72.7183	60.496	64.7335	59.9405
Throughput	110.197	159.565	142.317	161.809

is maximum with the help of congestion control algorithm. The rendezvous node also increases the throughput than the normal scenario (Figs. 4, 5 and 6).

The node with higher energy consumption speed loses its energy earlier which causes the node to die earlier. Figure 7 shows the graph of energy consumption

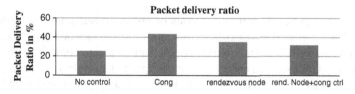

Fig. 4　Packet delivery ratio in four different test cases

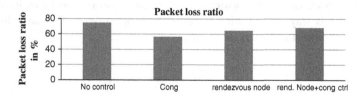

Fig. 5　Packet loss ratio in four different test cases

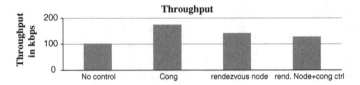

Fig. 6 Throughput in four different test cases

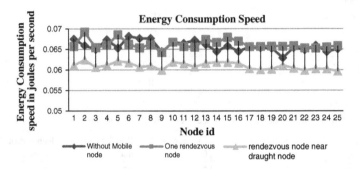

Fig. 7 Energy consumption speed of each node

speed per node. Without rendezvous node, the consumption speed variation is more than the use of rendezvous node. The node 2 is having energy consumption speed more than other nodes. So when rendezvous node reaches to the node 2, the energy consumption speed of node 2 gets decreased.

5 Conclusion

This paper has presented a result analysis of power consumption and congestion control. The energy of wireless biosensor nodes in the sensor network is a very precious resource. It is crucial to recharge the battery and to control the congestion of the sensor devices. More energy saving and speed of a single node can imbalance the power consumption. The congestion control and power consumption can directly affect the entire lifetime of the entire network. The power consumption of the network definitely increases the network lifetime. The use of wireless nodes that can be sink node or simple forwarder node in the network also helps to transfer data. So in this way, use of wireless nodes or cooperative nodes can reduce the energy consumption and increase the network lifetime.

References

1. Abusaimeh, H., & Yang, S. H. (2008). Balancing the power consumption speed in flat and hierarchical WSN. The *International Journal of Automation* and *Computing, 5*, 366–375.
2. Tao, L. Q., & Yu, F. (2010). ECODA: Enhanced congestion detection and avoidance for multiple class of traffic in sensor networks. *IEEE Transactions on Consumer Electronics, 56*, 1387–1394.
3. Azad, A. K. M., & Kamruzzaman, J. (2011). Energy-balanced transmission policies for wireless sensor networks. *IEEE Transactions on Mobile Computing, 10*(7), 927–940.
4. Wei, D., Vural, S., Moessner, K., & Tafazolli, R. (2011). An energy-efficient clustering solution for wireless sensor networks. *The IEEE Transactions on Wireless Communications, 10*(11), 3973–3983.
5. Alippi, C., Anastasi, G., Di Francesco, M. (2009). Energy management in wireless sensor networks with energy-hungry sensors. *IEEE Instrumentation and Measurement Magazine, 12*, 16–23.
6. Fang, W., Liu, F., Yang, F., Shu, L., & Nishio, S. (2010). Energy-efficient cooperative communication for data transmission in wireless sensor networks. *IEEE Transactions on Consumer Electronics, 56*, 2185–2192.
7. Yin, X., Zhou, X., Huang, R., Fang, Y., & Li, S. (2009). A fairness-aware congestion control scheme in wireless sensor networks. *IEEE Transactions on Vehicular Technology, 58*(9), 5225–5234.
8. Thanigaivelu, K., & Murugan, K. (2009). Impact of sink mobility on network performance in wireless sensor networks. *First International Conference on Networks Communications.*
9. Kalyani, A., & Despande, B. V. (2012). Energy Management to Optimize the Lifetime of Wireless. *Sensor Network., 59*(8), 26–29.
10. Park, S., Lee, W., & Cho, D.-H. (2012). Fair clustering for energy efficiency in a cooperative wireless sensor network. *IEEE Vehicular Technology Conference.*
11. Karenos, K., & Kalogeraki, V. (2010). Traffic management in sensor networks with a mobile sink. *IEEE Transactions on Parallel and Distributed Systems, 21*(10), 1515–1530.
12. Park, S., Lee, E., Yim, Y., Yu, F., & Kim, S. H. (2011). Novel strategy for data dissemination to mobile sink groups in wireless sensor networks. *IEEE International Symposium on Personal, Indoor and Mobile Radio Communications (PIMRC), 14*(3), 2259–2263.

Comparison of Different Similarity Functions on Hindi QA System

Bagde Sneha, Dua Mohit and Virk Zorawar Singh

Abstract This paper discusses a comparative analysis of different similarity measures for Hindi question answering system using machine learning approach from information retrieval and classification perspectives. Many machine learning tasks require similarity functions that evaluate likeness between examinations. Similarity computations are particularly important for clustering that depends on precise estimate of the distance between data points. This framework is considered for data matching for multiphrase words and misspelled words.

Keywords Hindi question answering system · Machine learning · Data mining · Similarity functions · Text similarity measure · N-gram approach · Jaccard coefficient similarity · Euclidean similarity measure · Jaro–Wrinkler

1 Introduction

A question answering system includes a process of data matching that aims to interpret whether two data occurrences represent the same entity. This approximate data matching process is relying on similarity functions [1]. Similarity measures have become an extremely popular tool in machine learning. One of the problems that occur in QA system using machine learning is data mining. Data is an essential entity or fact of our concern, but we should know how to retrieve or extract useful

B. Sneha (✉)
Department of Computer Science and Engineering, Banasthali Vidyapith, Banasthali, India
e-mail: snehabagde91@gmail.com

D. Mohit · V. Zorawar Singh
Department of Computer Engineering, National Institute of Technology, Kurukshetra, India
e-mail: er.mohitdua@gmail.com

V. Zorawar Singh
e-mail: zoravirk@gmail.com

© Springer Science+Business Media Singapore 2016
S.C. Satapathy et al. (eds.), *Proceedings of International Conference
on ICT for Sustainable Development*, Advances in Intelligent Systems
and Computing 408, DOI 10.1007/978-981-10-0129-1_68

657

entity from the large volumes of raw data. Data mining techniques help us in accomplishing this [1].

Data mining depends upon distance estimate between observations. The concept of similarity can be different depending on particular domain, task, or dataset available. It is desirable to learn similarity functions from training data to seize the correct notion of distance for a particular task available in a given domain. Another key application that can be benefit from using learnable similarity functions is clustering [2].

2 Different Similarity Functions

A text document can be modeled in many ways, "bag-of-words" being the most prominent representation [3] in IR and data mining. A phrase count is maintained in a bag and each word is made to correspond to an aspect in the followed data space. Consequently, the word appearing in the document with a high frequency, contributes a high weight. This weight can be raised if stemming is applied as N-variants of a base word add up.

Accurate clustering requires an error-free definition of the closeness between a pair of topic, concerning of either the pairwise comparison. In our work, first we use N-gram approach on dataset.

In [4], A.K. Patidar et al., arrived on a conclusion that Euclidean function works best with SNN clustering approach in contrast to Cosine, Jaccard, and correlation distance measure function. Conditional random field (CRF) model based on morpheme features for Tamil question classification is discussed in [5]. The CRF model to find the phrase which contains the information about EAT is trained with tagged question corpus. Another survey in [6] presented the question answering task from an information retrieval perspective and emphasized on the importance of retrieval models.

2.1 N-gram Approach

In this approach, the similarity rating is calculated, that is, based on the number of characters that is in the similar place in each gram [3]. The steps included are:

- Split the text into phrases or tokens consisting only words not digits
- Evaluate all possible N-gram.
- Evaluate frequencies of occurrences of each N-gram.
- Categorize the N-gram according to their frequencies from most frequent to least frequent.
- This gives the N-gram form for a document; these forms are saved in the text files.

But N-gram approach is the maximum time-taking approach in which each text or phrases are compared against all documents in the training classes in terms of similarity.

So, therefore, different similarity functions are used to determine the efficient string matching similarity measure. A variety of similarity measures are tested in the work for better string matching purpose such as cosine similarity, Jaro–Wrinkler, Jaccard coefficient similarity, Euclidean similarity.

2.2 Euclidean Distance

In this approach, the two text documents $D1$ and $D2$ are represented by the vectors t_{D1} and $t_{D2,}$ respectively. The Euclidean distance is given simply as

$$D_E(t_{D1}, t_{D2}) = \left(\sum |W_{tD1} - W_{tD2}|^2 \right)^{1/2} \tag{1}$$

where W_{tD1} = term frequency * index document frequency $(D1, t_{D1})$ and W_{tD2} = term frequency * index document frequency $(D2, t_{D2})$.

2.3 Cosine Similarity

This similarity function is applied to text documents. In this each document is treated as the vector in the space and the cosine between the two vectors gives us measure of similarity. The cosine similarity is

$$SIM_c(t_{D1}, t_{D2}) = (t_{D1} \cdot t_{D2}) \div (|t_{D1} * t_{D2}|) \tag{2}$$

2.4 Jaccard Coefficient

The Jaccard coefficient is another similarity function that provides the range between 0 and 1, where 1 means $t_{D1} = t_{D2}$ and 0 means t_{D1} and t_{D2} are disjoint. Therefore the distance measure is.

$$D_j = 1 - SIM_j \tag{3}$$

2.5 Jaro–Wrinkler

This similarity function depends on the sequence and number of common characters between the two strings [3].

The higher the distance of Jaro–Wrinkler measure between two strings, the more the two strings are similar. D_j denotes Jaro distance between two strings T_1 and T_2.

Therefore

$$D_j = \begin{cases} 0 & \text{if } s = 0 \\ 1/3(s/|T_1| + s/|T_2| + s - m/s) & \text{otherwise} \end{cases} \qquad (4)$$

where S is the number of matching characters. M is half of the number of transpositions. Two characters from T_1 and T_2 are considered matching only if they are the same and not farther than

$$[\max(|T_1|, |T_2|)] - 1 \qquad (5)$$

3 Text Similarity for Multiphrase on Hindi Text

The problem under consideration in the paper is multiphrase words and misspelled words text matching. For two strings, it describes the similarity of two strings to be the length of the longest prefix common to both strings. When the user types the query in the question field sometimes it just writes half the word it actually is or does the spelling mistakes. Then, from the string similarity measure working at the back-end, the string matching function gets the actual word and retrieves the answer according to the question. The data is used to evaluate the method as shown in Tables 1 and 2.

For example, for the string S1 = जम्मू कश्मीर की राजधानी क्या है?? For this sentence जम्मू कश्मीर is a multiphrase word. The user can type this query like S2 = जम्मू की राजधानी क्या है?? Or can like S3 = कश्मीर की राजधानी क्या है?? Or can do spelling mistakes.

To get the proper word or for retrieving the appropriate answer according to the given query, the similarity measure function matches the inappropriate query to the

Table 1 Value of different similarity functions for different string modes

Similarity functions	S2 = जम्मू कश्मीर	S2 = जम्मू	S2 = कश्मीर	S2 = जम्म	S2 = कश्मी
Euclidean	1	0.707	0.707	0	0
Cosine	1	0.552	0.552	0.225	0.225
Jaccard	1	0.531	0.531	0	0
Jaro	1	0.9	0.448	0.866	0.472

Table 2 Values of different similarity functions for multiphrase words

Similarity functions	$S2 = S1$	$S2 =$ केन्द्रीय	$S2 =$ केन्द्रय उत्पाद	केन्द्रीय $S2 =$ उत्पाद शुल्क
Euclidean	1	0.5	0.707	0.86
Cosine	1	0.579	0.683	0.88
Jaccard	1	0.25	0.5	0.75
Jaro	1	0.91	0.95	0.97

queries in the training database and finds the maximum similarity value to which this inappropriate query is matching maximum. A different similarity measure gives a different value and this text matching mathematical value depends on the threshold rate fixed. If one of the string similarity functions gives the string matching rate utmost close to the threshold value then that result is the best result and vice versa. If the threshold value is chosen too maximum, then there is a possibility of not getting any suitable result. And if the threshold value is selected too minimum, then the immaterial data may get generated. To calculate the threshold value for given similarity function for getting the optimal result, the threshold value is chosen between the intervals of $[S^{max},\ S^{min}]$.

4 Experimental Results

The similarity functions used in the experiment are Euclidean similarity function, Jaccard similarity function, cosine similarity function, and Jaro–Wrinkler function. Results according to these similarity functions for multiphrase word and misspelled words matching are based on the experiment.

For string $S1$ = ""जम्मू कश्मीर" की राजधानीक्या है?"? When another string $S2 = S1$ and when $S2$ = ""जम्मू"" (single phrase from the word) and when the $S2$ = ""कश्मीर"" (second half phrase from the word), when the string $S2$ = ""जम्म"" and $S2$ = ""कश्मी"" that are spelling mistaken words, when these different strings $S2$ in different string mode match with the String $S1$with the use of similarity measure. The results of different similarity functions are shown in Table 1.

For another string where $S1$ = ""केन्द्रीय उत्पाद शुल्क दिवस" कब मनाया जाता है" match with the different mode of strings gives different values using different similarity functions.

As shown in Table 2, when $S2$ = ""केन्द्रीय उत्पाद शुल्क"" all similarity functions give the same result; when $S2$ = ""केन्द्रीय"" Euclidean and cosine give the average result while Jaro–Wrinkler gives the best result; when $S2$ = ""केन्द्रीय उत्पाद"" and when $S2$ = ""केन्द्रीय उत्पाद शुल्क"" functions give the same average and best results, respectively, in all cases of string modes.

As shown in Tables 1 and 2, Jaccard similarity measuring function performs worst for multiphrase string matching case and also performs worst for the misspelled matching words. Cosine performs average for both cases and Euclidean performs good for the case multiphrase matching words but worst for the case spelling mistake words.

Measuring the impact of different similarity functions on different mode of texts, the experiment is using the Hindi language context. The implementation assigns manually classification labels that are usually used as guideline criteria to access the objective. The main objective is to match multiphrase words and misspelled words with their actual Hindi word.

In the implemented work, many comparisons on different strings with their different mode of texts are done but the two results shown from all of the above are the conclusion of the experiment. These experiment results shown in Tables 1 and 2 are illustrated by Figs. 1 and 2, respectively. The similarity measure values are represented by *Y*-axis and different mode of strings represented by *X*-axis.

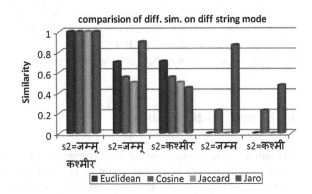

Fig. 1 Comparison for dataset 1

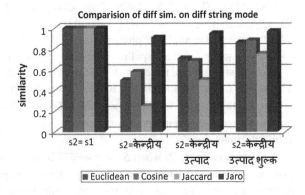

Fig. 2 Comparison for dataset 2

5 Conclusion and Future Work

The objective of the discussed work is to propose a procedure for measuring the best similarity functions for multiphrase word. To access the presented approach, several tests are carried out. As shown by the results, the best similarity measures are obtained by the Jaro–Wrinkler dataset. The future work is to use the concluded results for Hindi question answering system.

References

1. Huang, A. (2008). *Similarity measures for text document clustering*. Hamilton, New Zealand: Department of Computer Science the University of Waikato. In *New Zealand Computer Science Research Student Conference*.
2. Khalid, S. M. A., Jijkoun, V,. & de Rijke, M. (2007). Machine learning for question answering from tabular data. In *18th International Workshop on Database and Expert Systems Applications*, 1529-4188/07 © IEEE.
3. da Silva, R., Stasiu, R., Orengo, V. M., & Heuser, C. A. (2006). Measuring quality of similarity functions in approximate data matching. *Journal of Informatics*.
4. Patidar, A. K., Agarwal, J., & Mishra, N. (2012). Analysis of different similarity measure functions and their impacts on shared nearest neighbor clustering approach. *International Journal of Computer Application, 40*(16).
5. Lakshmana Pandian, S., & Geetha, T. V. (2008). *Tamil question classification using morpheme features*. Berlin Heidelberg: © Springer.
6. Kolomiyets, O., & Moens, M.-F. (2011). A survey of question answering technique from an information retrieval perspective. *Information Science, 181*, 5412–5434. Elsevier Inc.

5. Conclusion and Future Work

References

Preventing Faults: Fault Monitoring and Proactive Fault Tolerance in Cloud Computing

Anu Wadhwa and Anju Bala

Abstract Fault tolerance is about functioning of resources without any impact of faults occurring in them. Cloud computing has emerged as a revolutionary technology with pricing-per-use, scalability, and on demand availability of computing resources as its prominent features. But at the same time, reliability and security are some of the current issues in this technology. Fault tolerance can be used to resolve and handle the issues like reliability and availability. Different types of faults and fault tolerance techniques in cloud computing have been discussed. Further, various faults have been analyzed through Nagios monitoring tool. The experimental results have been validated through monitoring of various faults in cloud environment which can further improve the reliability and availability of cloud services.

Keywords Cloud computing · Fault tolerance · Fault monitoring · Nagios

1 Introduction

Cloud computing is one of the remarkable advances in information and communication technology (ICT). Like other computing technologies, cloud computing has also become essential to meet the day-to-day needs of general and technical communities [1]. Cloud computing can be recognized with some prominent features as accessing the computational resources over the Internet without reference to the service provider [1], paying for the resources as per the usage on the basis of predefined metrics, virtualization of the resources and hence high availability.

A. Wadhwa (✉) · A. Bala
Computer Science Department, Thapar University, Patiala, India
e-mail: wadhwa.anu26@gmail.com

A. Bala
e-mail: anjubala@thapar.edu

© Springer Science+Business Media Singapore 2016
S.C. Satapathy et al. (eds.), *Proceedings of International Conference on ICT for Sustainable Development*, Advances in Intelligent Systems and Computing 408, DOI 10.1007/978-981-10-0129-1_69

The applications, hardware resources, virtualized environment, system software all are delivered as services over the Internet in cloud computing [2]. IaaS, PaaS, SaaS are the services which cloud computing is providing to its customers. IaaS lies above the data centers in the cloud computing hierarchy. Virtualization is the principle concept of this service. A virtualized infrastructure is delivered from the service provider to the customer on demand. This infrastructure includes computational device, storage devices, network devices, etc. [3]. Examples of IaaS are Amazon EC2, Microsoft Azure, etc. PaaS provides a development platform which includes a set of services for developing and hosting applications on the cloud. From application designing to its implementation, verification, validation, and then finally its maintenance every phase of software development lifecycle is implemented at this service level [4]. Examples are Google App Engine, etc. In SaaS, developing tools are used over the Internet instead of installing them. It is also known as AssS (Application as a Service) [4]. Multitenancy feature of the cloud computing can be recognized at this service level. Multiple customers can have access to the single instance of an application at the same time [5]. Examples are NetSuite, etc. Performance and scalability are the prominent features of cloud computing are essences of SaaS. In spite of involvement of experts, this technology is facing many obstacles in its complete adoption by the users. Reason behind these hindrances is the use of the resources in the environment other than for which they were actually designed to work. Due to this number of faults occur in network, servers, storage space, etc. Fault tolerance can be used to resolve and handle the issues like reliability and availability.

Fault tolerance requires understanding of faults occurring in the system as well as impact of faults on the system, running application. As cloud computing architecture has different layers each of them is having their own set of services. Fault at any one layer causes hindrance in functionality of the layers above it.

Two different categories of faults are identified:

- Crash faults: shut down the system completely (hard disk failure) [6].
- Byzantine faults: affect certain parts of the system but application hosted on the system keeps on running, however, incorrectly [6].

The purpose of this paper is to discuss the types of faults, types of fault tolerance, techniques for handling them, and fault monitoring using Nagios. The rest of the paper is organized as follows. Section 2 is the related work. Section 3 describes the type of faults in cloud computing, challenges of fault tolerance, and fault tolerance levels. Section 4 discusses the fault monitoring. In Sect. 5, experimental results are given.

2 Related Work

Increasing demands for cloud computing are facing performance and security challenges, which are resulting in growing interests for cloud management [7]. Nagios is one of the highly acceptable monitoring tools for gathering information from remote nodes so that proactive fault tolerance can be implemented [7, 8]. Nagios generated results and other system related data are stored in hard disk. The data in hard disk contains information about CPU load statistics, hard disk memory status, network configuration, service status, etc. [9]. A monitoring taxonomy is designed in [10] to determine the monitoring characteristics and cloud scenarios for collecting relevant data. In [11, 12], comparative analyses of certain monitoring tools have been done. Table 1 depicts the characteristics of different.

Some requirements and characteristics of a monitoring system are discussed in [12] such as scalability, security, robustness, interoperatibiltiy [13], etc. The task of monitoring is equally important for both the service provider and the customers. Because of the concept of virtualization, jobs can migrate from one virtual node to the other hence it becomes troublesome [14, 15] to monitor the performance and at the same time it becomes necessary to assess the functioning of the resources.

Salient features of Nagios have been underlined including reliability and distribution of monitoring load in [16]. A GridICE is designed for which Nagios is serving the purpose of monitoring. At one of its layers Nagios is serving the purpose of analyzing the metrics and notifying. Analysis for different sites running in grid environment has been done and observations made helped in identifying issues at different levels [16]. Need for monitoring and its benefits have been identified in [17]. Monitoring is described as a three-step process in which collection of metrics is initial step then analyzing them and finally making the results as mentioned in [17]. Monitoring helps in making observations in virtualized environment. Performance fluctuation is mentioned as an important criterion for making observations. Different tools for monitoring have been studied for determining compatibility of tools with the environment for which they can serve the motive of examination. Fault tolerance has been discussed as one of the resulted benefit of monitoring in [17]. In [18] rule-based fault tolerance environment has been developed with the help of monitoring process. Various hardware parameters like temperature and voltage have been considered for making observations and collecting relevant information. Monitored content helps in predicting faults and finally rule-based engine helps in preventing failure.

Table 1 Monitoring tools

Tools	Type	Alerts
Nagios	N/W, service, host	Email, SMS
Ganglia	Distributed system	No
Zabbix	N/W, sensors, app	XMPP

3 Fault Tolerance

In cloud computing, one side is the cloud provider and the other side is the customer, who rents the cloud for running his application. Therefore, management of the faults can be done at different levels where both the providers and the customers may or may not be involved.

3.1 Levels of Fault Tolerance

Application fault tolerance: It is a customer-level fault management; sensors are deployed to monitor the liveliness of the customer application running on the cloud. Virtual machine fault tolerance: Both customers and the cloud providers are involved in handling faults at this level. Customers use the sensors in the cloud for sensing the faults and cloud providers use different techniques to detect and correct faults in the cloud. Physical machine fault tolerance: Faults at physical level are hard to detect at customer side as the hardware is present at the service provider side; therefore, the hardware faults are analyzed by the host. A monitoring system with sensors is deployed at the provider's side for handling physical machine-level faults.

3.2 Challenges of Fault Tolerance

Although a number of efficient techniques are available for handling faults in cloud computing system, but there are many challenges in implementing these technologies.

- Layered architecture of cloud computing itself is a big challenge as fault in single layer impacts the functioning of layers above it.
- Taking benefits from the resources more than for which they are actually meant causes hindrance in solving any issue.
- Some techniques are customer satisfactory while some are service provider based; hence, a technique which satisfies both is hard to be implemented.
- Carefully designed data centers are prone to the big faults because of the size of the data, location of the resources which is segregated all over the world.
- Increase in number of users and rising demand of high computation power is another prominent thing to be kept in mind while applying any of the fault tolerance techniques.

3.3 Handling the Faults

Faults can be handled at different points in time of their occurrence, on this basis three types of fault tolerance policies [19] are there.

- Proactive fault tolerance: Prevention is better than cure kind of strategy is followed here. Fault conditions are detected earlier and with the help of migration technique faulty VM is replaced and system keeps on running.
- Reactive fault tolerance: This method is applicable after the fault has occurred and is affecting application running on the cloud. A kind of on demand cloud fault tolerance service is provided to handle the error.
- Adaptive fault tolerance: Technological advancement is the basis of this approach; fault is handled automatically as per the circumstances when fault occurs.

4 Fault Monitoring

Fault tolerance can be achieved by following two basic things one is preventing errors to occur in the system and second is taking appropriate steps so that one fault could not cause entire system to shut down. Before taking any preventive step, we need to find when and how faults can occur in the system. Here, the first step of preventing system failure has been taken that is of monitoring so that faults can be prevented in advance.

Error cannot be eliminated without knowing the root cause of its occurrence or we cannot handle system failures without having any idea about system's performance. One of the available monitoring tools has been selected for this purpose called NAGIOS. It tells the current status of the hosts, devices, and services which are added to its configuration files.

Monitoring is done by monitoring agents on the basis of metric values obtained from the monitored components. Monitoring is done in the following steps:

- Collect the metrics from the resources.
- Send values to the monitoring server which stores and analyzes them.
- After server analysis reports are generated and notifications are sent to the contact groups (Fig. 1).

Results from the analysis are generated in different forms: Reports, summary, histogram, graphs, and maps. Further, faults occurring in the system can be easily recognized. The basic state labels [19] for a service or a host are "OK" or "up," "warning," "critical" or "down," "unknown," and "unreachable," which signifies functioning status.

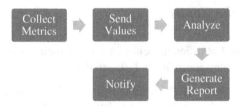

Fig. 1 Fault monitoring process

5 Experimental Results

The experiment is conducted for three different host machines out of which one is Windows host and remaining two are VM's with Ubuntu as operating system. We have monitored different hosts with the perspective of faults occurring in them. Virtual machines are analyzed by Nagios server for finding fault causing situations in them, which can be further corrected to prevent fault occurrences in the host machines. Scripts have been written for host which determines VM's identification details which include its name, IP address, services which Nagios will analyze (CPU utilization, disk check, memory status, etc.). Host and service availability are examined which will help in allocating VM's to the requested user or application. Nagios collects the metrics from the resources which are sent to the monitoring server where they are stored and analyzed. After analysis is done corresponding reports are generated, which determines the status of the resources, host itself, and services running on the host.

Figures below depict the host availability status, service check in the virtual machine and the local host, and finally the disk criticality. Figure 3 shows that out of three machines two are available and one is down which means if any of the application is demanding for some host machine then we can provide it as per the number of requests and if its demand is more than the availability then failure of the application can be prevented, which can occur due to lack of resources. Figure 2 is the summary of the all the host groups together which is again determining the availability of the host and service in the host. Figure 4 represents the faulty condition of the machine that is the criticality of the disk space. This information may prevent the allocation of further requests to the machine so that service failures will not be there.

Nagios configuration files can be edited easily as per the requirements. Check schedule, number of attempts, number of hosts, types of services on different hosts, etc., can be manually determined for monitoring. One of the Nagios plugin used here is called NRPE. It tracks services on remote system like disk space or CPU load. It allows Nagios server to query remote system as if it is local and to check services on the remote system (Table 2).

Status Summary For
All Host Groups

Host Group	Host Status Summary		Service Status Summary	
All Servers (all)	3 UP		4 OK	
	1 DOWN	1 Unhandled	3 CRITICAL	3 Unhandled
Debian GNU/Linux Servers (debian-servers)	1 UP		4 OK	
			2 CRITICAL	2 Unhandled
HTTP servers (http-servers)	1 UP		4 OK	
			2 CRITICAL	2 Unhandled
SSH servers (ssh-servers)	1 UP		4 OK	
			2 CRITICAL	2 Unhandled

Fig. 2 Host status summary

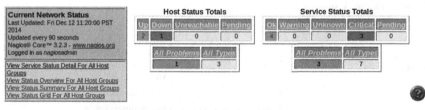

Current Network Status
Last Updated: Fri Dec 12 11:20:00 PST 2014
Updated every 90 seconds
Nagios® Core™ 3.2.3 - www.nagios.org
Logged in as nagiosadmin

View Service Status Detail For All Host Groups
View Status Overview For All Host Groups
View Status Summary For All Host Groups
View Status Grid For All Host Groups

Host Status Totals

Up	Down	Unreachable	Pending
2	1	0	0

All Problems	All Types
1	3

Service Status Totals

Ok	Warning	Unknown	Critical	Pending
4	0	0	3	0

All Problems	All Types
3	7

Host Status Details For All Host Groups

Host	Status	Last Check	Duration	Status Information
localhost	UP	2014-12-12 11:17:38	6d 10h 57m 50s	PING OK - Packet loss = 0%, RTA = 0.05 ms
store.fun.local	DOWN	2014-12-12 11:14:38	2d 2h 35m 24s	PING CRITICAL - Packet loss = 100%
windows	UP	2014-12-12 11:15:48	0d 9h 52m 55s	PING OK - Packet loss = 0%, RTA = 1.17 ms

3 Matching Host Entries Displayed

Fig. 3 Host status

After analysis is done corresponding reports are generated which determines the status of the resources, host itself, and services running on the host. Following analysis helped in preventing the occurrence of the fault as it determines criticality of the machine's working conditions.

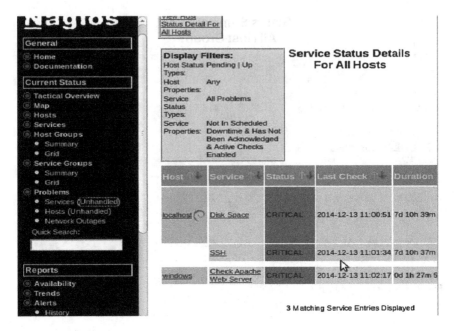

Fig. 4 Critical disk space and service

Table 2 Result interpretation by Nagios

Code	State
0	Up
1	Up or down
2	Down/unreachable
3	Down/unreachable

6 Conclusion and Future Work

An existing fault monitoring application Nagios core has been realized for analyzing the fault occurrence conditions in system. This application aids in proactive fault tolerance by determining the status of the host and services running on them. Fault tolerance can be easily achieved after understanding the existing faults and their root cause. By monitoring physical machine, virtual machine with Nagios it becomes more convenient to understand the behavior of the system corresponding to running hosts on it.

Future plan for implementing fault tolerance is to analyze available migration policies and developing a new algorithmic approach for migration of virtual machine proactively on the basis of results generated from the monitoring tool. For this cloud simulation needs to be done, CloudSim can be a good alternative. It helps in simulating data centers, number of users, etc., for cloud computing.

References

1. Buyyaa, R., Yeoa, C. S., Venugopala, S., Broberg, J., & Brandic, I. (2008). Cloud computing and emerging IT platforms: Vision, hype, and reality for delivering computing as the 5th utility. *Journal of Future*, 599–616.
2. Armbrust, M., Fox, A., Griffith, R., Joseph, A. D., Katz, R., & Konwinski, A., et al. (2010). A view of cloud computing. *Communications of the ACM*, 50–58.
3. Sultan, N. (2010). Cloud computing for education: A new dawn? *International Journal of Information Management*, 109–116.
4. Tsai, W. T., Sun, X., & Balasooriya, J. (2010, April). Service-oriented cloud computing architecture. In *Information Technology: New Generations (ITNG), 2010 Seventh International Conference* (pp. 684–689).
5. Xu, X. (2012). From cloud computing to cloud manufacturing. *Robotics and Computer-Integrated Manufacturing*, 75–86.
6. Jhawar, R., Piuri, V., & Santambrogio, M. (2012). A comprehensive conceptual system-level approach to fault tolerance in cloud computing. In *Systems Conference (SysCon), 2012 IEEE International* (pp. 19–22).
7. Povedano-Molina, J., Lopez-Vega, J. M., Lopez-Soler, J. M., Corradi, A., & Foschini, L. DARGOS: A highly adaptable and scalable monitoring architecture for multi-tenant clouds. *Future Generation Computer Systems*, 2041–2056.
8. Nagios Enterprises LLC. *Nagios-the industry satandard in IT infrastructure monitoring*. http://www.nagios.org.
9. Imamagic, E., & Dobrenic, D. Grid infrastructure monitoring system based on nagios. In *Proceedings of the 2007 Workshop on Grid Monitoring* (pp. 23–28). ACM.
10. Montes, J., Sánchez, A., Memishi, B., Pérez, M. S., & Antoniu, G. (2013). GMonE: A complete approach to cloud monitoring. *Future Generation Computer Systems*, 2026–2040.
11. Fatema, K., Emeakaroha, V. C., Healy, P. D., Morrison, J. P., & Lynn, T. (2014). A survey of Cloud monitoring tools: Taxonomy, capabilities and objectives. *Journal of Parallel and Distributed Computing*, 2918–2933.
12. Gogouvitis, S. V., Alexandrou, V., Mavrogeorgi, N., Koutsoutos, S., Kyriazis, D., & Varvarigou, T. (2012). A monitoring mechanism for storage clouds. In *Cloud and Green Computing (CGC), 2012 Second International Conference of IEEE* (pp. 153–159).
13. Aceto, G., Botta, A., De Donato, W., & Pescapè, A. (2013). Cloud monitoring: A survey. *Computer Networks*, 2093–2115.
14. Dinh, H. T., Lee, C., Niyato, D., & Wang, P. (2013). A survey of mobile cloud computing: Architecture, applications, and approaches. *Wireless Communications and Mobile Computing*, 1587–1611.
15. Park, J., Yu, H., Chung, K., & Lee, E. (2011). Markov chain based monitoring service for fault tolerance in mobile cloud computing. In *Advanced Information Networking and Applications (WAINA), IEEE Workshops of International Conference* (pp. 520–525).
16. Andreozzi, S., De Bortoli, N., Fantinel, S., Ghiselli, A., Rubini, G. L., & Tortone, G., et al. (2005). GridICE: A monitoring service for Grid systems. *Future Generation Computer Systems*, 559–571.
17. Ward, J. S., & Barker, A. (2014). Observing the clouds: A survey and taxonomy of cloud monitoring. *Journal of Cloud Computing*, 3(1), 1–30.
18. Rajachandrasekar, R., Besseron, X., & Panda, D. K. (2012, May). Monitoring and predicting hardware failures in HPC clusters with FTB-IPMI. In *Parallel and Distributed Processing Symposium Workshops & PhD Forum (IPDPSW), 2012 IEEE 26th International* (pp. 1136–1143).
19. Bala, A., & Chana, I. (2012). Fault tolerance-challenges, techniques and implementation in cloud computing. *IJCSI International Journal of Computer Science Issues*, 1694–0814.

Evolving the Reliability for Cloud System Using Priority Metric

Abishi Chowdhury, Kriti Agrawal and Priyanka Tripathi

Abstract The evolution of cloud in the past several years has shifted the IT industry toward utilizing the cloud as software, as a platform, or even as an entire infrastructure. As the demand for different cloud services is growing rapidly, it becomes a challenge for the cloud service providers to ensure the quality of the delivery of services. Cloud service quality refers to the ability to guarantee an intended level of performance or to provide different priorities to different users or applications. There are many ways to provide quality of cloud services, such as scheduling, admission control, traffic control, dynamic resource provisioning, etc. The essential criteria of QoS (quality of service) are reliability, availability, latency, price, etc. Reliability is the quality changing over time, i.e., failure-free service within a specific period of time. Therefore, reliability of cloud services is one of the most important issues in today's scenario. While scheduling the cloud resources to the users in the form of virtual machines, it is important to allocate the virtual machines according to their reliability and the users' priority. The authors have proposed an algorithm and also done simulation work regarding this problem. The authors also intend to deploy this algorithm in a controlled environment with 100 virtual machines.

Keywords Cloud computing · Data center · Virtual machine · Reliability · Priority · Failure

A. Chowdhury (✉) · K. Agrawal · P. Tripathi
Computer Engineering and Applications, National Institute of Technical
Teachers' Training and Research, Bhopal, India
e-mail: abishi.chowdhury@gmail.com

K. Agrawal
e-mail: akriti191@gmail.com

P. Tripathi
e-mail: ptripathi@nitttrbpl.ac.in

© Springer Science+Business Media Singapore 2016
S.C. Satapathy et al. (eds.), *Proceedings of International Conference
on ICT for Sustainable Development*, Advances in Intelligent Systems
and Computing 408, DOI 10.1007/978-981-10-0129-1_70

1 Introduction

Cloud computing brings a new era in the computational world by affording different services on demand to its global users. The significance of these services is displayed in a report from Berkeley as: "Cloud computing, the long-held dream of computing as a utility has the potential to transform a large part of the IT industry, making software even more attractive as a service [1]." Cloud computing uses the concept of pooling actual physical resources and granting them as virtual resources. The cloud services are used as pay-per-use basis; therefore, it is important that the performance should be well enough and it depends on efficient utilization of resources and as well as the reliability of the resources. As per [2], the reliability of a system that consists of n redundant components is: $r = 1 - (1 - r) n$. So, if the reliability of a data center reaches to 99 %, then the reliability of two redundant data centers will be 99.99 %, and the reliability of three redundant data centers will be 99.9999 %. Therefore, it can be concluded that large cloud providers will achieve higher reliability rates than private systems. Cloud services are very critical and complex; also it has a series of dependencies. So, it is important that all the involved members of a service provider should perform their role as efficient as possible to provide the services as reliable as possible. A reliable cloud service should function as the designer destined, when the customer needs it to perform, and wherever the linked customer is situated [3]. Reliability can help in measuring the fault tolerance ability of a system. A system is less reliable if its hardware and software dependencies are more and it is more reliable if the failure of hardware does not affect much on the performance of the software and vice versa [4]. Cloud services have a huge software, hardware, and infrastructure support; each of which can encounter failure at any time. And these failures make cloud less reliable.

The authors had proposed an algorithm to improve cloud service reliability with proper resource utilization [5] using the statistics done by the Microsoft Research [6, 7]. The authors had also simulated the same proposed algorithm with the help of CloudSim simulator and shown the results in the paper [8].

In this paper, the authors have simulated the proposed algorithm in large-scale with 100 virtual machines and 1000 cloudlets, and also proposed the idea of controlled experiment in which the proposed algorithm will be tested and the reliability of the virtual machines will be checked. The authors intend to perform the controlled experiment with 100 virtual machines and 1000 cloudlets.

2 Related Work

Resources are the root of cloud computing as these are provided to the users on demand. Therefore, it is necessary to manage and utilize the resources properly. In [9] an innovative approach for batch jobs had been proposed. Here, the jobs are allocated non-preemptively. The main advantage of this approach is that it avoids network and storage resources for checkpointing. In [10, 11] the allocation of virtual machines for real-time tasks and power management of data centers has been explained. Sealed-bid auction mechanism can be used for cloud resource allocation [12], which simplifies the decision rule of the cloud service provider. Several resource management techniques use different metrics for decision-making [13]. The reliability of these resources should be improved to provide failure-free services to the users. A fault tolerance model for virtual infrastructure has been proposed in [14, 15]. A disadvantage of this model is that it does not deal with real-time application. To overcome this, a model for real-time cloud application has been proposed in [16] which uses two sets of nodes; one is a set of virtual machines and the other is the set of adjudication nodes. There are five modules; acceptance test, time checker, reliability assessor, decision mechanism, and recovery cache. After implementing the proposed model, the authors have concluded that the increase in reliability after 10 computing cycles is 0.3439 and decrease in reliability is 0.8439. An algorithm-based fault tolerance method has been proposed in [17], which can be used for error detection and error correction in matrix operation. Another method has been proposed in [18] to improve the fault tolerance and reliability of matrix multiplication. There may be several types of failure [19] in a cloud environment; timeout, overflow, computing resource missing, data resource missing, network failure, etc. Other types of faults are crash fault and byzantine fault [20] and to resolve these faults replication, checkpoint, and monitoring can be used.

3 Algorithm to Improve Cloud Service Reliability with Proper Resource Utilization

The authors had proposed an algorithm [5] in which the reliability of virtual machines is evaluated dynamically, and according to the priority of the cloudlets higher reliable virtual machines will be allocated. The reliability of the virtual machine has been calculated using the formula 1:

$$R_S = \prod_1^N R_i (\text{i.e. } R_S = R_1 \cdot R_2 \cdot R_3 \ldots R_N) \ldots \tag{1}$$

where, N is the number of elements and R_i is the reliability of ith element of the system. If all the R_i are equal (to R say), then, $R_S = R^N$ when the system is a series system [21].

3.1 Data Center Reliability Assignment

At first, reliability of the data center will be assigned according to the Microsoft statistics [6]. And a failure counter will be set to calculate the numbers of failures occurring in the data center.

3.2 Virtual Machine Reliability Assignment

In the second step, reliability of the elements of virtual machines will be set. And then, reliability of virtual machine will be calculated according to (1).

3.3 Reliability Evaluation

Now, if failure occurs in the components of virtual machine then modified reliability will be calculated for the component in which failure occurred and also new reliability will be calculated for that particular virtual machine to which that fault element belongs.

3.4 Removal

Now, if number of failures reaches to a certain threshold then replace that element and also destroy that particular virtual machine and introduce new virtual machine.

3.5 Allocation of VMs to the Cloudlets

The user's request in the form of cloudlets will be allocated to the virtual machines according to the size.

3.6 Selection of Virtual Machines

In the last step, cloudlets according to their priority will be assigned to the highest reliable virtual machines.

4 Simulation and Results

The first step in the simulation work is to generate an adjacency matrix for allocation of virtual machines to the cloudlets. This matrix will produce a bipartite graph and the intuitive problem will then become a matching problem. The idea here is to find a perfect matching using the reliable scheduling algorithm, which will increase the reliability of the whole system. Figure 1 shows the output for the bipartite graphs. The values at the top row show the virtual machines and the red dots show the cloudlets. An edge between these two shows that the particular cloudlet can be allocated to that virtual machine.

The authors had done three types of simulations in [8] with the help of CloudSim simulator. In this paper the authors have done the same in large-scale with 100 virtual machines and 1000 cloudlets. Three types of simulations that had been done are:

(1) **Without considering reliability**
   Here the allocation of virtual machines to the cloudlets had been done without considering the reliability. It is a mismanaging way of scheduling. As a result,

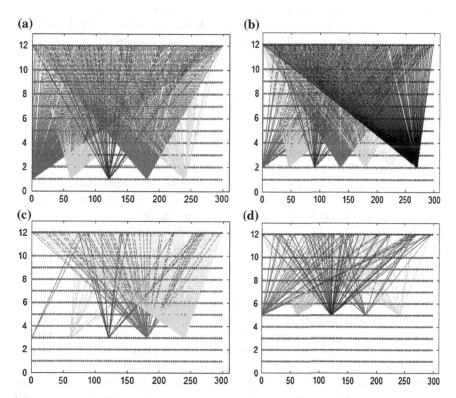

Fig. 1 Allocation of cloudlets to virtual machines in the form of bipartite graph

it can be seen from Graph 1 that an irregular allocation of virtual machine had taken place.

(2) **Reliability consideration before allocation**
In this technique, the simulation has been done without considering the intermediate failures (Graph 2). Though it is an ideal case, it is not possible in reality as fault may occur at any time during working.

(3) **Reliability consideration and failures during simulation**
This is the actual simulation of the proposed method. Here, failures had been considered during the simulation process (Graph 3).

5 Proposed Controlled Experiment to Check the Reliability of Virtual Machines

We will perform a small, controlled, and well-planned experiment [22, 23] to evaluate the reliability of virtual machines in their operational phase. Nature of the experiment is described in the following steps:

Step 1   Factor: Subject (computing skill and previous experience)
Method: M. Tech. third semester students, who will be briefed about the cloud technology and the experiments we are performing.

Step 2   Factor: Server load
Method: A single server hosting a particular website will be used.

Step 3   Factor: Internet speed
Method: Internet speed will be kept the same for all the computers.

Step 4   Factor: Environment and location for the experiment
Method: A single laboratory (M. Tech. cloud computing lab, NITTTR, Bhopal) will be used.

Step 5   Factor: Date, time, and place of evaluation
Method: All the experiments will be performed on specific dates. Time and place will remain the same for the experiments. Also, the duration for the experiment will be the same for all the experiments.

Step 6   Factor: Computers that will be used for virtual machines (processor speed, display units, input methods)
Method: All computers having the same configuration and speed will be used for this experiment.

Figure 2 shows the proposed model for the controlled experiment. In this model, a server containing resources for virtualization is created and on this server with the help of the hypervisors resources will be virtualized. There is a direct connection between the lab and this server. Students will use a thin client or a simple system for connecting to the server. Students request for resources and resources will be allocated to them. Throughout this process, a continuous monitoring of the system will be done for checking the throughput, network traffic, number of failures, and

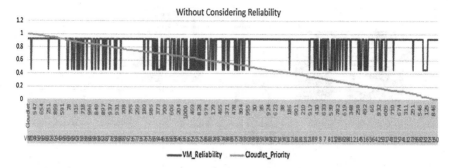

Graph 1 Without considering reliability

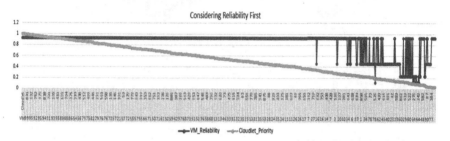

Graph 2 Reliability consideration before allocation

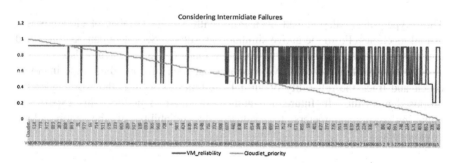

Graph 3 Reliability consideration and failures during simulation

availability of the system. On the basis of these parameters, evaluation of the system will be done. Also, during the whole experiment each time the reliability of the virtual machines will be checked and noted. And based on the updated reliability whether the virtual machines are allocated properly or not that will also be checked during this controlled experiment.

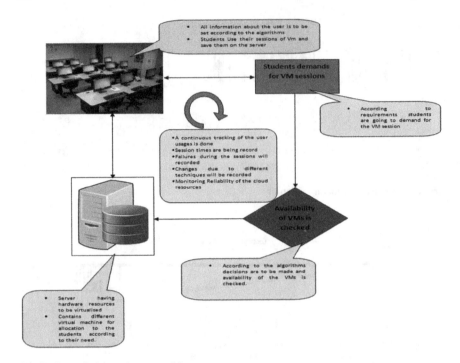

Fig. 2 Controlled experiment model

6 Conclusion and Future Work

Three types of simulation had been done with a large number of cloudlets and virtual machines. From the above simulation results in Sect. 4, it can be concluded that the proposed algorithm works up to the mark in the large-scale also.

Furthermore, the authors have proposed an idea of controlled experiment to check and track the reliability of virtual machines as well as the working of the proposed algorithm. In future, the authors destined to perform this experiment including more parameters to evaluate the reliability of the entire cloud environment.

References

1. Electrical Engineering and Computer Sciences University of California at Berkeley. In *Technical Report No. UCB/EECS-2009-28*. Retrieved February 10, 2009, from http://www. eecs.berkeley.edu/Pubs/TechRpts/2009/EECS-2009-28.html.
2. *Cloud Computing Bible*. Wiley Publishing, Inc. 10475 Crosspoint Boulevard Indianapolis, IN 46256 www.wiley.com Copyright © 2011 by Wiley Publishing, Inc., Indianapolis, Indiana Published by Wiley Publishing, Inc., Indianapolis, Indiana Published simultaneously in Canada. ISBN:978-0-470-90356-8.

3. Trustworthy Computing. *An introduction to designing reliable cloud services.* © 2014 Microsoft Corp. All rights reserved.
4. Yadav, N., Singh, V., & Kumari, M. (2014, February). Generalized reliability model for cloud computing. *International Journal of Computer Applications, 88*(14), 13–16.
5. Chowdhury, A., & Tripathi, P. (2014, December). Enhancing cloud computing reliability using efficient scheduling by providing reliability as a service. In *International Conference on Parallel, Distributed and Grid Computing (PDGC),* 978-1-4799-7683-6/14/$31.00©2014 IEEE.
6. Vishwanath, K., & Nagappan, N. (2010, June). *Characterizing cloud computing hardware reliability.* Microsoft Research One Microsoft Way, Redmond WA 98052 {kashi.vishwanath,nachin}@microsoft.com, SoCC'10, Indianapolis, Indiana, USA, Copyright 2010 ACM 978-1-4503-0036-0/10/06$10.00.
7. Chowdhury, A., & Tripathi, P. (2014, November). Quantifying the cloud computing reliability using a randomizer. In *CICN,* 978-1-4799-6929-6/14 © 2014 IEEE.
8. Chowdhury, A., & Tripathi, P. (2014, November). A novel attempt towards effective scheduling based on reliability in cloud environment. In *ACM Conference on ICTCS '14,* Udaipur, Rajasthan, India, Copyright 2014 ACM 978-1-4503-3216-3/14/11 $15.00 http://dx.doi.org/10.1145/2677855.2677891.
9. Jain, N., Menache, I., Joseph, & Yaniv, J. A truthful mechanism for value-based scheduling in cloud computing. Redmond, WA: Extreme Computing Group, Microsoft Research. Haifa, Israel: Computer Science Department, Technion.
10. Liu, S., Quan, G., & Ren, S. (2010). Line scheduling of real time services for cloud computing. In *World Congress on Services* (pp. 459–464).
11. Tsai, W., Shao, Q., Sun, X., & Elston, J. (2010). Service-oriented cloud computing. In *World Congress on Services* (pp. 473–478).
12. Lin, W., et al. (2010). Dynamic auction mechanism for cloud resource allocation. In *IEEE/ACM 10th International Conference on Cluster, Cloud and Grid Computing* (pp. 591–592).
13. Chowdhury, A., & Tripathi, P. (2014). A metrics based analysis of cloud resource management techniques. In *IEEE International Conference on Advanced Communication Control and Computing Technologies (ICACCCT).* ISBN:978-1-4799-3914-5/14/$31.00 ©2014 IEEE.
14. Kong, X., Huang, J., Lin, C., & Ungsunan, P. (2009, August). Performance, fault-tolerance and scalability analysis of virtual infrastructure management system. In *IEEE International Symposium on Parallel and Distributed Processing with Applications,* Chengdu, China.
15. Kong, X., Huang, J., & Lin, C. (2009, August). Comprehensive analysis of performance, fault-tolerance and scalability in grid resource management system. In *Eighth International Conference on Grid and Cooperative Computing,* Lanzhou, China.
16. Malik, S., & Huet, F. (2011). Adaptive fault tolerance in real time cloud computing. In *World Congress on Services IEEE.*
17. Huang, K., & Abraham, J. (1984). Algorithm-based fault tolerance for matrix operations. *IEEE Transactions on Computer, 33*(6), 518–528.
18. Deng, J., Huang, S., Han, Y., & Deng, J. Fault-tolerant and reliable computation in cloud computing. In *Department of Computer Science, University of North Carolina at Greensboro, Greensboro, NC 27412, USA.* Rockville, MD, USA: Intelligent Automation, Inc..
19. Dai, Y., Yang, B., Dongarra, J., & Zhang, G. Cloud service reliability: Modeling and analysis. In *Innovative computing laboratory, Department of Electrical Engineering & Computer Science.* Knoxville, TN, USA: University of Tennessee.
20. Jhawar, R., & Piuri, V. (2013). Fault tolerance and resilience in cloud computing environments. In J. Vacca (Ed.), *Computer and information security handbook* (2nd ed.). Morgan Kaufmann. ISBN:978-0-1239-4397-2.

21. *Applied R&M manual for defence systems Part D—supporting theory: Probabilistic R&M parameters and redundancy calculations.*
22. Mendes, E., Mosley, N., & Counsell, S. (2003). Early web size measures and effort prediction for web costimation. In *9th International Software Metrics Symposium* (pp. 18–39). IEEE Computer Society Press.
23. Mendes, E., & Mosley, N. (2005). *Web cost estimation: An introduction.* Idea Group Inc.

Multilevel Priority-Based Task Scheduling Algorithm for Workflows in Cloud Computing Environment

Anju Bala and Inderveer Chana

Abstract Task scheduling is the important concern for the execution of scientific workflow applications. An effective scheduling can increase the performance of the workflow applications and task scheduling is used to assign the cloud resources to workflow tasks. Hence, in this, priority-based task scheduling approach has been proposed that prioritizes the workflow tasks based on the length of the instructions. The proposed scheduling approach prioritize the tasks of cloud applications according to the limits set by six sigma control charts based on dynamic threshold values. Further, the proposed algorithm has been validated through the CloudSim toolkit. The experimental results validate that the proposed approach is effective for handling multiple task lists from workflows and in considerably reducing makespan and execution time.

Keywords Priority-based scheduling · VM allocation · Cloud computing · Task scheduling

1 Introduction

Various computing paradigms have assured to deliver utility-based computing such as cluster computing, grid computing, and currently cloud computing, which is used to share variety of resources, softwares, and information like a public utility [1]. It is a technology that allows consumers and businesses to use applications through the Internet without installation. Cloud computing resources can be utilized for applications such as scientific, business, e-commerce, and health care, etc.

A. Bala (✉) · I. Chana
Computer Science and Engineering, Thapar University, Patiala, India
e-mail: anjubala@thapar.edu

I. Chana
e-mail: inderveer@thapar.edu

© Springer Science+Business Media Singapore 2016   685
S.C. Satapathy et al. (eds.), *Proceedings of International Conference on ICT for Sustainable Development*, Advances in Intelligent Systems and Computing 408, DOI 10.1007/978-981-10-0129-1_71

The scientific applications can be represented in the form of workflows that stipulate a process or computation to be executed in the form of data flow and task dependencies. For the successful execution of the scientific applications on clouds, cloud platform should be able to handle the issues like dynamic provisioning of cloud resources to applications in order to meet quality of service parameters. However, there are some more open challenges that need to be resolved such as efficient scheduling of scientific workflows, handling applications and resource failures, heterogeneity of data, failure prediction of tasks, fault tolerant scheduling, etc. Workflow applications often require the scheduling efficiency in terms of execution time and makespan for performing large-scale experiments [2].

Efficient scheduling and resource provisioning are significant issues in both grid and cloud environments. Thus, there is a need to design priority-based task scheduling approach in cloud environment so as to reduce the makespan and execution time. Hence, this paper prioritized various independent tasks of a workflow using six sigma control charts. The tasks have been divided into various levels and further, the resources have been allocated according to the resource requirements of the various task levels.

The rest of the paper is structured as follows. Section 2 discusses the related work. The proposed scheduling algorithm and resource selection procedure is described in Sect. 3. Section 4 illustrates the experiment details and simulation results. Finally, Sect. 5 concludes the paper.

2 Related Works

This section presents existing workflow scheduling algorithms, priority scheduling algorithms along with our contributions.

2.1 Workflow Scheduling Algorithms

Workflow scheduling is the method of mapping the tasks to various cloud resources that can be implemented either by best-effort or QoS constraint-based approach [3]. Various scheduling heuristics and algorithms for workflows have been proposed by many authors. Topcuoglu et al. [4] have calculated the rank value for every task using HEFT, whereas Prodan and Fahringer [5] have employed GAs to schedule workflow applications on grid environment. Binato et al. [6] have shown that the GRASP can resolve job shop scheduling problems successfully. Recently, the GRASP has been explored by Blythe et al. [7] through the comparison of Min–Min heuristic with other heuristics for both compute and data-intensive applications but GARSP does not consider the scalability. Pandey et al. [8] offered a particle swarm optimization (PSO) heuristic as to schedule workflow applications on cloud resources by considering both computation cost and data transmission cost. They have not implemented the offered heuristic for real-time applications. YarKhan and Dongarra [9] have used simulated annealing (SA) to pick a suitable size of VMs for scheduling wherein Young et al. [10] has examined the performances of SA algorithms. Chetepen et al. [11] offered some scheduling heuristics, which are based on task replication and rescheduling of failed jobs. One of the authors Chen et al. [12] have partitioned large-scale scientific workflows. To solve the workflow-mapping problem, various scheduling heuristics have been used such as HEFT [6], Min–Min [13], Max–Min [14], MCT [15], etc. Sonmez [6] extended the scheduling heuristics for multiple workflows for grids but they have not implemented them in cloud environment. None of the said works have implemented priority-based scheduling approach for scientific workflow applications.

2.2 Priority-Based Scheduling Algorithms

Garg et al. [16] compared two allocation scheduling policies such as random overlap and round-robin verlap. In case of the round-robin overlap resource allocation and scheduling policy tasks are executed in a round-robin manner, while for random overlap policy tasks are randomly assigned to VMs.

They have shown that the response time of random overlap scheduling policy has increased. Dakshayini et al. [17] have also proposed priority-based and admission control-based scheduling schemes. They have provided the scheduling with the highest precedence for highly paid user service-requests, but they have not considered other priority levels. Ghanbari et al. [18] also proposed multidecision priority-driven algorithm, but they have not applied the proposed algorithm to set the priorities of tasks for cloud applications. None of the above said works have used priority-based task scheduling with statistical tools akin to six sigma control charts, etc. Six sigma control charts have been used to find three priority levels of multiple task lists in case of cloud applications such as planet lab workflow [19].

2.3 Our Contributions

In contrast to the existing work, a multilevel priority-based task scheduling algorithm has been proposed for workflow a type application that helps to handle and schedule various tasks effectively. The approach proposed in this paper is an extension of the work by Beloglazov et al. [20] and Kumar et al. [16]. They have also not used any priority levels for scheduling independent tasks in random overlap scheduling algorithm. Therefore, we have proposed priority-based task scheduling algorithm that prioritizes the task into different levels by using six sigma control charts.

3 Proposed Priority-Based Task Scheduling

In this section, various symbols have been utilized to schedule workflow tasks and provision the resources in order to attain task scheduling. The evaluated parameters have also been discussed. Table 1 describes the symbols that are used in this paper.

3.1 Task List Priority-Based Scheduling Algorithm

A priority-based autonomic task scheduling algorithm is presented as Algorithm 1. The proposed algorithm works as follows:

- First, the user submits a list of ready tasks to the broker.
- Independent task list is obtained from the list of unscheduled tasks. Once the task list has been obtained, initialize the values of control lists, i.e., UCL, MCL, and LCL to zero.
- For each independent task from x_i to x_n, find the average mean and standard deviation by getting millions of instructions per second (MIPs) values of each task. The task that has the minimum MIPS value will take less time.
- Set control limits using six sigma X-bar control charts, UCL = $\mu + 3s_i$, MCL = μ, set LCL = $\mu - 3s_i$, after the calculation of various control lists, calculate the current MIPS value of the first task.
- If the C_{value} <= LCL then put that task into high priority task list else if C_{Value} >= UCL then lay it into the low priority task list. If C_{Value} >= LCL and C_{Value} < = UCL then store it into the medium priority task list.

Table 1 Symbols and their meanings used

Symbols	Meanings
n	Total number of independent tasks
UCL	Upper control limit
MCL	Medium control limit
s_i	Standard deviation
LCL	Lower control limit
T	The set of tasks in the workflow w
T_s	The execute start time of task T's on cloud resource
T_f	The finish time of task T's on cloud resource
T_e	The execution time T on cloud resources
Tr	The total number of available resources
thl[i]	High priority list of tasks
tll[i]	Low priority list of tasks
tml[i]	Medium priority list of tasks

- Then submit the task lists based on the priority levels to various data centers for execution. Then data center will allocate the resources to submitted tasks by using resource allocation policy.
- Select the minimum migration time resource selection policy which is defined in Algorithm 2. Update the task dependency list after the execution of tasks. Task level priority scheduling is performed till there are no unscheduled independent jobs in the queue.

Algorithm 1: Priority Based Autonomic Task Scheduling

PROCEDURE: Workflow Submit
1. Input: Workflow W
3. Begin
4. Get a task list of unscheduled ready tasks.
5. end
6. PROCEDURE: Task Priority
7. Input: independent task list
8. Set UCL $\leftarrow$ 0, MCL $\leftarrow$ 0, LCL $\leftarrow$ 0
9. for each task x_i in the ready task list i =1 to n do
10. find the average mean $\mu = \sum$ get MIPS $(x_i) / n$

$$s_i = \sqrt{\sum_i \frac{(x-\mu)^2}{n-1}}$$

11. Set UCL = μ + 3s_i, MCL = μ, set LCL = μ - 3s_i
12. Calculate the current MIPS value $C_{value(Xi)}$ for each task
13. **if** C_{value} <= LCL **then** add task (x_i) to thl[i] // High Priority List
    Else if C_{Value} >=UCL **then** add task (x_i) to tll[i] // Low Priority
        List **Else** add task (x_i) to tml[i] // Medium Priority List
14. Submit the task lists to different virtual clusters according to the priorities
15. Select the Resources by Resource Selection Procedure (2)
16. Update task dependency list in workflow
17. end for

3.2 Resource Selection

When the task lists are submitted to data center, the resource selection algorithm used by data center has been described in Algorithm 2.

- First, select the resource which takes minimum time to execute the task. Then schedule and execute the highest order priority list until it becomes empty.
- In the similar way, select the resources for the medium priority and low priority lists.

Algorithm 2: Resource Selection
1. Input: Task Lists, Resource list (Number of Resources)
2. Begin i=0
3. While thl[i] is not empty do
4. Select a Resource which takes minimum time
5. Schedule the task on to the resource and execute it
6. Update the status of the resource
7. i = i+1
8. end while
9. Repeat this process from 3 to 8 for other Medium Priority and Low Priority Lists
10. end

3.3 Parameter Evaluation

Min–Min heuristic has been used to schedule those tasks which have the shortest execution time. The execution time can be defined in the terms of MIPs value of each task. The tasks which have the least MIPS value would be executed first.

- *ET (t, r) Execution Time* The amount of time taken by resource r to accomplish task t which is also defined as the difference between finish time and start time. $T_e = T_f - T_s$
- *Makespan* Makespan is defined as the difference between the submission time of the entry task and the output arrival time of the exit task.

4 Experimental Results

To evaluate the experimental results, CloudSim 3.0 has been utilized which supports the modeling and simulating one or more VMs on a data center [21, 22] as it is extremely difficult to validate the results on a real infrastructure. Therefore, to evaluate the usefulness of the proposed approach, an experiment was designed and conducted. The scheduling algorithm has been compared with random overlap algorithm, which is also implemented for workflows. We have simulated the data center that encompasses 25 heterogeneous nodes and 25 virtual machines. 12 of the nodes are HP ProLiant ML110 G4 servers, and the other 13 involves of HP ProLiant ML110 G5 servers. We have evaluated our results using five experiments with different tasks (Ccoudlets) 20, 40, 60, 80, 100 for workflows.

Case 1 The proposed scheduling algorithm reduces the execution time in comparison to random overlap algorithm, which is shown in Fig. 1. From the above line curve, we can also infer that the proposed algorithm is performing better in all the cases of the different cloudlet list. If the size of cloudlet increases then execution

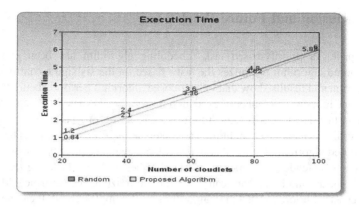

Fig. 1 Execution time (s)

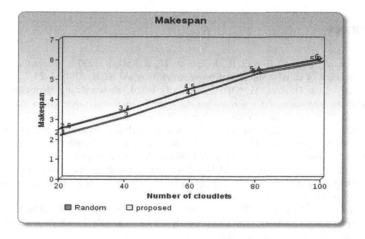

Fig. 2 Makespan

time will decrease; therefore, it will enhance the performance for more number of tasks.

Case 2 Figure 2 shows the comparative analysis of makespan of both the algorithms. Makespan reduces by prioritizing the tasks into various levels then these levels are submitted according to the priorities. High priority tasks which take less time have been submitted first, then medium priority task, and then low priority tasks. The tasks without prioritization take more time to complete the job.

5 Conclusion and Future Work

In this paper, a priority-based task scheduling algorithm has been presented. The proposed algorithm prioritizes the tasks based on the six sigma control charts. The experimental results show that our proposed algorithm performs better than the existing one in terms of number of execution time and makespan. This paper mainly focuses on task-level priority scheduling by considering one parameter. In future, we would like to incorporate multiple parameters at the time of priority scheduling. The proposed algorithm discusses only the independent task priority scheduling but it can also be extended for dependent tasks. Further, various clustering techniques such as K-means and K-medians can be used to optimize the performance. Multiple parameter-based priority scheduling algorithm can be implemented. It can be tested under a real testbed like Pegasus workflow management system.

References

1. Armbrust, M., Fox, A., Griffith, R., Joseph, A. D., & Katz, R. (2009). Above the clouds: A berkeley view of cloud computing. *Communications of the ACM, 53*(4), 1–25.
2. Ghanbari, S., & Othman, M. (2012). A priority based job scheduling algorithm in cloud computing. *International Conference on Advance Science and Contemprory Engineering, Shamsollah Ghanbari and Mohamed Othman/Procedia Engineering, 50,* 778–785.
3. Santos, D. L. (2009). Beyond six sigma—A control chart for tracking defects per billion opportunities (dpbo). Invited but unpublished presentation. *13th Annual International Journal of Industrial Engineering Conference, Las -Vegas, NV, 16,* 227–233.
4. Topcuoglu, H., & Hariri, S. (2002). Performance-effective and low-complexity task scheduling for heterogeneous computing. *IEEE Transactions on Parallel and Distributed System, 13,* 89–97.
5. Prodan, R., & Fahringer, T. (2005). Dynamic scheduling of scientific workflow applications on the grid using a modular optimisation tool. In *A Case Study, the 20th Symposium of Applied Computing* (pp. 687–694). Santa Fe, New Mexico, USA.
6. Binato, S. (2001). GRASP for job shop scheduling. In *Essays and surveys on meta-heuristics* (pp. 59–79). Kluwer Academic Publishers.
7. Blythe, J., Jain, S., Deelman, E., & Gil, Y. (2005). Task scheduling strategies for workflow-based applications in grids. *IEEE International Symposium on Cluster Computing and the Grid, 2,* 759–767.
8. Pandey, S., Wu, L., Guru, S. M., & Buyya, R. (2010). A particle swarm optimization-based heuristic for scheduling workflow applications in cloud computing environments. In *24th IEEE International Conference on Advanced Information Networking and Applications* (pp. 400–407).
9. YarKhan, A., & Dongarra, J. J. (2002). Experiments with scheduling using simulated annealing in a grid environment. In *3rd International Workshop on Grid Computing* (pp. 232–242), Baltimore, MD, USA.
10. Young, L., et al. (2003). Scheduling architecture and algorithms within the ICENI grid middleware. In *UK e-Science All Hands Meeting* (pp. 5–12). Bristol, UK, Nottingham, UK: IOP Publishing Ltd.

11. Rahman, M., Hassan, R., Ranjan, R., & Buyya, R. (2013). Adaptive workflow scheduling for dynamic grid and cloud computing environment. *Concurrency and Computation: Practice And Experience Concurrency Computer: Practice and Experience, 25,* 1816–1842.

12. Topcuoglu, H., & Hariri, S. (2002). Performance-effective and low-complexity task scheduling for heterogeneous computing. *IEEE Transactions on Parallel and Distributed System, 13,* 260–274.

13. Jia, Yu., & Buyya, R. (2005). Taxonomy of scientific workflow systems for grid computing. *ACM SIGMOD Record, 34,* 44–49.

14. Braun, T. D., & Siegel, H. J. (2001). A comparison of eleven static heuristics for mapping a class of independent tasks onto heterogeneous distributed computing systems. *Journal of Parallel and Distributed Computing, 61,* 810–837.

15. Blythe, J., Jain, S., & Deelman, E. (2005). Task scheduling strategies for workflow-based applications in grids (pp. 759–767). *In CCGrid, IEEE.*

16. Chen, W., & Deelman, E. (2012). Integration of workflow partitioning and resource provisioning. In *12th IEEE/ACM International Symposium on Cluster, Cloud and Grid Computing* (Vol. 133, pp. 764–768). Ottawa, Canada: IEEE.

17. Garg, S. K., & Buyya, R. (2011). NetworkCloudSim: Modeling parallel applications in cloud simulations. *Fourth IEEE International Conference on Utility and Cloud Computing., 37,* 105–113.

18. Dakshayini, D. M., & Guruprasad, D. H. S. (2011). An optimal model for priority based service scheduling policy for cloud computing environment. *International Journal of Computer Applications, 32,* 23–29.

19. Calheiros, R.N., Ranjan, R., Beloglazov, A., Rose, C. A. F. D., & Buyya, R. (2011). CloudSim: A toolkit for modeling and simulation of cloud computing environments and evaluation of resource provisioning algorithms. *Software: Practice and Experience, 41,* 23–50.

20. Beloglazov, A., & Buyya, R. (2012). Optimal online deterministic algorithms and adaptive heuristics for energy and performance efficient dynamic consolidation of virtual machines in cloud data centers. *Journal of concurrency Computation: Practices and Experiences, 24,* 1397–1420.

21. Chtepen, M., Dhoedt, B., Cleays, F., & Vanrolleghem. (2006). Evaluation of replication and rescheduling heuristics for gird systems with varying resource availability. In *Proceedings of 18th International Conference on Parallel and Distributed Computing Systems* (pp. 622–27). Anaheim, CA, USA.

22. Beloglazov, A., Abawajy, J., & Buyya, R. (2012). Energy-aware resource allocation heuristics for efficient management of data centers for cloud computing. *Future Generation Computer Systems, 28,* 755–768.

23. Bala, A., & Chana, I. (2011). A survey of various workflow scheduling algorithms in cloud environment. *International Journal of Computer Applications (IJCA), ncict*(4), 26–30.

24. Bala, A., & Chana, I. (2015). Autonomic fault tolerant scheduling approach for scientific workflows in cloud computing. *Concurrent Engineering: Research and Applications, 22,* 1–13.

Efficient DNA-Based Cryptographic Mechanism to Defend and Detect Blackhole Attack in MANETs

E. Suresh Babu, C. Nagaraju and M.H.M. Krishna Prasad

Abstract This paper addresses a novel method to detect and defend against the blackhole attack and cooperative blackhole attack using hybrid DNA-based cryptography (HDC) mechanism. Moreover, the proposed method upsurge the security issue with the underlying AODV routing protocol. Eventually, this HDC is one of the high potential candidates for advanced wireless ad hoc networks, which require less communication bandwidth and memory in comparison with other cryptographic systems. The simulation results of this proposed method provide better security and network performances as compared to existing schemes.

Keywords AODV · DNA-based cryptography · Blackhole attack · MANETS

1 Introduction

There is a necessity to design and develop a secure wireless mobile ad hoc network (SWMANETs), particularly useful for battlefield applications in order to perform security-sensitive operations. One of the primary concerns of these networks requires resilient security service, which are more vulnerable to limited physical insecurity of mobile nodes, as these nodes are disposed to attacks. These attacks are performed in both reactive and proactive routing protocols. The blackhole attack is one such type of assaults that can be performed against both these reactive and

E. Suresh Babu (✉)
Research Scholar, JNTUK, & K L University, Kakinada, AP, India
e-mail: sureshbabu.erukala@gmail.com

C. Nagaraju
Department of CSE, YV University, Proddutur, AP, India
e-mail: nagaraju.c@gmail.com

M.H.M. Krishna Prasad
Department of CSE, UEC, JNTUK, Kakinada, AP, India
e-mail: krishnaprasad.mhm@gmail.com

© Springer Science+Business Media Singapore 2016
S.C. Satapathy et al. (eds.), *Proceedings of International Conference on ICT for Sustainable Development*, Advances in Intelligent Systems and Computing 408, DOI 10.1007/978-981-10-0129-1_72

proactive routing protocols. Here, the malicious nodes pretend to be as a trusted router that impersonates the source and destination node by sending an imitated path request to the destination node and imitated path reply to the source node that was taking place in route discovery phase to claim that, it has the optimal route information. Finally, the blackhole node consumes the packet, and simply drops the packets, that reduce the network performance.

On the other hand, it is essential to design a secure protocol that can defend the blackhole attack and cooperative blackhole attack against on-demand routing protocol [1, 2]. The following are some of the challenging issues to secure against blackhole attack and cooperative blackhole attack. The first challenging issue is to secure the routing protocols against the blackhole attack. This problem has not properly addressed in most of the existing secure routing protocols or if addressed, there are very expensive in terms of bandwidth and limited computational capabilities. The second challenging issue is to defend the blackhole attack against ad hoc routing protocols that dynamically changes the topology, (i.e., what kind of key management and authentication schemes are needed? Unlike of Wireless networks, MANET cannot use any certificate authority (CA) server). The third challenging issue is the existing secure routing protocols may not be efficient or feasible to scale, as these protocols produce heavy traffic load and requires intensive computations. This paper mainly addresses all the above issues using hybrid DNA-based cryptographic mechanism to defend and detect a blackhole attack and cooperative blackhole attack against AODV routing protocol. We call this protocol, as secure routing protocol using hybrid DNA-based cryptography (SRP-HDC) that establishes cryptographically secure communication links among the communicating mobile nodes. The outline for the remainder of the paper is as follows: Sect. 2 specifies the related work. Section 3 enumerates the cryptographic mechanism and attack detection algorithm. Section 4 specifies the simulation results. Finally, we discuss the conclusion with future work in Sect. 5

2 Related Work

The proposed SRPHDC protocol for mobile ad hoc network may fall broadly into two categories, first, integrating the inspired pseudo biotic DNA predicated cryptographic approach into the existing AODV routing protocol and second, detecting the blackhole attack against AODV in MANETs. This section will discuss the revolution of the blackhole problem, which is a genuine security issue in mobile ad hoc network that affect its performance. Recently, many proposals had been proposed in defending, avoiding and detecting blackhole nodes in mobile ad hoc networks [1, 2]. In this paper, we collect and introduce the mechanisms that are proposed in recent years. In [3], Lu et al. proposed a secure AODV routing protocol against blackhole attack for MANETS by showing the security limitations of AODV. Isaac Woungang et al. [2] introduced a mechanism to identify a blackhole attack against the DSR routing protocol. However, their solutions cannot handle cooperative blackhole attacks and computation overhead is present. Most of the solutions discussed above are used to detect or avoid blackhole attacks on reactive routing protocols in the mobile ad hoc networks. However, most of the methods are used to either detect or prevent the blackhole attacks. In this paper, a novel method is presented based on the AODV protocol in which the adversaries are detected based upon the close neighborhood of the range and avoiding the blackhole using multiple paths between source and destination and finally defending by integrating inspired pseudo biotic DNA-based cryptographic mechanisms to the existing AODV routing protocol.

3 Secure Routing Discovery and Attack Detection Algorithm Using Closest Neighbor Mechanism

This section gives the prevention of the blackhole attacks against on-demand routing protocols can be performed using authentication and encryption mechanism.

3.1 Overview of AODV Routing Protocol

In brief, to summarize the AODV [4] routing protocol. AODV is the predominant on-demand routing protocol that offers low processing, low network utilization, ability to adapt the dynamic conditions and low memory overhead. We used this AODV as an underlying protocol to protect from the blackhole attack. The functionality of AODV is usually initiated with the route discovery process, whenever a valid route is not present, and another mechanism is route maintenance, as AODV fails to maintain lifelong route between the sending node and receiving node, due to the high mobility by nature. During the route discovery process, if the originator needs of a route, broadcasts route request (RREQ) packet (with regular information and security related information) to its neighboring nodes, which is described in the next Sect. 3.3. Once the neighbor node obtains a RREQ message from the originator, it broadcast the same RREQ message to its next hop with its current route. This process will be continual until it acquires the actual destination. This proposed work modifies the original AODV protocol. The slight modification is done at the destination side. To be more specific, the destination node broadcast the route reply (RREP) packet back to all its neighbor nodes with the current route until reaches to the source node to create multiple routes, instead of unicasting as in the original AODV routing protocol. To adapt the dynamic topology environment in the route discovery process of AODV in ad hoc networks, we used multiple, possibly disjoint, routes/path between source, and destination. This modified AODV routing protocol responds instead of unicasting with single RREP packet, it broadcasts

RREPs Reply (RREP) packet back to all its neighbor nodes with the current route until reaches to the source node. Once the first RREP message received by the source node, then it can begin sending the data to the destination, late arrived RREPs will be reserved or saved for future purpose. The alternative multiple paths between source and destination can be used for two purpose, first if the primary path fails to send the packets to the destination. Second, after detection of blackhole node (both single and cooperative blackhole node), source node will diverted the traffic with alternate route to the destination. Finally, once the path is entrenched, the nearest neighbor nodes will monitor the link status for the active routes. The nodes that do not conform the neighbor rating based on neighbor profile will be eliminated from the route as described from Sect. 3.3.4.

3.2 Background and Overview of DNA

In order to understand the rudimentary principles of DNA Cryptography [5–7] in a emerge area of DNA Computing, it is necessary to address the background details of central dogma of molecular biology that is how a DNA sequence is actually transcript and translated into a protein sequence as shown in Fig. 1. DNA (Deoxyribo Nucleic Acid) is the fundamental hereditary material that stores genetic · information found in almost every living organisms ranging from very small viruses to complex human beings. It is constituted by nucleotides which forms polymer chains. These chains are also known as DNA strands. Each DNA nucleotides contains a single base and usually consists of four bases, specifically, Adenine (A), Guanine (G), Cytosine (C), and Thymine (T) represent genetic code. These bases reads from the start promoter which forms the structure of DNA strand by forming

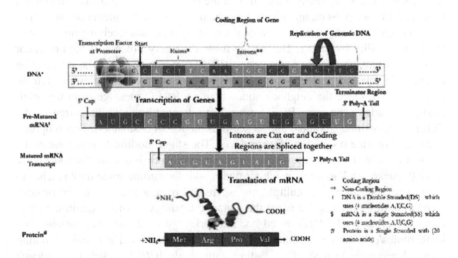

Fig. 1 Central dogma of molecular biology

two strands of hydrogen bonds, one is A with T and another is C with G; These DNA sequences are eventually transcript and interpreted into chains of amino acids, which constitutes proteins.

3.3 Node Authentication Using Hybrid DNA-Based Cryptosystem

This section describes the hybrid DNA-based cryptosystem, which is used to verify the data integrity and authenticate the mobile nodes. Moreover, this hybrid approach makes use of both the public and private key-based schemes. In other words, symmetric encryption will be used to achieve integrity and confidentiality, while asymmetric encryption will provide to authenticate the members of mobile nodes. Subsequently, the above method can be succeeded with the following assumptions.

- First, clearly, according the characteristics of MANETS, initial trustees must exist among the mobile nodes.
- Second, pairwise DNA-based mutual secret keys must exist between the nodes.
- Third, during the initialization phase of the network, we embed the unique ID, an initial key pair of DNA-based private key, and DNA-based public key for every node.
- Finally, the shared secret keys are used to reboot from PKI [8], which can distribute and generate a key pair of DNA-based private key and DNA-based public key for every node.

3.3.1 Secure Route Discovery Process on AODV Routing Protocol

Whenever, source node wants to send the data to the destination. It initiates the path discovery process only, when no valid route is present to the destination. Consequently, it broadcasts RREQ packet by creating pairwise DNA private key/shared key (the procedure of symmetric DNA-based cryptography is discussed in Sect. 3.3.3 with neighbor nodes, until RREQ reaches to the destination node. To achieve this, first, the source node generates pseudo random number and signs that number to create the certificate with its DNA secret key using asymmetric cryptosystem, subsequently the RREQ packet is secured by a Message Digest (MD5) algorithm, finally, the signature and generated hash value is attached to the RREQ control packet, and forwards to its intermediate nodes. Second, the intermediate node having source node DNA public key will verify the signed certificate and then decrypt the message that contains the shared secret key that can be summarized as.

$$cm_{q,i} + h\big(cm_{q,i} + K_{(sk,j)}\big) + \mathrm{E}\big(\mathrm{E}\big(K_{(sk,j)}, K_{(i,\mathrm{pub})}\big)K_{(i,\mathrm{pri})}\big) \tag{1}$$

where cm_q is RREQ control message that contains original message (M), identity Number (IN) of a node that forwards the original message, and sequence number (SeqNo) of the message, which can be written in (2). $h\big(cm_{q,i} + K_{(sk,j)}\big)$ represents the keyed hash Message algorithm with a shared key $K_{(sk,s)}$ on message $cm_{q,i}$ which can be written in (3). Finally '+' denotes the concatenation of strings; the suffix 'i' is the number of intermediate nodes form source to destination $i = 1, 2, 3, 4, \ldots n$ and suffix 'j' is the number of keys created the source node 's' which is usually represented as $j = k_1, k_2, k_3, \ldots, k_n$.

$$cm_{q,i} = \mathcal{M} + \big(IN_f\big) + \mathrm{SeqNo} \tag{2}$$

$$h\big(cm_{q,i} + K_{(sk,j)}\big) \tag{3}$$

Once RREQ control packet reaches with the valid route to the destination, then it verifies the signature, and decrypts the shared key with its private key and reply with a RREP control packet that transmits back to the source which can be summarized as

$$cm_{p,i} + h\big(cm_{p,i} + K_{(sk,j)}\big) + \mathrm{E}\big(\mathrm{E}\big(K_{(sk,j)}, K_{(i,\mathrm{pub})}\big)K_{(i,\mathrm{pri})}\big) \tag{4}$$

3.3.2 To Authenticate the Communications Between Nodes Using Pseudo DNA Asymmetric Cryptography

Suppose, a source node 's' want to send the data to the destination. It initiates with path discovery process with one of the 1-hop intermediate node 'r', then it generates the random key (Here, the Key will number of the splices, the starting code of the frame and removed length of the pattern codes) which will provided as shared DNA private key between source 's' and destination 'r' and then source node encrypts the DNA shared secret key 's_k' by utilizing its neighbor DNA public key K (pub, r) and then once again source node encrypts the encrypted DNA-based shared secret key 's_k' by utilizing its own private key K(pub, s). The output of encrypted shared key 's_k' provides signature for the RREQ control packet and secured by a Message Digest (MD5) algorithm. Once the RREQ control packet reaches with the valid route to the destination, then it verifies the integrity of message (signature) and decrypts the shared key with its private key and reply with a RREP control packet that transmits back to the source node. Hence, source node 's' can authenticate all intermediate node up to the destination node by creating the shared key 's_k' and distribute to all the nodes. The architectural diagram as shown in Fig. 2.

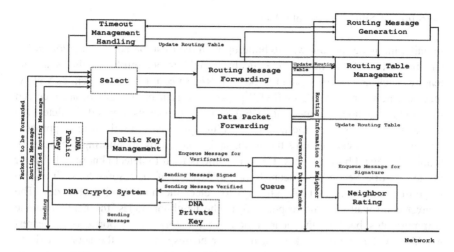

Fig. 2 Architectural diagram of asymmetric cryptography of SRPHDC

3.3.3 Communications Model for Exchanging the Shared Key Using Pseudo DNA-Based Symmetric Cryptography

In [9] we proposed a novel method called pseudo symmetric DNA-based cryptographic mechanisms. Particularly, whenever a node needs to transmit the packet, first, the node should share the unique DNA-based secret key with the source, i.e., private key are shared by the two participating nodes. Moreover, to protect the control or data traffic, the source node 's' can simple use the DNA-based shared key and sent to the destination 'd.' The complete format is shown below:

$$M + h(M + \text{SK}_{sd}) \tag{5}$$

where SK_{sd} is the shared key which is the part of the message M that is shared between source 's' and destination 'd.' Here pseudo DNA-based symmetric cryptography is mainly used to achieve data integrity and confidentiality.

3.3.4 Blackhole Attack Detection

This subsection performs the next task, after developing the secure route discovery process on AODV routing protocol, the detection scheme against blackhole attack should be incorporated into secure route discovery procedures.

Detection of Single Blackhole Attack

Initially, the nodes are allowed to build up with the trust, based on neighbor profile and behaviors of the nodes. Specifically, neighborhood profile includes all the features such as number of data packets or RREQ control packets sent as well as received, number of ACK, or RREP packets are sent or forwarded and received from/to the neighbors, number of RREQ/RREP/data packets dropped. To detect the misbehaving nodes, first, the neighbor profile approach is used. In this approach, the nearest node can identify the misbehaving nodes by monitoring the network traffic of its neighboring nodes. Here, a profile is used to detect abnormal behavior in the network. Second, the nearest node makes use of the responses, such as (No response, Shutdown, and Blacklisting) to detect the misbehavior of nodes. The nearest node waits for the response from the neighbor nodes, the neighbor node communicates back to the nearest node in the form of response (i.e., nearest neighbor will response back (shut down or no responses) to the source node or intermediate nodes). To achieve this k-nearest neighbor algorithm [10] is used to compute the k-nearest neighbor by calculating the distance between the nodes. Finally, the source node gives the neighbor rating [11] based on neighbor profile. After detecting the malicious node based on neighbor direct rating, the source node will diverted the traffic with different route to the destination. Hence, intrusion of malicious node effect to network becomes weaker. Therefore, we can conclude that more paths reduce malicious node intrusion to network.

Detection of Cooperative Blackhole Attack

The cooperative blackhole attack is more challenging attack to detect. In our design, source node has transferred its shared secret key with all its k-hop-neighbors using key management schemes. To detect the cooperative misbehaving nodes, first, the nearest foreign neighbor profile approach is used. In this approach, each node finds its closest nearest foreign neighbor (2-hop-neighbor) by establishing DNA shared keys between the sender and neighbor nodes. Hence, the nearest foreign neighbor can identify the misbehaving nodes by monitoring the network traffic of its neighboring nodes. After detecting the malicious node based on foreign neighbor indirect rating, the source node will diverted the traffic with different route to the destination. Eventually, intrusion of malicious node effect to network becomes weaker. Once again, we can conclude that more paths reduce malicious node intrusion to the network. In summary, the single blackhole and cooperative blackhole intrusion can be identified without the need of expensive signatures, as these signatures, which can be used to defend the route from end to end.

4 Simulation Results and Performance Analysis

To study the feasibility of our theoretical work, we have implemented and evaluated the secure routing protocol using hybrid biotic DNA-based cryptography method using network simulator [NS2], which is a software program running in Ubuntu-13.04 and conducted a series of experiments to evaluate its effectiveness. The experiment results show that this method is more efficient and increase the power against blackhole attacks. Our simulations are mainly used to compare between SRPHDC with AODV routing protocol with and without the presence of malicious nodes. Moreover, we also compare SRPHDC with modified AODV with and without the presence of malicious nodes and ARAN [12], respectively. The following Table 1 gives the simulation parameters, which are used to compare SRPHDC with AODV and ARAN respectively. Particularly, in order to calculate the performance of this proposed work, we had collect the simulated data by running the simulation up to 500 s with an input of 10, 25, 50, 75, 100 nodes.

Specifically, Fig. 3 depicts the throughput of SRPHDC and AODV routing protocol in the presence of 2, 4, 6, 8, 10 blackhole nodes out of 10, 25, 50, 75, and

Table 1 NS-2 parameters

Propagation model	Two ray ground
No. of nodes	10, 25, 50, 75, 100
Transmission range	250 m
Simulation time	500 s
Simulation area	750 m × 750 m
Node mobility	Model random waypoint
Traffic type	FTP/TCP
Data payload size	512 bytes/packet
Node pause time	0–20 s
Maximum node speed	1–20 m/s
Key size	1024 bits key

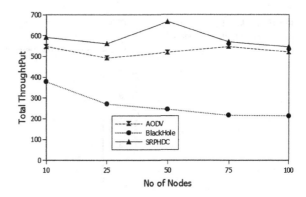

Fig. 3 Throughput of AODV versus SRPHDC in the presence of blackhole nodes

Fig. 4 Routing load of
AODV versus SRPHDC with
blackhole nodes

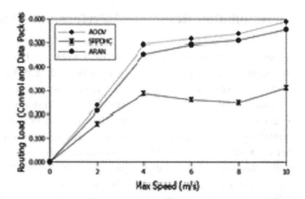

100 nodes, respectively. It is observed that AODV drastically decreases the throughput in the presence of 10–40 % blackhole nodes, the black nodes will be part of the route in AODV that result in a low packet delivery ratio (PDR) or high packet drop. While, SRPHDC still delivers the data packets with almost the same amount of data. Figure 4 depicts the total overhead of SRPHDC in comparison with AODV and SRPHDC. Here, it is observed that SRPHDC gives slight overhead than AODV due to the security mechanism and less overhead than modified AODV in the presence of different numbers of blackhole nodes. The reason is that SRPHDC can detect the blackhole nodes based on neighbor rating and remove them from routing. Moreover, SRPHDC gives less overhead than ARAN, because SRPHDC make use of DNA computing which provides less communication overhead and efficient storage capacity.

Figure 5 depicts the packet delivery ratio of ARAN and SRPHDC with different pause times. While, SRPHDC has a higher PDR than ARAN. Since, it delivers more data packets than ARAN to the destination. Figure 6 shows the end-to-end packet latency of SRPHDC and ARAN. In this case, it is observed that SRPHDC has lower latency than ARAN, Since, SRPDHC make use of DNA computing which provides parallel computation that causes less overhead of cryptographic

Fig. 5 Packet delivery ratio
of AODV, SRPHDC, ARAN

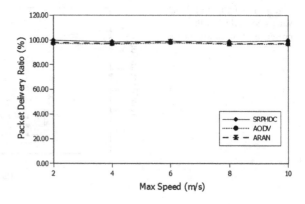

Fig. 6 End-to-end delay of
SRPHDC and ARAN

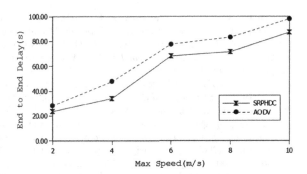

operations than ARAN. In ARAN, while processing the control packets, each and every node has to verify the signature and replace with its own digital signature that cause more overhead using conventional cryptography and additional delays at each node. Therefore, the end-to-end packet latency increases.

5   Conclusion

In this paper, we have proposed a scheme for defending and detecting the blackhole attack against AODV routing protocol in mobile ad hoc wireless networks. This proposed protocol establishes cryptographically secure communication among the nodes. The simulation results of this work provides better security, less computation overhead, and network performances as compared to existing ARAN schemes. As a conclusion, DNA cryptography is an emerging new idea and very promising field where research can be done in great innovation and development but this DNA cryptography lacks the related theory which is nevertheless still an open problem to model the good DNA cryptographic schemes.

References

1. Dasgupta, M. (2012). Network modelling of a blackhole prevention mechanism in mobile ad-hoc network. In *Fourth International Conference on Computational Intelligence and Communication Networks*.
2. Woungang, I. (2012). Detecting blackhole attacks on DSR-based mobile ad hoc networks. *IEEE International Conference on Computer, Information and Telecommunication Systems (CITS)*.
3. Lu, S., Li, L., Lam, K.-Y., & Jia, L. (2009). SAODV: A MANET routing protocol that can withstand black hole attack. *Proceedings of International. Conference on Computational Intelligence and Security*, December 11–14, Beijing, China (pp. 421–425).
4. Perkins, C. E., & Royer, E. M. (2001). The ad hoc on-demand distance-vector protocol. In C. E. Perkins (Ed.), *Ad Hoc Networking* (Chap. 6, pp. 173–220). Reading, MA: Addison-Wesley.
5. Ning, K. (2009). A pseudo DNA cryptography method CoRR 2009, abs/0903.2693.

6. Capkun, S., Buttyan, L., & Hubaux, J. (2003). Self-organized public-key management for mobile ad hoc networks. *IEEE Transactions on Mobile Computing, 2*(1), 1–13.
7. Gehani, A., LaBean, T. H., & Reif, J. H. (1999). DNA-based cryptography. In Winfree & Gifford, (Eds.), *Proceedings 5th DIMACS Workshop on DNA Based Computers* (Vol. 54, pp. 233–249). MIT, Cambridge, MA, USA.
8. Chen, J. (2003). A DNA-based, biomolecular cryptography design. In *IEEE International Symposium on Circuits and Systems (ISCAS)* (pp. 822–825).
9. Suresh Babu, E., Nagaraju, C., & Krishna Prasad, M. H. M. (2015). Light-weighted DNA based cryptographic mechanism against chosen cipher text attacks. *2nd International Doctoral Symposium on Applied Computation and Security Systems*, Kolkata, India. LNCS, Springer.
10. Andoni, A. (2009). Nearest neighbor search. In PhD Thesis. MIT, September 2009.
11. Hod, B. (2005). *Cooperative and reliable packet-forwarding on top of AODV*. Israel: School of Engineering and Computer Science The Hebrew University of Jerusalem.
12. Sanzgiri, K., LaFlamme, D., Dahill, B., Levine, B. N., Shields, C., & Belding-Royer, M. (2005). Authenticated routing for ad hoc networks. *IEEE Journal on Selected Areas in Communications, 23*(3), 598–610.

Sentiment Classification of Context Dependent Words

Sonal Garg and Dilip Kumar Sharma

Abstract With the increase in the use of Web 2.0, there are lot of opinions on the web about any product. Most of the opinions contain opinion words which has same polarity in all contexts. But there are some opinion words called context dependent words which have different polarity in different context. So there is a need to determine the polarity of ambiguous words (context dependent words) efficiently and effectively. This task is also known as word polarity disambiguation (WPD). This literature survey is done to familiarize with general applications and approaches of opinion mining then it presents the context dependent word polarity problem in depth by explaining the existing literature of sentiment classification of context dependent words and finally, some open problems, conclusion, and future directions are discussed.

Keywords Word polarity disambiguation · Opinion mining · Ambiguity · Context dependent word

1 Introduction

Today is an internet era. If anyone wants to purchase any product then he or she is interested in knowing the opinions about that product which helps in deciding whether to accept or reject that product. There are different sources of opinions. Opinions may be collected from newspapers, blogs, forum, discussion board, or review sites. Opinion mining is a technique to extract the opinion from the web

S. Garg (✉) · D.K. Sharma
Department of Computer Engineering & Applications, GLA University, Mathura, India
e-mail: Sonugarg174@gmail.com

D.K. Sharma
e-mail: dilip.sharma@gla.ac.in

© Springer Science+Business Media Singapore 2016
S.C. Satapathy et al. (eds.), *Proceedings of International Conference on ICT for Sustainable Development*, Advances in Intelligent Systems and Computing 408, DOI 10.1007/978-981-10-0129-1_73

about particular entity. There is huge volume of unstructured reviews so opinion mining is used to extract the sentiment from these reviews and present them in structured format.

There is lot of opinion on the web and some opinions contain context dependent words which have different polarity in different context. For instance consider the following sentences. This phone has long battery life. Here, the word long has positive polarity but in next sentence, this phone has long processing time here the same word long indicates negative polarity. So, to classify the sentiment correctly we need to remove ambiguity from the sentences with the help of tool called word polarity disambiguation.

Word polarity disambiguation is the computational identification of polarity of a word in given context [1]. It is a central challenge in opinion mining and it is an open research area in opinion mining [2].

Unfortunately, the problem of identification of specific polarity of given word seem to be easy. As human easily detect a sentiment of a word in a given context. But for a machine it is difficult. The goal of WPD is to obtain a sentiment of context dependent word from the given reviews using contextual information and to improve the accuracy of existing sentiment classification methods.

2 Sentiment Classification Levels

We can perform sentiment classification at three levels [2].

2.1 Document Level Classification

In document level sentiment classification it is assumed that opinions remain the same throughout the whole document. It is mostly used in review classification and buzz analysis. It is inappropriate for forums and blogs sentiment classification if it contain comparative sentences. To extract the irrelevant sentences from given documents is still a research challenge.

2.2 Sentence Level Classification

It is more grained approach in comparison to document level sentiment classification having same assumption that one-sided opinion remains throughout the whole sentence [1]. In this, sentiment classification can be done in two ways first one is grammatical syntactic approach and second one is using semantic approach [3].

Table 1 Sentiment level classification

Method	Advantage	Disadvantage
Document level	It is useful in the computation of overall polarity of a document	Difficulties in the extraction of irrelevant sentences from a document
		It is not possible to separately extract the different sentiments about different aspect of an entity
Sentence level	Used for subjectivity/objectivity classification	Inappropriate for complex sentences
		Fails when use express different views in same sentence
Aspect level	Used in product review mining where the sentiments occurred on particular aspect matters	Do not consider long range dependency

2.3 Aspect Level Classification

Most fine-grained approach is aspect-based sentiment classification which is concerned with sentiments expressed toward certain aspects of a topic. Its major a application in product review mining (Table 1).

3 Applications of Opinion Mining

The opinion mining can be used in many fields [4].

3.1 Question Answering [5, 6]

Opinion mining can be used in question answering system. It also proves useful in rationale questions which require more information about related entity.

3.2 Online Advertising [7]

Opinion mining can be used to show advertisements on WebPages according to user's interest and promotes the products with higher ratings.

3.3 Detection of Flames [5, 6]

Detection of flames, arrogant words, overheated words, or hatred language used in emails or websites can be helpful in finding the places to post the advertisement. The sentiments reflected from the text written on the websites or mail can be analyzed to find the overall content inclination. So helps in determining sensitive content.

3.4 Individual Consumers [7]

Sentiment analysis helps in decision making whether to accept that product or reject that product based on the pros and cons of that product. Sentiment analysis plays an important role in recommendation system by comparative analysis of opinions of different products.

3.5 Organizations and Business [7]

Opinion mining plays a very important role in businesses and organizations by reducing a huge amount of money and time spent in perceiving consumers opinions about its products and competitors products using survey or other techniques but also it helps manufacturer to limit the risk of overproduction and underproduction. It also helps in marketing of business and product benchmarking.

4 Context Dependent Word Sentiment Classification

There are lots of opinion words which have same orientation in any context, e.g., good, bad. But some ambiguous words exist which has different orientation in different context. It is impossible to determine the context dependent word polarity without prior knowledge about that product.

General sentiment classification model considering context dependent word can be given as (Fig. 1).

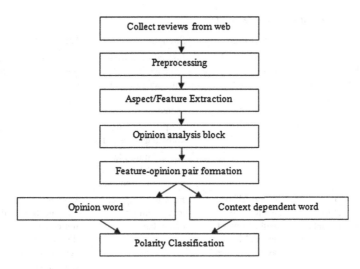

Fig. 1 General context dependent word sentiment classification

5 State of the Art

In [8], to handle context dependent word first linguistic or convention rule is used then for identifying opinion words both the opinion words and features are used. Sentiment orientation of domain opinion word is finding online.

ΔTF_IDF approach is used in [9] to assign weight score but the limitation of this approach is that it does not consider the distributions of terms across various documents for classification it only considers the total number of documents in which term occurs. In this paper [10], authors adopt mutual information method to predict the sentiment of term in a document. This paper considers the distributions of terms across various documents. In this authors, assumed that opinion word have same polarity in same domain. Authors [11] consider that the same opinion word has different polarity in same domain so uses the combination of opinion word and feature to calculate the polarity. Here, they calculate the polarity by using conditional mutual information (CMI).

Aspect-based sentiment analysis and summarization system was proposed [12] which deal with the context dependent opinion words. For constructing the training dataset, first authors used an online dictionary for identifying context dependent words and assign the polarity using natural linguistic rules. After that a combination of opinion words and feature was used for classification of remaining. For this task interaction information method was used.

In this work [13] integrated WSD algorithm to improve the task of polarity classification. In this paper, extended gloss overlap algorithm was used with the help of this voxpop system accuracy increased to 60 %.

This paper [14] aims to determine the most related sense of word according to context then determined the sentiment of word using SentiWordNet. A graph-based

Table 2 Summary of papers under WPD

Paper	Method used	Strength	Limitations
A holistic lexicon-based approach to opinion mining [8]	Lexicon-based approach	Instead of current sentence it also consider external information in other sentence	Dictionary definitions are generally small
		No prior domain knowledge needed	
		Handle multiple conflicting opinion words in a sentence	
Delta tf-idf: an improved feature space for sentiment analysis [9]	Delta tf-idf is used to assign a weight to word then SVM classifier is used	Easy to compute, implement and understand	For sentiment classification, it does not consider distribution of terms across various documents
Word sense disambiguation of opinionated words using extended gloss overlap [13]	Lexicon-based approach (SentiWordNet 3.0)	Improve the accuracy from 50.5 to 60 %	Does not consider contextual valence shifters
			Domain specific
Sentiment classification using graph based word sense disambiguation [14]	Lexicon-based approach	Gives high accuracy	Rely on dictionary
An information theoretic approach to sentiment polarity classification [10]	Mutual information	Works on distribution of terms across the document	Fail when same opinion word have different polarity in same domain
Revised mutual information approach for German text sentiment classification [11]	Revised mutual information	Works well when same opinion word have different polarity in same domain	Domain dependent dataset, sentence level classification
Aspect based summarization of context dependent opinion words [12]	Interaction information method	Domain independent. Aspect level classifications	Does not consider noun and adverb as opinion words
Word polarity disambiguation using bayesian model and opinion-level features [15]	Bayesian model	Consider both inter-level and intra-level opinion features	It does not work well when opinion level features are not used

WSD algorithm is used that builds a graph corresponding to words sequence from Word Net and then find the stronger link as a related meaning of target word.

In this work [15], Bayesian model was used to disambiguate word sentiment polarity. Authors use both the intra-opinion features and inter-opinion features (Table 2).

6 Open Problems [7]

In this section, we give brief discussion of few problems occurred during handling of context dependent words.

6.1 Selection of Feature Word

The most important problem prevailing in the field of WPD is the proper selection of feature word. Generally, noun is used to represent aspect but sometimes verbs and adjectives also used as a feature word.

6.2 Grouping the Synonym

It is possible that different customers express their reviews on same feature using different wordings. For example, in mobile domain, voice and sound represent same feature.

6.3 Handling Abbreviations

Customer can give comments using abbreviations or shorthands. They can use Moto g for Motorola, pic for picture, gud in place of good. So there is a difficulty in handling these words.

6.4 Lexicon-Based Method

Lexicon-based method is not suitable for handling context dependent words. Here, the same word represents different polarity in different context.

6.5 Selection of Opinion Words

Generally, adjective expressed opinion in opinion mining but there are some adverbs and noun which are opinion word and context dependent in nature so there is a need to include more opinion words.

6.6 Detection of Fake Review

There is a huge amount of fake reviews used to promote their product or criticize other products. So there is a need to identify these fake reviews by detecting duplicates. Finding fake or spam review is still a challenge in opinion mining.

6.7 Standard Dataset Problem

There is a lack of standard dataset of context dependent words.

7 Conclusion

In this paper, we have described a basic concept of opinion mining, its applications, different levels and different types of approaches used for the sentiment classification. We have also compared lexicon-based approaches. Comparison is done on the basis of approach used, its strength, and weakness. This paper mainly focuses on sentiment classification using context dependent words. Supervised approaches give higher performance in opinion mining but they require large manually labeled data. Creating large dataset is challenging as it requires intense manual effort. It works well in handling context dependent words. A very little work has been done in sentiment classification of context dependent words. In future, we aspire to include more opinion words. Furthermore, unsupervised approaches can be utilized and work can be done to disambiguate other part of speech (POS) other than adjectives. There is a need to consider noun and adverb as context dependent words also with different feature extraction algorithms to improve the accuracy of the system.

References

1. Liu, B. (2010). Sentiment analysis and subjectivity. In N. Indurkhya & F. Damerau, (Eds.), *Handbook of Natural Language Processing* (2nd edn.).
2. Pang, B., & Lee, L. (2008). Opinion mining and sentiment analysis. *Foundations and Trends in Information Retrieval., 2*(1/2), 1–135.
3. Kaur, A., & Duhan, N. (2015). A survey on sentiment analysis and opinion mining. *International Journal of Innovations & Advancement in Computer Science,* 107–116.
4. Feldman, R. (2013). Techniques and applications for sentiment analysis. *Communications of the ACM, 56*(4), 82–89.
5. Liu, B., & Zhang, L. (2012). A survey of opinion mining and sentiment analysis. In C. C. Aggarwal & C. X. Zhai (Eds.), *Mining Text Data* (pp. 415–163).New York: Springer.
6. Chandrakala, C. S., & Sindhu, C. (2012). Opinion mining and sentiment classification: A survey. *ICTACT Journal on Soft Computing, 3*(1), 420–425.
7. Rashid, A., et al. (2013). A survey paper: areas, techniques and challenges of opinion mining. *International Journal of Computer Science (IJCSI), 10*(6), 2.
8. Ding, X., Liu, B., & Yu, P. S. (2008). A holistic lexicon-based approach to opinion mining. In *Proceedings of the 2008 International Conference on Web Search and Data Mining* (pp 231–240).
9. Matineau, J., & Finin, T. (2009). Delta tfidf:an improved feature space for sentiment analysis. In *Proceedings of the Third International ICWSM Conference* (pp. 258–261).
10. Yuming, L., Zhang, J., Wang, X., & Zhou, A. (2012). An information theoretic approach to sentiment polarity classification. In *Proceedings of the 2nd Joint WICOW/AIRWeb Workshop on Web Quality* (pp. 35–40). ACM.
11. Farag, S., & Mathiak, B. (2013). Revised mutual information approach for german text sentiment classification. In *Proceedings of the 22nd International Conference on World Wide Web Companion* (pp. 579–586).
12. Kansal, H., & Toshniwal, D. (2014). Aspect based summarization of context dependent opinion words. *18th International Conference on Knowledge-Based and Intelligent Information & Engineering Systems—KES* (pp. 166–177).
13. Rosario, B., Rajon, C., & Cheng, C. K. (2011). Word sense disambiguation of opinionated words using extended gloss overlap. In *Proceedings of the 8th Natural Language Processing* (pp 1–5).
14. Jalilvand, A., & Salim, N. (2012). Sentiment classification using graph based word sense disambiguation. *AMLTA 2012, CCIS, 322,* 351–358.
15. Xia, Y., Cambria, E., Hussain, A., & Zhao, H. (2014). Word polarity disambiguation using bayesian model and opinion-level features. *Cognitive Computation.*

Opinion Mining Classification Based on Extension of Opinion Mining Phrases

Shivam Rathi, Shashi Shekhar and Dilip Kumar Sharma

Abstract Opinion mining is the field of study that analyses people's thoughts, sentiments, emotions and attitude towards entities, product, services, issues, topics, events and their attributes. There are many different tasks such as opinion extraction, sentiment mining, emotional analysis, review mining etc. The important aspect of opinion minion is to gather the information from reviews, blogs, etc. and then finding out the behavior of that information, i.e. the information is related to either positive or negative context. The positive and negative reviews or blogs deal with a numerical value. The value is to be calculated using SentiWordNet 3.0. The opinion words are mainly adjective words such as "good," "better," "awesome." But there arises several problems because identifiers negation words and the extension of the opinion words such as "very very good" are not considered. In this paper, details about opinion mining, how the polarity value deals with positive and negative and how to deal with Roman language reviews and blogs is discussed.

Keywords Word polarity disambiguation · Opinion mining · Ambiguity · Natural language processing

1 Introduction

The bang of public media has created extraordinary opportunities for citizens to publicly accent their opinions, but has made severe bottlenecks when it comes to making sense of these opinions. In these types of reviews, users express their

S. Rathi (✉) · S. Shekhar · D.K. Sharma
Department of Computer Engineering & Applications,
GLA University, Mathura, India
e-mail: shivamrathi13@gmail.com

S. Shekhar
e-mail: shashi.shekhar@gla.ac.in

D.K. Sharma
e-mail: dilip.sharma@gla.ac.in

© Springer Science+Business Media Singapore 2016
S.C. Satapathy et al. (eds.), *Proceedings of International Conference on ICT for Sustainable Development*, Advances in Intelligent Systems and Computing 408, DOI 10.1007/978-981-10-0129-1_74

opinions straight forward in terms of the reviews of any product about the product's features that they liked it or not. For example, "this phone (model no) has good battery backup," clarifies that the battery backup review given by the user is positively inclined.

An object can be defined as "an object O is an entity which can be a product, topic, person, event, or organization. It is associated with a pair, O: (T, A), where T is a hierarchy or taxonomy of components or parts and sub-components of O, and A is a set of attributes of O [1]."

There are lots of opinions on the web and some opinions contain context dependent words which have different polarity in different context [2]. For instance consider the following sentences- "This phone has long battery life." Here, the word "long" has positive polarity but in other sentence say, "this phone has long processing time." Here the same word "long" indicates negative polarity. So, to classify the sentiments correctly we need to remove ambiguity from the sentences with the help of tool called word polarity disambiguation.

Word Polarity disambiguation is the computational identification of polarity of a word in given context [3]. It is a central challenge in opinion mining and it is an open research area in opinion mining [4, 5].

The general framework of sentiment classification can be given as (Fig. 1).

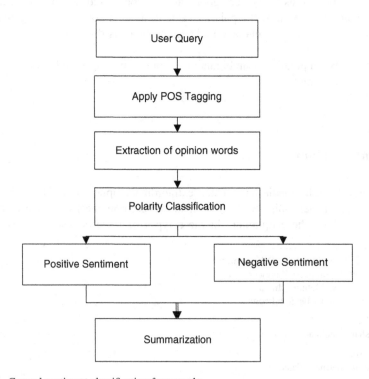

Fig. 1 General sentiment classification framework

2   Sentiment Classification Levels

Sentiments of a review title are based on the opinion of any product, and its issues. There are three types of levels:

1. Document level
2. Sentence level
3. Entity and aspect level

2.1   User Query

User query is the review or a blog of any product with its features. A user gives review of any product on the basis of its features like its battery, size, shape.

2.2   POS Tagger

Part-of-Speech tagger is used to classify each word of the sentence. It is classified in terms of noun, adjective, verb. After POS tagger, it is easy to extract the opinion value words and calculate the score with the help of SentiWordNet (Fig. 2).

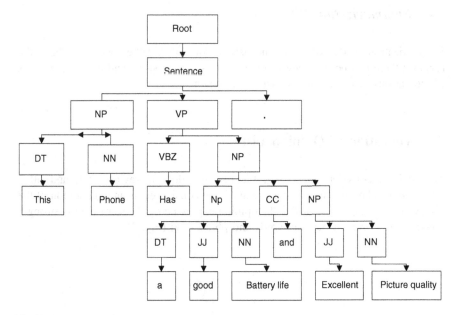

Fig. 2 Phrase extraction tree

Table 1 Specified target

POS tag	Description	Example
JJ	Adjective	Good
JJS	Adjective, superlative	Best
NN	Common noun	Quality
CC	Coordinating conjunction	And
DT	Determiner	The

POS tagger is classified each and every word in the sentence, for example, "the phone has a good battery life and excellent picture quality" [6, 7] through his POS tagger phone can be define by the noun phrase, good define by adjective which give the opinion of that phone. Furthermore, all specification related to the sentence shown in Table 1.

2.3 Polarity Classification

Moreover, extraction of opinion words is calculated by the dataset and the score is calculated as the binary values, i.e., 1 and −1. If the score is in −1, then it is clear that the word which is extracted by the POS tagger has negative value otherwise positive.

2.4 Summarization [8]

Summarization is described by the association rules. When we talk about the reviews; there can be two views to the reviews, i.e., positive and negative reviews of the various products on the basis of its features.

3 Generation of Opinion List

The reviews consider star rating, the review title, the review text, and other information. The title and the star rating can be considered as an aggregation of the review text. Thus, it is clear that the opinion expressed by the star rating is strongly related to the opinion bearing words and phrases [9] (Fig. 3).

Fig. 3 Overview of the
opinion list generation

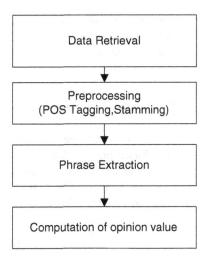

3.1 Data Retrieval and Preprocessing

3.1.1 Language Detection

First, in social media websites there are many reviews which give the opinion either
positive or negative. The review is to be classified with the help of polarity score
using SentiWordNet [10]. The score is to be calculated in the binary form either 1
or −1. In preprocessing, the language is to be detected, i.e., in which language the
review is written needs to be found out. It can either be English or any other
language.

3.1.2 Filtering

The star rating or the reviews corresponds to the polarity of an opinion and the
generation of opinion value is described later in this paper. In a sentence, few words
can be removed or filtered or omitted so that the polarity of the opinion gives the
accuracy that is positive or negative.

There are several issues which can be filtered before calculating the polarity:

- Few reviews are to be written in subjunctive form, i.e., "could have been a bad
  film." The subjunctives words are "could," "would," "should." Therefore,
  review titles with subjunctive are discarded.
- Some titles are questioning based, i.e., "why this movie is good." Therefore,
  these types of interrogative sentences are excluded from dataset of reviews.

- Many reviews consider "but" which is an indicator of bipolar opinion. For example, "good phone but poor battery backup." Such reviews which contain "but" are omitted.

3.1.3 Calculation of Opinion Value

In a dataset, there are millions of reviews and each review consists of adjective-based phrases. The adjective-based phrases are "good," "bad," etc. The polarity value is calculated with the help of 'n' defined as the no of reviews of product. The SentiWordNet 3.0 dataset is used. The formulae which are used to calculate the opinion value are:

$$SR_i = \frac{\sum_{j=1}^{n} S_j^i}{n} \tag{1}$$

$$OV_i = \frac{SR_i - 3}{2} \tag{2}$$

From (1), we have to calculate the star rating where 'n' is the no of reviews and the star rating is to be used in the calculation of opinion value. Then, (3) is subtracted in (2) because positive, negative, and neutral words tell about the opinion of the review on the basis of star rating.

3.2 Opinion Value for Missing Phrase

The problem in negation words and identifiers is to know the opinion value of the review. Let us take an example related to adjective "good." If any review uses "good," then it is easy to find the opinion. But if the reviews use words like "very very good," then finding the polarity becomes challenging (Fig. 4).

$$OV_{VG} = n * OV_V + OV_G \tag{3}$$

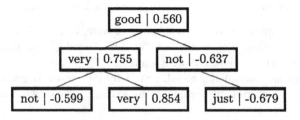

Fig. 4 Missing phrase except very good

Through this formula, opinion value of "very good" is to be calculated by 'n' where n is the no of times "very" comes and the sum of "very" to the polarity value of "good".

4 State of Art

A lot of work has been done in the area of extension of opinion phrases which tells about the polarity of the reviews. These works have mainly focused on adjective-based phrases, verb-based phrases, association rule mining, summarization, and aspect-based phrases. Each of these methods has their own strengths and weaknesses mentioned in Table 2.

Table 2 Comparison of previous works

S. no.	Paper	Methodology	Strength	Weakness
1.	A method for opinion mining of product reviews using association rules [8]	POS tagging association rule mining	Deals with NLP	Implicit opinion sentence
2	A generic approach to generate opinion lists of phrases for opinion mining applications [11]	Calculate score opinion words, consider missing phrase	Fast generation of opinion list and algorithm for missing phrase	Not considered "but" and negation words. Considered only adjective based phrase
3	Verb oriented sentiment classification [12]	Bag of words approach and linguistic knowledge	Extract opinion structure from the text	Enrich the dictionary
4	Identifying customer preferences about tourism products using an aspect-based opinion mining approach [1]	Aspect orientation rule and summarization	Bipolar sentences	Poor precision and performance related to frequent noun and NPs
5	Enriching semantic knowledge bases for opinion mining in big data application [13]	Identify ambiguous sentiment. Integrate context term	Work on two different resources WordNet and ConceptNet	Avoid the inclusion of context terms. Use crowd sourcing to annotate large sentiment corpora

5 Conclusion and Future Directions

In this paper, we have described the basic concept of opinion mining, different levels and different types of approaches used for the sentiment classification. In the mentioned papers, mainly reviews are considered in English language and not in a roman language. So, there is a future scope to create a database which gives the opinion value of Roman words such as "acha," "bekar." Thereafter, these words can be compared with "good," "bad," and other words of English language dataset. Thus, it will lead to increase in the performance rating of the websites related to any product that includes reviews in Roman language. We work on bipolar opinion also which considered such as good phone but battery poor then these type of sentences cannot be excluded from the data set and find out the polarity of these bipolar opinions words.

References

1. Marrese-Taylora, E., Velasquez, J. D., Bravo-Marquezb, F., & Matsuoc, Y. (2013). Identifying customer preferences about tourism products using an aspect-based opinion mining approach. In *17th International Conference in Knowledge Based and Intelligent Information and Engineering Systems—KES* (pp. 182–191).
2. Pang, B., & Lee, L. (2008). Opinion mining and sentiment analysis. *Foundations and Trends in Information Retrieval, 2*(1/2), 1–135.
3. Liu, B. (2010). Sentiment analysis and subjectivity. In N. Indurkhya & F. Damerau (Eds.), *Handbook of Natural Language Processing* (2nd edn.).
4. Miller, G. A. (1995). WordNet: A lexical database for English. *Communications of the ACM, 38*, 39–41.
5. Mukherjee, A., Liu, B., & Glance, N. (2012). Spotting fake reviewer groups in consumer reviews. In *Proceedings of the 21st International World Wide Web Conference* (pp. 191–200).
6. Klein, D., & Manning, C. D. (2003). Fast exact inference with a factored model for natural language parsing. *Advances in Neural Information Processing Systems 15 (NIPS 2002)* (pp. 3–10). Ma: MIT Press Cambridge.
7. Liu, B., & Zhang, L. (2012). A survey of opinion mining and sentiment analysis. In C. C. Aggarwal & C. Zhai (Eds.), *Mining Text Data* (pp. 415–463). US: Springer.
8. Kim, W. Y., & Kim, K. I. (2009). A method of opinion mining of product reviews using association rules. *ACM ICIS*.
9. Jindal, N., Liu, B., & Lim, E.-P. (2010). Finding unusual review patterns using unexpected rules. In *Proceedings of the 19th ACM International Conference on Information and Knowledge Management*, pp. 1549–1552.
10. Karamibekr, M., & Ghorbani, A. A. (2012). Verb oriented sentiment classification. In *IEEE/WIC/ACM International Conferences on Web Intelligence and Intelligent Agent Technology (WI-IAT)* (Vol. 1, pp. 327–331).
11. Rill, S., & Drescher, J. (2012). A generic approach to generate opinion list of phrases for opinion mining application. *WISDOM'12*, August 12, 2012.
12. Kaur, A., & Duhan, N. (2015). A survey on sentiment analysis and opinion mining. *International Journal of Innovations & Advancement in Computer Science, 4*, 107–116.
13. Weichselbraun, A., Gindlb, S., & Schar, A. (2014). Enriching semantic knowledge bases for opinion mining in big data application. *Knowledge-Based Systems, 69*, 78–85.

Facial Expression Recognition Using Variants of LBP and Classifier Fusion

Sarika Jain, Mishra Durgesh and Thakur Ramesh

Abstract Automatic facial expression analysis is an interesting and challenging problem; we empirically evaluate facial expression based on ensemble methodology, which builds a classification model by integrating multiple classifiers, for improving prediction performance. This method is adopted by many researchers in field of statistics, pattern recognition, and machine learning. This paper presents an ensemble-based facial expression recognition system using local binary pattern, it is used for feature extraction. The extracted feature histogram represents the local texture and global shape of face images. Three different classifiers which are Euclidian distance, neural network, and support vector machine are used for classification. The proposed ensemble classifier approach has demonstrated superior performance compared to individual classifiers. The ensemble-based classifier yielded an accuracy of 97.20 %; the best accuracy obtained from all other single classifier schemes tested using the Cohn-Kanade database.

Keywords Facial representation · Human computer interaction · Local binary pattern · Ensemble · Support vector machine · Neural network

1 Introduction

Facial expressions provide an important behavioral measure for studies of emotions, cognitive processes, and social interaction. Facial expression recognition (FER) has recently become a promising research area. Facial expressions constitute

S. Jain (✉) · M. Durgesh
Computer Science Department, Sri Aurobindo Institute of Technology,
Indore, India
e-mail: sarika_jain04@hotmail.com

T. Ramesh
International Institute of Professional Studies, Devi Ahilya Vishwavidyalaya,
Indore, India

© Springer Science+Business Media Singapore 2016   725
S.C. Satapathy et al. (eds.), *Proceedings of International Conference
on ICT for Sustainable Development*, Advances in Intelligent Systems
and Computing 408, DOI 10.1007/978-981-10-0129-1_75

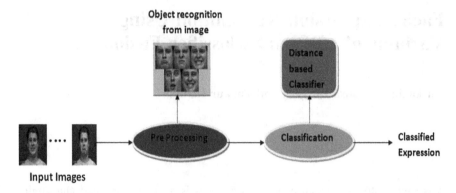

Fig. 1 Facial expression recognition system

an essential part of nonverbal communications, which consists of all the messages other than words that are used in communication. Most of the facial expression recognition systems attempt to recognize a set of prototypic expressions; those are happiness, surprise, anger, sadness, fear, and disgust. This practice follows from the studies of Ekman and Friesen [1] and Ekman [2]. These studies propose that basic emotions have corresponding prototypic facial expressions.

Facial expression recognition requires more subtle and discriminative feature extraction as compare to other recognition methods. Facial expression recognition has three stages of preprocessing, feature extraction, and classification as shown in Fig. 1.

Preprocessing includes removal of noise and unwanted information from input image. We are using standard dataset which does not contain noise in images. To reduce the unwanted data we have used object detection techniques which resulted as detected face. Dimensionality reduction was our first choice to extract features. We have used different feature extraction methods. Local binary patterns extracts features from the input images which are used as an input to the classifier. The set of features is formed to describe the facial expression. To increase the recognition rate we have applied various extraction methods on processed face image in place of whole images [3, 4]. We have used various classifiers Euclidian distance, Support Vector Machine, Neural Network as classifier. The experiments are performed on Cohn-Kanade dataset [5].

2 Local Binary Pattern

The local binary pattern operator is an image operator which transforms an image. It can be considered as small-scale appearance of the image, where an array or sequence of pixels with integer labels used as metadata for small-scale image. These labels are commonly the histogram, used in image analysis. LBP is

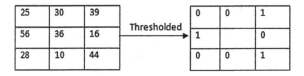

25	30	39
56	36	16
28	10	44

Thresholded →

0	0	1
1		0
0	0	1

Fig. 2 Calculation of LBP

compatible with almost all images ranging from monochrome still images, color images, videos, and volumetric data. LBP operator selects a group of pixels around each pixel in image, thresholds is used find P neighbor gray values with respect to the center pixel [6]. All values are concatenated to get the binary values (Fig. 2).

$$\text{LBP}\,(x_c, y_c) = \sum_{n=0}^{7} s\,(i_n - i_c)2^n \tag{1}$$

where n runs over the n neighbors of the central pixel, i_c and i_n are the gray-level values of the central pixel and the surrounding pixel, and $s(x)$ is 1 if $x \geq 0$ and 0 otherwise.

The pixels in this block are threshold by its center pixel value, multiplied by powers of two and then summed to obtain a label for the center pixel [6]. Neighborhood will consists of pixels; if it is comprise of n values than 2^n different labels can be obtained depending on values of the center and the pixels in the neighborhood [7].

3 Classification Method

3.1 Euclidian Distance

To compute the Euclidean distance the distance will be absolute value of the difference between their coordinates this absolute value should always be non-negative number.

Suppose we consider two points X and Y in two-dimensional Euclidean spaces. The distance between two points in the plane with coordinates (x, y) and (a, b) is given by

$$\text{dist}((x, y),\,(a, b)) = \sqrt{(x - a)^2 + (y - b)^2} \tag{2}$$

Now if we construct a line segment with the endpoints of X and Y. This line segment will form the hypotenuse of a right triangle. we note that the lengths of the

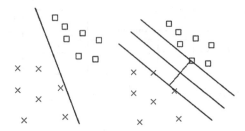

Fig. 3 Separating hyperplane (*Left*) a random one, (*Right*) one that maximizes margin of separability

legs of this triangle are given by $|x - a|$ and $|y - b|$. The distance between the two points will then be given as the length of the hypotenuse [8].

3.2 Support Vector Machines

The field of face processing, a sub discipline of image processing has successfully applied support vector machine. One of the reasons for that is its fast and efficient implementation. SVMs are basically binary classifiers from the perspective of statistical learning theory. SVM comes from theoretical bounds of generalization error.

First, the error bound is minimized by maximizing the margin, i.e., minimal distance between hyperplane separating two classes and the data points closest to the hyperplane. Second, the upper bound on generalization of error does not depend upon dimensionality of space [9] (Fig. 3).

3.3 Neural Network

An artificial neural network is a replica of biological neural system. An artificial neural network is an adaptive system. Adaptive in context of problem solving as it is flexible for all operational changes. These iteration wise modifications are done in training phase. The input and corresponding output for training data is essential for these networks as it conveys the information which helps in weight adjustment [10].

A neural network has to be configured such that the application of a set of inputs produces the desired set of outputs [11]. Weights can be adjusted explicitly, using a priori knowledge. Another way is to 'train' the neural network by providing the set of patterns as reference and some learning rules. These patterns are matched with another set of patterns based on similarities and error is calculated. The weights are optimized as the error becomes zero (Fig. 4).

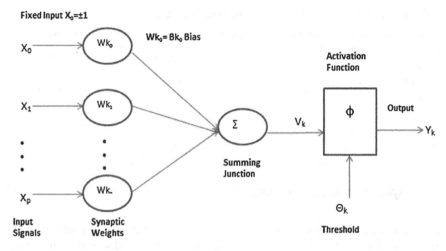

Fig. 4 Architecture of neural networks

4 Proposed Method

See Fig. 5.

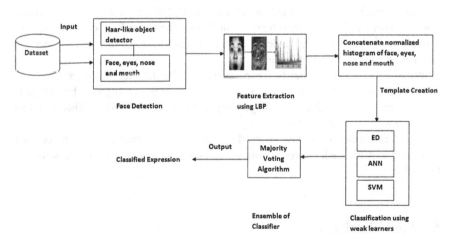

Fig. 5 Overview of proposed method

Table 1 Expression wise recognition rate using ensemble classifiers

Expression	Average recognition rate of three classifiers (%)	Recognition rate using ensemble classifiers (%)
Anger	100	100
Disgust	91.67	100
Fear	83.33	95.83
Happy	100	100
Sad	66.67	95.83
Surprise	91.67	100

5 Experiments and Results

Experiment performed using various future selection and classifiers landed us with final selection as Uniform LBP and three classifiers as Distance metric, i.e., Euclidian distance, artificial neural network, and support vector machine to create an Ensemble Classifier. Finally, to improvise the prediction performance ensemble methodology is used, which builds a classification model by integrating multiple classifiers. This method is adopted by many researchers in field of statistics, pattern recognition, and machine learning. An ensemble-based facial expression recognition system using local binary pattern for feature extraction and three different classifiers which are Euclidian distance, neural network, and support vector machine for classification. The proposed ensemble classifier approach has demonstrated superior performance compared to individual classifiers. The ensemble-based classifier yielded an accuracy of 97.2 %; the best accuracy obtained from all other single classifier schemes tested using the Cohn-Kanade database. The comparison is shown in the table. Our motive was to reduce the misclassification rate with reasonable time complexity.

We also got to know that few expressions can be detected easily compared to other expression, such as Fear and Sad are more complex to recognize. Even human may experience difficulties in recognizing ambiguous expression. That is major issue leads to misclassification.

In our experiments we have worked on improving overall recognition accuracy of system and for each expression its accuracy is also calculated and presented in the Table 1.

6 Analysis of Results

We started with distance measure and the results were good with one of them was 80.95 %. We used back propagation method and the performance was good but the convergence time was quite high. To improve on time parameter we used faster version of back propagation, i.e., Conjugate Gradient method that gives an

Table 2 Expression wise recognition rate w.r.t classifiers

S. no	Pre processing	Feature extraction	Classifier	Recognition rate
1	Face region detection using manual cropping	LBP	Bray Curtis	80.95
2			Chi square	71.40
3			Euclidian	69.04
4			Taxi cab	69.40
5			Chess board	45.23
6	Face detection using viola jones	LBP + PCA	Euclidian distance	95.13
7	Face detection using viola jones	LBP + 2DPCA	Euclidian distance	95.83
8	Face and facial part detection using viola jones	Uniform LBP	Artificial neural network	95.83
9			Support vector machine	95.83
10			Euclidian distance	95.83
11			Ensemble classifier	97.20

approximate increase in the rate of convergence of about an order of magnitude as compared to conventional back-propagation method. The results were 95.83 %. We tried Support vector Machine (SVM) that makes binary decisions, so the multiclass classification is achieved by using the one against-rest technique, best recognition performance is obtained by using RBF kernel, i.e., 95.83.

The expressions were misclassified with many classifiers, only two expressions which were never incorrectly classified: happy and anger. Reason behind working on multiple classifiers was to find the best one. Another reason is to find expression and a classifier pair which may tell which classifiers is able to recognize a particular expression. Finally, with ensemble of classifiers misclassification rate reduced greatly and the rate of recognition was also improved to 97.20 % (Table 2).

7 Conclusion and Future Work

The key points of our work can be summarized as follows:

- Viola Jones method is a powerful object recognition technique and gave well results in face and facial parts detection.
- We used LBP for feature extraction; our experiments suggest that LBP features are effective and efficient for face expression recognition.

- Misclassification of expressions is reduced and the recognition accuracy was increased up to highly satisfactory level.

We will extend our work for real time images. We will apply these methods for micro-expression recognition. In classifier fusion to find best classifier we can use genetic algorithms.

References

1. Ekman, P., & Friesen, W. (1976). Pictures of facial affect. *Consulting Psychologists.*
2. Ekman, P. (1993). Facial expression and emotion. *American Psychologist, 48,* 384–392.
3. Zaker, N., & Mahoor, M. H., Mattson, W. I., Messinger, D. S., & Cohn, J. F. (2013). A comparison of alternative classifiers for detecting occurrence and intensity in spontaneous facial expression of infants with their mothers. In *Automatic Face and Gesture Recognition* (pp. 22–26).
4. Bafandehkar, A., Nazari, M., & Rahat, M. (2011). Pictorial structure based keyparts localization for facial expression recognition using Gabor filters and Local Binary Patterns Operator. *2011 International Conference of Soft Computing and Pattern Recognition (SoCPaR)* (pp. 429, 434), 14–16 Oct. 2011. doi:10.1109/SoCPaR.2011.6089282.
5. Kanade, T., Cohn, J. F., & Tian, Y. Comprehensive database for facial expression analysis. *Proceedings of the Fourth IEEE International Conference on Automatic Face and Gesture Recognition,* Grenoble, France (pp. 46–53).
6. Ojala, T., Pietikainen, M., & Harwood, D. (1996). A comparative study of texture measures with classification based on featured distribution. *Pattern Recognition, 29*(1), 51–59.
7. Anitha, M., Venkatesha, K., & Adiga, B. (2010). A survey on facial expression databases. *International Journal of Engineering Science and Technology, 2*(10), 5158–5174.
8. Turk, M., & Pentland, A. (1991). Eigenfaces for recognition. *Journal of Cognitive Neuroscience, 3*(1).
9. Valstar, M., & Pantic, M. (2006). Fully automatic facial action unit detection and temporal analysis. In *Proceedings of Conference on Computer Vision and Pattern Recognition* (pp. 149–158).
10. Chen, L. S. (2000). Joint processing of audio–visual information for the recognition of emotional expressions in human–computer interaction. Ph.D. Thesis, University of Illinois at Urbana-Champaign, Department of Electrical Engineering, 2000.
11. Rama, L. R., Babu, G. R., & Kishore, L. (2012). Face recognition based on eigen features of multi scaled face components and artificial neural network. *International Journal of Security and Its Applications (IJSIA), 5*(3), SERSC, 23–44.

FLSU-Based Energy Efficient Protocol Design for WSN with Mobile Sink

Nitika Vats Doohan and Sanjiv Tokekar

Abstract Data transmission from source to sink is the basic concept of wireless sensor networks (WSNs). If the sink is stationary then the nodes nearer to the sink will deplete their energy faster and put the sink out of reach to other nodes in the network. To address the problem mobile sinks are preferred so that energy available used judiciously, consequently increases the overall uptime of the network. Challenges lies with the sinks location updation to all nodes in the network using minimal communication overheads. In this paper, we proposed a mechanism where sink updates its position and movement in fixed local area with time. The proposed protocol takes care of minimal communication overhead compared with LURP (Link Updation Register Protocol), and BBM (Basic Broadcast Method). Additionally proposed protocol is simple, reliable, scalable, and supports multiple sinks with minimal communication overhead.

Keywords Wireless sensor networks · Mobile sinks · Data transmission · LURP · Energy-Efficiency

1 Introduction

Wireless sensor networks (WSNs) has emerged as a new and challenging research area. Sensor is a device that produces a measurable response to a change in a physical condition, such as temperature, pressure, etc. Sensors frequently used for measurement of basic physical phenomena including acceleration, shock, vibration, humidity, flow rate, force, magnetic field, and wind speed, etc. [1, 2]. Depending upon the application requirement, sensor nodes can have integrated with special onboard capabilities such as relaying, sensing, and data aggregation. Essentially, wireless sensor networks are information-retrieval networks and are constraint in

N.V. Doohan (✉) · S. Tokekar
Institute of Engineering and Technology, DAVV, Indore, MP, India
e-mail: nitika.doohan@gmail.com

© Springer Science+Business Media Singapore 2016
S.C. Satapathy et al. (eds.), *Proceedings of International Conference on ICT for Sustainable Development*, Advances in Intelligent Systems and Computing 408, DOI 10.1007/978-981-10-0129-1_76

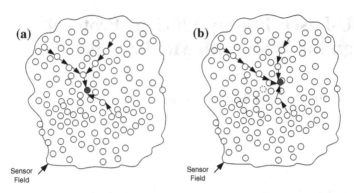

Fig. 1 **a** Single stationary sink with nodes. **b** Single mobile sink with nodes

energy resources, memory, and computation power. These networks are not having point-to-point communication network [3, 4]. In more simplified term, WSN is conglomeration of nodes that senses, processes, and communicates the information within the network. The gathered information is further passed on to the sink for further analysis via multiple hop paths. Sink(s) can be stationary or mobile to suit the application requirement. Routing the data from source node to sink is challenging but it become even more difficult when the sink(s) is mobile [4]. Figure 1a depicts a typical scenario where single sink is stationary. The nodes, nearer to the sink deplete their energy faster owing to the traffic from the whole network use these nodes to reach to sink. Once sink become unreachable, WSN just become uninformative network. Thus, the stationary sink setup suffers from the shorter overall network lifetime and degraded overall performance. Figure 1b depicts the scenario, where the sink is mobile and the nodes acting as a gateway to mobile sink/s are change as the sink changes its position, which results in increase of overall network lifetime, and hence improving the performance.

Mobile sink is application-driven novel concept to enhance the network lifetime by avoiding the excessive overhead at the node, which is closest to the sink as compared with static sink. For seamless data transmission, mobile sink needs to be synchronous with the nodes in the network. The researchers have proposed several data dissemination protocols for static sinks [5]. Our proposed approach not only support dynamic sinks but also scales well with network size and its lifetime. The strategy of mobile sinks is implemented in TTDD, LURP, BBM, etc. and is discussed in following paragraphs [6].

TTDD (Two-Tier Data Dissemination) design uses a proactively build two-tier grid structure so that only sensors located at grid points need to acquire the forwarding information. Mobility of sink is very low and communication overhead is more [7]. In Location update-based routing protocol (LURP), when the sink node moves, it only needs to broadcast its location information within a local area rather than among the entire network [8]. However, LURP approach does not support with multiple mobile sinks and need some augmentation. BBM uses the broadcast approach. It supports the mobility of sink and when the sink moves, broadcast takes place. This leads to very high communication overhead and hence network lifetime decreases [9]. One hop neighbour [10] information is important for sending the data in WSN network. In the paper, we propose a scheme the communication, where overhead is lesser compared to LURP and BBM approaches. In addition, it will support multiple mobile sinks along with network scalability and reliability. Here, we simulate all the results via using the network simulator version 2 i.e. NS2 [10].

2 Proposed Protocol in WSN

Mobility of sink node brings more challenges in data dissemination as the sink moves from one location to another. Many researchers suggested that if sink is mobile then the location of the sink should be updated on regular basis so that node can transmit the data to the sink. In addition, if the sink's location is updated on regular basis then the nodes that are continuously updating the location will be drained out energy swiftly. In our approach, there is a node named as fixed location sink updation (FLSU) preferred to is placed at the center so that it will be reachable in minimum hop count by the occurring event anywhere in the network. Initially FLSU, will broadcast its location to the entire network so that sink and other nodes in the network know the FLSU location. Any node with high power backup can also act as FLSU, without affecting the proposed protocol mechanism and its features. Whenever the sink moves, it will update its location to the FLSU. This way FLSU maintains current location of sink, all the time in its database. The uptime life of FLSU is decided by the predefined threshold value of energy level. At the end of FLSU life, it passes it services to the nearest most powerful node. The decision of powerful node is based on the remaining energy level in the node. This way presence of the FLSU will be assured, until the end of useful network life.

2.1 Protocol Design

FLSU-based protocol is an improvement over LURP. In this protocol; a node FLSU; has to manage the current location of the sink in the rendezvous region.
   Assumptions:

 i. FLSU is assumed to be with high battery level. Initial location of the FLSU may be located in any part of the network, preferable at center of the rendezvous region.
 ii. Initially FLSU broadcasts its location to the entire network and hence the entire nodes in the network will learn the location of the FLSU. Consequently, FLSU learns the current location of the sink in the network.
iii. Broadcasting is required only once at the beginning.

```
Pseudo code for FLSU: -
FLSU is stationary and will change its position only
when drained out.
   a)  At initial phase: -
FLSU broadcast its initial location in the network and
Sink will give its position to FLSU periodically.

   b)  In working phase: -
If (node transmits data)
      then
          if (cache > threshold) or (cache = empty)
             then
{
  request_FLSU            //request FLSU for sink location
            FLSU_reply               // sink location reply
            Send_data_to_sink
            }
Update_cache        //node  updates   its   own   cache  with
current location & broadcast to its one hop neighbor so
that  request  to  FLSU  to  obtain  sink  location  can  be
minimized
                  else
                     Send data to sink directly   //      with
available sink location information in its cache
               fi
   fi
```

2.2 Working of Protocol

FLSU works in two-phase protocol. In the first phase, when the network has to set up, first the FLSU will broadcast its position in the entire network so that the sink will update its cache. The initial position of sink known to FLSU as shown in the Fig. 2.

After getting the initial position of the sink, FLSU will update its cache with sink's location information. When sensor node wants to send the data, initially it will send the request to FLSU. Then FLSU forward its data to the sink and in response of that FLSU sends the location of the sink to the respective sensor node as shown in Fig. 3. Then in series of the next transmission, the particular node will send the data directly to the sink. It overall increases the lifetime of the network and reduces the complexity as well as the transmissions in the network. The second phase of the protocol starts, when sink changes its position from one location to another as depicted in Fig. 4. In that case, sink will update its position to the nodes in the rendezvous region as shown in Fig. 5.

Now, if any sensor node wants to send the data, they will send it to the FLSU first and then FLSU will send the data to the sink and the same time FLSU will tell

Fig. 2 FLSU broadcast its location through the network

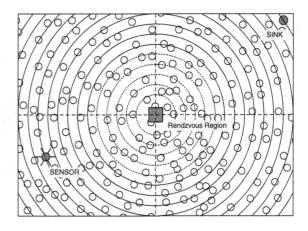

Fig. 3 Event triggered and it update information to FLSU

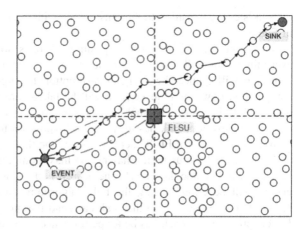

Fig. 4 Sink has moved to new location and update its position

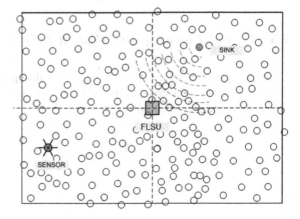

Fig. 5 FLSU redirects the
source node data to new
sink's position

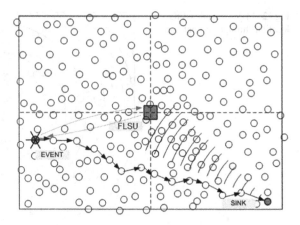

Fig. 5 FLSU redirects the source node data to new sink's position

the sink's current position to the sensor node with the reverse acknowledgement as in Fig. 5.

Whenever the sink changes its position, there is no need to do the broadcast every time. Sink will update its location to the limited nodes in rendezvous region. Hence, it will reduce the energy consumption to large extent. And overall the computation power is reduced as it will be taken care by the participating nodes itself. While in distributed mobile sink support approach, every time when sink changes its position, sink has to multicast its position to all the former APs.

In case, if power of FLSU depleted below the threshold power, then the nearest node in the rendezvous region having the maximum available power will take the lead and will become the FLSU. The nearest node, which is having the maximum energy will be chosen using the bully algorithm. Now, the new FLSU has to broadcast in the network so that sink can send the query or data can be retrieved. Hence, propose approach has better performances over other proposed approaches.

2.3 Cost of Updating the Location Information of the Sink

During the period of time T, we assume that the data packets from the node S is sent to the sink. As the sink changes its location, it will update its position to the FLSU. The cost of updating the sink location is:

$$E = mnh + \left(\frac{T}{t}\right)Nh. \tag{1}$$

$$n = d \cdot \frac{\sqrt{N}}{L}. \tag{2}$$

where $\frac{N}{L^2}$ denotes the density of the network; d is travel distance by sink. The total updating cost of the sink's location among the entire network is $\left(\frac{T}{t}\right)Nh$.

When putting (2) into (1), we get:

$$E_1 = md \cdot \frac{\sqrt{N}}{L}h + \left(\frac{T}{t}\right)Nh \qquad (3)$$

Total updating cost in LURP is

$$E_2 = m\frac{\pi R^2}{L^2}Nh + \left(\frac{T}{t}\right)Nh \qquad (4)$$

and, the total updating cost of BBM is

$$E_3 = mNh \qquad (5)$$

2.4 For Better Efficiency of FLSU

Following assumptions were made for analysis the better efficiency of FLSU (as shown in Fig. 6). Assumptions are:-

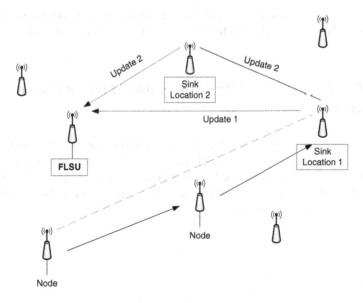

Fig. 6 FLSU updating new location data to FLSU

i. Multiple nodes in rendezvous region are deployed in master-slave or in distributed configuration.
ii. FLSU powered with long backup battery, recharging by portable solar panel can also be considered based on the environmental conditions.
iii. Update at node cache based on time factor or distance travel.

2.5 Advantages of FLSU

1. Minimal communication overhead.
2. Based on simple logic, which requires less complex algorithm to implement and thus less computing power requirement.
3. There is no data relay except first packet for keeping the information. Therefore, very less computation power is required.
4. Efficiently supports multiple sinks.

3 Challenges of FLSU

3.1 Loss of Data Between Sink Updates at FLSU

In the proposed approach, it is possible that during the sink movement the sink location updation will result the loss of packet-stream reach to the sink.

3.2 Continues Flow of Data When Sink Moves at Uniform Speed

As can be seen from Fig. 7, if sink is moving all the time at variable speed and not sending the acknowledgement to each packet received, data may be loss during updating process. The chances of loss of data are more when large data has to be transmitted, which is normally not true in the WSN environment. Data is always a burst in nature for short period.

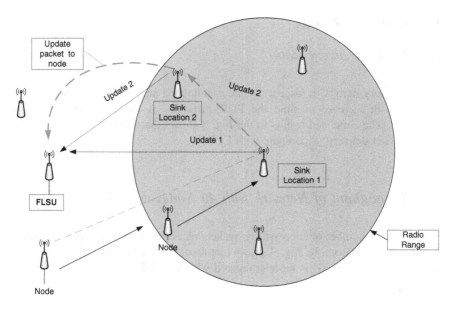

Fig. 7 New location information updation to sink for data transmission

Table 1 Parameters used for simulation as shown below:

Parameter name	Value
Number of nodes	10, 20, 30, 40
Number of mobile sinks	02
Number of stationary sinks	02
Number of sources	02
Energy per node	1 J
Packet size	1024 bytes
Queue size	50 packets
Simulation time	50 s
Transmission range	10 m
Trans transmission stop time	45 s
Sink's velocity	3 m/s

3.3 Parameters Used for Simulation

Extensive simulation for the FLSU approach using the Network Simulator version 2 i.e. NS2 [11] were performed and results show that proposed approach consistently performs well as compared with LURP approach.

4   Simulation Results

4.1   Packet Delivery Ratio

Packet delivery ratio with respect to time is simulated and compared. It was found out that with increase of time the PDR increases as the event generation is increasing. The simulated results in the Fig. 8 shows that the FLSU approach has higher packet delivery as compare with LURP.

4.2   Throughput of Network with 40 Nodes w.r.t. Time

Throughput is number of successfully packets received by the sinks per unit time. Simulated results from the Fig. 9 shows that amount of data transferred in a given amount of time in FLSU is more compared with the LURP.

4.3   Routing Overhead of Network with 10 Nodes w.r.t. Time

Routing overhead is the number of packets traversed in the network to establish the path divided by total no of packets traveled in the network up to that instant of time.

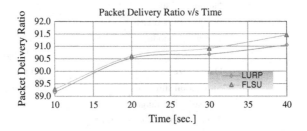

Fig. 8 Packet delivery ratio comparison of FLSU with LURP w.r.t. time

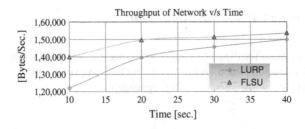

Fig. 9 Throughput of Network of FLSU with LURP w.r.t. time

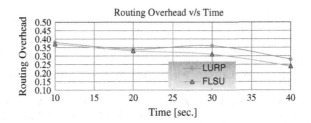

Fig. 10 Routing overhead of FLSU and LURP w.r.t. time

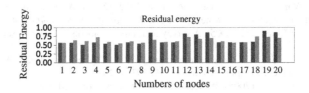

Fig. 11 Overview of the energy residual level w.r.t. time

Figure 10 shows the routing overhead of FLSU and LURP protocols. Routing overhead in FLSU is found to be less than compared approach.

4.4 Energy Residual of 20 Nodes in the Network

Figure 11 shows an overview of the energy residual level in different. The average residual energy of the nodes in FLSU is more than the compared existing approach.

5 Conclusions

Present study is based on the related work done to achieve the sink mobility at low communication overheads. Results were simulated and compared with standard protocols and proposed approach found to be simple, scalable, and more efficient. The performance of the FLSU protocol is comparable or surpassing with the standard approaches. In the proposed work, we have tried to minimize communication overhead, broadcasting, and other handshakes to the great extent to achieve the location updates on all the nodes continuously. The proposed approach is more efficient as having reduced transmissions and fully support to multiple mobile sinks. The said approach is very simple and easy to manage, robust, and scalable.

References

1. Karl, H., & Willig, A. (2005). *Protocols and architectures for wireless sensor networks*. West Sussex, England: John Wiley and Sons Ltd.
2. Pottie, G. J., & Kaiser, W. J. (2000). Wireless integrated network sensors. *Communications of the ACM, 43*(5), 51–58.
3. Ye, F., Zhong, G., Lu, S., & Liu, Z. (2005). Gradient broadcast: A robust data delivery protocol for large scale sensor networks. *ACM Wireless Networks (WINET), 11*(2), 285–298.
4. Stallings, W. (2005). *Wireless communications and networks* (2nd ed.). Upper Saddle River, NJ, US: Pearson Prentice Hall.
5. Luo, H., Ye, F., Cheng, J., Lu, S., & Zhang, L. (2005). TTDD: Two-tier data dissemination in large-scale wireless sensor networks. *Wireless Networks, 11*, 161–175.
6. Wang, G., Wang, T., Jia, W., Guo, M. G., & Li, J. (2009). Adaptive location updates for mobile sinks in wireless sensor networks. *The Journal of Supercomputing, 47*, 127–145.
7. Akkaya, K., & Younis, M. (2005). A survey on routing protocols for wireless sensor networks. *Ad Hoc Networks, 3*(3), 325–349.
8. Akyildiz, I. F., Su, W., Sankasubramaniam, Y., & Cayirci, E. (2002). Wireless sensor networks: A survey. *Computer Networks, 38*, 393–422.
9. Bulusu, N., & Jha, S. (2005). *Wireless sensor networks*. Norwood, USA: Artech House.
10. Ding, G., Sahinoglu, Z., Orlik, P., Zhang, J., & Bhargava, B. (2006). Tree-based data broadcast in IEEE 802.15.4 and ZigBee networks. *IEEE Transactions on Mobile Computing, 5*, 1561–1574.
11. The VINT Project. (2013). The network simulator—ns-2. http://www.isi.edu/nsnam/ns/.

Goal-Based Constraint Driven Dynamic RESTful Web Service Composition Using AI Techniques

Digvijasinh Rathod, Satyen Parikh and M.S. Dahiya

Abstract SOAP and REST are two architectural styles to develop web service where SOAP follows operation centric and REST follows resource centric approach. Objective of dynamic web service modeling is to compose existing web services published on internet and generate new value-added service to satisfy client request; and it receives great response from many researchers. Even though widespread adoption of web service, there are some situation where we need automatic composition of existing web services from service repository and generate workflow automatically to fulfill client objective. After studying many research papers we find that Automatic RESTful Web Service Composition is still unexplored as compare to SOAP/WSDL-based web service. In this research paper, we proposed goal based constraint-driven internal web service composition and automatic workflow (GCDIWBC) algorithm Using AI planning which generates new value-added service by considering nested automatic service composition and generate workflow also. We also considered case where one automatic web service composition can call other automatic web service composition internally in the composition and generate workflow dynamically which will be run on workflow execution engine like WS-BPEL or WSMO. In other words service dynamically generated by GCDIWBC can nested to other generated service by other. GCDIWBC has implemented using prototype of travel scheduling scenario and performance evaluation with RESTful and SOAP/WSDL web services using state-of-the-art JSHON2 AI planner. Result shows that proposed algorithm with RESTful web service using JSHON2 outperforms SOAP/WSDL standard both in its response time and scalability.

D. Rathod (✉) · M.S. Dahiya
Institute of Forensic Science, Gujarat Forensic Sciences University, Gandhinagar, India
e-mail: todigvijay@yahoo.com

M.S. Dahiya
e-mail: msdahiya49@rediffmail.com

S. Parikh
Department of Computer Application, Ganpat University, Mehsana, India
e-mail: satyen.parikh@ganpatuniversity.ac.in

© Springer Science+Business Media Singapore 2016
S.C. Satapathy et al. (eds.), *Proceedings of International Conference on ICT for Sustainable Development*, Advances in Intelligent Systems and Computing 408, DOI 10.1007/978-981-10-0129-1_77

Keywords REST · RESTful · SOAP · WSDL · Web service · Automatic composition · HTTP · Workflow · WS-BPEL · AI planning

1 Introduction

REST and SOAP/WSDL are two architectural styles to build web services. SOAP/WSDL web service follows operation centric and RESTful web service follows resource centric approach. RESTful HTTP-based and SOAP/WSDL-based services are two architectural styles for building distributed system and web services can be described as software entities that are capable of delivering certain functionalities over a network as per WWW consortium. Representational state transfer (REST) is an architectural style that first defined by Fielding [1] for distributed system such as World Wide Web. This architectural style describes six constraints: client-server, stateless, cacheable, layered system, code on demand (optional), and uniform interface. Resources and representations are key concepts of RESTful web services where resources are abstraction of information and each of these resources are named with unique identifiers like URIs. When browser sends HTTP GET request to web application for targeted resources, it receives response in form of HTML representation of the resource's state. RESTful web services uses HTTP methods, i.e., GET, PUT, POST, and DELETE for CRUD [2] (Create, Read, Update, and Delete) to access and manipulate resources. Today's over changing business environment single service not enough to satisfy requirements of users and need to compose more than one web service and generate value added service. Composition of SOAP/WSDL web service received great attention form research community [3–9]. Web service composition enables combining existing services and creating more valuable services that satisfy user requirements. It encourages reuse of existing services by discovering available interfaces; existing web services can be reused to create new services that are able to provide more functionality. However, as the number of available web services proliferates, the problem of finding and combining multiple Web services takes much time and costs in the development, integration, and maintenance of complex services. For this reason, automated composition becomes interesting as it can reduce costs and time in the software development cycle. Automated composition is the task of generating automatically a new Web service that achieves a given goal by interacting with a set of available Web services [5]. Recently RESTful web service gain momentum to develop distributed computing and few attempts have been taken by research in area of composition of RESTful web services. As far as our survey concern there is no serious attempt has been taken by researcher in area of automated composition of RESTful web service. In this research paper, we proposed novel approach termed as goal-based constraint-driven internal web service composition and automatic workflow (GCDIWBC) algorithm Using AI planning. We generated workflow based on user recruitment using JSHON—AI planner. AI planner generated

```
16. Planning Domain pd[ ]←——Get_Planning_Domain(uc[ ] ,
                             fg [ ] ,ps[ ] ,pls [ ]).
17. Raw_Work_Flow [ ] ←—— Get_Raw_Work_Flow(pd[]).
18. Loop: initial i =0 ; i <= Raw_Work_Flow[i].length;
          i++
19. Service_Type[ ]←—— Raw_Work_Flow[i].get_Service_Type();
20. Loop: initial j =0 ; j<=Service_Type[ j ] .length ; j++
21.Loop:Service_Instance[]←—— Discover_Service(Service_Type[j])
22 Condition: if  Service_Instance[ ] equals Null then
23. // here Service_Type regarded as dynamic service
24. pd1[ ]←—— Create_ Planning_Domain(Service_Type[j])
25. this.Service_Instance[ ]←—— DWM (pd1 [ ], uc[ ])
  26. end if
  27. Service_Type[ j ].set_Service_Instances(
                      this.Service_Instance[ ])
  28. End Loop
  29. Raw_Work_Flow[i].set_Service_Type(Service_Type[ ])
  30. End Loop
  31. Pre_Final_Work_Flow[ ]←—— Optimal_Service_Instance( pd[]
                             , uc [ ]Raw_Work_Flow [ ])
  32. Final_Work_Flow[ ] Create_Executable_Service(pd[] ,
                             Pre_Final_Work_Flow[ ])
  33. Loop: intical i=0; i<=Final_Work_Flow[ ].length
  34. Final_Work_Flow[i].publishWebService()
  35. End Loop
  36. Return Final_Work_Flow[ ]
```

Fig. 1 Proposed goal-based constraint-driven internal web service composition and automatic workflow (GCDIWBC) algorithm using AI planning

workflow of service type that algorithm search in web service registry to find respective RESTful web service. If any service type not found in web service registry will be considered as value added registry which need to generate by composing existing web services. We discussed algorithm in detail in Fig. 1.

The remainder of the paper is organized as follows: Sect. 2 discuss recent research work related to automatic RESTful web service composition, which provides the foundation for our research. Section 3 discuss about Travel scheduling motivating scenario which will be used to develop benchmark application. In Sect. 4 we discuss about the problem formulation of proposed approach. Section 5 discussion about proposed algorithm: goal-based constraint-driven internal web service composition and automatic workflow (GCDIWBC) Sect. 6 describes experiment setup and configuration to develop benchmark application Sect. 7 discussion of results. We conclude our research work and presents future research directions in Sect. 8.

2 Related Work

During last decade following researchers contributed in composition of RESTful-and SOAP-based web services.

Rauf et al. [10] introduce UML based modeling approach for RESTful web service composition. However, there is no detail implementation of RESTful web service composition is given. Zhao and Doshi [11] represented initial efforts toward automated RESTful web services modeling using online shopping scenario. Researcher presented sate transition system (STS)-based situation calculus to automate the RESTful web services composition. Rosenberg et al. [12] explored Bite: process composition model for Web applications, but bite programming language does not support HTTP PUT method. Automated RESTful web service composition: Primary requirement of automated web service composition is to generate composition schema semi/fully automated which will minimize user interaction and eliminates human errors and reduces the overall cost of process. There are four group of approached examined 1. Workflow-based [13–14] 2. Model based [15, 16] 3. Mathematics based [17, 18] and 4. AI planning based [19, 20]. Our proposed model used Quality of Service aware Workflow-based AI—planning approach for fully automation of RESTful web services.

3 Motivating Scenario: Traveling Scheduling Scenario

We used travel scheduling motivating scenario to describe our proposed approach. In this scenario, customer wants to travel from departure location to destination location with hotel booking. It is time consuming and tedious job for customer to find different ways like traveling by bus/by train to intermediate destination and also finding travel option from intermediate destination to goal city with hotel booking. There are many services are available which will provide travel information (either by bus or by train or by airplane), hotel information, and payment information but these individual services are not enough to give complete travel plan. This is a perfect example where travel schedule plan with hotel booking and payment can be generated automatically based on customer need. But single services are not enough to satisfy requirement of customer and need to composite them to generate value-added service dynamic web service composition (DWSC) which will satisfy request of the customer. Some time it is also need to use one DWSC to other DWSC. Our approach provides complete travel plan with travel schedule, hotel details with booking and travel reservation (bus/train/airline) with payment facility where customer need to provide departure time, departure location, destination location, and user constraints. In this scenario, we need to answer these questions 1. How to generate workflow of services based on need of customer? 2. How to find candidate service from repository based on workflow? 3. If candidate service not found, how to generate value added service by composing available services? 4. How to select services suitable for composition? 5. How composed services will be added to generated workflow? 6. How generated workflow represent by available workflow language? 7. How to execute generated workflow?

4   Problem Formulation

In this section, we describe RESTful web service composition problem by definition and then present automated composition planning problem from AI planning perspective.

Definition 1 (*Web service*) web service can be defined with 2-tuple WS = (Iw, Ow) where Iw = {Iw1, Iw2, ...} is an input parameter set, Iwi $\in$ Iw and Ow = {Ow1, Ow2 ...} is an output parameter set, Owi $\in$ Ow. In the motivating scenario we developed following web service with specification. In our motivating scenario, we developed more than 50 web services as shown in Fig. 1 with specification. These web services are register in RESTRegistery.

Definition 2 (*Service Type*) Service type is defined with 1-tuple service type St1 = {ws1} St2 = {ws2} ... where ws1, ws2 ... is instance of service. We developed following (Table 2) Service type in our motivation scenario.

Definition 3 (*Service URL*) Service URL defined with 1-tuple; Service URL SU1 = {wsurl1}, SU2 = {wsur2} ... where wsurl1, wsurl2 ... is URL of resources of RESTful web services. http://www.machotel.com/ ANShatabdiTrainService/

Definition 4 (*Web Service Repository*) Web service repository is defined as WSR = {St1, Ws1, Qos1, St2, Ws2, Qos2, ...} where St1, St2 ... is service type; Ws1, Ws2 ... is instance of web service; Qos1, Qos2 ... is quality of service factor.

Definition 5 (*Workflow*) Workflow is defined as Wf = {wt1, wt2, ...} where wt1, wt2 is available service type.

Definition 6 (*RESTful Composition Request*) is defined as RCR = {Wrin, Wrout} where Wrin = {Wrin1, Wrin2, ...} is input interface parameter and Wrout = {Wrout1, Wrout2, ...} is output interface parameter.

Definition 7 (*Operator*) Operator: {name(op), precond(op), add(op), del(op)} where name(op) is operator name and it will be expressed as n{x1, x2, ...} where x1, x2 ... represent parameter of operator. precond(op) is precondition of the operator op, i.e., properties necessary for its execution. add(op) and del(op) defined two sets of properties describing, respectively the facts to be added and the facts to be deleted of the world state after the execution of op.

Definition 8 (*Action*) An action is Instance of operator If a is an action and Si is a state such as precond + (a) Si and precond − (a) $\cap$ Si = $\varnothing$ then a is applicable in Si, and the result of this application is the state: $Si + 1 = \gamma (Si, a) = (Si\text{-effect} - (a))$ U effect + (a)

Definition 9 (*Planning space*) Problem space PS of Domain DO is triplet PS = (Op, Sinit, C, Goal) where Op is set of operators, Sinit is the initial state and Sinit $\in$ S. Goal is the goal defines a coherent set of instantiated predicates, i.e., world properties must be reached.

Definition 10 (*Parameter matching*) Let IR = (ir1, ir2, ...) is an interface parameter and web service Ws = (Iw, Ow) then fully parameter matching defined as IR ^ Ws when Iw <= IR. If IR ^ Ws then all output parameter can be added to IR and it is denoted as IR $\in$ Ws = IR U Ow.

Definition 11 (*Value-added service*) value added service is defined as Vas = {ws1, ws2, ...} where any of the web service type {wt1, wt2, ...} of work flow (Wf) doesn't belongs to WSR, i.e., that web service need to generate by composition existing web service.

Definition 12 (*WSC problem*) RESTful WSC is defined as 2—tuple RWSC = {RCR, WSR} where RCR = {Wrin, Wrout} and WSR = {wt1, ws1, q1, wt2, ws2, q2, ...}, then IO = {ws1, ws2, ...} is ordered sequence of web services associated web service type found from WSR such that ws1 $\in$ ws2 ... $\in$ wsn}.

Definition 13 (*Planning problem*) there are variety of applications are available of AI planning, such as robots, testing, automated code synthesis etc., It stats from initial state and generate workflow of web service type (sequence of actions) automatically. Each service type have instance of web service which need to be search in web service repository to find candidate web service type. If web service type not found in web service repository, it need to be generate (value added web service) by composing existing web services.

Planning domain defines all the operators who can apply to the world. A problem has to specify the initial state as well as the purpose to achieve. A domain D of L is a restricted state transition system $\Sigma = (S, A, \gamma)$ such as: S = 2 {instantiated atom of L} and A = {the set of instantiated operators of op} where op is the set of operators.

5 Proposed Novel Algorithm: Automated RESTful Web Service Composition Using AI Planning

Novel approach to solve automated RESTful web service composition problem is as follows and proposed algorithm shows in Fig. 1 and steps is as follow,

1. Customer need to provide departure location, departure time, destination location, and user constraint.
2. AI planner generates workflow of travel routes plan (sequence of service types) between departure location and destination location with hotel options.
3. Services will be discovered from RESTRegistry based on generated service types.
4. Optimal services are selected from discovered services based on Quality-of-Service (QoS).
5. If services not found from RESTRegistry, value-added service need to be generate based on available services.

6. Finally operator of selected services are combined and invoked by the order of the service type in the workflow.

6 Experiment Setup and Configuration

To validate the feasibility and efficiency of proposed algorithm, we measured response time and average composition time of RESTful and SOAP/WSDL web service using the state-of-the-art automated planners JSHON2—Java implementation of Simple Hierarchical Ordered Planner show in Table 1. We have conducted extensive experiments on 54,000 Web services that are distributed in the 18 groups of large-scale Web service repositories. The datasets are published on ICEBE05 and can be freely downloaded from the website of Web service challenge. These 18 groups of Web service repositories are categorized into Composition1 and

Table 1 The absolute increase and its rate of response time of RESful wand SOAP/WSDL with JSHON2 along with the change of service number

|P|	|W|	Proposed algorithm with RESTful web service using JSHOP2			Proposed algorithm with RESTful web service using SOAP/WSDL		
		ART	RTI	Rate	ART	RTI	Rate
4–8	2156	0.363	–	–	2.345	–	–
	2656	0.381	0.018	0.05	2.675	0.33	0.14
	4156	0.412	0.031	0.08	3.768	1.093	0.41
16–20	2156	1.14	–	–	4.153	–	–
	2656	1.376	0.236	0.21	12.12	7.967	1.92
	4156	2.36	0.984	0.72	15.897	3.777	0.31
32–36	2156	3.285	–	–	16.156	–	–
	2656	3.422	0.137	0.04	17.124	0.968	0.06
	4156	4.657	1.235	0.36	18.451	1.327	0.077
4–8	3356	0.548	–	–	9.436	–	–
	5356	0.981	0.433	0.79	10.897	1.461	0.15
	8356	1.515	0.534	0.54	13.679	2.782	0.26
16–20	5356	3.888	–	–	18.376	–	–
	6712	4.536	0.648	0.17	18.843	0.467	0.025
	8356	5.611	1.075	0.24	18.964	0.588	0.03
32–36	3356	5.936	–	–	15.908	–	–
	5356	6.691	0.755	0.13	21.435	5.527	0.35
	8356	9.314	2.623	0.39	21.946	0.511	0.02

\|P\| gives the parameter size of input or output of a service. \|W\| is the number of services in a repository. ART is the average response time on all of the 10 composition requests in a Web service repository. RTI is the absolute increase of average response time compared with the preceding ART. Rate is the increase rate of average response time

Composition2 in the service repository distributions, the number of services involved in a dataset ranges from 2156 to 8356, and the size of input or output parameters in a service ranges from 4–8, 16–20 to 32–36. In terms of the number of services and parameter size in a Web service repository, the easiest dataset to be dealt with is Composition1-20-4. On the contrary, the most difficult dataset is Composition2-100-32. Each service repository either in the Composition1 or Composition2 has 10 composition requests for test. All of the experiments are performed on a PC with Intel(R) Core(TM) i3—2310 M CPU@ 2.10 GHz with 4G RAM. Apart from AI planners tested on Windows 7.

7 Result Analysis

GCDIWCAW using JSHON2 can best satisfy all of the composition requests because response time of GCDIWCAW (RESTful web service) using JSHON2 to the easiest dataset (Composition1-20-4) ranges from 0.363 to 0.412 s and for hardest dataset (Composition2-100-32) ranges from 5.963 to 6.314 s; still remains within a short period of time. For GCDIWCAW with SOAP/WSDL standard for the easiest dataset (Composition1-20-4) response time ranges from 2.345 to 3.768 s and for hardest dataset (Composition2-100-32) ranges from 15.908 to 21.946 s. This shows that GCDIWCAW (RESTful web service) using JSHON2 is significantly faster than the rest of the approach discuss above when a dataset becomes more difficult to handle.

8 Conclusion

The ability to automatically and efficiently compose Web services can potentially simplify the implementation of business processes. This research work presents an efficient approach for dynamic composition of RESTful Web services using the JSHON2 state-of-the-art automated planners. Our research work focus on recent challenges of RESTful web service composition and also provides optimal solution using Travel Scheduling motivating scenario. In our research work, we proposed GCDIWCAW algorithm and proposed framework to implement proposed algorithm. RESTful web service composition is very active research area among research community and our research is focus of very important aspects of automatic composition of RESTful web service composition. The extensive experiments conducted on large-scale Web service repositories indicate that our proposed algorithm with RESTful web service using JSHON2 outperforms SOAP/WSDL standard both in its response time and scalability.

References

1. Fielding, R. T. (2000). Architectural styles and the design of network-based software architectures. In *PhD thesis* (pp. 89–88). Irvine: University of California.
2. Rathod, D. M., Parikh, S. M., & Buddhadev, B. V. (2013). Structural and behavioral modeling of RESTful web service interface using UML: Proceedings of 2013. In *IEEE International Conference on Intelligent Systems and Signal Processing (ISSP)* (pp. 28–33). doi:10.1109/ISSP.2013.6526869. ISBN:978-1-4799-0316-0.
3. Hamadi, R., & Benatallah, B. (2003). A petri net-based model for web service composition. In *Proceedings of Australasian Database Conference (ADC 2003)*, Adelaide, Australia.
4. Waldo, J., Wyant, G., Wollrath, A., & Kendall, S. (1994). A note on distributed computing. In *Technical report, Sun Microsystems Laboratories* http://researchsun.com/techrep/1994/smli_tr-94-29.pdf.
5. Fielding, R. T., & Taylor, R. N. (2002). Principled design of the modern web architecture. *ACM Transactions on International Technology, 2*(2), 115–150.
6. Bellwood, T., et al. *Universal description, discovery and integration specification (UDDI) 3.0* http://uddi.org/pubs/uddi-v3.00-published-20020719.htm.
7. McIlraith, S., Son, T. C., & Zeng, H. (2001, March/April). Semantic web services. *IEEE Intelligent Systems, 16*(2), 46–53.
8. Casati, F., Ilnicki, S., & Jin, L. (2000). Adaptive and dynamic service composition in EFlow. In *Proceedings of 12th International Conference on Advanced Information Systems Engineering(CAiSE), Stockholm, June 2000*. Sweden: Springer.
9. Casati, F., Sayal, M., & Shan, M. C. (2001). Developing e-services for composing e-services. In *Proceedings of 13th International Conference on Advanced Information Systems Engineering(CAiSE), Interlaken, June 2001*. Switzerland: Springer.
10. Rauf, I., Ruokonen, A., Systa, T., & Porres, I. (2010). Modeling a composition RESTful web service with UML. In *Proceedings of the 4th European Conference on Software Architecture (ECSA 2010)* (pp. 253–260). Copenhagen.
11. Zhao, H., & Doshi, P. (2009). Towards automated RESTful web service composition. In *Proceedings of the 2009 IEEE International Conference on Web Services (ICWS '09)* (pp. 189–196). Washington.
12. Rosenberg, F., Curbera, F., Duftler, M. J., & Khalaf, R. (2007) Bite: Workflow composition for the web. In *Service Oriented Computing: Fifth International Conference* (pp. 94–106). Springer.
13. Agarwal, V., Chafle, G., Mittal, S., Srivastava, B. (2008). Understanding approaches for web service composition and execution. In *COMPUTE'08: Proceedings of the 1st Bangalore Annual Compute Conference, ACM 2008* (pp. 1–8). New York, USA.
14. Peltz, C. (2003, January). *Web services orchestration a review of emerging technologies, tools, and standards*. Hewlett Packard & Co.
15. Srivastava, B., & Koehler, J. (2003). Web service composition—current solutions and open problems. In *Proceedings of ICAPS 2003 Workshop on Planning for Web Services*.
16. Traverso, P., & Pistore, M. (2004). Automated composition of semantic web services into executable processes. In *Proceedings of 3rd International Semantic Web Conference* (pp. 380–394).
17. Ter Beek, M. H., Bucchiarone, A., & Gnesi, S. (2007). Formal methods for service composition. *Annals of Mathematics, Computing & Tele Informatics, 1*(5), 1–10.
18. Bartalos, P., & Bielikova, M. (2010). QoS aware semantic web service composition approach considering pre/postconditions. In *International Conference on Web Services, IEEE CS 2010. Automatic Dynamic Web Service Composition* (Vol. 82, pp. 345–352).
19. Alrifai, M., Risse, T., Dolog, P., & Nejdl, W. (2009). A scalable approach for Qos-based web service selection. In *Service-Oriented Computing 2008 Workshops* (pp. 190–199). Springer.

20. DiBernardo, M., Pottinger, R., & Wilkinson, M. (2008). Semi-automatic web service composition for the life sciences using the biomoby semantic web framework. *Journal of Biomedical Informatics, Semantic Mashup of Biomedical Data, 41*(5), 837–847. ISSN:1532-0464.
21. Bhatnagar, R., & Patel, J. (2014). Scady: A scalable & dynamic toolkit for enhanced performance in grid computing. In *IEEE International Conference on Pervasive Computing ICPC—2015*, Jan 8–10, 2015, Pune. *API Specification for Small Grid Middleware case study, IEEE Xplore, India Conference (INDICON), 2014 Annual IEEE* (pp. 1–5). doi:10.1109/INDICON.2014.7030545.

Author Index

© Springer Science+Business Media Singapore 2016
S.C. Satapathy et al. (eds.), *Proceedings of International Conference
on ICT for Sustainable Development*, Advances in Intelligent Systems
and Computing 408, DOI 10.1007/978-981-10-0129-1

Printed in the United States
By Bookmasters